The Virulence of
Escherichia coli

Special Publications of the Society for General Microbiology

Publications Officer: Colin Ratledge, 62 London Road, Reading, UK

1. Coryneform Bacteria,
    eds. I. J. Bousfield and A. G. Callely

2. Adhesion of Microorganisms to Surfaces,
    eds. D. C. Ellwood, J. Melling and P. Rutter

3. Microbial Polysaccharides and Polysaccharases,
    eds. R. C. W. Berkeley, G. W. Gooday and D. C. Ellwood

4. The Aerobic Endospore-forming Bacteria:
    Classification and Identification,
    eds. R. C. W. Berkeley and M. Goodfellow

5. Mixed Culture Fermentations,
    eds. M. E. Bushell and J. H. Slater

6. Bioactive Microbial Products: Search and Discovery,
    eds. J. D. Bu'Lock, L. J. Nisbet and D. J. Winstanley

7. Sediment Microbiology,
    eds. D. B. Nedwell and C. M. Brown

8. Sourcebook of Experiments for the Teaching of Microbiology,
    eds. S. B. Primrose and A. C. Wardlaw

9. Microbial Diseases of Fish,
    ed. R. J. Roberts

10. Bioactive Microbial Products, Volume 2,
    eds. L. J. Nisbet and D. J. Winstanley

11. Aspects of Microbial Metabolism and Ecology,
    ed. G. A. Codd

12. Vectors in Virus Biology,
    eds. M. A. Mayo and K. A. Harrap

13. The Virulence of *Escherichia coli*,
    ed. M. Sussman

This book is based on a Symposium of the SGM held at Newcastle upon Tyne, 3 January 1983.

The Virulence of *Escherichia coli*

Reviews and Methods

Edited by

M. Sussman

*Department of Microbiology, Medical School,
University of Newcastle upon Tyne,
Newcastle upon Tyne, UK*

1985

Published for the

Society for General Microbiology

by

ACADEMIC PRESS

(Harcourt Brace Jovanovich, Publishers)

London   Orlando   San Diego   New York
Toronto   Montreal   Sydney   Tokyo

ACADEMIC PRESS INC. (LONDON) LTD.
24–28 Oval Road
LONDON NW1 7DX

United States Edition published by
ACADEMIC PRESS, INC.
Orlando, Florida 32887

British Library Cataloguing in Publication Data

The Virulence factors of escherichia coli : reviews
    and methods.—(Special publications of the
    Society for General Microbiology)
    1. Escherichia coli
    I. Sussman, M.   II. Series
    596'.0234   QR201.E82

Library of Congress Cataloging in Publication Data
Main entry under title:

The Virulence factors of Escherichia coli.

    Based on the Society for General Microbiology's
Symposium on Virulence Markers of Escherichia Coli...
Newcastle–upon–Tyne...Jan. 1983
    Includes bibliographical references and index.
    1. Escherichia coli––Congresses. 2. Virulence
(Microbiology)––Congresses. I. Sussman, Max. II. Society
for General Microbiology. III. Symposium on Virulence
Markers of Escherichia Coli (1983 : Newcastle–upon–Tyne,
Tyne and Wear)
    QR82.E6V57 1984   616'.0145   84–12421
    ISBN 0–12–677520–6 (alk. paper)

For Jean
Proverbs xxxi : 27–29

Contributors

W. D. ALLEN   *Department of Immunology, Unilever Research, Colworth Laboratory, Sharnbrook, Bedford MK44 1LQ, U.K.*

A. W. ASSCHER   *Department of Renal Medicine, Welsh National School of Medicine, KRUF Institute, Royal Infirmary, Cardiff CF2 1SZ, Wales.*

J. BAILEY   *Pathology Department, Central Veterinary Laboratory, Weybridge, Surrey KT15 3NB, U.K.*

N. H. CARBONETTI   *Department of Genetics, University of Leicester, Leicester LE1 7RH, U.K.*

SUSAN CHICK   *Department of Renal Medicine, Welsh National School of Medicine, KRUF Institute, Royal Infirmary, Cardiff CF2 1SZ, Wales.*

PAMELA B. CRICHTON   *Department of Medical Microbiology, University of Dundee Medical School, Ninewells Hospital, Dundee DD1 9SY, U.K.*

I. M. FEAVERS   *Department of Microbiology, University of Sheffield, Sheffield S10 2TN, U.K.*

E. GRIFFITHS   *National Institute for Biological Standards and Control, Hampstead, London NW3 6RB, U.K.*

R. J. GROSS   *Division of Enteric Pathogens, Central Public Health Laboratory, London NW9 5HT U.K.*

M. J. HARBER   *Department of Renal Medicine, Welsh National School of Medicine, KRUF Institute, Royal Infirmary, Cardiff CF2 1SZ, Wales.*

C. R. HARWOOD   *Department of Microbiology, Medical School, University of Newcastle upon Tyne, Newcastle upon Tyne NE2 4HH, U.K.*

J. HOLMGREN   *Department of Medical Microbiology, University of Göteborg, S-413 46 Göteborg, Sweden.*

BARBARA JANN   *Max-Planck-Institut für Immunbiologie, D-7800 Freiburg-Zähringen, Federal Republic of Germany.*

K. JANN   *Max-Planck-Institut für Immunbiologie, D-7800 Freiburg-Zähringen, Federal Republic of Germany.*

MARGARET A. LINGGOOD   *Department of Immunology, Unilever Research, Colworth Laboratory, Sharnbrook, Bedford MK44 1LQ, U.K.*

RUTH MACKENZIE   *Department of Renal Medicine, Welsh National School of Medicine, KRUF Institute, Royal Infirmary, Cardiff CF2 1SZ, Wales.*

N. MACKMAN   *Department of Genetics, University of Leicester, Leicester LE1 7RH, U.K.*

J. A. MORRIS   *Central Veterinary Laboratory, Weybridge, Surrey KT15 3NB, U.K.*

D. C. OLD   *Department of Medical Microbiology, University of Dundee Medical School, Ninewells Hospital, Dundee DD1 9SY, U.K.*

S. H. PARRY   *Department of Microbiology, Medical School, University of Newcastle upon Tyne, Newcastle upon Tyne NE2 4HH, U.K. (Present address: Department of Immunology, Unilever Research, Colworth Laboratory, Colworth House, Sharnbrook, Bedford MK44 1LQ, U.K.)*

P. PORTER   *Department of Immunology, Unilever Research, Colworth Laboratory, Sharnbrook, Bedford MK44 1LQ, U.K.*

J. R. POWELL   Department of Immunology, Unilever Research, Colworth Laboratory, Sharnbrook, Bedford MK44 1LQ, U.K.

DIANA M. ROOKE   Department of Microbiology, Medical School, University of Newcastle upon Tyne, Newcastle upon Tyne NE2 4HH, U.K.

B. ROWE   Division of Enteric Pathogens, Central Public Health Laboratory, London NW9 5HT, U.K.

SYLVIA M. SCOTLAND   Division of Enteric Pathogens, Central Public Health Laboratory, London NW9 5HT, U.K.

D. J. SHEEHAN   Pathology Department, Central Veterinary Laboratory, Weybridge, Surrey KT15 3NB, U.K.

H. R. SMITH   Division of Enteric Pathogens, Central Public Health Laboratory, London NW9 5HT, U.K.

W. J. SOJKA   Central Veterinary Laboratory, Weybridge, Surrey KT15 3NB, U.K.

PAULINE STEVENSON   National Institute for Biological Standards and Control, Hampstead, London NW3 6RB, U.K.

M. SUSSMAN   Department of Microbiology, Medical School, University of Newcastle upon Tyne, Newcastle upon Tyne NE2 4HH, U.K.

P. W. TAYLOR   Bayer AG, Pharma-Forschungszentrum, Institut für Chemotherapie, D-5600 Wuppertal 1, Federal Republic of Germany. (Present address: Molecular Diagnostics, Inc., 400 Morgan Lane, West Haven, Connecticut 06516, U.S.A.)

C. J. THORNS   Central Veterinary Laboratory, Weybridge, Surrey KT15 3NB, U.K.

G. A. H. WELLS   Pathology Department, Central Veterinary Laboratory, Weybridge, Surrey KT15 3NB, U.K.

P. H. WILLIAMS   Department of Genetics, University of Leicester, Leicester LE1 7RH, U.K.

Foreword

Escherichia coli is an ubiquitous and versatile bacterium. Non-pathogenic strains have formed the work-horses of biochemists and geneticists for elucidating general biological phenomena such as metabolic pathways, protein and nucleic acid synthesis, enzyme induction and repression and the role of DNA in heredity. As a member of the commensal flora of the gut, *E. coli* does not cause harm and indeed may contribute to the flora's protective action against extraneous infection. Unfortunately, some types of *E. coli* produce severe disease in man and farm animals: enteritis, urethritis, septicaemia, meningitis and other syndromes. Why are these particular types of *E. coli* pathogenic and what are the determinants concerned? These questions were discussed at a symposium held early in January 1983 at the University of Newcastle upon Tyne during a meeting of the Society for General Microbiology. The symposium was arranged by Professor M. Sussman and I had the pleasure of being Chairman. This book incorporates the contributions to the symposium together with 17 additional chapters on methods.

After two chapters have set the scene by describing the many and varied diseases caused by *E. coli* in man and farm animals, other contributions deal with the mechanisms and determinants that might be involved in the five main steps of pathogenesis. Adhesion in the primary infection of mucous surfaces receives considerable attention with up-to-date information on both the bacterial adhesins and host receptors. The ability of some strains to penetrate surface epithelial cells and enter the tissues is noted and potential determinants are discussed. The third fundamental requirement for virulence—ability to grow in the tissues of the host—is stressed in relation to the bacterial derivation of enough iron for growth from the environment *in vivo*. The two main determinants of interference with humoral and phagocytic defence—complete (i.e. with a full polysaccharide side chain) lipopolysaccharides and the K acid polysaccharide antigens of encapsulated *E. coli*—are discussed at the level of the relation of chemical structure to biological function. The various toxins produced by *E. coli*, the heat-labile and stable enterotoxins, the endotoxin, the haemolysin and a toxin possibly responsible for pig oedema disease, receive the attention they deserve both at a sophisticated level of molecular biology and in relation to the disease syndromes they might cause.

Penultimately, the genetics of virulence in *E. coli* is discussed, especially the plasmids that code for virulence factors such as adhesins, toxins and the determinants of iron usage and cell penetration. Finally, the immunological reaction of the pig to infection by *E. coli* and to killed organisms is described, and the

ix

potential of vaccines is explored in relation to both protection of individual animals and elimination of virulence and drug resistance plasmids from the *E. coli* strains in pig populations.

My predominant impression from the symposium and the book is that, despite the advances in knowledge of virulence factors in the last decade, we are still far from explaining completely the many and varied disease syndromes wrought by *E. coli*. To get nearer to this explanation we shall have to look at the impact of combinations and variants of known virulence factors and also for additional factors. Indeed, if non-pathogenic *E. coli* is the microbiologist's friend, the pathogenic varieties are his enigma.

The book provides valuable information and an outlook for the future to researchers and clinicians interested in infectious disease as a whole, and especially those interested in *E. coli* infections of medical and veterinary importance. My only regret is that it does not include some of the many interesting points that emerged during the lively discussion at the symposium.

Harry Smith, F.R.S.
Department of Microbiology
University of Birmingham
Birmingham, U.K.

Preface

This book is based on the Society for General Microbiology Symposium on 'Virulence Markers of *Escherichia coli*', which took place in Newcastle upon Tyne in the wake of the New Year celebrations in January 1983. The subtle change of title of this book from that of the Symposium is deliberate. The reviews are more extensive than those presented at the Symposium and more than two-thirds of the chapters are invited contributions, in addition to those delivered at the Symposium. Most of these chapters describe methods for the study of *E. coli* virulence that are widely dispersed in the literature. It is hoped that they will be useful to those who are new to 'coli virulence studies'. There may also be something here for the old hands.

As this book is sent on its way through the press, memories of a happy and profitable one-day symposium come to mind and especially the willing co-operation of all the contributors. Professor Harry Smith, who has generously written the Foreword, also provided many penetrating observations during the discussions. I am deeply grateful to him. My sincere thanks are also due to Sterilin Ltd., Feltham, Middlesex, England, for their generous assistance with part of the travelling expenses of the invited speakers at the symposium.

Publication of this book will take place during the centenary year of the discovery of *Bacterium coli commune* by Dr. Theodor Escherich. This happy coincidence constitutes these pages as a memorial to a great paediatrician and distinguished bacteriologist.

October 1984   M. Sussman

Contents

II Methods

The Virulence of
Escherichia coli

Theodor Escherich 1857–1911. (Reproduced with permission from
L. Schonbauer, Das Medizinische Wien, Urban & Schwarzenberg,
Vienna, 1947.)

1

Theodor Escherich (1857–1911): A Biographical Note

M. SUSSMAN

Department of Microbiology, Medical School, University of Newcastle upon Tyne, Newcastle upon Tyne, UK

The name of Dr. Theodor Escherich has become well known through the naming for him of the tribe *Escherichiae,* the genus *Escherichia* and its type species *Escherichia coli,* which he had first described as *Bacterium coli commune* (Escherich, 1885a,b). His achievements, however, extend far beyond this important description.

Escherich was born on 29 November 1857 in Ansbach, Bavaria, in southern Germany into a respected Frankonian family. His father, Dr. Ferdinand Escherich, was a noted medical statistician who was honoured with the then significant title of '*Medizinalrat*' (Medical Counsellor). It seems that Theodor was a lively schoolboy given to pranks and practical jokes, which attracted adverse opinion and led to his being sent to the Jesuit School 'Stella matutina' at Feldkirch in Austria.

In 1876 Escherich went up to university and, since it was the practice in Germany at the time to attend several universities, he spent semesters at the Universities of Strassbourg, Kiel, Berlin, Würzburg and Munich, where he was awarded his doctorate in 1881. In the following year he began to work at the Würzburg University Clinic under Prof. Karl Gerhardt, editor of the first German 'Handbook of Paediatrics', who seems to have had a considerable influence on Escherich, and he was eventually appointed First Assistant. Though his first papers were concerned with the diagnostic methods propounded by Gerhardt, his later papers from Würzburg showed evidence of broadening interests which included biochemistry. These interests were later to be extended in the laboratories of Hoppe-Seyler, Voit, Pettenkofer and Soxhlet.

Escherich's interest in paediatrics had been aroused by Gerhardt but the state of the specialty in the German universities at the time made a thorough training in

1

paediatrics possible only abroad. So Escherich went to Paris and later Vienna where, working at the renowned St. Anna Children's Hospital, of which twenty years later he was to become the Director, he decided to devote his life to paediatrics.

From Vienna Escherich returned to Munich where Frobenius, a pupil of Robert Koch, introduced him to the then still relatively new techniques for the pure culture and identification of bacteria. With this knowledge Escherich conceived the plan to clarify the significance of bacteria in the physiology and pathology of digestion in infants. As a first step he showed that human milk is normally sterile but that pyogenic cocci may be present in certain pathological conditions. There followed a series of studies of the bacteriology of infant faeces in the course of which *Bacterium coli commune* (now *Escherichia coli*) and *Bacterium lactis aerogenes* (now *Klebsiella pneumoniae*) were discovered and their properties described, including the composition of the gas produced when these organisms ferment carbohydrate and the observation that under anaerobic conditions their growth is strictly dependent on the fermentation of carbohydrate (Escherich, 1885a,b). Though it was widely believed at the time that the intestinal flora plays a significant role in nutrition, Escherich concluded from his studies that if this was so, the contribution was slight.

In 1886 Escherich published a monograph '*Die Darmbakterien des Säuglings und ihre Beziehung zur Physiologie der Verdauung*' ('The intestinal bacteria of the infant and their relation to the physiology of digestion') and was appointed Docent in Paediatrics (roughly equivalent to Senior Lecturer) at the University of Munich. Here he continued his work in infant nutrition and pointed out the hazards of the high sodium content of cow's milk and encouraged a return to breast feeding.

In 1890 the Austrian Ministry of Education issued a call to the 33-year-old Escherich to take up the post of extraordinary Professor when the Chair of Paediatrics became vacant at the University of Graz. Here, taking up this post and the Directorship of the St. Anna Children's Hospital, he demonstrated his remarkable organizational and administrative abilities. The hospital was renovated and extended to create one of the most attractive and best equipped children's hospitals in Austria. In particular, the hospital was furnished with incubator rooms for the newborn and wards were set aside for diphtheria patients. In addition, laboratories and a clinical lecture theatre were built. Escherich's reputation and that of the hospital he directed led to a massive increase in the numbers of both in-patients and out-patients treated there.

Apart from his administrative activities at Graz, Escherich engaged in a wide range of researches. Between 1891 and 1893 he investigated the aetiology and pathogenesis of diphtheria and, following on the experimental work in animals by Behring and Kitasato (1890), he was able, as early as 1893, to demonstrate the presence of antitoxin in the sera of children convalescing from the disease.

Escherich played his part in establishing the value of the treatment of diphtheria with antitoxin by treating children in his wards and he did so against substantial opposition from his colleagues. Another topic of his research during this period was the condition now known as idiopathic infantile tetany, of which he described the physical signs in 1890. Escherich eventually reached the conclusion that hypoparathyroidism played a part in the pathogenesis of this condition and in 1909 he summarized knowledge on the subject in a monograph.

Escherich then returned to his earlier interest in intestinal bacteria with the object of investigating their possible pathogenicity in infants. In the course of these studies he suggested that coliform bacteria are associated with acute contagious intestinal infections in the first year of life. There is evidence that at this time he repeatedly isolated dysentery bacilli in pure culture and that he may just have missed being the first to describe these organisms, which were described at this time by Shiga (1898). Escherich and his pupils noted the frequency of isolation of *E. coli* from the urine of young girls and they were the first to recognise its significance in urinary tract infection.

Soon after the general release of tuberculin Escherich used it on his young patients and showed, as others did, that it had no therapeutic application but that it was valuable as a diagnostic tool. It was in the course of this work that the specific 'prick reaction' to tuberculin was recognized.

In 1894 Escherich was appointed Ordinary Professor and Examiner at Graz and when, in 1902, the Chair of Paediatrics at the University of Vienna became vacant, the Viennese Faculty unanimously called Escherich to the Chair and the Directorship of the St. Anna Children's Hospital of Vienna, which he accepted in spite of the difficult problems he knew would face him there.

On arrival in Vienna he set to work with energy and determination to reduce infant mortality and, among other projects, he encouraged the school doctor service, the movement to provide nursing care for tuberculous children, and the establishment of the Association for Paediatricians in Vienna. Escherich, as so often far ahead of his time, organised a paediatric advisory service for the obstetric clinics in Vienna.

These and other administrative preoccupations prevented Escherich from personal research. He nevertheless took an active interest in the work of his pupils, who worked on subjects including scarlet fever, dysentery, allergy and tuberculosis.

Among his pupils were Moro, who described the well-known reflex in the newborn, and Clemens von Pirquet (1874–1929), the discoverer of the cutaneous tuberculin reaction, who also conceived the notion of allergy and who succeeded Escherich in the Chair in Vienna.

A count of Escherich's papers to 1909 numbers 158, while those written by his pupils and associates number 271. He died from a cerebrovascular accident on 15 February 1911.

4 M. SUSSMAN

References

Behring, E. and Kitasato, S. (1890). Ueber das Zustandekommen der Diphtherie—Immunität und der Tetanus-Immunität bei Thieren. *Deutsche Medizinische Wochenschrift* **16**, 1113, 1145.

de Rudder, B. (1957). Zu Theodor Escherichs 100. Geburtstag. *Deutsche Medizinische Wochenschrift* **82**, 1620–1621.

Escherich, T. (1885a). Die Darmbacterien des Neugeborenen und Säglings. *Fortschritte der Medicin* **3**, 515–522.

Escherich, T. (1885b). Die Darmbacterien des Neugeborenen und Säuglings. *Fortschritte der Medicin* **3**, 547–554.

Pfaundler, M. (1911). Theodor Escherich. *Muenchener Medizinische Wochenschrift* **58**, 521–523.

Shiga, K. (1898). Uebur den Erreger der Dysenterie in Japan. *Centralblatt für Bakteriologie, Parasitenkunde und Infektionskrankheiten, Abteilung 1* **23**, 599–600.

Zappert (1911). Hofrat Prof. Theodor Escherich. *Wiener Medizinische Wochenschrift* **61**, 498–499.

Part I
Reviews

2

Escherichia coli in Human and Animal Disease

M. SUSSMAN

*Department of Microbiology, Medical School, University of Newcastle upon Tyne,
Newcastle upon Tyne, UK*

Introduction

The organism now known as *Escherichia coli* was first described by Escherich (1885) under the name *Bacterium coli commune*. It had been isolated from normal infant faeces. Though it was at first regarded as non-pathogenic, Escherich (1894) showed that it was present in the urine of young girls suffering from urinary tract infection and suggested that it reached the bladder by the ascending route. In failing to distinguish between his ''*B. coli*'' and the then undescribed genus *Shigella*, Escherich (1899) considered that *E. coli* was the cause of dysentery but it is thought (von Pirquet, 1910–1911) that he discarded this notion in later years.

In spite of the search for a bacterial cause for many diseases, which on occasion led to the erroneous identification of certain bacteria as pathogens, *E. coli* was probably generally regarded as of low virulence. This is hardly surprising when *E. coli* is compared with the virulent pathogens such as *Corynebacterium diphtheriae*, *Streptococcus pyogenes*, *Vibrio cholerae*, *Mycobacterium tuberculosis* and others which were then the centre of interest. *E. coli* was thought of as the major commensal present in faeces. Such a view continued for some 60 years from its discovery. The breakthrough to the discovery of the specific virulence of *E. coli* had to await the systematic study of its serology.

This introductory chapter will provide a brief review of the properties of *E. coli* and the part it plays in human and animal disease.

Characteristics of *E. coli*

Escherichia coli, which belongs to the family Enterobacteriaceae, is the only member of the genus *Escherichia*. Though no other species are at present officially recognized in the genus, *E. coli* can be subdivided into stable biotypes on the basis of a variety of properties (Chapter 11).

Morphology

E. coli is a short Gram-negative, non-sporing and, usually, peritrichous and fimbriate bacillus. A capsule or microcapsule is often present (Chapter 5) and a few strains produce a profuse polysaccharide slime.

Biochemical Characteristics

E. coli is a facultative anaerobe that grows readily on simple culture media and synthetic media with glycerol or glucose as the sole source of carbon and energy. When growing anaerobically there is an absolute requirement for fermentable carbohydrate. Glucose is fermented to give acid and gas.

Some of the characteristics of *E. coli* are listed in Table 1.

Table 1.   *Characteristics of Escherichia coli (after Ørskov, 1974)*

Optimum growth temperature	37°C
Catalase	+
Oxidase	−
Indole production	+
Methyl Red reaction	+
Voges–Proskauer reaction	−
Citrate utilization	−
NO_3^- reduction	+
Growth in KCN	−
Fermentation	Mixed acid
Gas from glucose (37°C)	+
Acid from: Arabinose	
Lactose	
Maltose	May be delayed
Mannitol	+
Sorbitol	
Trehalose	
Adonitol	
Inositol	−
Dulcitol	
Esculin	
Salicin	Varies for different strains
Sucrose	
Xylose	
Mole % G + C	50–51

Cultural Characteristics

In broth smooth (S) strains of *E. coli* produce a uniform turbidity, while rough (R) strains tend to leave a clear supernatant medium over a granular deposit of growth. On solid media, colonies are circular and smooth with an entire edge; some strains produce mucoid colonies (Macone *et al.*, 1981). On media containing washed erythrocytes a soluble α-haemolysin can be demonstrated and there is a cell-associated β-haemolysin which is released from cells by lysis (Smith 1963; see also Chapter 23). Lactose-fermenting organisms with the appearance of *E. coli* are frequently referred to as 'coliforms' but they cannot, without further definitive identification, be regarded as *E. coli*.

Serological Characteristics

A variety of antigens may be used to type *E. coli* strains. The best known of these are the O somatic lipopolysaccharide, K capsular and H flagellar antigens, which form the basis of the typing scheme first introduced by Kauffmann (1944) and since expanded (see Chapter 14). The serotyping of *E. coli* fimbriae is also available to expand previous typing schemes (Ørskov and Ørskov, 1983; Parry *et al.*, 1982).

Other Typing Characteristics

Apart from serological properties other characteristics of *E. coli* have formed the basis of typing schemes. Thus, bacteriophage typing (reviewed by Milch, 1978) and colicine typing (reviewed by Gillies, 1978) have been employed but they have not come into general use.

Recent general accounts of the Enterobacteriaceae are given by Gross and Holmes (1983) and Holmes and Gross (1983).

Distribution

The primary habitat of *E. coli* is the gastro-intestinal tract, and principally the bowel, of mammals and birds. When found in nature, either in soil or water or elsewhere, it is derived from the primary habitat, usually by faecal contamination.

Colonization and Persistence of *E. coli*

Colonization by *E. coli* takes place soon after birth (Escherich, 1885; Bettelheim *et al.*, 1974) and its source is to be found in the mother and the inanimate environment (O'Farrell *et al.,* 1976). Little is known of the factors that might promote intestinal colonization but Hanson *et al.* (1983) have found a variety of adhesins including MS, P and X fimbriae on faecal and oral isolates from the newborn. The variety of fimbrial adhesins and their distribution among the strains provide no clues as to their possible significance in colonization.

Once established, *E. coli* remains part of the faecal flora. It is likely that in any individual the *E. coli* population consists of a majority serotype and a number of minority serotypes (Kauffmann and Perch, 1943; Wallick and Stuart, 1943). There is evidence that the serotype may change from time to time. Thus, the serotypes in hospital patients may be different from those in the population at large (Kennedy *et al.,* 1965; Winterbauer *et al.,* 1967; Cooke *et al.,* 1969) and it seems that the new strains are derived from contaminated food (Cooke *et al.,* 1970; Shooter *et al.,* 1970). Work with volunteers has, in some cases, suggested that ingested organisms only rarely persist (Sears *et al.,* 1950; Sears and Brownlee, 1953; Smith, 1969); in other cases persistence regularly followed ingestion of at least 10^6 organisms (Cooke *et al.,* 1971). It is likely that both the nature and the conditions of growth determine the likelihood of persistence. Thus, the strains used by Smith (1969) were derived from animals and carried R factors and such strains may not easily colonize the bowel (Cooke, 1974). Indeed, it is known that the prevalence of serogroups O1, O2, O4, O5, O18, O20, O25 and O75 in the human faecal flora is greater than that of other serogroups and it is thought that this is in some way due to the capacity of the various serogroups to persist in the bowel (Ewing and Davies, 1961; Guinée, 1963). There is evidence that persistence of *E. coli* carrying R factors depends on their O serogroup (Hartley and Richmond, 1975). Moreover, growth conditions may affect adhesin production *in vitro* (Parry *et al.,* 1982) and this in turn may affect colonization by and persistence of inocula fed to volunteers.

In animals, too, *E. coli* colonizes the bowel early in life and it is likely that the source is to be found in the faecal flora of the dam (Smith, 1965).

E. coli is a small minority member of the bowel flora of mammals and both individuals and groups may be found in whom it is absent. Dubos *et al.* (1963) have suggested that in the case of mice, which they studied, *E. coli* is not a normal member of the flora but should rather be regarded as a potential pathogen.

Normal Flora

E. coli is regarded as a member of the normal flora of man. As we have seen, colonization takes place very early in life. It appears rapidly in the saliva (Russell and Melville, 1978) but does not appear to colonize the normal mouth or pharynx. Before the significance or magnitude of the anaerobic bowel flora was recognized, *E. coli* was thought to be the predominant faecal organism. However, it is now recognized that, though it is universally present in the human bowel, *E. coli* is a small minority constituent of the flora.

The function of *E. coli* in the faecal flora is difficult to assess. There are suggestions that it may have a nutritional significance by providing a source of vitamins in some animals. The normal flora of the alimentary tract has been extensively reviewed (Drasar and Hill, 1974; Skinner and Carr, 1974; Clarke and Bauchop, 1977; Wilson, 1983).

Virulence Factors

Colonization Factors

Colonization of a mucous membrane is the prerequisite of most diseases due to *E. coli*. This presents particular problems because these surfaces have natural clearing mechanisms that tend to remove particles, including bacteria, and pathogens have evolved adhesins to overcome these mechanisms. Indeed, it is generally the case that stable microbial populations, such as those in rivers or soil, are anchored to a substratum. Populations not so attached are held at their preferred ecological location by other physical means, such as flotation in static waters, but such mechanisms would be distinctly hazardous in, say, a river. Detached microorganisms, unless they alight on a suitable alternative substratum, may rapidly be removed from their niche and cease to be part of the stable population from which they originate. The fate of such organisms is indeterminate. The general importance of microbial adhesion has been recognized in recent years and has been the subject of a number of monographs (Ellwood *et al.*, 1979; Beachey, 1980; Berkeley *et al.*, 1980; Bitton and Marshall, 1980; O'Connor *et al.*, 1981).

In the case of *E. coli* a number of fimbrial adhesins have been described and they are discussed in detail in Chapter 4. It should be noted, however, that non-fimbriate *E. coli* may be capable of attaching to surfaces (Ip *et al.*, 1981; Sussman *et al.*, 1982a) by mechanisms that have not yet been elucidated.

The heterogeneity of fimbriae, in terms of the receptors to which they bind, and their antigenicity, is significant for the sites that they colonize and may have importance for the host response to colonization.

Endotoxins

These are the cell envelope lipopolysaccharides (LPS) of Gram-negative organisms. The LPS molecule consists of three parts. Inserted into the outer membrane of the cell is (1) *lipid A* to which is attached (2) the *core oligosaccharide,* and the outermost part of the molecule is (3) the *O-specific polysaccharide.* The

Table 2. *Biological activities of lipid A or lipopolysaccharide*
(after Kabir et al., 1978; Westphal et al., 1983)

Lethal	Local Shwartzman reaction
Hypotensive	Platelet aggregation
Pyrogenic	Hagemann factor activation
Induction of prostaglandin synthesis	Induction of plasminogen activator
Leucopoenic	Adjuvant activity
Leucocytotic	Thymus-independent immunogen
Enhancement of phagocytosis	B lymphocyte polyclonal activation
Enhancement of non-specific resistance to infection	B lymphocyte mitogenicity
Macrophage activation	Interferon induction
Complement activation	Gelation of *Limulus* amoebocyte lysate

latter endows the organisms with its O-serogroup identity (Ørskov *et al.*, 1977), while lipid A endows the LPS with its toxicity. The endotoxicity of LPS (Table 2) is expressed when it is released in the body by the breakdown of infecting bacterial cells and the effects can be reproduced by parenteral administration of purified lipopolysaccharide or lipid A. Apart from the activation of the complement and coagulation cascades, the toxicity of LPS can probably be accounted for by its effects on cell membranes.

Organisms in which the LPS is complete are termed smooth (S) and they are virulent. Loss of parts or the whole of the repeating subunit structure of the O-antigenic polysaccharide side chains in mutant strains leads to a reduction of virulence. The toxicity of the LPS extracted from such mutants is similar, indicating that virulence of living whole organisms is unrelated to LPS toxicity but is a function of the O-antigenic side chains (Medearis *et al.*, 1968). This may be related to the greater ease of opsonization of strains of *E. coli* producing degraded LPS (Van Dijk *et al.*, 1981).

Endotoxins are directly involved in the production of the symptoms of disease, ranging from fever at one extreme to the potentially fatal endotoxic shock of septicaemia. The structure and activity of endotoxin are the subject of several recent reviews (Kabir *et al.*, 1978; Bradley, 1979; Westphal *et al.*, 1983) and they are discussed in Chapter 5.

Exotoxins

E. coli strains may produce a number of toxins. The best recognized of these are the plasmid-coded heat-labile (LT) and the heat-stable (ST) enterotoxins produced by strains causing diarrhoeal disease in man.

LT is a protein of molecular weight 86,500 which is antigenically related to, but not identical with, *Vibrio cholerae* enterotoxin (choleragen). It has a subunit structure and consists of one A fragment and five B fragments. The latter bind the toxin to GM_1 gangliosides in the mucosal cell membrane and create a functional pore through which the A fragment enters the cell. There it ATP-ribosylates a component of the adenylate cyclase enzyme system, which is thereby activated and increases the production of cyclic adenosine monophosphate (cAMP). The effect of this on the intact intestinal mucosa is to give rise to a net secretion of water, sodium, chloride and bicarbonate ions.

ST is a polypeptide of approximate molecular weight 5000. It is poorly antigenic, unrelated to choleragen and acts by stimulating the guanylate cyclase system of mucosal epithelial cells. The resulting increase in the production of cyclic guanosine monophosphate (cGMP) leads to secretion of water and electrolytes. It is this secretion induced by LT and ST enterotoxins which gives rise to the characteristic watery diarrhoea of small bowel origin associated with intestinal infection with organisms producing these toxins (Turnberg, 1979). The LT and ST enterotoxins are discussed in Chapter 6.

Konowalchuk *et al.* (1977) described a heat-labile cytotoxin for Vero cells present in filtrates of some strains. This toxin, which they termed VT, is distinct from LT and ST in having no effect on Y1 adrenal or CHO cell lines. It has been postulated that it is related to the cytotoxin of *Shigella dysenteriae* type 1 (Shiga) (O'Brien *et al.*, 1983). The genes for VT production in one strain are carried by a temperate phage (VT phage) and VT^+ organisms are lysogenic for this phage (Scotland *et al.*, 1983). These toxins and their genetics are discussed in Chapter 8 and methods for their demonstration are described in Chapter 19.

A few invasive animal pathogenic strains of *E. coli* produce a heat-labile toxin controlled by the *vir* plasmid. The toxin is lethal for rabbits, mice and chickens when given intravenously (Smith, 1978).

Many strains of *E. coli* produce haemolysins and such strains are far more common in animals than man (Smith, 1963). The significance of haemolysin production for virulence is uncertain but since uptake of haemoglobin breakdown products may be a means of scavenging essential iron, haemolysin production may be an accessory virulence factor (Sussman, 1974; Welch *et al.*, 1981; see also Chapter 7).

Invasiveness

Certain strains of *E. coli* are able to invade the mucosal cells of the colon just like *Shigella* spp. These enteroinvasive *E. coli* (EIEC) produce a dysentery-like disease in man which will be discussed below. The ability of these strains to invade colonic mucosal cells is reflected in the production of keratoconjunctivitis in guinea-pigs (Serény test; Serény, 1955) and the ability to invade and multiply in HeLa cells (Dupont *et al.*, 1971) and HEp-2 cells (Mehlman *et al.*, 1977) (see also Chapter 19). The genetics of invasiveness of EIEC is discussed in Chapter 8.

Other Virulence Factors

Many strains of *E. coli* isolated from systemically invasive disease in animals carry the plasmid for colicin V (ColV) production. This appears to be associated with ability to cause septicaemia (Smith, 1974) though colicin V activity is not essential for ColV plasmid-mediated virulence enhancement (Quackenbush and Falkow, 1979). Williams and Warner (1980) have presented evidence that the ColV plasmid mediates iron uptake independently of colicin V synthesis and activity. It is possible that this constitutes a factor for enhanced virulence (see Chapter 7).

Diseases of Man due to *E. coli*

E. coli is a multipotent pathogen that has evolved the ability to cause disease in several body systems and, at least in the bowel, there are several different mechanisms of pathogenesis. Moreover, it is possible that the causative role of *E. coli* in a number of diseases is not yet recognized.

Most of the diseases with which we are concerned are related to mucosal surfaces. What follows infection may be due to production of exotoxins or local invasion of and multiplication in mucosal cells. In some cases the mechanisms are not understood. Systemic invasive disease due to *E. coli* is unusual in man. When it occurs the septicaemia that develops may be accompanied by a distinct organotropism as in meningitis.

Mucosal colonization of the intestine and of the urinary tract may be asymptomatic. This may be because the host is immune to the pathogenicity mechanisms exhibited by the infecting strain. Asymptomatic infection in the urinary tract may be a more or less extended interlude between episodes of symptomatic infection and is more properly regarded as *covert*. Indeed, the covert nature of such infections is emphasized by their suspected implication in reduction of life expectancy in the aged but possibly at all ages.

Gastro-Intestinal Infections

The gastro-intestinal tract is easily accessible for pathogens ingested with food and drink. Bowel infections are, therefore, among the most common infections of man, their incidence depending on such environmental factors as personal and food hygiene and temperature. The part played by *E. coli* in such infections was suspected as early as Escherich himself and later, from time to time (Rowe, 1979), until the serology of *E. coli* began to be systematized (Kauffmann, 1944) and accurate definition and comparison of strains became possible.

The beginnings of modern knowledge about the involvement of *E. coli* in gastro-enteritis go back to Bray (1945), who found that 42 of 44 strains isolated from an outbreak of diarrhoea among infants in a hospital were serologically similar. A few years later several similar outbreaks were reported in which the serotype of the causative strain was identified (Giles and Sangster, 1948; Giles *et al.*, 1949; Taylor *et al.*, 1949). In the years that followed *E. coli* serology was refined (Kauffmann, 1954, 1966) and accurate identification of strains greatly facilitated.

The first indication of a specific virulence factor associated with *E. coli* was the report by Taylor *et al.* (1961) that certain strains associated with diarrhoea in children induced secretion of fluid and electrolytes into isolated loops of rabbit ileum. Several years later a serogroup O148 strain was shown to be associated with diarrhoea in British troops in Aden (Rowe *et al.*, 1970); it was later shown to be enterotoxigenic. At about the same time two different enterotoxins were identified in porcine strains (Smith and Gyles, 1970) and it was observed that there is a requirement for enterotoxigenic and enteropathogenic strains to colonize the small bowel (Gorbach *et al.*,1971; Thomson, 1955). In the case of enterotoxigenic strains fimbrial colonization factors were later identified (Evans *et al.*, 1977).

Finally, an unusual type of *E. coli* infection resembling dysentery due to *Shigella* spp. was identified in Japan (Ogawa *et al.*, 1968).

The *E. coli* strains associated with disease and briefly described above fall into three groups and are now separately designated. Strains associated with disease in infants are termed enteropathogenic *E. coli* (EPEC) while strains producing enterotoxins and principally associated with diarrhoea in travellers are termed enterotoxigenic *E. coli* (ETEC). Strains that cause the dysentery-like syndrome are called enteroinvasive *E. coli* (EIEC).

EPEC. Diarrhoeal disease, particularly that associated with the warmer seasons, has long been recognized and used to be given evocative names such as '*cholera infantum*', '*cholera nostras*' and, more recently, 'summer diarrhoea.' Most commonly a disease of infants and very young children, it was characterized by a very high mortality, a high incidence among the underprivileged and an apparent association with artificial feeding. Despite a behaviour that suggested that an infective agent was involved, none was identified. Between the 1920s and the 1950s there was an unexplained decline of summer diarrhoea in the general population. The history of EPEC infections has been detailed by Rowe (1979).

Outbreaks of gastro-enteritis in Aberdeen and London (Giles *et al.*, 1949; Giles and Sangster, 1948; Taylor *et al.*, 1949) were subsequently shown to be due to serotypes O55:B5 and O111:B4. Since then a number of other serotypes of EPEC have been identified (Taylor, 1961) (Table 3).

Since the decline of 'summer diarrhoea', outbreaks of infantile gastro-enteritis have been particularly associated with hospitals and day nurseries, though spo-

Table 3. *O serogroups of E. coli associated with normal faeces and various infections in humans[a]*

Normal faeces	EPEC[b] Outbreaks	EPEC[b] Sporadic	ETEC[c]	EIEC[d]	Neonatal meningitis	UTI[e]	Septicaemia
O1	O18	O1	O1	O28	O1	O1	O1
O2	O20	O2	O6	O112	O6	O2	O2
O4	O26	O4	O7	O115	O7	O4	O4
O5	O44	O6	O8	O124	O16	O6	O6
O6	O55	O8	O9	O136	O18	O7	O7
O7	O86	O15	O15	O143	O83	O8	O8
O8	O111	O21	O20	O144		O9	O9
O18	O112	O51	O25	O147		O11	O11
O20	O114	O75	O27	O152		O22	O18
O25	O119	O85	O60	O164		O25	O22
O45	O124		O63			O62	O25
O75	O125		O75			O75	O75
O81	O126		O78				
	O127		O80				
	O128		O85				
	O142		O88				
	O158		O89				
	O159		O99				
			O101				
			O109				
			O114				
			O115				
			O126				
			O128				
			O142				
			O148				
			O153				
			O159				

[a]Data from Ewing and Davies, 1961; Ørskov *et al.*, 1977; Rowe, 1979; Dupont, 1982.
[b]Enteropathogenic *E. coli*.
[c]Enterotoxigenic *E. coli*.
[d]Enteroinvasive *E. coli*.
[e]Urinary tract infection.

radic and community cases have been observed. Apparently infections in the community are more common in developing countries. Study of hospital outbreaks has facilitated the elucidation of the epidemiology and has permitted the identification of the causative *E. coli* and the serogroups responsible. Frequently cases in the community precede hospital outbreaks, which are due to admission to the hospital of a baby with established disease (Thomson *et al.*, 1956) or a carrier excreting the pathogenic strain. Bottle-feeding has been recognized as a particular risk factor.

The mechanisms of pathogenicity of EPEC are not fully understood. Oral challenge experiments indicate that a dose of 10^5 to 10^{10} organisms is required to induce diarrhoea (Ferguson and June, 1952; Levine *et al.*,1978). It seems likely that the rapid growth of *E. coli* facilitates development in contaminated food of the large numbers of organisms required to produce infection. Post-mortem studies have shown that EPEC proliferate in the upper small bowel. It has also been shown that some EPEC induce secretion of fluid into perfused segments of rat jejunum (Klipstein *et al.*, 1978) though they produce neither LT nor ST enterotoxins. The possible role of VT (Konowalchuk *et al.*, 1977) in the production of diarrhoea has yet to be established but it may not be necessary for the expression of enteropathogenicity by EPEC (Rothbaum *et al.*, 1982).

ETEC. Disease due to ETEC appears to be prevalent mainly in tropical and developing countries and affects those travelling to these places. The significance of ETEC in travellers' diarrhoea began to be clarified with the report by Rowe *et al* (1970) that a singel serotype (**O**148:H28), later shown to be enterotoxigenic, was responsible in 54% of cases of diarrhoea in a group of soldiers who had travelled from Great Britain to Aden. Similar observations were made by Dupont *et al.* (1971) in U.S. troops in the Far East from whom the same serotype was isolated. These findings led to a series of studies on groups of travellers to a number of developing countries and in each case ETEC were found to be responsible for a substantial proportion of cases of diarrhoea (Shore *et al.*, 1974; Gorbach *et al.*, 1975; Merson *et al.*, 1976; Sack *et al.*, 1977). ETEC serotypes have also been isolated from outbreaks of diarrhoea on cruise liners, where food was the source (e.g. Hobbs *et al.*, 1976), and from staff and visitors to a national park in the United States, where contaminated water was the source (Rosenberg *et al.*, 1977). The whole subject of travellers' diarrhoea has been reviewed by Nye (1979).

Apart from common-source food or water outbreaks, the spread of ETEC is probably by direct person-to-person contact. The relationship to travel is due to susceptible individuals from developed countries acquiring infection in tropical or developing countries where the prevalence of ETEC is high and where they are a significant cause of diarrhoeal disease at all ages but most particularly in infants and children (Guerrant *et al.*, 1975; Nalin *et al.*, 1975; Wadström *et al.*,

1976; Sack *et al.*, 1977b). Where standards of hygiene are high, ETEC infections are unimportant as a cause of diarrhoea.

Oral challenge experiments in human volunteers have shown that 10^8 to 10^{10} ETEC organisms are required to set up infection and diarrhoea. The incubation period and the effects are, to some extent, dose-dependent (Dupont *et al.*, 1971; Levine *et al.*, 1977). ETEC may produce either LT or ST (e.g. Gross *et al.*, 1976) or both (e.g. Rosenberg *et al.*, 1977). However, enterotoxin production alone does not appear to be sufficient to make an ETEC strain pathogenic. Evans *et al.* (1975) found the fimbrial colonization factor CFA/I present on an ETEC strain of serotype **O**78:**H**11. Another such fimbrial factor (CFA/II) is present on certain other ETEC serotypes (Evans and Evans, 1978). These colonization factors appear to permit ETEC to adhere to the ileal mucosa, where the enterotoxins are then secreted in close proximity to their target epithelial cells.

EIEC. These organisms may be atypical in that they may be non-motile, late or non-lactose fermenting and anaerogenic, though frequently not all these characteristics are present. In addition, certain EIEC O serogroups are antigenically related to *Shigella* (Edwards and Ewing, 1972; Ørskov *et al.*, 1977). Clinically the syndrome they produce is similar to or indistinguishable from *Shigella* infection.

The pathogenesis of EIEC-induced dysentery was studied by Dupont *et al.* (1971) by oral challenge of guinea-pigs and in rabbit ileal loops. They showed that epithelial cell invasion was the essential feature. After oral challenge human volunteers showed the characteristic features of mild dysentery. As in *Shigella* dysentery (Levine *et al.*, 1973) the challenge dose for human volunteers is comparatively low (Hobbs *et al.*, 1949). The enteroinvasion characteristic of EIEC is associated with the presence of a plasmid (Sansonetti *et al.*, 1982; Silva *et al.*, 1982; see also Chapter 8).

Disease due to EIEC is uncommon, but it may be confused with shigellosis and its prevalence underestimated. A large outbreak in the United States was traced to cheese imported from France (Marier *et al.*, 1973; Tulloch *et al.*, 1973). However, little is known about the mechanism or route of transmission of EIEC but they may be assumed to be similar to those of *Shigella*.

Urinary Tract Infection

The urinary tract consists of the kidneys, ureters, bladder and urethra. Usually urine flows from the kidneys into the bladder from which it is voided by way of the urethra and constitutes an effective wash-out system (O'Grady *et al.*, 1968). If the functional valves between the ureters and the bladder are incompetent, urine may reflux from the bladder back into the renal pelvis as a result of the pressure set up in the bladder during micturition. Indeed, if the pressure is high enough the reflux may extend into the collecting ducts of the kidney (intrarenal reflux) (Rolleston *et al.*, 1974). Backward flow along the urethra, if it occurs, can be of significance only in the short female urethra. In a sense, therefore, the urinary tract is a 'closed' system protected from the gross entry of micro-organisms. Nevertheless, urinary tract infection (UTI) is extremely common. It has a special age and sex distribution, being most common in females and increasing in incidence with age in both sexes. The clinical and related aspects of UTI have been reviewed by Sussman and Asscher (1979).

Laboratory diagnosis of UTI depends on the isolation of the infecting organism from the urine but a potential complication is contaminating organisms present in urine passed in the normal way. Though this problem can be avoided by obtaining urine with a catheter, this is undesirable because of the risk of introducing infection into the bladder in this way (Beeson, 1958). Collection of urine by direct puncture of the bladder with a syringe, though quite safe, is not commonly carried out. A significant advance in the diagnosis and study of UTI was the statistical definition of the criteria for the diagnosis by Kass (1956). He demonstrated that the presence of more than 10^5 organisms per millilitre of a single species in a properly collected speciman of urine was evidence, with high probability, for the presence of infection in the bladder rather than contamination of the specimen during voiding. One such count indicates presence of infection with a probability of greater than 80% and three such counts bring the probability close to 100%.

Several different types of UTI are recognized. The most common is covert bacteriuria (CBU), which was originally called asymptomatic bacteriuria. As the name implies the patient is unaware of its presence but it is now recognized that it may be the silent phase of an infection punctuated by symptomatic episodes (Sussman *et al.*, 1969). The latter may be an infection limited to the bladder (acute cystitis) with mainly local symptoms or an infection of the kidney (acute pyelonephritis) with more or less severe constitutional symptoms in which bacterial invasion of the kidney parenchyma takes place. These kinds of UTI may also occur in individuals who have not previously been bacteriuric. Chronic pyelonephritis, in which scarring of the kidney occurs, is usually the result of kidney infection very early in life and certainly before the age of 5 years. Chronic pyelonephritis, if it is progressive, may destroy sufficient of the kidney substance to produce chronic renal failure with all that this implies. It is now clear that rarely, if ever, does chronic pyelonephritis arise *de novo* after early childhood in the structurally and functionally normal urinary tract. For this reason we shall be concerned here mainly with acute infections in the normal urinary tract. In the structurally and/or functionally abnormal urinary tract bacterial virulence factors are probably not as important as host factors in establishing and maintaining infection.

The organism most commonly isolated from all types of UTI is *E. coli* (Kunin *et al.*, 1964; Sussman *et al.*, 1969; McAllister *et al.*, 1971; Newcastle Asymptomatic Bacteriuria Research Group, 1975) and they mostly belong to a restricted range of serogroups (Kunin, 1966; Grüneberg *et al.*, 1968; and Table 3). The prevalence of these serogroups in UTI is similar to that in the faeces of normal individuals (Turck *et al.*, 1962; Grüneberg and Bettelheim, 1969; Grüneberg *et al.*, 1968; Roberts *et al.*, 1975). This type of evidence gave rise to the *prevalence theory* of *E. coli* pathogenicity in UTI, which postulates that the strains responsible for UTI are simply those that predominate in the faeces. It has become apparent, however, that there are distinct differences not only between the strains in faeces and UTI but also between the strains responsible for different types of UTI. This gave rise to the notion that special pathogenicity features are associated with uropathogenic *E. coli*. This *special pathogenicity theory* is now generally accepted and several characteristics have been postulated to be involved in the pathogenesis of UTI by *E. coli*.

Capsular (K) antigens. These are important *E. coli* virulence factors (see Chapter 5). Only a limited range of these acidic polysaccharide antigens (K1, K2, K3, K12, K13) is found with significant frequency in UTI (Mabeck *et al.*, 1971; Kaijser *et al.*, 1977). Though there appears to be no difference in the frequency with which K antigens are found in cystitis and pyelonephritis, presence of K antigens favours kidney invasion in a mouse model (Kalmanson *et al.*, 1975) particularly if they are present in large amounts (Nicholson and Glynn, 1975). Similarly, strains causing acute pyelonephritis contain more K antigen than those causing cystitis (Glynn *et al.*, 1971; Kaijser, 1973; McCabe *et al.*, 1975).

K antigens are anti-phagocytic and K1, as we shall see, is poorly immunogenic. Such properties may be of considerable significance in UTI accompanied by tissue invasion. Supporting evidence for this comes from the observation that in experimental pyelonephritis anti-K antibodies are more protective than anti-O antibodies (Kaijser and Olling, 1973; Kaijser and Ahlstedt, 1977; Kaijser *et al.*, 1978).

Somatic (O) antigens. Roberts *et al.* (1975) have shown that the concordance of O serogroup between the predominant faecal strain and that causing UTI may be absent in CBU. They suggested that CBU may be of such long standing that the faecal O serogroup has changed. It is likely, therefore, that symptomatic UTI frequently occurs when the urinary tract is first colonized with a given strain of *E. coli*. Similarly, Lindberg *et al.* (1975) found that the eight most common faecal O serogroups were more prevalent in symptomatic infection than in CBU. Several other differences between strains of *E. coli* isolated from symptomatic UTI and CBU have been summarized by Hanson *et al.* (1975). They showed that rough strains are more common in CBU than in symptomatic infection. This suggests that in the course of prolonged colonization, as in CBU, strains with altered cell envelopes are selected. Such strains also tend to be more serum-sensitive and they yield extracts that are less efficient in sensitizing erythrocytes for indirect haemagglutination (IHA) than are strains from symptomatic infections. These differences are reflected in the host response, in that IHA titres of sera from patients with CBU are lower when the patient's own strain is the source of antigen than when the antigen is prepared from standard strains of the same serogroup. Similarly, strains from patients with CBU are less immunogenic than those from patients with acute UTI (Sohl-Akerlund *et al.*, 1977). In chronic UTI *E. coli* may undergo serological degradative changes (Bettelheim and Taylor, 1969) and it has been suggested that these are due to prolonged exposure to specific urinary secretory immunoglobulin A (sIgA) (Hanson *et al.*, 1978). All this suggests that intact O antigens may be one of the virulence factors of *E. coli* in UTI. Complementary evidence is provided by the observation that when CBU is eliminated by treatment, symptoms frequently occur when reinfection with a strain of different serotype takes place (Asscher *et al.*, 1969).

Haemolysin production. We have already noted the suggestion (Sussman, 1974) that haemolysin production may be a virulence factor in that haemolysis makes available a source of iron where it would otherwise be scarce. Haemolytic strains of *E. coli* were first observed by Dudgeon *et al.* (1921, 1923) in normal faeces, UTI and extra-intestinal infections. Since then other workers have found that haemolytic strains are more prevalent among those isolated from UTI and other extra-intestinal infections than they are in normal faeces (Cooke and Ewins, 1975; Minshew *et al.*, 1978a,b). Half of the strains isolated from UTI produce both α- and β-haemolysins, whereas few of those from normal faeces do so (Cooke and Ewins, 1975).

Sufficient iron appears to be present in urine (Diem and Lentner, 1970) to support the growth of *E. coli*. The significance of haemolysin production for virulence is, therefore, likely to be principally in tissue, including kidney, invasive infections.

Adherence. As we have seen, it seems likely that K antigens and haemolysin production are virulence factors for invasive infection. The nature and state of O antigens of infecting *E. coli* strains determine their immunogenicity, antigenicity and serum sensitivity. Though immunogenicity and antigenicity may be of some significance for UTI limited to the lumen of the urinary tract, and the degree of serum sensitivity is probably a highly significant feature in determining tissue invasion, O antigens are unlikely to be of primary importance as virulence factors. Moreover, none of the above appear to be capable of endowing *E. coli* with the capacity to colonize the urinary tract against the wash-out effects of urine flow. It is, therefore, significant that the capacity of *E. coli* to adhere to uroepithelial cells (UEC) correlates with the severity of UTI from which they are isolated (Svanborg Edén *et al.*, 1976). This observation was extended by Svanborg Edén *et al.* (1978) and the use of UEC obtained from urinary sediment for the experimental study of adhesion has been described in detail (Svanborg Edén, 1978).

The properties of the adhesion and colonization factors of *E. coli* are considered in detail in Chapter 4 and this account will deal principally with the part they play in the pathogenesis of UTI and its consequences.

Since adhesion was first implicated in the pathogenesis of UTI confirmatory evidence has continued to accumulate. Though only a small proportion of faecal *E. coli* adhere to UEC, some 70% of those isolated from patients with acute pyelonephritis do so (Svanborg Edén *et al.*, 1978) and the predominant faecal strain of patients with UTI has adherence properties similar to those of the urinary strain isolated early in the infection (Svanborg Edén *et al.*, 1979b). Similarly, *E. coli* colonizing the distal urethra and introitus of normal women are non-adhesive (Sussman, 1982a). Moreover, *E. coli* adheres preferentially to periurethral and vaginal cells from women and girls who are prone to UTI (Kallenius and Winberg, 1978; Fowler and Stamey, 1977). This is highly significant since colonization of the periurethral area and introitus is a prelude to UTI (O'Grady *et al.*, 1970; Bollgren and Winberg, 1976; Stamey, 1980).

It is generally agreed that the adhesive properties of *E. coli* can, in most cases, be accounted for by the presence of fimbriae and that haemagglutination (HA) is a valuable model for the analytical study of fimbrial adhesins. By this and other means several different HA types of fimbrial adhesin have been recognized. The first of these were the so-called mannose-sensitive haemagglutinins (MS-HA), which give an agglutination of guinea-pig erythrocytes that is inhibitable by D-mannose, and mannose-resistant haemagglutinins (MR-HA) which give an agglutination of human erythrocytes that is not inhibited by D-mannose. The first indication that the latter were of significance in adhesion in the urinary tract was provided by Svanborg Edén and Hanson (1978), who showed that adhesion of *E. coli* to UEC was not inhibited by mannose. In fact, both MS fimbriae and MR fimbriae adhere to UEC (Parry *et al.*, 1982) but the presence of MR fimbriae is associated with strains isolated from patients with UTI (Sussman *et al.*, 1982a; Parry *et al.*, 1983).

The receptors to which MS-HA fimbriae attach is unknown but it is recognized that they attach to uromucoid and the Tamm-Horsfall glycoprotein which it contains (Ørskov *et al.*, 1980; Parry *et al.*, 1982). It has been pointed out that urinary mucus may be a source of confusion in *in vitro* tests of the capacity of *E. coli* to adhere to UEC (Chick *et al.*, 1981). The possible significance of MS-HA fimbriae in the pathogenesis will be considered below.

Adhesion of uropathogenic *E. coli* to human periurethral cells correlates with the presence of MR-HA for human erythrocytes (Kallenius and Möllby, 1979; Kallenius *et al.*, 1980c) and this has been confirmed with purified fimbriae (Korhonen *et al.*, 1980). The adhesion was shown by Kallenius *et al.* (1980b) to be due to fimbriae that bind to the carbohydrate moiety of the P blood group antigen. They also showed that uropathogenic *E. coli* do not agglutinate erythrocytes from individuals with the rare p̄ blood group phenotype, which do not have P antigens. The fimbriae that bind to the P antigen have been termed P fimbriae (Kallenius *et al.*, 1981) and the minimal receptor has been identified as the disaccharide α-D-Gal*p*-(1→4)-β-D-Gal*p* (Kallenius *et al.*, 1980a; Svenson *et al.*, 1983). Independent studies by Leffler and Svanborg Edén (1980, 1981) provided evidence that globoseries glycosphingolipids present in cell membranes are the receptors on UEC for the fimbriae of uropathogenic *E. coli*. The identity of the receptor on erythrocytes with that on UEC was demonstrated by Kallenius *et al.* (1981). At about the same time it was shown that not all uropathogenic *E. coli* with MR-HA adhesins possess P fimbriae. Such strains, which do not react in a P-specific manner, were termed X-specific and it was shown that adhesins of these two kinds could occur separately or together on the same organism. Although one X-specific strain of *E. coli* has been shown to have M blood group-specific HA activity (Väisänen *et al.*, 1982) the receptors for other X-specific adhesins are probably sialoglycoproteins.

Of the candidate virulence factors for *E. coli* in the upper urinary tract, P fimbriae are found with greatest frequency. In three studies they were found in at least 81% of strains from girls with pyelonephritis giving MR-HA of human erythrocytes (Kallenius *et al.*, 1981; Leffler and Svanborg Edén, 1981; Väisänen *et al.*, 1981). The figures remain very impressive even when the prevalence of P-fimbriate organisms is calculated as a percentage of the total number of *E. coli* isolated rather than only of those giving MR-HA. P-fimbriate *E. coli* are far less common in acute cystitis and less common again in CBU (Kallenius *et al.*, 1981; Leffler and Svanborg Edén, 1981). The prevalence of P-fimbriate *E. coli* in faeces appears to be less than 10% but they may represent as much as 50% of the MR-HA strains isolated (Leffler and Svanborg Edén, 1981).

Adult UTI also yield a high proportion of MR-HA strains (Parry *et al.*, 1983). In pregnant women with UTI there is a significant correlation between infection with MR-HA strains and a past history of UTI. Thus, a pregnant woman with an *E. coli* UTI and a past history of UTI has a seven-fold greater chance that this infection is due to an MR-HA strain than if she had no such history. Similarly, the risk of a pregnant patient with symptomatic UTI having an infection with an MR-HA strain is six-fold greater than for a patient with CBU (Parry *et al.*, 1983).

The significance of P fimbriae in lower UTI in children and in adults is uncertain but studies to establish these points are in progress. Moreover, the significance of X-specific adhesins in all types of UTI remains to be established.

Since attachment of P fimbriae to UEC depends on the expression of the P

antigen on the cell surface, such expression may be expected to affect susceptibility to infection both qualitatively and quantitatively. Indeed, in the study by
Parry *et al.* (1983) the predictive character of past history for the nature of the
infecting organism in UTI of pregnancy may be explainable in this way. The
fluorescence-activated cell sorting technique for the study of the density and
localization of P receptors on cells, as described by Svenson and Kallenius
(1983), will be of value in examining the relationship between susceptibility and
receptor expression. Lomberg *et al.* (1983) have shown that in the absence of
vesicoureteric reflux, P_1 blood group expression on the patient's cells contributes
to susceptibility to recurrent pyelonephritis due to P-fimbriate *E. coli*. When
reflux is present no advantage is conferred on P-fimbriate organisms, presumably
because the reflux carries the organisms to the kidney. One might also predict
that in the absence of reflux, individuals with the rare p̄ phenotype would be
highly resistant to upper UTI due to *E. coli*.

Whether adhesins are produced in the urinary tract is of some importance in
formulating convincing models for the function of virulence factors in the pathogenesis of UTI. Direct examination of *E. coli* in urine without subculture suggests that MS-HA adhesins are not produced *in vivo* while MR-HA adhesins may
be produced *in vivo* (Sharon *et al.*, 1981). Production of P fimbriae *in vivo* has
been demonstrated with bacteria from centrifuged fresh urine by a receptor-
specific particle agglutination test (Svenson *et al.*, 1982; Möllby *et al.*, 1983).
The report by Harber *et al.* (1982) which purports to demonstrate a lack of
adherence to UEC by freshly isolated urinary *E. coli* is, therefore, difficult to
understand. The methods used have been the subject of criticism (Sussman *et al.*,
1982b) and the matter is further discussed by Parry and Rooke in Chapter 4.

A two-phase hypothetical model for colonization of the urinary tract has been
formulated in general terms by Ørskov *et al.* (1980). They propose that MS-HA
fimbriae play a role as colonization factors in the colon, where large amounts of
mucus are produced. At other sites where mucus is produced and where there is a
tendency to emptying, as in the urinary tract, mucus may have a dual role. Either
organisms are captured by free mucus and are removed during micturition or, if
additional adhesins such as MR-HA fimbriae are produced, their expression,
presumably after a phase switch (Eisenstein, 1981), may determine colonization
of the urinary tract.

Experimental UTI. A variety of animal experimental models of UTI have been
proposed. Most of these are models of pyelonephritis produced either by the
haematogenous or the ascending route. It has proved extremely difficult to produce kidney infection by the ascending route so long as the bladder is normal and
in the absence of marked reflux. If the bladder is modified surgically or by the
introduction of a foreign body, kidney invasion is more readily obtained. The
available models have been briefly reviewed by Gorrill (1968) and Miraglia
(1970).

The relevance of animal models of pyelonephritis for infection in man is doubtful. Indeed, the difficulty of producing kidney infection experimentally when the urinary tract is normal appears exactly to parallel the situation in man, where infection of the normal tract does not lead to kidney invasion unless the pathogen has special virulence factors. Thus, unless suitable receptors for colonization are present on their UEC, animals do not provide suitable models for infection in man. Receptors for *E. coli* MR-HA adhesins are not present in the rat or pig (Parry *et al.*, 1982), both of which have been used for the study of experimental UTI. The necessary receptors do, however, appear to be present in the mouse (Hagberg *et al.*, 1983a) but until recently the most commonly used mouse models involved haematogenous infection (e.g. Harle *et al.*, 1975). The recent report by Hagberg *et al.* (1983b) of an ascending UTI model in the mouse without obstruction is, therefore, of great interest. Monkey UEC also possess the necessary receptors (Parry *et al.*, 1982) and an important model of UTI in *Macaca fascicularis* has been described by Kallenius *et al.* (1983).

Indirect consequences of UTI. Several indirect results of UTI have been reported. Their mechanism is obscure but the possibility that the effects of bacterial infection may be far-reaching and indirect should not be lost on those with an interest in the mechanisms of microbial pathogenicity.

A relationship between CBU and raised blood pressure in adults has been reported in a number of surveys (Kass *et al.*, 1965, 1978; Freedman *et al.*, 1965; Sussman *et al.*, 1969) but the increase is small, of marginal statistical significance and probably of little clinical significance (Kunin, 1979; Stamey, 1980).

The presence of CBU in pregnancy is associated with a high risk of progression to acute pyelonephritis (Kass, 1960). The reasons are not clearly understood but may be related to abnormalities of the urinary tract (Williams *et al.*, 1968; Grüneberg *et al.*, 1969), some of which may be temporary and associated with pregnancy, and the properties of urine in pregnancy that enhance its qualities as a bacterial growth medium (Asscher *et al.*, 1966). It has also been claimed that CBU in pregnancy is associated with an excess risk of toxaemia of pregnancy (pre-eclampsia), prematurity, low birth weight and stillbirth (Kass, 1960; Norden and Kass, 1968; Beard and Roberts, 1968) but it seems unlikely that CBU alone is responsible (Asscher, 1980). However, it is interesting that in a pregnant mouse experimental model *E. coli* was found to have a profound effect on foetal development (Coid *et al.*, 1978).

Perhaps the most challenging indirect effect of CBU to have been reported is that it reduces life expectancy in old age (Dontas *et al.*, 1981; Dontas, 1983) and at other ages (Evans *et al.*, 1982).

Meningitis

Meningitis due to *E. coli* is predominantly an infection of the newborn. About 80% of the strains responsible carry the K1 antigen (McCracken *et al.*, 1974) and such strains are carried by up to 30% of normal individuals (Schiffer *et al.*, 1976). Rectal swabs from 35% of these give an almost pure culture of K1-bearing organisms and in an additional 31% they represent the majority of the organisms cultured (Sarff *et al.*, 1975). Up to 38% of the newborn may carry such organisms in their stool (Schiffer *et al.*, 1976) and the origin in the majority of these is to be found in the maternal stool (Sarff *et al.*, 1975). Some two-thirds of the strains responsible for meningitis belong to serogroups O1, O7, O16 and O18ac (Schiffer *et al.*, 1976).

Grados and Ewing (1970) found that the *E. coli* isolated from the cerebrospinal fluid of a newborn infant with meningitis was agglutinated by antiserum specific for *Neisseria meningitidis* groups B and Kasper *et al.* (1973) showed that the capsules of the two organisms shared antigenic specificity. The acidic K1 polysaccharide is a polymer of *N*-acetylneuraminic acid (Barry and Goebel, 1957; Barry, 1958) and it is probably structurally identical with the meningococcal group B polysaccharide (Liu *et al.*, 1971).

As a rule serum antibody is protective against blood-borne organisms that cause meningitis but K1 antibodies are infrequently found in normal adults (Schiffer *et al.*, 1976). Even adult patients recovering from infections with group B meningococci may have only low levels of antibody (Brandt *et al.*, 1972) and children after urinary tract infections with K1$^+$ organisms produce no serum antibody against this antigen (Kaijser *et al.*, 1973). It would appear, therefore, that K1 antigen is poorly immunogenic. However, antibody produced in animals by injection of whole organisms is protective (Wolberg and DeWitt, 1969) and anti-K1 sIgA is present in human colostrum (Schiffer *et al.*, 1976) and may be the usual protection for the newborn infant.

Only some 40% of infants with *E. coli* septicaemia in the absence of meningitis are infected with K1$^+$ organisms (Schiffer *et al.*, 1976), that is, about half the prevalence in meningitis. It may be, therefore, that K1$^+$ organisms are particularly invasive and likely to give rise to septicaemia and that, because of a tropism for the meninges, the organisms in the blood stream set up meningitis. The nature of the special tendency to attack the meninges is unknown but may be related to the organotropism described by Buddingh and Polk (1939). The first line of defence against invasion is likely to be colostral and milk sIgA but once invasion has taken place the infant is unprotected in the absence of maternal circulating IgG (Wilfert, 1978).

Wound Infection

Wound infections due to *E. coli* have long been recognized (Keighley and Burdon, 1979). They most commonly follow surgical operations, such as appendicectomy, in the course of which the alimentary tract is entered. The strain isolated from the wound infection is usually the same as can be obtained from the patient's pre-operative rectal swab (Quick and Brogan, 1968). Nothing is known of the virulence factors of *E. coli* that are operative in wound infection. However, the severity of the infection appears to be inoculum-dependent. Thus, if a very severely infected (gangrenous) appendix is delivered through the operation wound, subsequent wound infection is more likely to occur (Annotation, 1970). However, it is now apparent that this type of wound infection is probably due to a mixture of *E. coli* and non-sporing anaerobes, in particular *Bacteroides fragilis,* which is also part of the intestinal flora. There is experimental evidence that they may be acting synergistically (Onderdonk *et al.,* 1976). Thus, when sub-infective doses of *E. coli* and *B. fragilis* are inoculated together into surgical incisions in guinea-pigs, infection with pus formation takes place (Kelly, 1978). The amounts and ratios of the organisms inoculated are critical. It may be that interference with neutrophil phagocytosis by *B. fragilis* (Ingham *et al.,* 1977) permits *E. coli* to establish infection, though it normally becomes established with difficulty in experimental wounds (Quick and Brogan, 1968). Such a mechanism is supported by the observation that metronidazole, which has significant activity against anaerobes, will prevent experimental *E. coli* sepsis (Onderdonk *et al.,* 1979). Once infection is established virulence factors such as K antigens and iron uptake may play a significant role in maintaining it but evidence is lacking.

Septicaemia

This may be defined as the entry of bacteria into the blood stream with a variety of clinical effects. Generally speaking, for septicaemia to result, organisms must enter the circulation in large numbers, usually over a period of time. The source may be either a discrete lesion, such as a collection of pus, or a more diffuse infected tissue lesion or, most commonly, a urinary tract infection. Septicaemia is to be distinguished from bacteraemia, in which small or moderate numbers of organisms enter the circulation over a very short time span and are removed by the phagocytic system without causing symptoms. In septicaemia the organisms are also more or less rapidly removed from the circulation but it is endotoxin (LPS) released from organisms in the blood stream that causes the symptoms; except when the function of the phagocytes is depressed or their circulating number is reduced (less than 10^8/litre) or, possibly, in the agonal state, bacteria do not remain long enough in the blood stream to multiply there. Nevertheless, there is a continuum between bacteraemia and septicaemia and it is the symptoms that are the hallmark of the latter.

Little is known of the virulence factors, other than endotoxin, associated with the capacity of *E. coli* to cause septicaemia in man. In the neonate presence of K1 is thought to be important in producing invasion and meningitis (see above). The significance of the ColV-associated iron uptake system (Williams, 1979) in human disease is unknown. However, in view of the circumstances in which septicaemia occurs in man, namely as an uncommon complication of very common infections, and then usually in compromised individuals, it may be that host factors rather than microbial virulence factors are the determinants. Of the many Gram-negative organisms that can cause septicaemia *E. coli* most commonly does so (McCabe, 1981) probably because infections due to it are so frequent.

The symptoms of septicaemia are due to the effects of endotoxin released from organisms. As has been noted above, endotoxin has a wide range of biological activities (Table 2), such as activation of the interacting physiological amplification cascades including the complement, coagulation and fibrinolytic systems (Gewurz and Lint, 1977; Ulevitch and Cochrane, 1977). This leads to disseminated intravascular coagulation which, apart from other effects, damages blood vessels (Wardle, 1980). The activation of the kallikrein system increases vascular permeability and dilatation, leading to some of the haemodynamic changes that together constitute endotoxic shock (Stoner, 1972). Endotoxin is also a cell toxin and a variety of organs may be affected including the lung (Corrin, 1980), kidney and liver (Coalson *et al.*, 1979).

Haemolytic-Uraemic Syndrome (HUS)

This condition, which is the commonest cause of acute renal failure in children, was first described in 1955 and consists of an acute febrile illness followed by acute renal failure and intravascular haemolysis. Though usually a disease of infancy, the peak age incidence varies in different places and there is a seasonal variation. In addition, case clustering has been observed and all this points to infection as a cause.

Karmali *et al.* (1983) have reported evidence of infection with VT$^+$ *E. coli* in 73% of sporadic cases of HUS in Canada. There is some evidence for a similar association in the United Kingdom.

Haemorrhagic Colitis

This condition, characterized by sudden severe abdominal colic and grossly bloody diarrhoea, was described by Riley *et al.* (1983). There had been previous reports of sporadic cases associated with antibiotic administration. They studied two food-related outbreaks in Oregon and Michigan in each of which serotype O157:H7 organisms were isolated from affected individuals. A similar outbreak has been reported from Canada (cited by Day *et al.*, 1983) and it has been reported that the strains responsible produce VT (O'Brien *et al.*, 1983).

Diseases of Animals due to *E. coli*

Several types of disease can be produced by *E. coli* in animals. These include enteric colibacillosis, in which the organism remains localized to the intestine and in which either diarrhoea or toxaemia may be produced, and systemic coli-bacillosis, which is characterized by invasion and septicaemia. *E. coli* can also give rise to mastitis in cows. These conditions are considered in detail in Chapter 3.

In addition, *E. coli* may be responsible in some cases for the metritis-mastitis-agalactia syndrome (Ross *et al.,* 1969), possibly as an event secondary to an endocrine disturbance (Martin, 1970). The strains responsible belong to various serotypes and tend to produce a distinct capsule. The route by which organisms reach the udder is uncertain and both haematogenous spread during parturition or shortly thereafter and ascending infection from the bowel via the genital mucosa have been proposed. Since the syndrome is accompanied by adrenal hypertrophy and a raised plasma cortisol level, it has been suggested that it is a phenomenon secondary to an endocrine disturbance.

In young rabbits *E. coli* may produce a dysentery-like diarrhoea secondary to infection with coccidia (Löliger *et al.,* 1969a,b). There is extensive colonization of the colon and caecum and frequently also of the jejunum, apparently facilitated by epithelial damage produced by the coccidia (Matthes, 1969). The causative organisms adhere to the intestinal epithelium but neither LT or ST production has been demonstrated.

E. coli may cause enteric colibacillosis and systemic colibacillosis in domestic animals including foals and puppies. The pathogenesis of these infections is unknown. In dogs *E. coli* is associated with UTI, especially in the presence of outflow obstruction caused by prostatic hypertrophy, and in bitches it is frequently isolated from infections of the womb (pyometra). Haemolytic strains of *E. coli* are said to be associated with diarrhoea in puppies.

The intense rearing of chickens, turkeys, geese, ducks and pigeons is associated with septicaemia and granulomatous disease due to *E. coli.* Septicaemia in chickens and turkeys is associated particularly with serotypes O1:**K**1, O2:**K**1, O78:**K**80 but others may be found (Nivas *et al.,* 1977). Infection is due to inhalation of faecal dust and chicks in the first week of life are particularly susceptible, though infection may occur later either if environmental contamination is sufficiently severe or secondary to virus infection (Goren, 1978). The organisms reach the air sacs, multiply there and set up septicaemia. Coli-granulomatosis (Hjärre's disease) is an uncommon sporadic infection due to capsulate strains of *E. coli* of various serogroups. It is characterized by the development of granulomata in lung, liver, kidney, appendix and skin. The capsule is thought to be the relevant virulence factor.

The Immune Response to *E. coli*

The immune response to *E. coli* is mounted against O, H, K and fimbrial antigens. Whether there are responses to other bacterial antigens, such as the

protoplast membrane and internal constituents, is unknown. However, with prolonged exposure to organisms such responses are likely to take place.

Bacterial O antigens (lipopolysaccharides) are rather resistant to degradation
and have long been regarded as poor immunogens. Neither macrophage nor T
lymphocyte participation is necessary in the antibody response to free O antigen,
which is due to direct stimulation of B lymphocytes in a thymus-independent
manner. At low concentrations B lymphocyte activation appears to be specific,
while at higher concentrations polyclonal activation takes place. The thymus-
independent immunogenicity of O antigens is due to the polysaccharide side
chains with their repeating oligosaccharide epitopic groups and the antibody
response is generally limited to IgM production (Humphrey, 1982; Munro and
Brenner, 1982) but under some conditions there is production of IgG.

The immune response to flagella (H antigens) of *E. coli* appears not to have
been studied. However, flagellins are usually highly immunogenic proteins and
it may be assumed that the response is characteristic of protein immunogens in
general and similar to the response to comparable antigens of other enterobacteria.

The immunogenicity of capsules (K antigens) depends on their chemical structure. Thus K1 and K5 are poorly immunogenic in man, while others give rise to
good responses (see Chapter 5).

Fimbriae are protein structures and their immunogenicity may be regarded as
typical of proteins.

Natural (normal) antibodies. Views have differed as to whether these are truly
'pre-formed' antibodies, unrelated to previous exposure to antigenic stimulation,
or antibodies produced in response to cross-reacting antigens, but the latter is
now the accepted view. Such antibodies against *E. coli* may be found in germfree
animals (Ikari, 1964). It has been suggested that anti-A and anti-B blood group
isoantibodies are due to immunization with bacteria including *E. coli* (Springer,
1970).

Response to colonization and infection. Serum antibody to *E. coli* present in
infants at birth is maternal IgG but at this age mucosal surfaces are unprotected.
Therefore, passive protection by maternal secretory IgA derived from colostrum
and milk plays an important role in mucosal immunity, particularly because sIgA
formation is the slowest of the specific immunity mechanisms to develop (Burgio
et al., 1980; Hanson *et al.,* 1983). The sIgA produced by infants and children
appears only rarely to include anti-fimbrial antibody (Hanson *et al.,* 1983) but
such antibodies are present in breast milk (Svanborg Eden *et al.,* 1979a) together
with anti-K1 antibody (Carlsson *et al.,* 1982) and anti-O antibodies (Gindrat *et
al.,* 1972). Anti-fimbrial antibody is probably necessary to prevent colonization
of neonates with *E. coli,* since neither anti-O antibody (Carlsson *et al.,* 1976) nor

anti-K antibody (Carlsson *et al.*, 1982) inhibits colonization. In infants intestinal colonization induces production of serum antibodies (Lodinova *et al.*, 1973).

Recovery from *Salmonella* gastroenteritis (La Brooy *et al.*, 1980) and cholera (Waldman *et al.*, 1971) in man is accompanied by the appearance of specific sIgA and the same may be taken to occur in analogous *E. coli* infections. The same occurs in intestinal infections in animals and the antibodies produced are protective (Porter *et al.*, 1977; also see Chapter 9).

Normal urine contains both IgG and IgA, which appear to originate from the glomerular filtrate and from tissue fluid (Burdon, 1973b) and their concentration increases in UTI (Burdon, 1970). Some of these antibodies are specific for *E. coli* (Tourville *et al.*, 1968) but they are unlikely to play any part in the local antibacterial defence of the urinary tract. Such a role is reserved for sIgA (Burdon, 1973b, 1976). The local immune response in the urinary tract has been reviewed by Holmgren and Smith (1975).

Systemic infections may be regarded as those in which tissue invasion is sufficient to produce general rather than purely local symptoms. Fever is, perhaps, the best recognized of these general symptoms. Systemic infections can be separated into those in which infection is limited to the tissues and those in which organisms enter the blood stream. In either case serum IgM and IgG are produced.

In its immune response kidney-invasive UTI may be regarded as typical of tissue-invasive infection. Thus, whereas lower UTI, which is usually limited to superficial infection, gives rise to local production of sIgA, which can be found in the urine, kidney invasion by *E. coli* in acute pyelonephritis is accompanied by the appearance of serum IgM (Needell *et al.*, 1955; Percival *et al.*, 1964) and in recurrent or chronic infection IgG appears (Hanson and Winberg, 1966; Hanson *et al.*, 1969). The bacteria appearing in the urine in upper UTI are frequently coated with antibody (Thomas *et al.*, 1975). In addition, there are cell-mediated immune responses in pyelonephritis and it has been reported that in early pyelonephritis in childhood these may be depressed (Ahlstedt *et al.*, 1983). Other aspects of the immune response to UTI have been reviewed by Holmgren and Smith (1975).

The immune response to septicaemia is difficult to distinguish from that of tissue invasion since in man the latter almost always follows the former. The nature of the response to septicaemia without previous tissue invasion, as may follow instrumentation of the infected urinary tract, has not been recorded. An analogous situation may be the chronic leakage of bacterial antigens into the circulation in cirrhosis of the liver, where the detoxication functions of the liver are depressed and in which there is a marked increase in anti-*E. coli* antibodies (Wright, 1982). In any case, the dominant features of septicaemia are the toxic effects of endotoxin release (Young *et al.*, 1977), probably caused by pre-existing IgM antibody and complement in the circulation. The immune response to *E. coli* septicaemia in poultry has been reviewed by Parry and Porter (1981).

Protective (effector) mechanisms. In mucosal infection sIgA acts in several ways. It inhibits adhesion of the infecting organism (Svanborg Edén *et al.*, 1976), neutralizes toxins (Rowley, 1983) and may have a bactericidal action (Burdon, 1973a).

In an animal model of intraperitoneal infection with *E. coli* it has been shown that the first event is immigration of phagocytic cells. It appears that under these conditions both anti-O and anti-K antibodies are protective but the effect of the anti-K antibodies is more powerful (Ahlstedt, 1983). Similar observations have been made with a rabbit pyelonephritis model (Kaijser *et al.*, 1983) and a mouse model (Kaijser and Ahlstedt, 1977). The greater effectiveness of anti-K antibodies is probably related to their opsonic effect in the presence of K antigen (Van Dijk *et al.*, 1977) and anti-O antibody is bactericidal (see Chapter 25). Polyclonal IgM is not protective, probably because of its low specificity. Antibody-dependent cell-mediated cytotoxicity against *E. coli* has been demonstrated and may be an effector mechanism in immunity but it may also hold potential for the production of host tissue damage (Hagberg *et al.*, 1982).

Protection. A consideration of effector mechanisms in immunity to *E. coli* makes it clear that protection may be achieved in a number of ways depending on the pathogenicity mechanism involved. Thus, in surface infections specific sIgA production would be protective but neither the antigens nor the best routes of antigen administration necessary to obtain such responses have been clearly defined. The use of purified fimbrial vaccines for this purpose has been reviewed by Korhonen and Rhen (1982) and the use of oral vaccines of whole bacteria is discussed in Chapter 9. Where toxin-dependent pathogenicity mechanisms are involved, the use of toxoids or toxin subunits as immunization antigens are a possibility (see Chapter 6). The problems involved in the choice of antigen and its route of administration have been considered by Rowley (1983).

The generation of bactericidal O and opsonic K antibodies by immunization to combat systemic infection in man is probably impractical for several reasons. The variety of O antigens on virulent *E. coli* is too large for the preparation of a suitable vaccine and, since O antigenic vaccines would be unavoidably endotoxic, their administration would in any case be unacceptable. Similar *Pseudomonas* vaccines have, however, been used (Jones, 1983). The problem in the case of K antigens is the poor immunogenicity of the important serotypes.

In experimental animals the toxic effect of endotoxin can be neutralized by administration of antibody to core LPS (Braude and Douglas, 1972; Braude *et al.*, 1973). Similarly, the presence in patients with septicaemia of high titres of antibody to core LPS determinants is protective (McCabe *et al.*, 1972). It has also been shown that passive administration of antiserum to core LPS, produced in human volunteers by immunization with a heat-killed *E. coli* J5 vaccine, reduces the death rate from Gram-negative septicaemia (Ziegler *et al.*, 1982).

References

Ahlstedt, S. (1983). Host defence against intraperitoneal *Escherichia coli* infection in mice. *Progress in Allergy* **33**, 236–246.

Ahlstedt, S., Hagberg, M., Jodal, U. and Maåcrild, S. (1983). Cell-mediated immune parameters in children with pyelonephritis caused by *Escherichia coli*. *Progress in Allergy* **33**, 289–297.

Annotation (1970). Wound infection after appendicectomy. *Lancet* **1**, 930–931.

Asscher, A. W. (1980). "The Challenge of Urinary Tract Infection." Academic Press, London.

Asscher, A. W., Sussman, M., Waters, W. E., Davis, R. H. and Chick, S. (1966). Urine as a medium for bacterial growth. *Lancet* **2**, 1037–1041.

Asscher, A. W., Sussman, M., Waters, W. E., Evans, J. A. S., Campbell, H., Evans, K. T. and Williams, J. E. (1969). Asymptomatic significant bacteriuria in the non-pregnant woman. II. Response to treatment and follow-up. *British Medical Journal* **1**, 804–806.

Barry, G. T. (1958). Coliminic acid, a polymer of *N*-acetylneuraminic acid. *Journal of Experimental Medicine* **107**, 507–521.

Barry, G. T. and Goebel, W. F. (1957). Coliminic acid, a substance of bacterial origin related to sialic acid. *Nature (London)* **179**, 206–208.

Beachey, E. H. (1980). "Bacterial Adherence." Chapman & Hall, London.

Beard, R. W. and Roberts, A. P. (1968). Asymptomatic bacteriuria during pregnancy. *British Medical Bulletin* **24**, 44–49.

Beeson, P. B. (1958). The case against the catheter. *American Journal of Medicine* **24**, 1–3.

Berkeley, R. C. W., Lynch, J. M., Melling, J., Rutter, P. R. and Vincent, B. (1980). "Microbial Adhesion to Surfaces." Ellis Horwood, Chichester.

Bettelheim, K. A. and Taylor, J. (1969). A study of *Escherichia coli* isolated from chronic urinary tract infection. *Journal of Medical Microbiology* **2**, 225–236.

Bettelheim, K. A., Breadon, A., Faiers, M. C., O'Farrell, S. M. and Shooter, R. A. (1974). The origin of O-serotypes of *Escherichia coli* in babies after normal delivery. *Journal of Hygiene* **72**, 67–78.

Bitton, G. and Marshall, K. C. (1980). "Adsorption of Microorganisms to Surfaces." Wiley, New York.

Bollgren, I. and Winberg, J. (1976). The periurethral aerobic flora in girls highly susceptible to urinary infections. *Acta Paediatrica Scandinavica* **65**, 81–87.

Bradley, S. B. (1979). Cellular and molecular mechanisms of action of bacterial endotoxins. *Annual Review of Microbiology* **33**, 67–94.

Brandt, B. L., Wyle, F. A. and Artenstein, M. S. (1972). Radioactive antigen-binding assay for *Neisseria meningitidis* polysaccharide antibody. *Journal of Immunology* **108**, 913–920.

Braude, A. I. and Douglas, H. (1972). Passive immunisation against the local Shwartzman reaction. *Journal of Immunology* **108**, 505–512.

Braude, A. I., Douglas, H. and Davis, C. E. (1973). Treatment and prevention of intravascular coagulation with antiserum to endotoxin. *Journal of Infectious Diseases* **128**, Supplement, S157–S164.

Bray, J. (1945). Isolation of antigenically homogeneous strain of *Bact. coli Neapolitanum* from summer diarrhoea of infants. *Journal of Pathology and Bacteriology* **57**, 239–247.

Buddingh, G. J. and Polk, A. D. (1939). Experimental meningococcus infection of the chick embryo. *Journal of Experimental Medicine* **70**, 485–497.

Burdon, D. W. (1970). Quantitative studies of urinary immunoglobulins in hospital patients, including patients with urinary tract infection. *Clinical and Experimental Immunology* **6**, 189 196.

Burdon, D. W. (1973a). The bactericidal action of immunoglobulin A. *Journal of Medical Microbiology* **6**, 131–139.

Burdon, D. W. (1973b). Immunoglobulins in the urinary tract. Discussion on a possible role in urinary tract infection. *In* "Urinary Tract Infection" (Eds. W. Brumfitt and A. W. Asscher), pp. 148–158. Oxford University Press, London and New York.

Burdon, D. W. (1976). Immunological reactions to urinary infection: The nature and function of secretory immunoglobulins. *In* "The Scientific Foundations of Urology" (Eds. D. L. Williams and G. D. Chisolm), pp. 192–196. Heinemann, London.

Burgio, G. R., Lanzavecchia, A., Plebani, A., Jayakar, S. and Ugazio, A. G. (1980). Ontogeny of secretory immunity: Levels of secretory IgA and natural antibodies in saliva. *Pediatric Research* **14**, 1111–1114.

Carlsson, B., Gothefors, L., Ahlstedt, S., Hanson, L. A. and Winberg, J. (1976). Studies of *Escherichia coli* O antigen. Specific antibodies in human milk, maternal serum and cord blood. *Acta Paediatrica Scandinavica* **65**, 216–224.

Carlsson, B., Kaijser, B., Ahlstedt, S., Gothefors, L. and Hanson, L. A. (1982). Antibodies against *Escherichia coli* capsular (K) antigens in human milk and serum—their relation to the *E. coli* gut flora of the mother and neonate. *Acta Paediatrica Scandinavica* **71**, 313–318.

Chick, S., Harber, M. J., Mackenzie, R. and Asscher, A. W. (1981). Modified method for studying bacterial adhesion to isolated uroepithelial cells and uromucoid. *Infection and Immunity* **34**, 256–261.

Clarke, R. T. J. and Bauchop, T. (Eds.) (1977). "Microbial Ecology of the Gut." Academic Press, London.

Coalson, J. J., Archer, L. T., Benjamin, B. A., Beller-Todd, B. K. and Hinshaw, L. B. (1979). A morphologic study of live *Escherichia coli* organism shock in baboons. *Experimental and Molecular Pathology* **31**, 10–22.

Coid, C. R., Sanderson, H., Slavin, G. and Altman, D. G. (1978). *Escherichia coli* infection in mice and impaired fetal development. *British Journal of Experimental Pathology* **59**, 292–297.

Cooke, E. M. (1974). "*Escherichia Coli* and Man." Churchill-Livingstone, Edinburgh and London.

Cooke, E. M. and Ewins, S. P. (1975). Properties of strains of *Escherichia coli* isolated from a variety of sources. *Journal of Medical Microbiology* **8**, 107–111.

Cooke, E. M., Ewins, S. P. and Shooter, R. A. (1969). The changing faecal population of *Escherichia coli* in hospital medical patients. *British Medical Journal* **4**, 593–595.

Cooke, E. M., Shooter, R. A., Kumar, P. J., Rousseau, S. A. and Foulkes, A. (1970). Hospital food as a possible source of *Escherichia coli* in patients. *Lancet* **1**, 436–437.

Cooke, E. M., Hettiaratchy, F. G. T. and Buck, A. C. (1971). Fate of ingested *Escherichia coli* in normal persons. *Journal of Medical Microbiology* **5**, 361–369.

Corrin, B. (1980). Lung pathology in septic shock. *Journal of Clinical Pathology* **33**, 891–894.

Day, N. P., Scotland, S. M., Cheasty, T. and Rowe, B. (1983). *Escherichia coli* O157:H7 associated with human infection in the United Kingdom. *Lancet* **1**, 825.

Diem, K. and Lentner, C. (1970). "Documenta Geigy Scientific Tables," 7th ed., 664. Geigy, Basel.

Dontas, A. S. (1983). The effect of bacteriuria on survival in old age. *Geriatric Medicine Today* **2**, 74–82.

Dontas, A. S., Kasviki-Charvati, P., Papanayiotou, P. C. and Marketos, S. G. (1981). Bacteriuria and survival in old age. *New England Journal of Medicine* **304**, 939–943.

Drasar, B. S. and Hill, J. J. (1974). "Human Intestinal Flora." Academic Press, London.

Dubos, R., Schaedler, R. W. and Costello, R. (1963). Composition, alteration and effects of intestinal flora. *Federation Proceedings, Federation of American Societies for Experimental Biology* **22**, 1322–1329.

Dudgeon, L. S., Wordley, E. and Bawtree, F. (1921). On *Bacillus coli* infections of the urinary tract, especially in relation to haemolytic organisms. *Journal of Hygiene* **20**, 137–164.

Dudgeon, L. S., Wordley, E. and Bawtree, F. (1923). On *Bacillus coli* infections of the urinary tract, especially in relation to haemolytic organisms. Second communication. *Journal of Hygiene* **21**, 168–198.

Dupont, H. L. (1982). *Escherichia coli* diarrhoea. *In* ''Bacterial Infections of Humans: Epidemology and Control'' (Eds. A. S. Evans and H. A. Feldman), pp. 219–234. Plenum Medical, New York.

Dupont, H. L., Formal, S. B., Hornick, R. B., Snyder, M. J., Libonati, J. P., Sheehan, D. G., Labrec, E. H. and Kalas, J. P. (1971). Pathogenesis of *Escherichia coli* diarrhea. *New England Journal of Medicine* **285,** 1–9.

Edwards, P. R. and Ewing, W. H. (1972). ''Identification of Enterobacteriaceae.'' Burgess, Minneapolis, Minnesota.

Eisenstein, B. (1981). Phase variation of type 1 fimbriae in *Escherichia coli* is under transcriptional control. *Science* **214,** 337–339.

Ellwood, D. C., Melling, J. and Rutter, P. (1979). ''Adhesion of Microorganisms to Surfaces.'' Academic Press, London.

Escherich, T. (1885). Die Darmbakterien der Sänglings und Neugeborenen. *Fortschritte der Medizin* **3,** 515–522.

Escherich, T. (1894). Über colicystitis un Kinderalter. *Jahrbuch für Kinderheilkunde* **44,** 268.

Escherich, T. (1899). Zur Ätiologie der Dysenterie. *Zentralblatt für Bacteriologie, Parasitenkunde und Infektionskrankheiten, Abteilung 1* **25,** 385.

Evans, D. A., Kass, E. H., Hennekens, C. H., Rosner, B., Miao, L., Kendrick, M.l., Miall, W. E. and Stuart, K. L. (1982). Bacteriuria and subsequent mortality in women. *Lancet* **1,** 156–158.

Evans, D. G. and Evans, D. J. (1978). New surface-associated heat-labile colonisation factor antigen (CFA/II) produced by enterotoxigenic *Escherichia coli* of serogroups O6 and O8. *Infection and Immunity* **21,** 638–647.

Evans, D. G., Silver, R. P., Evans, D. J., Chase, D. G. and Gorbach, S. L. (1975). Plasmid-controlled colonisation factor associated with virulence in *Escherichia coli* enterotoxigenic for humans. *Infection and Immunity* **12,** 656–666.

Evans, D. G., Evans, D. J., Jr. and Dupont, H. L. (1977). Virulence factors of enterotoxigenic *Escherichia coli*. *Journal of Infectious Diseases* **136,** Supplement, 118–123.

Ewing, W. H. and Davies, B. R. (1961). ''The O-antigen groups of *Escherichia coli* cultures from different sources.'' Centre for Disease Control, Atlanta, Georgia.

Ferguson, W. W. and June, R. C. (1952). Experiments on feeding adult volunteers with *Escherichia coli* O111, **B**4, a coliform organism associated with infant diarrhoea. *American Journal of Hygiene* **55,** 155–169.

Fowler, J. E., Jr. and Stamey, T. (1977). Studies of introital colonisation in women with recurrent urinary tract infection. VII. The role of bacterial adherence. *Journal of Urology* **117,** 472–476.

Freedman, L. R., Phair, J. P., Seki, M., Hamilton, H. B., Nefziger, M. D. and Hirata, M. (1965). The epidemiology of urinary tract infections in Hiroshima. *Yale Journal of Biology and Medicine* **37,** 262–282.

Gewurz, H. and Lint, T. F. (1977). Alternative modes and pathways of complement activation. *In* ''Comprehensive Immunology,'' Vol. 2, Biological Amplification Systems in Immunology (Eds. N. K. Day and R. A. Good), pp. 17–45. Plenum Medical, New York.

Giles, C. and Sangster, G. (1948). An outbreak of infantile gastroenteritis in Aberdeen. The association of a special type of *Bact. coli* with the infection. *Journal of Hygiene* **46,** 1–9.

Giles, C., Sangster, G. and Smith, J. (1949). Epidemic gastroenteritis of infants in Aberdeen during 1947. *Archives of Disease in Childhood* **24,** 45–53.

Gillies, R. R. (1978). Bacteriocin typing of Enterobacteriaceae. *In* ''Methods in Microbiology,'' Vol. 11 (Eds. T. Bergan and J. R. Norris), pp. 78–86. Academic Press, London.

Gindrat, J.-J., Gothefors, L., Hanson, L. A., and Winberg, J. (1972). Antibodies in human milk against *Escherichia coli* of the serotypes most commonly found in neonatal infections. *Acta Paediatrica Scandinavica* **61,** 587–590.

Glynn, A. A., Brumfitt, W. and Howard, C. J. (1971). K antigens of *Escherichia coli* and renal involvement in urinary tract infections. *Lancet* **1,** 514–516.

Gorbach, D. L., Banwell, J. G., Chatterjee, B. D., Jacobs, B. and Sack, R. B. (1971). Acute undifferentiated human diarrhoea in the tropics. I. Alterations in intestinal microflora. *Journal of Clinical Investigation* **50**, 881–889.

Gorbach, S. L., Kean, B. H., Evans, D. G., Evans, D. J. and Bessudo, D. (1975). Travellers' diarrhoea and toxigenic *Escherichia coli*. *New England Journal of Medicine* **292**, 933–935.

Goren, E. (1978). Observations on experimental infection of chicks with *Escherichia coli*. *Avian Pathology* **7**, 213–224.

Gorrill, R. H. (1968). Susceptibility of the kidney to experimental infection. *In* "Urinary Tract Infection" (Eds. F. O'Grady and W. Brumfitt), pp. 24–36. Oxford University Press, London and New York.

Grados, O. and Ewing, W. H. (1970). Antigenic relationship between *Escherichia coli* and *Neisseria meningitidis*. *Journal of Infectious Diseases* **122**, 100–103.

Gross, R. J. and Holmes, B. (1983). The Enterobacteriaceae. *In* "Principles of Bacteriology, Virology and Immunity," 7th ed., Vol. 2 (Eds. G. S. Wilson, A. A. Miles and M. T. Parker), pp. 272–284. Edward Arnold, London.

Gross, R. J., Scotland, S. M. and Rowe, B. (1976). Enterotoxin testing of *Escherichia coli* causing epidemic enteritis in the United Kingdom. *Lancet* **1**, 629–630.

Grüneberg, R. N. and Bettelheim, K. A. (1969). Geographical variation in serological types of urinary *Escherichia coli*. *Journal of Medical Microbiology* **2**, 219–224.

Grüneberg, R. N., Leigh, D. A. and Brumfitt, W. (1968). *Escherichia coli* serotypes in urinary tract infections. Studies in domicilliary antenatal and hospital practice. *In* "Urinary Tract Infection" (Eds. F. O'Grady and W. Brumfitt), pp. 68–79. Oxford University Press, London and New York.

Grüneberg, R. N., Leigh, D. A. and Brumfitt, W. (1969). Relationship of bacteriuria in pregnancy to acute pyeloncphritis, prematurity and fetal mortality. *Lancet* **2**, 1–3.

Guerrant, R. L., Moore, R. A., Kirschenfold, B. A. and Sande, M. A. (1975). Role of toxigenic and invasive bacteria in acute diarrhoea of childhood. *New England Journal of Medicine* **293**, 567–573.

Guinee, P. A. M. (1963). Preliminary investigations concerning the presence of *E. coli* in man and various species of animals. *Zentralblatt für Bakteriologie, Parasitenleunde, Infektionskrankheiten und Hygiene, Abteilung 1: Originale* **188**, 201–218.

Hagberg, L., Hull, R., Hull, S., Falkow, S., Freter, R. and Svanborg Edén, C. (1983a). Contribution of adhesion to bacterial persistence in the mouse urinary tract. *Infection and Immunity* **40**, 265–272.

Hagberg, L., Engberg, L., Preter, R., Lam, J., Olling, S. and Svanborg Edén, C. 1983b). Ascending, unobstructed urinary tract infection in mice caused by pyelonephritogenic *Escherichia coli* of human origin. *Infection and Immunity* **40**, 273–283.

Hagberg, M., Ahlstedt, S. and Hanson, L. A. (1982). Antibody-dependent cell-mediated cytotoxicity against *Escherichia coli* O antigens. *European Journal of Clinical Microbiology* **1**, 59–65.

Hanson, L. A. and Winberg, J. (1966). Demonstration of antibodies of different immunoglobulin types to the O-antigen of the infecting strain in infants and children with pyelonephritis. *Nature (London)* **212**, 1495–1496.

Hanson, L. A., Holmgren, J., Jodal, U. and Lomberg, J. (1969). Precipitating antibodies to *Escherichia coli* O antigens: A suggested difference in the antibody response of infants and children with first and recurrent attacks of pyelonephritis. *Acta Paediatrica Scandinavica* **58**, 506–512.

Hanson, L. A., Ahlstedt, S., Jodal, U., Kaijser, B., Larsson, P., Lidin-Janson, G., Lincoln, K., Lindberg, U., Mattsby, I., Olling, S., Peterson, H. and Sohl, A. (1975). The host–parasite relationship in urinary tract infection. *Kidney International* **8**, S28–S34.

Hanson, L. A., Ahlstedt, S., Carlsson, B., Kaijser, B., Larsson, P., Mattsby-Baltzer, I., Sohl-Akerlund, A., Svanborg Edén, C. and Svennerholm, A.-M. (1978). Secretory IgA antibodies to enterobacterial virulence antigens: their induction and possible relevance. *In* "Secretory Immunity and Infection" (Eds. J. R. McGhee, J. Mestecky and J. L. Babb), pp. 165–176. Plenum, New York.

Hanson, L. A., Söderström, T., Brinton, C., Carlsson, B., Larson, P., Mellander, L. and Svanborg-Eden (1983). Neonatal colonisation with *Escherichia coli* and the ontogeny of the antibody response. *Progress in Allergy* **33**, 40–52.

Harber, M. J., Chick, S., Mackenzie, R. and Asscher, A. W. (1982). Lack of adherence to epithelial cells by freshly isolated urinary pathogens. *Lancet* **1**, 586–568.

Harle, E. M. J., Bullen, J. J. and Thomson, D. A. (1975). Influence of oestrogen on experimental pyelonephritis caused by *Escherichia coli*. *Lancet* **2**, 283–286.

Hartley, C. L. and Richmond, M. H. (1975). Antibiotic resistance and survival of *E. coli* in the alimentary tract. *British Medical Journal* **4**, 71–74.

Hobbs, B. C., Rowe, B. and Taylor, J. (1949). School outbreak of gastro-enteritis associated with a pathogenic paracolon bacillus. *Lancet* **2**, 530–532.

Hobbs, B. C., Rowe, B., Kendall, M., Turnbull, P. C. B. and Ghosh, A. C. (1976). *Escherichia coli* O27 in adult diarrhoea. *Journal of Hygiene* **77**, 393–400.

Holmes, B. and Gross, K. J. (1983). Coliform bacteria; various other members of the Enterobacteriaceae. *In* "Principles of Bacteriology, Virology and Immunity," 7th ed., Vol. 2 (Eds. G. S. Wilson, A. A. Miles and M. T. Parker), pp. 285–292. Edward Arnold, London.

Holmgren, J. and Smith, J. W. (1975). Immunological aspects of urinary tract infection. *Progress in Allergy* **18**, 289–352.

Humphrey, J. H. (1982). The fate of antigens. *In* "Clinical Aspects of Immunology" (Eds. P. J. Lachmann and D. K. Peters), pp. 161–186. Blackwell, Oxford.

Ikari, N. S. (1964). Bactericidal antibody to *Escherichia coli* in germ-free mice. *Nature (London)* **202**, 879–881.

Ingham, H. R., Sisson, P. R., Tharagonnet, D., Selkon, J. B. and Codd, A. A. (1977). Inhibition of phagocytosis *in vitro* by obligate anaerobes. *Lancet* **2**, 1252–1254.

Ip, S. M., Crichton, P. B., Old, D. C. and Duguid, J. P. (1981). Mannose-resistant and eluting haemagglutinins and fimbriae in *Escherichia coli*. *Journal of Medical Microbiology* **14**, 223–226.

Jones, R. J. (1983). Infection in the burned patient. *In* "Infections in the Immunocompromised Host" (Eds. C. S. F. Easmon and H. Gaya) pp. 7–11. Academic Press, London.

Kabir, S., Rosenstreich, D. L. and Mergenhagen, S. E. (1978). Bacterial endotoxins and cell membranes. *In* "Bacterial Toxins and Cell Membranes" (Eds. J. Jeljaszewicz and T. Wadström), pp. 59–87. Academic Press, London.

Kaijser, B. (1973). Immunological studies on some *Escherichia coli* strains with special reference to K antigen and its relation to urinary tract infection. *Journal of Infectious Diseases* **127**, 670–677.

Kaijser, B. and Ahlstedt, S. (1977). Protective capacity of antibodies against *Escherichia coli* O and K antigens. *Infection and Immunity* **17**, 286–289.

Kaijser, B. and Olling, S. (1973). Experimental haematogenous pyelonephritis due to *Escherichia coli* in rabbits: The antibody response and its protective capacity. *Journal of Infectious Diseases* **128**, 41–49.

Kaijser, B., Jodal, U. and Hanson, L. A. (1973). Studies on the antibody response and tolerance to *E. coli* K antigens in immunised rabbits and children with urinary tract infection. *International Archives of Allergy* **44**, 260–273.

Kaijser, B., Hanson, L. A., Jodal, U., Lidin-Janson, G. and Robbins, J. B. (1977). Frequency of *Escherichia coli* K antigens in urinary tract infections in children. *Lancet* **1**, 663–664.

Kaijser, B., Larsson, P. and Olling, S. (1978). Protection against ascending *Escherichia coli* pyelonephritis in rats and significance of local immunity. *Infection and Immunity* **20**, 78–81.

Kaijser, B., Larsson, P., Nimmich, W. and Söderstrom, T. (1983). Antibodies to *Escherichia coli* K and O antigens in protection against acute pyelonephritis. *Progress in Allergy* **33**, 275–288.

Kallenius, G. and Möllby, R. (1979). Adhesion of *Escherichia coli* to human periurethral cells is correlated to mannose-resistant agglutination of human erythrocytes. *FEMS Microbiology Letters* **5**, 295–299.

Kallenius, G. and Winberg, J. (1978). Bacterial adherence to periurethral epithelial cells in girls prone to urinary tract infection. *Lancet* **2**, 540–543.

Kallenius, G., Möllby, R., Svenson, S. B., Winberg, J. and Hultberg, H. (1980a). Identification of a carbohydrate receptor recognised by uropathogenic *Escherichia coli*. *Infection* **8**, Supplement 3, S288–S293.

Kallenius, G., Möllby, R., Svenson, S. B., Winberg, J., Lindblat, A. and Svenson, S. (1980b). The p^k antigen as receptor of pyelonephritic *Escherichia coli*. *FEMS Microbiology Letters* **7**, 297–302.

Kallenius, G., Möllby, R. and Winberg, J. (1980c). *In vitro* adhesion of *Escherichia coli* to human periurethral cells. *Infection and Immunity* **28**, 972–980.

Kallenius, G., Möllby, R., Hultberg, H., Svenson, S. B., Cedergren, B. and Winberg, J. (1981). Structure of carbohydrate part of receptor on human uroepithelial cells for pyelonephritogenic *Escherichia coli*. *Lancet* **2**, 604–606.

Kallenius, G., Svenson, S. B., Hultberg, H., Möllby, R., Winberg, J. and Roberts, J. A. (1983). P-fimbriae of pyelonephritogenic *Escherichia coli*: Significance for reflux and renal scarring—A hypothesis. *Infection* **11**, 73–76.

Kalmanson, G. M., Harwick, H. J., Turck, M. and Guze, L. B. (1975). Urinary tract infection; localisation and virulence of *Escherichia coli*. *Lancet* **1**, 134–136.

Karmali, M. A., Steele, B. T., Petrie, H. and Lim, C. (1983). Sporadic cases of haemolytic-uraemic associated with faecal cytotoxin and cytotoxin-producing *Escherichia coli* in stools. *Lancet* **1**, 619–620.

Kasper, D. L., Winkelhake, J. L., Zollinger, W. D., Brandt, B. C. and Artenstein, M. S. (1973). Immunochemical similarity between polysaccharide antigens of *Escherichia coli* O7:K1(L):NM and group B *Neisseria meningitidis*. *Journal of Immunology* **110**, 262–268.

Kass, E. H. (1956). Asymptomatic infections of the urinary tract. *Transactions of the Association of American Physicians* **69**, 56–63.

Kass, E. H. (1960). Bacteriuria and pyelonephritis in pregnancy. *Archives of Internal Medicine* **105**, 194–198.

Kass, E. H., Savage, W. and Santamarina, B. A. G. (1965). The significance of bacteriuria in preventive medicine. *In* "Progress in Pyelonephritis" (Ed. E. H. Kass), pp. 3–10. Davis, Philadelphia.

Kass, E. H., Miall, W. E., Stuart, K. L. and Rosner, B. (1978). Epidemiologic aspects of infections of the urinary tract. *In* "Infections of the Urinary Tract" (Eds. E. H. Kass and W. Brumfitt), pp. 1–7. University of Chicago Press, Chicago.

Kauffmann, F. (1944). Zur serologie der Coli-Gruppe. *Acta Pathologica et Microbiologica Scandinavica* **21**, 20–45.

Kauffmann, F. (1954). "Enterobacteriaceae," 2nd ed. Munksgaard, Copenhagen.

Kauffmann, F. (1966). "The Bacteriology of Enterobacteriaceae." Munksgaard, Copenhagen.

Kauffmann, F. and Perch, B. (1943). Über die Coliflora des gesunden Menschen. *Acta Pathologica et Microbiologica Scandinavica* **20**, 201–220.

Keighley, M. R. B. and Burdon, D. W. (1979). "Antimicrobial Prophylaxis in Surgery." Pitman Medical, Tunbridge Wells.

Kelly, M. J. (1978). The quantitative and histological demonstration of pathogenic synergy between *Escherichia coli* and *Bacteroides fragilis* in guinea-pig wounds. *Journal of Medical Microbiology* **11**, 573–523.

Kennedy, R. P., Florde, J. J. and Petersdorf, R. G. (1965). Studies on the epidemiology of *E. coli* infections. IV. Evidence for nosocomial flora. *Journal of Clinical Investigation* **44**, 193–201.

Klipstein, F. A., Rowe, B., Engert, R. F., Short, H. B. and Gross, R. J. (1978). Enterotoxigenicity of enteropathogenic serotypes of *Escherichia coli* isolated from infants with epidemic diarrhoea. *Infection and Immunity* **21**, 171–178.

Konowalchuk, J., Speirs, J. I. and Stavric, S. (1977). Vero response to a cytotoxin of *Escherichia coli*. *Infection and Immunity* **18**, 775–779.

Korhonen, T. K. and Rhen, M. (1982). Bacterial fimbriae as vaccines. *Annals of Clinical Research* **14**, 272–277.

Korhonen, T. K., Edén, S. and Svanborg Edén, C. (1980). Binding of purified *Escherichia coli* pili to human urinary tract epithelial cells. *FEMS Microbiology Letters* **7**, 237–240.

Kunin, C. M. (1966). Asymptomatic bacteriuria. *Annual Review of Medicine* **17**, 383–406.

Kunin, C. M. (1979). "Detection, Prevention and Management of Urinary Tract Infection," 3rd ed. Lea & Febiger, Philadelphia.

Kunin, C. M., Deutscher, R. and Paquin, A. (1964). Urinary tract infection in schoolchildren: An epidemiological, clinical and laboratory study. *Medicine (Baltimore)* **43**, 91–130.

La Brooy, J. T., Davidson, G. P., Shearman, D. J. C. and Rowley, D. (1980). The antibody response to bacterial gastroenteritis in serum and secretions. *Clinical and Experimental Immunology* **41**, 290–296.

Leffler, H. and Svanborg Edén, C. (1980). Chemical identification of a glycosphingolipid receptor for *Escherichia coli* attaching to human urinary tract epithelial cells and agglutinating human erythrocytes. *FEMS Microbiology Letters* **8**, 127–134.

Leffler, H. and Svanborg Edén, C. (1981). Glycolipid receptors for uropathogenic *Escherichia coli* on human erythrocytes and uroepithelial cells. *Infection and Immunity* **34**, 920–929.

Levine, M. M., Dupont, H. L., Formal, S. B., Hornick, R. B., Takeuchi, A., Gangarosa, E. J., Snyder, M. J. and Libonati, J. P. (1973). Pathogenesis of *Shigella dysenteriae* 1 (Shiga) dysentery. *Journal of Infectious Diseases* **127**, 261–270.

Levine, M. M., Caplan, E. S., Waterman, D., Cash, R. A., Hornick, R. B. and Snyder, M. J. (1977). Diarrhoea caused by *Escherichia coli* that produce only heat-stable enterotoxin. *Infection and Immunity* **17**, 78–82.

Levine, M. M., Bergquist, E. J., Nalin, D. R., Waterman, D. H., Hornick, R. B., Young, C. R., Sotman, S. and Rowe, B. (1978). *Escherichia coli* strains that cause diarrhoea but do not produce heat-labile or heat-stable enterotoxins and are not invasive. *Lancet* **1**, 1119–1122.

Lindberg, U., Hanson, L. A., Jodal, U., Lidin-Janson, G., Lincoln, K. and Olling, S. (1975). Asymptomatic bacteriuria in schoolgirls. II. Differences in *Escherichia coli* causing asymptomatic and symptomatic bacteriuria. *Acta Paediatrica Scandinavica* **64**, 432–436.

Liu, T. Y., Gotschlich, E. C., Dunne, F. T. and Jonssen, E. K. (1971). Studies on the meningococcal polysaccharides. II. Composition and chemical properties of the group B and group C polysaccharide. *Journal of Biological Chemistry* **216**, 4703–4712.

Lodinova, R., Jonja, V. and Wagner, V. (1973). Serum immunoglobulins and coproantibody formation in infants after artificial intestinal colonisation with *Escherichia coli* O83 and oral lysozyme administration. *Pediatric Research* **7**, 659–669.

Löliger, H.-C., Matthes, S., Schubert, H. J. and Heckmann, F. (1969a). Die akuten Dysenterien der Jungkaninchen. *Deutsche Tierärztliche Wochenschrift* **76**, 16–20.

Löliger, H.-C., Matthes, S., Schubert, H. J. and Heckmann, F. (1969b). Die akuten Dysenterien der Jungkaninchen. *Deutsche Tierärztliche Wochenschrift* **76**, 38–41.

Lomberg, E., Hanson, L. A., Jacobsson, B., Jodal, U., Leffler, H. and Svanborg Edén, C. (1983). Correlation of P blood group, vesicoureteral reflux and bacterial attachment in patients with recurrent pyelonephritis. *New England Journal of Medicine* **308**, 1189–1192.

Mabeck, C. E., Ørskov, F., and Ørskov, I. (1971). *Escherichia coli* serotypes and renal involvement in urinary tract infections in children. *Lancet* **1**, 1312–1314.

McAllister, T. A., Alexander, J. G., Dulake, C., Percival, A., Boyce, J. M. H. and Wormald, P. J. (1971). The sensitivity of urinary pathogens—a survey. *Postgraduate Medical Journal* **47**, 7–18.

McCabe, W. R. (1981). Gram-negative bacteremia. *In* "Medical Microbiology and Infectious Diseases" (Ed. A. I. Braude), pp. 1387–1394. Saunders, Philadelphia.

McCabe, W. R., Carlings, P. C., Bruins, S. and Greely, A. (1975). The relation of K antigen to the virulence of *Escherichia coli*. *Journal of Infectious Diseases* **131**, 6–10.

McCabe, W. R., Kreger, B. E. and Johns, M. (1972). Type-specific and cross-reactive antibodies in Gram-negative bacteraemia. *New England Journal of Medicine* **287**, 261–267.

McCracken, G. H., Jr., Sarff, L. D., Glode, M. P., Mize, S. G., Schiffer, M. S., Robbins, J. B., Gotschlich, E. C., Ørskov, I. and Ørskov, F. (1974). Relation between *Escherichia coli* K1 capsular polysaccharide antigen and clinical outcome in neonatal meningitis. *Lancet* **2**, 246–250.

Macone, A. B., Pier, G. B., Pennington, J. E., Matthews, W. J. and Goldman, D. A. (1981). Mucoid *Escherichia coli* in cystic fibrosis. *New England Journal of Medicine* **304**, 1445–1449.

Marier, R., Wells, J. G., Swanson, R. C., Callahan, W. and Mehlman, I. J. (1973). An outbreak of enteropathogenic *Escherichia coli* foodborne disease traced to imported cheese. *Lancet* **2**, 1376–1378.

Martin, C. E. (1970). Status of the mastitis–metritis–agalactia syndrome of sows. *Journal of the American Veterinary Medical Association* **157**, 1519–1521.

Matthes, S. (1969). Die Darmflora gesunder und dysenteriekranker Kaninchen. *Zentralblatt für Veterinärmedizin, Reihe B* **16**, 563–570.

Medearis, D. N., Jr., Camitta, B. M. and Heath, E. C. (1968). Cell wall composition and virulence in *Escherichia coli*. *Journal of Experimental Medicine* **128**, 399–414.

Mehlman, I. J., Eide, E. L., Sanders, A. C., Fishbein, M. and Alusso, C. C. G. (1977). Methodology for recognition of invasive potential of *Escherichia coli*. *Journal of the Association of Official Analytical Chemists* **60**, 546–562.

Merson, M. H., Morris, G. K., Sack, D. A., Wells, T. G., Feeley, J. C., Sack, R. B., Creech, W. B., Kapikian, A. Z. and Gangarosa, E. J. (1976). Travellers' diarrhoea in Mexico. *New England Journal of Medicine* **294**, 1299–1305.

Milch, H. (1978). Phage typing of *Escherichia coli*. *In* "Methods in Microbiology," Vol. 11 (Eds. T. Bergan and J. R. Norris), pp. 87–155. Academic Press, London.

Minshew, B. H., Jorgensen, J., Counts, G. W. and Falkow, S. (1978a). Association of hemolysin production, hemagglutination of human erythrocytes and virulence for chicken embryos of extraintestinal *Escherichia coli* isolates. *Infection and Immunity* **20**, 50–54.

Minshew, B. H., Jorgensen, J., Swanstrum, M., Grootes-Reuvecamp, G. A. and Falkow, S. (1978b). Some characteristics of *Escherichia coli* strains isolated from extraintestinal infections in humans. *Journal of Infectious Diseases* **137**, 648–654.

Miraglia, G. J. (1970). Model infections in the evaluation of antimicrobial agents. *Transactions of the New York Academy of Sciences* **32**, 337–347.

Möllby, R., Kallenius, G., Korhonen, T. K., Winberg, J. and Svenson, S. B. (1983). P-fimbriae of pyelonephritogenic *Escherichia coli*: Detection in clinical material by a rapid receptor-specific agglutination test. *Infection* **11**, 68–72.

Munro, A. J. and Brenner, M. K. (1982). Cell interactions in the induction of the allergic response. *In* "Clinical Aspects of Immunology" (Eds. P. J. Lachmann and D. K. Peters), pp. 187–198. Blackwell, Oxford.

Nalin, D. R., McLaughlin, J. C., Rahaman, M., Yunus, M. and Curlin, G. (1975). Enterotoxigenic *Escherichia coli* and idiopathic diarrhoea in Bangladesh. *Lancet* **2**, 1116–1119.

Needell, M. H., Neter, E., Staubitz, E. and Bingham, W. A. (1955). The antibody (haemagglutination) response of patients with infection of the urinary tract. *Journal of Urology* **74**, 674–682.

Newcastle Asymptomatic Bacteriuria Research Group (1975). Asymptomatic bacteriuria in schoolchildren in Newcastle upon Tyne. *Archives of Disease in Childhood* **50**, 90–102.

Nicholson, A. M. and Glynn, A. A. (1975). Investigation of the effect of K antigen in *Escherichia coli* urinary tract infection by use of a mouse model. *British Journal of Experimental Pathology* **56**, 549–553.

Nivas, S. C., Peterson, A. C., York, M. D., Pomeroy, B. S., Jacobson, L. D. and Jordan, K. A. (1977). Epizootiological investigations of colibacillosis in turkeys. *Avian Diseases* **21**, 514–530.

Norden, C. W. and Kass, E. H. (1968). Bacteriuria of pregnancy—a critical appraisal. *Annual Review of Medicine* **19**, 431–470.

Nye, F. J. (1979). Travellers' diarrhoea. *Clinics in Gastroenterology* **8**, 767–781.

O'Brien, A. D., Lively, T. A., Chen, M. E., Rothman, S. W. and Formal, S. B. (1983). *Escherichia coli* O157:H7 strains associated with haemonhagic colitis in the United States produce a *Shigella dysenteriae* 1 (Shiga) like cytotoxin. *Lancet* **1**, 702.

O'Connor, M., Whelan, J. and Elliott, K. (1981). "Adhesion and Microorganism Pathogenicity." Pitman Medical, London.

O'Farrell, S. M., Lennox-King, S. M. J., Bettelheim, K. A., Shaw, E. J. and Shooter, R. A. (1976). *Escherichia coli* in a maternity ward. *Infection* **4**, 146–152.

Ogawa, H., Nakaumura, A. and Sakazaki, R. (1968). Pathogenic properties of enteropathogenic *Escherichia coli* from diarrheal children and adults. *Japanese Journal of Medical Science and Biology* **21**, 339–349.

O'Grady, F., Gauci, C. L., Watson, B. W. and Hammond, B. (1968). *In vitro* models simulating conditions of bacterial growth in the urinary tract. *In* "Urinary Tract Infection" (Eds. F. O'Grady and W. Brumfitt), pp. 80–90. Oxford University Press, London and New York.

O'Grady, F., Richards, B., McSherry, M. A., O'Farrell, S. M. and Cattell, W. R. (1970). Introital enterobacteria, urinary infection and the urethral syndrome. *Lancet* **2**, 1208–1210.

Onderdonk, A. B., Bartlett, J. G., Louie, T., Sullivan-Seigler, N. and Gorbach, S. L. (1976). Microbial synergy in experimental intra-abdominal abscess. *Infection and Immunity* **13**, 22–26.

Onderdonk, A. B., Louie, T. J., Tally, F. P. and Bartlett, J. G. (1979). Activity of metronidazole against *Escherichia coli* in experimental intra-abdominal sepsis. *Journal of Antimicrobial Chemotherapy* **5**, 201–210.

Ørskov, F. (1974). *Escherichia. In* "Bergey's Manual of Determinative Bacteriology" (Eds. R. E. Buchanan and N. E. Gibbon), pp. 293–296. Williams & Wilkins, Baltimore, Maryland.

Ørskov, I. and Ørskov, F. (1983). Serology of *Escherichia coli* fimbriae. *Progress in Allergy* **33**, 80–105.

Ørskov, I., Ørskov, F., Jann, B. and Jann, K. (1977). Serology, chemistry and genetics of O and K antigens of *Escherichia coli*. *Bacteriological Reviews* **41**, 667–710.

Ørskov, I., Ørskov, F. and Birch-Andersen, A. (1980). Comparison of *Escherichia coli* fimbrial antigen F7 with type 1 fimbriae. *Infection and Immunity* **27**, 657–666.

Parry, S. H. and Porter, P. (1981). Immunity to *Escherichia coli* and *Salmonella In* "Avian Immunology" (Eds. M. E. Rose, L. N. Payne and B. M. Freeman), pp. 327–347. British Poultry Science, Edinburgh.

Parry, S. H., Abraham, S. N. and Sussman, M. (1982). The biological and serological properties of adhesion determinants of *Escherichia coli* isolated from urinary tract infections. *In* "Clinical, Bacteriological and Immunological Aspects of Urinary Tract Infection in Children" (Ed. H. Schulte-Wissermann), pp. 113–125. Thieme, Stuttgart.

Parry, S. H., Boonchai, S., Abraham, S. N., Salter, J. M., Rooke, D. M., Simpson, J. M., Bint, A. J. and Sussman, M. (1983). A comparative study of the mannose-resistant and mannose-sensitive haemagglutinins of *Escherichia coli* isolated from urinary tract infections. *Infection* **11**, 123–128.

Percival, A., Brumfitt, W. and De Louvois, J. (1964). Serum antibody levels as an indication of clinically inapparent pyelonephritis. *Lancet* **2**, 1027–1033.

Porter, P., Parry, S. H. and Allen, W. D. (1977). Significance of immune mechanisms in relation to enteric infections of the gastrointestinal tract in animals. *In* "Immunology of the Gut" (Eds. R. Porter and J. Knight), pp. 55–75. Elsevier, Amsterdam.

Quackenbush, R. C. and Falkow, S. (1979). Relationship between colicin V activity and virulence in *Escherichia coli*. *Infection and Immunity* **24**, 562–564.

Quick, C. A. and Brogan, T. D. (1968). Gram-negative rods and surgical wound infection. *Lancet* **1**, 1163–1167.

Riley, L. W., Remis, R. S., Helgerson, S. D., McGee, H. B., Wells, J. G., Davis, B. R., Herbert, R. J., Olcott, E. S., Johnson, L. M., Hargrett, N. T., Blake, P. A. and Cohen, M. L. (1983). Haemorrhagic colitis associated with a rare *Escherichia coli* serotype. *New England Journal of Medicine* **308**, 681–685.

Roberts, A. P., Linton, J. D., Waterman, M. M., Gower, P. E. and Koutsaimanis, K. G. (1975). Urinary and faecal *Escherichia coli* O-serotypes in symptomatic urinary tract infection and asymptomatic bacteriuria. *Journal of Medical Microbiology* **8**, 311–318.

Rolleston, G. L., Making, T. M. J. and Hodson, C. J. (1974). Intrarenal reflux and the scarred kidney. *Archives of Disease in Childhood* **49**, 531–539.

Rosenberg, M. L., Koplan, J. P., Wachsmuth, J. K., Wells, J. G., Gangarosa, E. J., Guerrant, R. L. and Sack, D. A. (1977). Epidemic diarrhoea at Crater Lake from enterotoxigenic *Escherichia coli*. *Annals of Internal Medicine* **86**, 714–718.

Ross, R. F., Christian, L. L. and Spear, M. L. (1969). Role of certain bacteria in mastitis–metritis–agalactia of sows. *Journal of the American Veterinary Medical Association* **155**, 1844–1852.

Rothbaum, R., McAdams, A. J., Gramella, R. and Partin, J. C. (1982). A clinicopathologic study of enterocyte-adherent *Escherichia coli*: A cause of protracted diarrhoea in infants. *Gastroenterology* **83**, 441–454.

Rowe, B. (1979). The role of *Escherichia coli* in gastroenteritis. *Clinics in Gastroenterology* **8**, 625–644.

Rowe, B., Taylor, J. and Bettelheim, K. A. (1970). An investigation of travellers' diarrhoea. *Lancet* **1**, 1–5.

Rowley, D. (1983). Immune responses to enterobacteria presented by various routes. *Progress in Allergy* **33**, 159–174.

Russell, C. and Mellville, T. H. (1978). Bacteria in the human mouth. *Journal of Applied Bacteriology* **44**, 163–181.

Sack, D. A., Kaminsky, D. C., Sack, R. B., Wamola, I. A., Ørskov, F., Ørskov, I., Slack, R. C. B., Arthur, R. R. and Kapikian, A. Z. (1977a). Enterotoxigenic *Escherichia coli* diarrhoea in travellers: A prospective study of American Peace Corps volunteers. *Johns Hopkins Medical Journal* **141**, 63–70.

Sack, D. A., McLaughlin, J. C., Sack, R. B., Ørskov, F. and Ørskov, I. (1977b). Enterotoxigenic *Escherichia coli* isolated from patients at a hospital in Dacca. *Journal of Infectious Diseases* **135**, 275–280.

Sansonetti, P. J., d'Hautville, H., Formal, S. B. and Toncas, M. (1982). Plasmid-mediated invasiveness of 'Shigella-like' *Escherichia coli*. *Annales de Microbiologie (Paris)* **132A**, 351–355.

Sarff, L. M., McCracken, G. H., Jr., Schiffer, M. S., Glode, M. P., Robbins, J. B., Ørskov, I. and Ørskov, F. (1975). Epidemiology of *Escherichia coli* K1 in healthy and diseased newborns. *Lancet* **2**, 1099–1104.

Schiffer, M. S., Oliviera, E., Glode, M. P., McCracken, G. H., Jr., Sarff, L. M. and Robbins, J. B. (1976). A review: Relation between invasiveness and the K1 capsular polysaccharide of *Escherichia coli*. *Pediatric Research* **10**, 82–87.

Scotland, S. M., Smith, H. R., Willshaw, G. A. and Rowe, B. (1983). Vero cytotoxin production in strains of *Escherichia coli* is determined by genes carried on bacteriophage. *Lancet* **2**, 216.

Sears, H. I. and Brownlee, F. (1953). Further observations on the persistence of individual strains of *E. coli* in the intestinal tract of man. *Journal of Bacteriology* **63**, 47–57.

Sears, H. I., Brownlee, I. and Uchiyama, J. K. (1950). Persistence of individual strains of *E. coli* in the intestinal tract of man. *Journal of Bacteriology* **59**, 293–301.

Serény, B. (1955). Experimental *Shigella* keratoconjunctivitis: A preliminary report. *Acta Microbiologica Academiae Scientarum Hungaricae* **2**, 293–296.

Sharon, N., Eshdat, Y., Silverblatt, F. J. and Ofek, I. (1981). Bacterial adherence to cell surface sugars. *In* "Adhesion and Microorganism Pathogenicity" (Eds. M. O'Connor, J. Whelan and K. Elliott), pp. 119–135. Pitman Medical, Tunbridge Wells.

Shooter, R. A., Cooke, E. M., Rousseau, S. A. and Breaden, A. L. (1970). Animal sources of common serotypes of *E. coli* in the food of hospital patients. Possible significance in urinary tract infection. *Lancet* **2**, 226–228.

Shore, E. G., Dean, A. G., Honk, K. J. and Davis, B. R. (1974). Enterotoxin-producing *Escherichia coli* and diarrhoeal disease in adult travellers: A prospective study. *Journal of Infectious Diseases* **129**, 577–582.

Silva, R. M., Toledo, M. R. F. and Trabulsi, L. R. (1982). Correlation of invasiveness with plasmid in enteroinvasive strains of *Escherichia coli*. *Journal of Infectious Diseases* **146**, 706.

Skinner, F. A. and Carr, J. G. (Eds.) (1974). "The Normal Microbial Flora of Man." Academic Press, London.

Smith, H. W. (1963). Haemolysins of *Escherichia coli*. *Journal of Pathology and Bacteriology* **85**, 197–211.

Smith, H. W. (1965). The development of the flora of the alimentary tract in young animals. *Journal of Pathology and Bacteriology* **90**, 495–513.

Smith, H. W. (1969). Transfer of antibiotic resistance from animal and human strains of *Escherichia coli* to resident *Escherichia coli* in the alimentary tract of man. *Lancet* **1**, 1174–1176.

Smith, H. W. (1974). A search for transmissible pathogenic characters in invasive strains of *Escherichia coli*: The discovery of a plasmid-controlled toxin and a plasmid-controlled lethal character closely associated, or identical, with colicin V. *Journal of General Microbiology* **83**, 95–111.

Smith, H. W. (1978). Transmissible pathogenic characteristics of invasive strains of *Escherichia coli*. *Journal of the American Veterinary Medical Association* **173**, 601–607.

Smith, H. W. and Gyles, C. L. (1970). The relationship between two apparently different enterotoxins produced by enteropathogenic strains of *Escherichia coli* of porcine origin. *Journal of Medical Microbiology* **3**, 387–401.

Sohl-Akerlund, A., Ahlstedt, S., Hanson, L. A., Lundberg, U. and Olling, S. (1977). Differences of antigenicity of *Escherichia coli* strains isolated from patients with various forms of urinary tract infections. *International Archives of Allergy and Applied Immunology* **55**, 458–467.

Springer, G. F. (1970). Importance of blood group substances in interactions between man and microbes. *Annals of the New York Academy of Sciences* **169**, Article 1, 134–152.

Stamey, T. (1980). "Pathogenesis and Treatment of Urinary Tract Infections." Williams & Wilkins, Baltimore, Maryland.

Stoner, H. B. (1972). Specific and non-specific effects of bacterial infection in the host. *Symposia of the Society for General Microbiology* **22**, 113–128.

Sussman, M. (1974). Iron and infection. *In* "Iron in Biochemistry and Medicine" (Eds. A. Jacobs and M. Worwood), pp. 649–679. Academic Press, London.

Sussman, M. and Asscher, A. W. (1979). Urinary tract infection. *In* "Renal Disease" (Eds. D. Black and N. F. Jones), pp. 400–436. Blackwell, Oxford.

Sussman, M., Asscher, A. W., Waters, W. E., Evans, J. A. S., Campbell, H., Evans, K. T. and Williams, J. E. (1969). Asymptomatic significant bacteriuria in the non-pregnant woman. I. Description of a population. *British Medical Journal* **1**, 799–803.

Sussman, M., Abraham, S. N. and Parry, S. H. (1982a). Bacterial adhesion in the host–parasite relationship of urinary tract infection. *In* "Clinical, Bacteriological and Immunological Aspects of Urinary Tract Infections in Children" (Ed. H. Schulte-Wissermann), pp. 103–112. Thieme, Stuttgart.

Sussman, M., Parry, S H., Rooke, D. M. and Lee, M. J. S. (1982b). Bacterial adherence and the urinary tract. *Lancet* **1**, 1352.

Svanborg Edén, C. (1978). Attachment of *Escherichia coli* to human urinary tract epithelial cells. *Scandinavian Journal of Infectious Diseases, Supplement* **15**, 1–54.

Svanborg Edén, C. and Hanson, L. A. (1978). *Escherichia coli* pili as possible mediators of attachment of human urinary tract epithelial cells. *Infection and Immunity* **21**, 229–237.

Svanborg Edén, C., Hanson, L. A., Jodal, U., Lindberg, U. and Sohl-Akerlund, A. (1976). Variable adhesion to normal urinary tract epithelial cells of *Escherichia coli* strains associated with various forms of urinary tract infection. *Lancet* **2**, 490–492.

Svanborg Edén, C., Eriksson, B., Hanson, L. A., Jodal, U., Kaijser, B., Lidin-Janson, G., Lindberg, U. and Olling, S. (1978). Adhesion to normal human uroepithelial cells of *Escherichia coli* from children with various forms of urinary tract infection. *Journal of Pediatrics (St. Louis)* **93**, 398–403.

Svanborg Edén, C., Carlsson, B., Hanson, L. A., Jann, K., Korhonen, T. K. and Wadström, T. (1979a). Anti-pili antibodies in breast milk. *Lancet* **2**, 1235.

Svanborg Edén, C., Lidin-Janson, G. and Lindberg, U. (1979b). Adhesiveness to urinary tract epithelial cells of fecal and urinary *Escherichia coli* isolates from patients with symptomatic urinary tract infections or asymptomatic bacteriuria of varying duration. *Journal of Urology* **122**, 185–188.

Svenson, S. B. and Källenius, G. (1983). Density and localization of P-fimbriae-specific receptors on mammalian cells: Fluorescence-activated cell analysis. *Infection* **11**, 6–12.

Svenson, S. B., Kallenius, G., Möllby, R., Hultberg, H. and Winberg, J. (1982). Rapid identification of P-fimbriated *Escherichia coli* by a receptor-specific particle agglutination test. *Infection* **10**, 209–214.

Svenson, S. B., Hultberg, H., Kallenius, G., Korhonen, T. K., Möllby, R. and Winberg, J. (1983). P-fimbriae of pyelonephritogenic *Escherichia coli*: Identification and chemical characterisation of receptors. *Infection* **11**, 61–67.

Taylor, J. (1961). Host specificity and enteropathogenicity of *Escherichia coli*. *Journal of Applied Bacteriology* **24**, 316–325.

Taylor, J. Powell, B. W. and Wright, J. (1949). Infantile diarrhoea and vomiting: A clinical and bacteriological investigation. *British Medical Journal* **2**, 117–125.

Taylor, J., Wilkins, M. P. and Payne, J. M. (1961). Relations of rabbit gut reaction to enteropathogenic *Escherichia coli*. *British Journal of Experimental Pathology* **42**, 43–52.

Thomas, V. L., Forland, M. and Shelokov, A. (1975). Antibody-coated bacteria in urinary tract infection. *Kidney International* **8**, Supplement 4, S20–S22.

Thomson, S. (1955). The role of certain varieties of *Bacterium coli* in gastro-enteritis of babies. *Journal of Hygiene* **53**, 357–367.

Thomson, S., Watkins, A. G. and Gray, O. P. (1956). *Escherichia coli* gastro-enteritis. *Archives of Disease in Childhood* **31**, 340–345.

Tourville, D., Bienenstock, J. and Tomasi, T. B. (1968). Natural antibodies of human serum, saliva and urine reactive with *Escherichia coli*. *Proceedings of the Society for Experimental Biology and Medicine* **128**, 722–727.

Tulloch, E. F., Ryan, K. J., Formal, S. B. and Franklin, F. A. (1973). Invasive enteropathic *Escherichia coli* dysentery. An outbreak of 28 adults. *Annals of Internal Medicine* **79**, 13–17.

Turck, M., Petersdorf, R. G. and Fournier, M. R. (1962). The epidemiology of non-enteric *Escherichia coli* infections: Prevalence of serological groups. *Journal of Clinical Investigation* **41**, 1760–1765.

Turnberg, L. A. (1979). The pathophysiology of diarrhoea. *Clinics in Gastroenterology* **8**, 551–568.

Ulevitch, R. J. and Cochrane, C. G. (1977). The chemistry and biology of the proteins of the Hageman factor-activated pathways. *In* "Comprehensive Immunology," Vol. 2, Biological Amplification Systems in Immunology (Eds. N. K. Day and R. A. Good), pp. 205–217. Plenum Medical, New York.

Väisänen, V., Tallgren, L. G., Mäkelä, P. H., Kallenius, G., Hultberg, H., Elo, J., Siitonen, A., Svanborg Edén, C., Svenson, S. B. and Korhonen, T. K. (1981). Mannose-resistant haemagglutination and P antigen recognition are characteristic of *Escherichia coli* causing primary pyelonephritis. *Lancet* **2**, 1366–1369.

Väisänen, V., Korhonen, T. K., Jokinen, M., Galimberg, C. G. and Enholm, C. (1982). Blood group M specific haemagglutination in pyelonephritogenic *Escherichia coli*. *Lancet* **1**, 1192.

Van Dijk, W. C., Verbrugh, H. A., Peters, R., Van Der Tol, M. E., Peterson, P. K. and Verhoef, J. (1977). *Escherichia coli* K antigen in relation to serum-induced lysis and phagocytosis. *Journal of Medical Microbiology* **10**, 123–130.

Van Dijk, W. C., Verbrugh, H. A., Van Erne-Van der Tol, M. E., Peters, R. and Verhof, J. (1981). *Escherichia coli* antibodies in opsonisation and protection against infection. *Journal of Medical Microbiology* **14**, 381–389.

van Pirquet, C. (1910–1911). Theodor Escherich. *Zeitschrift für Kinderheilkunde* **1**, 423–441.

Wadström, T., Aust-Keltis, A., Habte, D., Holmgren, J., Meeuwisse, G., Möllby, R. and Söderlind, D. (1976). Enterotoxin-producing bacteria and parasites in stools of Ethiopian children with diarrhoeal disease. *Archives of Disease in Childhood* **51**, 865–870.

Waldman, R. H., Bencic, Z., Sakazaki, R., Sinha, R., Ganguly, R., Deb, B. C. and Mukerjee, S. (1971). Cholera immunology. I. Immunoglobulin levels in serum, fluid from the small intestine and faeces from patients with cholera non-choleric diarrhoea during illness and convalescence. *Journal of Infectious Diseases* **123**, 579–586.

Wallick, H. and Stuart, C. A. (1943). Antigenic relationships of *E. coli* isolated from one individual. *Journal of Bacteriology* **45**, 121–126.

Wardle, E. N. (1980). Bacteraemic and endotoxic shock. *Journal of Clinical Pathology* **33**, 888.

Welch, R. A., Dellinger, E. P., Minshew, B. and Falkow, S. (1981). Haemolysin contributes to virulence of extra-intestinal *E. coli* infections. *Nature (London)* **294**, 665–667.

Westphal, O., Jann, K. and Himmelspach, K. (1983). Chemistry and immunochemistry of bacterial lipopolysaccharides as cell wall antigens and endotoxins. *Progress in Allergy* **33**, 9–39.

Wilfert, C. M. (1978). *E. coli* meningitis: K1 antigens and virulence. *Annual Review of Medicine* **29**, 129–136.

Williams, G. L., Davies, D. K. L., Evans, K. T. and Williams, J. E. (1968). Vesicoureteric reflux in patients with bacteriuria of pregnancy. *Lancet* **2**, 1202–1205.

Williams, P. H. (1979). Novel iron uptake system specified by ColV plasmids: An important component in the virulence of invasive strains of *Escherichia coli*. *Infection and Immunity* **26**, 925–932.

Williams, P. H. and Warner, P. J. (1980). ColV plasmid-mediated, colicin V-independent iron uptake system of invasive strains of *Escherichia coli*. *Infection and Immunity* **29**, 411–416.

Wilson, G. (1983). The normal bacterial flora of the body. *In* "Principles of Bacteriology, Virology and Immunity," Vol. 1. (Eds. G. Wilson and H. M. Dick), pp. 230–246. Edward Arnold, London.

Winterbauer, R. H., Turak, M. and Petersdorf, R. G. (1967). Studies on the epidemiology of *E. coli* infections. V. Factors affecting acquisition of specific serological groups. *Journal of Clinical Investigation* **46**, 21–29.

Wolberg, G. and DeWitt, D. (1969). Mouse virulence of K(L) antigen-containing strains of *Escherichia coli*. *Journal of Bacteriology* **100**, 730–737.

Wright, R. (1982). The liver. *In* "Clinical Aspects of Immunology" (Eds. P. J. Lachmann and D. K. Peters), pp. 878–902. Blackwell, Oxford.

Young, L. S., Martin, W. J., Meyer, R. D., Weinstein, R. J. and Anderson, E. T. (1977). Gram-negative rod bacteraemia: Microbiologic, immunologic and therapeutic considerations. *Archives of Internal Medicine* **86**, 456–471.

Ziegler, E. J., McCutcheon, J. A., Fierer, J., Glauser, M. P., Sadoff, J. C., Douglas, H. and Braude, A. I. (1982). Treatment of Gram-negative bacteremia and shock with human antiserum to a mutant *Escherichia coli*. *New England Journal of Medicine* **307**, 1225–1230.

3

Escherichia coli as a Pathogen in Animals

J. A. MORRIS AND W. J. SOJKA

Central Veterinary Laboratory, Weybridge, Surrey, UK

Introduction

Although *Escherichia coli* is a normal inhabitant of the intestinal tract of animals and man the organism can also cause disease. Most of these diseases are each associated with relatively few serological groups of *E. coli,* even though a very large number exist in nature. These serogroups tend to be host-specific. Thus, the *E. coli* associated with neonatal diarrhoea in calves are not usually associated with calf septicaemia and *E. coli* found in enteric diseases in pigs are rarely encountered in cattle, sheep or poultry. There are some exceptions, notably strains of *E. coli* belonging to OK serogroup **O**78:**K**80, that are frequently isolated from poultry, calves and lambs affected with colisepticaemia but these organisms are rarely found in pigs.

E. coli infections are usually referred to as colibacillosis and frequently occur in young animals. Two main forms are recognized: (1) systemic colibacillosis caused by bacteraemic strains of *E. coli,* and (2) enteric colibacillosis caused by enteropathogenic strains of *E. coli.* Coliform mastitis is yet another form of *E. coli* infection but this occurs in adult cattle.

In the systemic form of colibacillosis, the bacteraemic strains of *E. coli* pass through the mucosa of the alimentary or respiratory tract and enter the blood stream. This invasion may result in a generalized infection (colisepticaemia) or a localized infection manifested as meningitis and/or arthritis in calves and lambs. In poultry air-sacculitis and pericarditis are frequently observed.

Enteric colibacillosis may be subdivided into two main syndromes: (1) colibacillary diarrhoea and (2) colibacillary toxaemia, which includes oedema disease of pigs. In enteric colibacillosis enteropathogenic *E. coli* colonize the small intestine and elaborate toxin(s). Although invasion may occur in the terminal stages of disease, the bacteria are generally confined to the intestinal tract. In colibacillary diarrhoea, which is by far the most common enteric colibacillosis, enterotoxigenic *E. coli* produce enterotoxins that induce the movement of fluid

47

from the tissues into the gut lumen, thereby causing diarrhoea. Endotoxin is principally involved in colibacillary toxaemia and this can be associated with either enterotoxigenic or non-enterotoxigenic *E. coli*. Oedema disease involves endotoxin and/or a third type of toxin often referred to as neurotoxin.

Poultry are normally only affected by the systemic form of colibacillosis, while in piglets enteric colibacillosis is more common. In calves systemic colibacillosis is a frequent sequel of colostrum deprivation and is related to the level of immunoglobulin in their sera. Diarrhoea is a common disease in calves but the evidence suggests that only a proportion of these outbreaks can be attributed to enterotoxigenic *E. coli* alone. In lambs both forms of colibacillosis occur at approximately equal frequency. Colibacillosis in calves and lambs usually occurs early in life, frequently within a few days of birth. In pigs, however, enteric colibacillosis can occur in newborn piglets and also in the older animal at weaning.

In contrast to both the systemic and enteric forms of colibacillosis, which are caused by a relatively small number of *E. coli* serotypes, the isolates from bovine coliform mastitis belong to a very large number of serological groups. *E. coli* from these OK groups are regularly present in the bovine gut and there is a close relationship between coliform contamination of bedding and the occurrence of *E. coli* mastitis. The organisms do not invade the mammary tissue but remain in the lumen of the teat canal and the lactiferous sinus. *E. coli* endotoxin is probably absorbed and causes a mild mastitis or an acute toxic reaction.

The earlier literature concerning various aspects of *E. coli* infections in farm animals and poultry has been reviewed by Sojka (1965, 1970, 1971). This will be summarized and some of the more recent accounts relating to the more important aspects of *E. coli* infections will be described under the main headings Enteric colibacillosis, Systemic colibacillosis and Coliform mastitis.

Enteric Colibacillosis

Enteric diseases of animals are a complex syndrome involving a variety of infectious agents, secondary opportunists and precipitating factors. However, there is increasing evidence that certain strains of *E. coli* are frequently involved in disturbances of the intestinal tract of newborn piglets, calves and lambs. Enteric colibacillosis probably involves the following sequence of events: (1) infection with enteropathogenic *E. coli,* (2) proliferation of the organism in the small intestine, (3) formation and release of toxin(s) and (4) development of lesions (Nielsen *et al.,* 1969). The syndromes will be described according to the principal toxins thought to be involved: colibacillary diarrhoea, involving enterotoxin; and colibacillary toxaemia, involving endotoxin.

Colibacillary Diarrhoea

The enterotoxins responsible for colibacillary diarrhoea increase the movement of electrolyte from the body tissue into the intestinal lumen. The diarrhoea occurs most frequently in newborn calves, lambs and piglets (1–3 days old), where it is usually referred to as neonatal *E. coli* diarrhoea. Colibacillary diarrhoea also occurs in pigs in the immediate post-weaning period (usually 1 wk after weaning) and this condition is often known as weanling colibacillary diarrhoea.

Neonatal E. coli diarrhoea. Neonatal diarrhoea in which *E. coli* was implicated comprised 35% of all the diagnoses made in pigs between 1975 and 1980 by the Ministry of Agriculture, Fisheries and Food Laboratories in England and Wales. In calves the proportion was 26% and in lambs 17% (J. Bell, personal communication). In neonatal diarrhoea, disease occurs in the absence of both enteritis and bacteraemia. However, invasion may sometimes occur in the terminal stages of the disease or as a result of post-mortem changes. Thus in many respects *E. coli* neonatal diarrhoea resembles cholera in man: in both diseases (1) the apparent cause of death is fluid and electrolyte loss resulting from diarrhoea, (2) the organism is confined to the lumen of the small intestine, (3) the increase in net fluid movement occurs across an intact epithelium and (4) this increase in net secretion is caused by enterotoxins produced by these bacteria.

Stevens (1963a,b) reported that piglets are usually born healthy but illness occurs abruptly about 12 h after birth. One or two piglets in the litter may be found dead, others may be moribund even before the onset of diarrhoea while the remainder show varying degrees of diarrhoea and listlessness. Affected piglets continue to suck until just before death. Although the whole litter occasionally dies there are usually a few survivors that make a complete recovery.

At post-mortem examination specific lesions are not generally found (Saunders *et al.*, 1960; Stevens, 1963a,b). There are signs of dehydration, the stomach is distended with clotted milk and the intestine is distended with fluid, mucus, gas and particles of digesta. Some parts appear congested while others are pale. However, in most animals examined immediately after death there is no congestion or other evidence of inflammation. The absence of macroscopic abnormality in the alimentary tract and internal organs of piglets affected by *E. coli* neonatal diarrhoea has been reported (Smith and Jones, 1963; Moon, 1969) and the ultrastructure of the villous epithelial cells of piglets appears normal (Dress and Waxler, 1970).

A constant bacteriological finding is the intense proliferation of certain serotypes of *E. coli* in the small intestine while the numbers of other bacteria in the gut (*Lactobacillus, Streptococcus, Clostridia*) are normal (Smith and Jones, 1963). Abnormally large numbers of enterotoxigenic *E. coli* are present in the small and large intestine ($\geq 10^8$ organisms/g ileal mucosal scrapings) in piglets,

Table 1. Colibacillosis in calves and lambs: serological OK groups of E. coli isolates from the systemic and enteric forms of colibacillosis and virulence factors found in isolates from the latter form

| Systemic colibacillosis | | Enteric colibacillosis[a]: neonatal E. coli diarrhoea | | | | | |
| Isolate | Serological OK groups[b] | Isolate | Serological OK groups[c] | Adhesins | | Heat-stable enterotoxin (STa) | Endotoxin |
				K99	F41		
Bacteraemic E. coli	O78:K80	Enterotoxigenic E. coli	O8:K85	+		+	+
	O137:K79		O8:K25	+		+	+
	O15:K		O8:K28	+		+	+
	O35:K		O20:K17	+		+	+
	O115:K		O20:K	+		+	+
	O117:K98		O9:K30[d]	+	+	+	+
	O119:K		O9:K35	+	+	+	+
	O26:K60		O9:K	+	+	+	+
	O86:K61		O101:K30[d]	+	+	+	+
	(other but less frequently		O101:K28	+	+	+	+
	reported O groups: O1,		O101:K32	+	+	+	+
	O2, O8, O20, O55,		O101:K	+	+	+	+
	O73 and O114)		O101:K30		+	+	+
			O101:K		+	+	+

[a] In addition to neonatal E. coli diarrhoea, enteric colibacillosis may be manifested as colibacillary toxaemia (enteric toxaemia) in which endotoxin appears to be involved.

[b] Based on Sojka (1965), Renault (1979) and Ørskov (1978).

[c] Based on Ørskov et al. (1975), Ørskov and Ørskov (1978) and our own results.

[d] Most commonly encountered OK groups.

calves and lambs. These *E. coli* belong to relatively few serological groups (Sojka, 1971). In both calves and lambs usually the same *E. coli* serological OK groups are involved (Table 1) and these are often referred to as calf–lamb ETEC strains. In piglets, however, the majority of serogroups associated with disease are rarely found elsewhere (Table 2). Nevertheless, some *E. coli* principally associated with disease in calves and lambs (e.g. **O**101:**K**30,**K**99,**F**41) do occasionally cause disease in piglets. The *E. coli* isolates from piglets are often haemolytic.

Neonatal E. coli diarrhoea: Virulence determinants. For an *E. coli* to cause diarrhoea in a susceptible host it must possess at least two virulence determinants. It must be able to colonize the small intestine, which is promoted by colonization antigens or adhesins (e.g. K88, K99), and it must be able to produce enterotoxin, which causes the diarrhoea.

Most *E. coli* from piglets affected by neonatal *E. coli* diarrhoea produce a common antigen, K88 (Sojka, 1971), as shown in Table 2. This antigen is a mannose-resistant haemagglutinin for guinea-pig erythrocytes and can be observed as very fine filaments (referred to as fimbriae or pili) that are often branched when examined by electron microscopy (Stirm *et al.,* 1967a,b; Gaastra and de Graaf, 1982). Ørskov *et al.* (1961), who first described the K88 antigen, reported that it can be demonstrated readily on bacteria grown at 37°C but not on organisms grown at 18°C. It was also observed that the ability to produce K88 antigen can be transmitted experimentally from one strain of *E. coli* to another in mixed cultures. It was confirmed later that the production of this antigen is plasmid-mediated (Ørskov and Ørskov, 1966; Smith and Halls, 1968). Three forms of the antigen are now recognized: K88ab, K88ac (Ørskov *et al.,* 1964) and K88ad (Guinée and Jansen, 1979).

Although enterotoxigenic *E. coli* do not invade the epithelial cells (Smith and Halls, 1967a,b), they either adhere to or become closely associated with the intestinal mucosa (Arbuckle, 1970, 1971) and can be seen along the villi from the tip to the base but they are not usually present within the crypts (Bertschinger *et al.,* 1972).

Adherence to the mucosal surface would help to overcome the mechanical clearance mechanism of peristalsis and would therefore promote colonization of the small intestine by *E. coli*. K88 antigen promotes colonization of the anterior small intestine (Smith and Linggood, 1971) and it has been demonstrated that the antigen is responsible for attachment of the bacterium to the enterocyte brush borders *in vivo* (Jones and Rutter, 1972) and *in vitro* (Jones and Rutter, 1972; Wilson and Hohmann, 1974; Sellwood *et al.,* 1975).

Smith and Linggood (1971) showed that removal of the K88 plasmid from an *E. coli* strain isolated from piglets affected by neonatal diarrhoea was accompanied by a loss of its ability to cause disease. Reintroduction of a K88 plasmid

Table 2. *Enteric colibacillosis in pigs: serological OK groups of E. coli isolates and their main virulence factors*

Age at which piglets are affected	Clinical disease	Serological OK groups	Virulence factors (E. coli isolates)										Type of E. coli
			Adhesins					Toxins					
								Enterotoxin		Endotoxin	Neurotoxin		
			K88ab	K88ac	987P	K99	F41	ST (heat-stable)	LT (heat-labile)				
Neonatal period (usually first few days of life)	Colibacillary diarrhoea / Neonatal E. coli diarrhoea	O149:K91[a]	+					+	+[b]	+		Enterotoxigenic E. coli (ETEC)
		O8:K87[a]		+				+	+	+		
		O8:K87[b]		+				+	+	+		
		O45:KE65[a]		+				+	+	+		
		O138:K91[a]		+				+	+	+		
		O141:K85ab[a]	+					+	+	+		
		O141:K85ac		+				+	+	+		
		O147:K89[a]		+				+	+	+		
		O157:KV17[a]		+				+	+	+		
		O9:K101			+			+		+		
		O9:K103			+			+		+		
		O9:K35				+		+		+		

52

Disease	Serotype							
	O20:K101	+				+		
	O20:K	+				+		
	O64:KV142		+	+		+		+
	O101:K30		+	+	+	+		+
	O101:K			+	+	+		+
Post-weaning (usually 1 wk after weaning) Weanling E. coli diarrhoea	O141:K85ab[a]		+	+		+		+
	O141:K85ac		+	+		+		+
	O138:K81[a]		+	+		+		+
Colibacillary toxaemia (including oedema disease)	O141:K85ab[a]		+	+		+		+
	O141:K85ac		+	+		+		+
	O138:K81[a]		+	+		+		+
	O139:K92			+		+		+
	O45:KE65			+		+		+

[a] OK groups more frequently encountered (O149:K91 most common).
[b] Strains possessing K88 adhesins usually produce both ST and LT enterotoxin.

53

from another isolate of *E. coli* restored its virulence. A non-enteropathogenic strain of *E. coli* was rendered enteropathogenic for piglets by introducing two plasmids: K88 and Ent (which mediates enterotoxin production). The K88 antigen enabled the organism to multiply in the anterior small intestine, while the Ent plasmid was responsible for the diarrhoea that followed (Smith and Linggood, 1971, 1972).

Some piglets are gentically resistant to infection with K88[+] *E. coli* (Sellwood *et al.*, 1975). At least two phenotypes exist, one in which K88[+] *E. coli* attach to the enterocyte brush borders in the small intestine and another in which attachment does not occur. The phenotype is the product of two alleles at a single locus inherited in a simple Mendelian manner, the adhesive allele being dominant over the non-adhesive allele. It was therefore possible to select for breeding pigs which possess the non-adhesive gene and thus to produce progeny resistant to neonatal *E. coli* diarrhoea caused by K88[+] *E. coli* (Rutter *et al.*, 1975). In a recent study involving the three known K88 variants, at least five different phenotypes of piglets could be distinguished in *in vitro* enterocyte brush-border tests (Bijlsma *et al.*, 1981). One phenotype is susceptible to all three variants, three phenotypes are susceptible to only one or two variants and one phenotype is entirely resistant to K88-mediated adhesion.

Although the majority of *E. coli* associated with neonatal diarrhoea in piglets produce K88, not all strains produce this antigen. Moon *et al.* (1977) examined K88[−] *E. coli* and found that some produced K99 (Sojka, 1972, cited in Smith and Linggood, 1972; Ørskov *et al.*, 1975), a fimbrial antigen frequently found on *E. coli* associated with neonatal diarrhoea in calves and lambs. Like K88, the K99 antigen is plasmid-mediated and appears to be responsible for colonization of the small intestine by the organism (Smith and Linggood, 1972). Moon *et al.* (1977) demonstrated that K99[+] *E. coli* isolated from piglets and four of five K99[+] *E. coli* isolated from calves and lambs colonized the lower small intestine and produced diarrhoea in experimentally infected colostrum-deprived piglets. Similar findings were reported by Smith and Huggins (1978) when they infected piglets with a K99[+] *E. coli* isolated from a calf with diarrhoea. Subsequent studies demonstrated that K88[−] K99[−] *E. coli* associated with neonatal diarrhoea in piglets produced a fimbrial antigen known as 987P (Nagy *et al.*, 1977; Isaacson *et al.*, 1977) which is responsible for the adhesion of these bacteria to the villous epithelium of the lower small intestine. Unlike K88 and K99, the 987P fimbriae do not haemagglutinate erythrocytes and the genetic determinants are not readily transmitted. The 987P fimbriae are usually demonstrated on bacteria that do not produce any of the other known fimbriae, but Schneider and To (1982) reported the isolation of two strains of *E. coli* that possessed the ability to produce 987P fimbriae or K88. This was observed *in vitro* and it was not known whether this occurred in the small intestine or whether the same bacterium produced both fimbriae.

Yet another fimbrial antigen associated with colonization of the small intestine was predicted when Moon *et al.* (1980) examined K88$^-$ K99$^-$ 987P$^-$ *E. coli* isolated from piglets with diarrhoea. These bacteria have now been shown to produce F41 fimbriae (Morris *et al.*, 1983). These fimbriae were originally demonstrated in association with K99 on certain strains of *E. coli* (Morris *et al.*, 1980, 1982a) associated with diarrhoea in calves.

The OK serogroups of *E. coli* most frequently isolated from newborn calves and lambs with diarrhoea are shown in Table 1. These bacteria produce K99 fimbrial antigen which functions in a similar way to K88 in piglets. In contrast to K88$^+$ bacteria, *E. coli* that produce K99 usually colonize the lower portion of the ileum. However, in common with K88, K99 is a mannose-resistant haemagglutinin (Burrows *et al.*, 1976; Morris *et al.*, 1977; de Graaf *et al.*, 1981). *E. coli* from OK serogroups O9 and O101 produce F41 fimbriae (de Graaf and Roorda, 1982) in addition to K99 (Morris *et al.*, 1978, 1980). The relative roles of K99 and F41 in the colonization of the small intestine have not been determined but K99$^+$ F41$^+$, K99$^+$ F41$^-$ and K99$^-$ F41$^+$ *E. coli* have each been isolated from calves with *E. coli* neonatal diarrhoea (J. A. Morris, unpublished) and these three types of *E. coli* have also been isolated from newborn piglets with diarrhoea.

With the possible exception of K88, the fimbrial antigens described have not been reported on *E. coli* isolated from animals more than a few weeks old. Furthermore, Smith and Huggins (1978) reported that field isolates of K88$^-$ K99$^-$ 987P$^-$ *E. coli* that were enteropathogenic for weaned pigs did not proliferate in the small intestine of newborn piglets. The *in vitro* studies of Runnels *et al.* (1980) suggest that the small intestine develops resistance to K99-mediated adhesion as the host ages. However, this was not evident with K88-mediated adhesion. Their data demonstrated that when age-related resistance to adhesion occurred, this resistance was innate rather than antibody-mediated. It was concluded that although some enterotoxigenic *E. coli* only adhere to enterocytes from the newborn piglet, this does not entirely explain the marked resistance of older animals to *E. coli* infections.

Although colonization of the small intestine is an essential prerequisite for strains to produce diarrhoea, colonization per se does not cause diarrhoea. The organism must also produce enterotoxin in a susceptible host. Two main classes of enterotoxin are recognized: (1) heat-stable toxin, ST (Smith and Halls, 1967b), and (2) heat-labile toxin, LT (Gyles and Barnum, 1969; Smith and Gyles, 1970). ST is produced by all enterotoxigenic strains of *E. coli* (Smith and Gyles, 1970) but some strains, particularly those possessing the K88 antigen, produce both ST and LT (Gyles, 1971). Production of these toxins is controlled by transmissible plasmids.

The ligated small intestinal loops of piglets were first used to study enterotoxigenic *E. coli* of porcine origin by Nielsen (1963) and their use has been developed by many research workers (e.g., Moon, 1965; Smith and Halls, 1967a, 1968; Truszczynski and Ciosek, 1972; Pesti and Semjen, 1973; Burgess *et al.*, 1978) to determine the enteropathogenicity of the *E. coli* isolates.

Representatives of *E. coli* from the OK serogroups shown in Table 2 were isolated by Smith and Halls (1967a) from the faeces and alimentary tract of pigs in epidemiologically unrelated outbreaks of diarrhoea. They found that each strain consistently dilated segments of porcine small intestine. Similarly, *E. coli* isolated from herds affected by both oedema disease and weanling diarrhoea (Table 2) also consistently dilated gut segments but no reaction occurred with *E. coli* from herds where oedema disease was not associated with diarrhoea. These workers also reported no reaction with *E. coli* belonging to OK serogroups associated with diarrhoea of non-infectious origin, bacteraemia in calves and lambs, or diarrhoea in man.

Non-enterotoxigenic *E. coli* that received the Ent plasmid from an enterotoxigenic *E. coli* in mixed culture transmission experiments dilated segments of piglet intestine (Smith and Halls, 1967a, 1968). Moon and Whipp (1970) reported that three piglet strains of *E. coli* (two belonging to OK group O64 and the other to OK group O101) consistently dilated ligated segments from piglets less than 1 wk old but not segments from older pigs. These were referred to as class 2 strains and were considered atypical since typical strains (class 1) dilated segments from both neonatal and older pigs. Smith and Linggood (1972) confirmed these findings and found that these strains also dilated ligated segments of calf and lamb intestine. A similar reaction was obtained with calf and lamb enterotoxigenic *E. coli*, suggesting that enterotoxins produced by the atypical piglet strains and by *E. coli* enteropathogenic for the calf and lamb may be identical. Cultures of typical piglet *E. coli* dilated segments of 3-day-old and 8-wk old pigs, calves and lambs. *E. coli* enteropathogenic for man and non-enteropathogenic *E. coli* from pigs and calves consistently gave negative results in all these tests. Smith and Linggood (1972) suggested that the enterotoxin produced by atypical *E. coli* enteropathogenic for piglets and the *E. coli* enteropathogenic for calves and lambs resembled ST rather than the LT produced by typical *E. coli*.

Results obtained with cell-free culture fluids in the ligated-loop tests are similar to those obtained with live organisms. Cell-free culture fluids from enterotoxigenic *E. coli* were used by Smith and Halls (1967b), who showed that these preparations dilated gut segments whereas similar preparations from non-enterotoxigenic *E. coli* did not. Serologically different *E. coli* enteropathogenic for piglets produce indistinguishable toxins which are distinct from endotoxin. The latter, prepared from both enteropathogenic and non-enteropathogenic *E. coli*, consistently failed to dilate intestinal segments (Smith and Halls, 1967a; Gyles and Barnum, 1969). The enterotoxin described by Smith and Halls was heat-stable (ST) while Gyles and Barnum (1969) demonstrated an additional enterotoxin which was heat-labile (LT).

The effects on piglets of crude ST (broth filtrates) and crude LT (whole cell lysates) administered by stomach tube were investigated by Kohler (1971). He reported that both preparations produced diarrhoea. With LT the diarrhoea developed later than with ST, it was more persistent and most piglets died. In contrast, none of the piglets given the ST preparation died. Smith and Gyles (1970) investigated the effects of oral administration of LT and ST preparations from various strains of *E. coli,* some of which had acquired the Ent plasmid from other strains in mixed culture transmission experiments. Both LT and ST preparations produced diarrhoea while similar preparations from non-enterotoxigenic *E. coli* were generally without effect. However, diarrhoea did occur in two of these 14 control piglets, suggesting that bacterial products other than enterotoxin might be capable of producing diarrhoea.

Burgess *et al.* (1978) studied the enterotoxins produced by a strain of *E. coli* isolated from piglets (*E. coli* strain P16). These toxins were examined in a number of animal models before and after treatment of the toxins with methanol and they demonstrated two heat-stable toxins which they referred to as STa and STb. STa differs from STb in that it is soluble in methanol, has a lower thermal stability and has a different spectrum of biological activity. Thus, in experiments with equivalent doses of STa and STb, activity in neonatal piglets and infant mice is due principally to STa while STb shows the greatest activity in the loops of weaned pigs. It is tempting to speculate that the class 1 *E. coli* described by Moon (i.e. Smith's 'typical' *E. coli*) produce STa and STb, thereby causing diarrhoea in both neonatal and weaned piglets, while class 2 *E. coli* (Smith's 'atypical' *E. coli*) produce only STa, thereby causing diarrhoea only in neonatal piglets (Burgess *et al.,* 1978). The results obtained by Whipp *et al.* (1981) confirmed that class 1 and class 2 *E. coli* produce different types of ST. However, methanol-insoluble enterotoxic activity (STb) was isolated from both classes of *E. coli* and it was concluded that the differences between the types of ST produced by each class did not correspond with their solubility in methanol. It was also concluded that the response of the pig intestine was a function not only of the ST subtype but also of the age of the pig.

When a strain of *E. coli* (strain B44) enteropathogenic for newborn calves was examined in ligated calf ileal loops, enterotoxic activity corresponding to STa was found (Newsome, 1980). In three surveys of *E. coli* isolated from calves affected by neonatal diarrhoea, the majority of strains produced STa (Sivaswamy and Gyles, 1976; Moon *et al.,* 1976; Lariviere *et al.,* 1979). LT was produced in four of the eight strains in which STa was not demonstrated. It was unclear whether the remainder produced STb or whether they simply failed to produce detectable levels of STa.

A report by Gyles (1979) suggests that calf strains may produce a second enterotoxin, which he refers to as ST_2. Burgess *et al.* (1980) reported that *E. coli* strain P16 produces two forms of STa. It is highly probable that terms such as

STa, STb and LT denote three biological classes of enterotoxin rather than three discrete molecular entities.

Weanling colibacillary diarrhoea. Diarrhoea of older piglets associated with enterotoxigenic *E. coli* is frequently referred to as weanling colibacillary diarrhoea. The sudden changes of environment and flora that occur at weaning may lead to the multiplication of *E. coli* in the small intestine, culminating in enteric colibacillosis which is manifested as colibacillary diarrhoea or colibacillary toxaemia. Both syndromes may be observed in the same herd. The enterotoxigenic *E. coli* associated with diarrhoea belong to the same OK serogroups as those that cause toxaemia (Table 2) and these organisms are frequently haemolytic.

In outbreaks of weanling colibacillary diarrhoea it is often the large thriving piglets in a herd that collapse and die. The clinical characteristics, post-mortem appearance and bacteriological findings resemble those of neonatal *E. coli* diarrhoea (Stevens, 1963a) but the mortality is low and sometimes only one or two piglets in a herd may be affected. Isolates of *E. coli* most frequently belong to OK group O141:K85ab, or to OK groups O138:K81 and O141:K85ac (Sojka, 1971). Neither K88 nor any of the other adhesive antigens have been detected on these *E. coli*, but they do produce enterotoxin in ligated intestinal loop experiments (Smith and Halls, 1967a).

Colibacillary Toxaemia

Disease associated with the absorption of endotoxin occurs infrequently in newborn calves, lambs and piglets but is more of a problem in piglets at weaning. Colibacillary toxaemia in the neonate is often referred to as enteric toxaemia or endotoxaemia.

Enteric toxaemia in neonatal calves, lambs and piglets. In outbreaks of neonatal *E. coli* diarrhoea, sudden death of newborn piglets, calves and lambs, showing no other clinical symptoms, is generally attributed to the effect of absorbed endotoxin produced by the enterotoxigenic *E. coli*. In these animals the concentration of endotoxin may become lethal before the enterotoxin induces diarrhoea. The enteric toxaemic form of colibacillosis in calves described by Gay *et al.* (1964) is characterized by sudden collapse and death. This was associated with a rapid proliferation of *E. coli* belonging to one or other of the serogroups O8, O9 or O101 and occurred during the first week of life, usually within 48 h of birth. There is a complete loss of muscle tone and marked flaccid paralysis. Clinically there may be signs of dehydration but calves usually die before scouring is evident. Wray and Thomlinson (1969) reported that day-old calves intravenously injected with phenol-extracted endotoxin from *E. coli* O78:K80 and O115:K showed mild shock and developed enteritis. Isolates belonging to these two OK groups, however, are usually considered to be bacteraemic rather than enteropathogenic. Shreeve and Thomlinson (1972) demonstrated rapid absorption of endotoxin from the intestinal tract of piglets. They also found that piglets were hypersensitive to *E. coli* endotoxin at birth and suggested that hypersensitivity may play a part in the pathogenesis of *E. coli* neonatal diarrhoea (Shreeve and Thomlinson, 1971).

Colibacillary toxaemia in pigs at weaning. In addition to colibacillary weanling diarrhoea, weaner piglets can be affected by colibacillary toxaemia. Three syndromes have been described: (1) shock in weaner syndrome, (2) haemorrhagic enteritis and (3) oedema disease, of which the last is the most important.

A syndrome affecting weaners and store pigs in which one or two piglets in a litter die suddenly was described as shock in weaner syndrome by Schimmelpfennig (1970). Oedema is observed in the intestine, lungs and kidneys as well as the central nervous system. Serous exudate is common in the large body cavities and, in severe cases, blood may pass into the lumen of the intestine, thus simulating classical haemorrhagic enteritis. Schultz *et al.* (1961) suggested that the lesions were an expression of acute shock and that the intestinal changes were due to permeability disturbances caused by shock rather than by enteritis.

Under the term haemorrhagic enteritis Stevens (1963a) described a form of enteritis in post-weaning piglets characterized by sudden death and inflammation of the intestine. Although diarrhoea sometimes occurs this is not a consistent feature of the disease. Haemorrhagic lesions are characteristic and these are usually present in the stomach (Thomlinson and Buxton, 1962) although they are occasionally confined to the intestine and sometimes only the jejunum and ileum. Typically, the intestinal mucosa is severely congested and haemorrhages occur in the lumen. The mesenteric lymph nodes are enlarged and haemorrhagic and subcutaneous oedema and serous fluid within the body cavities may also be present. Schimmelpfennig (1970) considered shock in weaner syndrome to be an acute form of haemorrhagic enteritis and Thomlinson and Buxton (1962) pointed out the similarity between the latter and oedema disease.

Oedema disease is an acute disease of young thriving pigs usually occurring about 1 wk after weaning (Sojka, 1965, 1970; Schimmelpfennig, 1970). The clinical features are characterized by (1) an apparent low morbidity and high case fatality rate, (2) oedema of subcutaneous tissues, usually over the frontal aspect of the head and involving the orbital tissues and (3) signs of neurological disturbances which are variable but generally include ataxia, convulsion and then death. In some herds diarrhoea may be a clinical feature when enterotoxigenic *E. coli* are involved. Sometimes animals may be found dead without having been observed ill (Nielsen and Clugston, 1971).

At post-mortem examination subcutaneous oedema is observed together with oedema in the submucosa of the cardiac region of the stomach and oedema of the mesentery and spiral colon. The most characteristic microscopic lesions are those of angiopathy affecting the small arterioles (Kurtz *et al.,* 1969; Clugston *et al.,* 1973). In piglets that have survived for several days there are lesions of focal encephalomalacia in the brain stem together with lesions in the small arteries and arterioles (Kurtz *et al.,* 1969). These malacia lesions are thought to be the result of vascular injury leading to oedema and ischaemia (Nielsen, 1981).

Bacteriologically, oedema disease is characterized by high concentrations of *E. coli* in the small and large intestine. These organisms can also be isolated from the mesenteric lymph nodes of affected piglets but there is no invasion of other tissues (Sojka *et al.,* 1957; Sojka, 1965). The *E. coli* belong to relatively few serological OK groups (Table 2) and the same OK groups are associated with the other syndromes of colibacillary toxaemia. *E. coli* from the OK types **O**141:**K**85ab and **O**138:**K**81 usually cause diarrhoea and oedema disease. Some isolates of the serological group **O**141:**K**85ac are enterotoxigenic while others are not. *E. coli* of the OK groups **O**139:**K**82 and **O**45:**K**″E65″ are usually non-enterotoxigenic.

Schimmelpfennig (1970) considered that shock in weaner syndrome and haemorrhagic enteritis are due to direct absorption of *E. coli* endotoxin and that both are a form of endotoxin poisoning. In contrast, he suggested that oedema disease is caused by a neurotoxin and he postulated that in the pig the action of endotoxin and neurotoxin would compete. Thomlinson (1963) suggested that the lesions are due to anaphylaxis. According to this theory, *E. coli* endotoxin polysaccharides are constantly absorbed in small quantities by the reticuloendothelial system, a process which results in the formation of tissue-sensitizing antibodies (Buxton and Thomlinson, 1961). If *E. coli* multiply rapidly as a result of dietary or other changes occurring at weaning, then a sudden absorption of *E. coli* endotoxin from the intestine would produce an anaphylactic reaction in the previously sensitized piglet. Whether this reaction produces shock in weaner syndrome, haemorrhagic enteritis or oedema disease would depend upon the degree of hypersensitivity of the animal and the amount of endotoxin absorbed.

Smith and Halls (1967b) considered that the enterotoxin produced by strains of *E. coli* associated with oedema disease is important only in the causation of the diarrhoea and plays no part in oedema disease. If an organism is capable of producing diarrhoea, shock in weaner syndrome and oedema disease (e.g. an *E. coli* of OK group **O**141:**K**85ab), the factors that determine which of these syndromes, if any, develop probably depend upon the conditions within the small intestine. It is reasonable to infer that diarrhoea results if the organism produces sufficient enterotoxin. Profuse diarrhoea would probably decrease the likelihood of oedema disease or other colibacillary toxaemias occurring, since the metabolic products responsible for these diseases would more likely be swept through the intestine before being absorbed. If diarrhoea is mild, or if no enterotoxin is produced at all (e.g. as in an infection of *E. coli* **O**139:**K**82), the accumulation of bacterial products might lead to oedema disease, shock in weaner syndrome or haemorrhagic enteritis.

Intravenous injection of cell-free extracts of *E. coli* belonging to those serotypes frequently associated with oedema disease results in a condition indistinguishable from field outbreaks, whereas similar extracts from other *E. coli* are without effect (Erskine *et al.*, 1957). Nielsen and Clugston (1971) prepared a partially purified freeze-thaw extract of an *E. coli* of serogroup **O**139:**K**82 and referred to this as EDP (edema disease principle). They also prepared a phenol–water extract from the same strain and referred to this as endotoxin. Paired animals were given a single intravenous dose of EDP or endotoxin and pairs were then killed and examined at regular intervals after infection. Both groups of piglets rapidly developed pyrexia and leucopenia but those injected with endotoxin were most severely affected. These animals vomited and passed profuse soft faeces. At 3 and 6 h lesions were similar in both groups and considered characteristic of endotoxaemia. At 24 and 48 h pigs injected with endotoxin were clinically normal and had no significant lesions. Animals given EDP had recovered from endotoxaemia by 12 h but by 24 h they had developed neurological symptoms and typical oedematous lesions. Thus, the clinical responses induced by EDP were, in part, similar to those of endotoxin in the period immediately after injection but much reduced in intensity. However, EDP produced additional lesions not associated with the shock-like syndrome and these delayed reactions were characteristic of oedema disease.

Schimmelpfennig and Weber (1978) isolated crude oedema disease neurotoxin from *E. coli* belonging to serogroups **O**138:**K**81, **O**139:**K**82 and **O**141:**K**85. The three strains each produced a common toxin which was not present in extracts of *E. coli* from serological groups not associated with oedema disease. The toxin caused fits and paralysis in mice 24 h after intravenous injection. It was thermolabile, sensitive to formalin and poorly immunogenic although neutralizing antisera could be prepared in rabbits after prolonged immunization. The toxin shared no antigenic relationship with O or K antigen and was not related to *Shigella* toxin.

Smith and Halls (1968) found that large numbers of *E. coli* from serogroup **O**141:**K**85 were present on scrapings from the walls rather than the contents of the small intestine in piglets affected by oedema disease. They concluded that this strain adhered to the intestinal epithelium. A similar conclusion was reached by Kenworthy (1970), using germ-free piglets infected with *E. coli* from the same serogroup. Gilka and Salajka (1968) examined piglets with oedema disease naturally infected with *E. coli* from serogroup **O**139:**K**82 and reported that the bacteria were not associated with the villous epithelium. Bertschinger and Pohlenz (1980, 1983), using pigs experimentally infected with pure cultures of *E. coli* serogroup **O**139:**K**82 or gut contents of a pig naturally infected with these organisms, demonstrated microcolonies or continuous layers of these bacteria

adherent to the mucosal surface in the lower small intestine. Maximum colonization of the ileum by the *E. coli* occurred between 1 and 7 days before clinical
symptoms of disease became visible.

Systemic Colibacillosis

Systemic colibacillosis occurs frequently in calves, lambs and poultry. Bacteraemic strains of *E. coli* pass through the mucosa of the alimentary or respiratory tract and enter the blood stream where they cause either a generalized infection
(colisepticaemia) or a localized infection such as meningitis and/or arthritis in
calves and lambs or air-sacculitis and pericarditis in poultry. In piglets systemic
colibacillosis is less frequently encountered.

Calves

Systemic colibacillosis commonly occurs in hypogammaglobulinaemic calves;
this deficiency occurs in colostrum-deprived animals but it may also occur in
some calves that have been fed colostrum but failed to absorb it. Colostrum must
be ingested within a few hours of birth if it is to be effective, for little or no
absorption occurs after 24–36 h and in some calves even after 6 h.

The generalized infection occurs in the first few days of life and follows an
acute, often fatal, course. Diarrhoea may be seen before death but it is not a
constant feature (Barnum *et al.*, 1967). Shock-like syndromes may develop in
septicaemic calves (Tennant *et al.*, 1978), possibly caused by endotoxin. Calves
that die are usually well hydrated and enlargement of the spleen and petechial
haemorrhages on the surfaces of the heart, peritoneum and blood vessels are
common. In some animals the course of systemic colibacillosis is more prolonged. *E. coli* localize in tissues with a predilection for joint cavities and
leptomeninges causing polyarthritis, meningitis or both (Barnum *et al.*, 1967).

In generalized infection pure cultures of *E. coli* can usually be isolated from
organs and tissues throughout the body. In localized infection *E. coli* can sometimes be isolated from the infected sites. These bacteraemic isolates belong to
one or another of relatively few serological groups as shown in Table 1 (based on
Sojka, 1965; Renault, 1979; Ørskov, 1978). The commonest OK group is
O78:K80, which is also frequently encountered among isolates from lambs and
poultry (Sojka, 1965).

A condition similar to that occurring in the field has been reproduced experimentally in colostrum-deprived calves by the administration of certain strains of
E. coli including organisms from serogroups O78:K80 (Fey and Margadant,
1962ab) and O137:K89 (Gay *et al.*, 1964; Schoenaers and Kaeckenbeeck,
1958). Fey and Margadant (1962b) showed that infection with certain strains of

E. coli was successful only when *E. coli* were administered immediately after birth and before the animal received colostrum. If colostrum was ingested and gammaglobulin absorbed, experimental reproduction of colisepticaemia failed. Hence the main factor which renders a calf susceptible to systemic colibacillosis appears to be a deficiency of gammaglobulin (Penhale *et al.,* 1970). Logan and Penhale (1971) found that the IgM fraction of serum gammaglobulin prevented colisepticaemia when the immunoglobulin was injected intravenously. In contrast, similar administration of IgM had no effect on enteric colibacillosis caused by enterotoxigenic *E. coli* (Penhale and Logan, 1971).

Lambs

As in other species of animals, systemic colibacillosis in lambs may be generalized or localized (e.g., Sojka, 1971). It may occur in lambs 1 day to 14 wk old but most commonly 2- to 3-wk-old lambs are affected (Jensen, 1974). The clinical and post-mortem characteristics are very similar to those described for calves. The serological OK groups of *E. coli* are also similar to those found in calves (Table 1).

Experimental reproduction of the disease has been reported by several workers (Roberts, 1957, 1958; Kater *et al.,* 1963; Terlecki and Sojka, 1965). Kater *et al.* (1963) were able to reproduce the disease in older lambs (5–8 wk old) using cultures of *E. coli* **O**78:**K**80. However, Terlecki and Shaw (1959) found that 3-month-old lambs were refractory to experimental infection with bacteraemic *E. coli* **O**20:**K**″X644″ isolated from lambs with meningo-encephalitis. This was thought to be age-related since Terlecki and Sojka (1965) were successful in reproducing the disease in very young lambs (1–3 days old) using the same isolates. Some experimental lambs died of septicaemia within 24 h, others developed meningitis and/or arthritis but none developed diarrhoea. Lambs inoculated with *E. coli* strain **O**141:**K**85ab, isolated from a pig with oedema disease, remained clinically healthy.

Piglets

Usually 24- to 48-h-old piglets are affected. Some piglets are found dead while others are comatose. Diarrhoea does not usually occur. In acutely affected piglets there may be no gross lesions but in less acute cases there may be subserous or submucosal haemorrhages. Meningitis, arthritis and pneumonia may also be observed.

Poultry

Many manifestations of *E. coli* infections in poultry have been described (Sojka, 1965) but the most common is systemic colibacillosis. This is frequently referred to as colisepticaemia and less frequently as coliform pericarditis or air-sacculitis. Colisepticaemia is primarily a disease of 5- to 12-wk-old broiler chicks with a maximum incidence at 6–9 wk (Gross, 1972) but it can also occur in newborn chicks and turkeys. The disease is more common in the winter months and this may be related to inadequate ventilation (Sojka, 1965). Most outbreaks of colibacillosis occur in flocks affected with other respiratory diseases, especially *Mycoplasma gallisepticum* (Gross, 1957).

Symptomatically, the disease is similar to *M. gallisepticum* infection in that there is respiratory distress and sneezing associated with lesions in the lower respiratory tract. Depression, severe green diarrhoea and often lameness are the main clinical characteristics. Most deaths occur during the first 5 days of the disease. At post-mortem examination, the most frequent lesion is fibrinous pericarditis. The pericardial sac is usually opaque, thickened, filled with a purulent exudate and usually firmly attached to the myocardium. Fibrinous pericarditis is often accompanied by perihepatitis, the liver being enlarged and covered by a gelatinous exudate, the air sacs are thickened and contain yellow caseous material. Synovitis is a common sequel to colisepticaemia and hock joints are most frequently affected (Sojka and Carnaghan, 1961). Panopthalmitis, usually of only one eye, can also develop in some birds but this is less common.

Bacteriological examination of the visceral organs of birds that died of colisepticaemia leads to the isolation of *E. coli* usually in profuse cultures. Such strains frequently belong to relatively few serological OK groups, the commonest being **O78:K80**, **O2:K1** and **O1:K1** (Sojka and Carnaghan, 1961). Other serological groups reported included **O3**, **O6**, **O8**, **O11**, **O15**, **O22**, **O55**, **O74**, **O88**, **O95** and **O109**. Many other O groups were isolated infrequently (Gross, 1972).

The infecting strain of *E. coli* can often be isolated in pure cultures from the pericardial lining of birds which have died, but in killed birds the lesions may be resolving and sterile. In chronic cases of arthritis the infecting organism may be localized in affected joints.

Harry and Hemsley (1965) reported that in normal broiler chickens *E. coli* was restricted mainly to the upper respiratory tract. Only in diseased birds were pathogenic *E. coli* isolated from the air sacs and trachea. These workers concluded that production of colisepticaemia depends on factors other than the mere presence of a bacteraemic strain of *E. coli* in the upper respiratory tract.

It is generally considered that certain predisposing factors are essential in the pathogenesis of colisepticaemia—*M. gallisepticum* and infectious bronchitis virus (IBV) are commonly implicated although in some outbreaks of colisepticaemia it appears that *E. coli* may be the sole cause (Gordon, 1959; Sojka and Carnaghan, 1961). The experimental infections described by Gross (1956, 1958) and by Sojka and Carnaghan (1961) indicate that certain serotypes of *E. coli* are capable of producing syndromes indistinguishable from those observed in the field. In contrast, there was no disease when each of two control strains was injected intravenously. One *E. coli* strain was of unknown antigenic serotype and had been isolated from the faeces of a healthy chicken, the other was from OK serogroup **O**141:**K**85ab and had been isolated from a pig with oedema disease.

Virulence determinants in bacteraemic E. coli. A high proportion of invasive strains of *E. coli* isolated from calves, lambs or chicks produce colicin V (ColV), a plasmid-determined antibacterial protein (Smith, 1974). By transferring the ColV plasmid from virulent field isolates into non-pathogenic strains of *E. coli* such as strain K-12, Smith found that the acquisition of the ColV phenotype was always accompanied by an increase in lethality. This was associated not with toxic activity but with a greater ability to survive in blood and peritoneal fluids. Furthermore, the virulence of pathogenic field strains was significantly reduced when the ColV plasmid was eliminated by treatment with curing agents such as sodium lauryl sulphate (Smith and Huggins, 1976). Reintroduction of the ColV plasmid by conjugation restored the *E. coli* to their original virulence. Although Smith demonstrated that the genes associated with the increased survival were present on the ColV plasmid, he pointed out that his results did not indicate whether it was the ColV genes themselves or other genes present on the ColV plasmid that were responsible. Whatever the products involved, it was concluded that they must function by remaining intimately associated with the bacterium as ColV$^+$ bacteria usually survive much better than isogenic ColV$^-$ bacteria when both forms are inoculated simultaneously into the same host (Smith and Huggins, 1976). ColV$^+$ organisms are less sensitive than ColV$^-$ forms to the host defence mechanisms dependent upon antibody and complement. The determinants for serum resistance and for colicin V production are closely linked (Binns *et al.*, 1979). These workers cloned a fragment of the ColV plasmid and found that it increased the virulence of the recombinant *E. coli* by approximately 100-fold. A genetic determinant for resistance to the bactericidal effect of serum was mapped within the fragment. It was concluded that the association between the genes specifying colicin V and virulence in ColV plasmids resembles the relationship between the genes specifying for K88 antigen production and raffinose utilization (Binns *et al.*, 1979). Most plasmids containing the K88 determinants also contain the determinants for raffinose utilization which appears to be unrelated to virulence (Smith and Parsell, 1975). Minshew *et al.* (1978) examined *E. coli* invasive for man and found that the ability to haemagglutinate human erythrocytes was a common feature of these bacteria. It was suggested that this property might reflect a specific common adherence factor. Clancy and Savage (1981) reported that *E. coli* K-12 into which a ColV plasmid had been introduced adhered *in vitro* to the small intestine of mice in two- to three-fold greater numbers than the isogenic strain, without the plasmid. These differences were statistically significant. Furthermore, the ColV strain produced fimbriae that absorbed male-specific phage, whereas its isogenic variant without ColV did not.

In addition to describing the ColV plasmid, Smith (1974) also described an independent plasmid, Vir, which was found in an *E. coli* strain isolated from a lamb with fatal bacteraemia. Vir$^+$ *E. coli* appear to be far less common in nature than ColV$^+$ *E. coli* and have only been isolated from calves and lambs (Lopez–Alvarez and Gyles, 1980). In contrast to the Vir$^-$ *E. coli,* isogenic bacteria containing the Vir plasmid were highly toxic when injected into chickens or mice. Culture filtrates and especially bacterial ultrasonicates of Vir$^+$ recombinant *E. coli* were also toxic. Since the toxin was highly susceptible to heat and low pH and was non-dialysable, Smith (1974) suggested that it was probably protein in nature. When injected into chickens the toxin produced characteristic lesions very similar to those caused by the chick lethal toxin produced by *E. coli* isolated from colisepticaemia in poultry (Truscott, 1973; Truscott *et al.*, 1974). Thus, post-mortem examination showed the main gross pathological changes to be fluid accumulation in the peritoneal cavity and pericardial sac, various degrees of a perihepatic lesion characterized by a fibrinous or gelatinous pellicle covering the liver, and congestion of the liver and spleen (Truscott *et al.*, 1974; Smith 1974).

The Vir phenotype is characterized not only by the synthesis of toxin but also by the production of a specific surface antigen. This antigen could be detected by absorbing OK antiserum raised to the Vir$^+$ *E. coli* with the Vir$^-$ isogenic strain. Smith (1974) showed that although Vir$^+$ organisms were agglutinated by this specific antiserum the toxic activity was not neutralized. He pointed out that the toxin and the surface antigen may be two different products of independent genes located on the same plasmid. Lopez-Alvarez and Gyles (1980) examined isolates of *E. coli* from colisepticaemia in calves and found that some strains produced the Vir toxin but not the Vir surface antigen. When the Vir plasmid was transferred to an *E. coli* strain that failed to produce fimbriae the recombinant organism produced these structures. Two types of fimbriae were observed; one type appeared to be sex fimbriae and the other was thought to be the surface antigen associated with the Vir phenotype. Morris *et al.* (1982b) demonstrated that Vir$^+$ recombinant *E. coli* attached *in vitro* to calf enterocyte brush borders, ciliated epithelial cells and squamous epithelial cells from the oesophagus whereas the Vir$^-$ isogenic *E. coli* failed to adhere to these cells. Furthermore, the adhesion of Vir$^+$ *E. coli* was inhibited by antiserum to an antigenically unrelated Vir$^+$ recombinant *E. coli* but antiserum to the Vir$^-$ form was without effect.

Bovine *E. coli* Mastitis

During recent years, bovine coliform mastitis has become very common. It continues to be a major problem and reflects the poor hygienic conditions of winter housing and milking parlours on many farms. Data from the Veterinary

Investigation Diagnosis Analysis Mark II (VIDA II), based on the reports from 24 Veterinary Investigation Centres in England, Wales and Scotland, indicated that during 1981 *E. coli* was most frequently implicated as a causal agent in bovine mastitis, accounting for 2904 diagnoses, followed by coagulase-positive staphylococci (2291), *Streptococcus uberis* (2227), *Str. dysgalactiae* (1419), *Corynebacterium pyogenes* (560) and *Str. agalactiae* (455).

Coliform mastitis affects lactating cattle, particularly during the winter months. *E. coli* mastitis is often severe in early lactation which, in the majority of herds, coincides with changes from grazing to winter housing and feeding. During the summer, when most cows are at grass, *E. coli* mastitis is rare. A close relationship has been demonstrated between coliform contamination of bedding and the occurrence of *E. coli* mastitis (Bramley and Neave, 1975). *E. coli* isolates from cattle with mastitis are similar to those found in the faeces of healthy cows (Linton *et al.*, 1979). *E. coli* will survive and even multiply in bedding (Bramley and Neave, 1975). Some evidence suggests that cows, especially those in early lactation, produce incubation temperatures in bedding when they lie down, thus contributing to increased bacterial bedding populations (Francis *et al.*, 1980) and increased teat end challenge. The same authors stated that during low ambient temperatures in winter, cow lying times lengthened, causing greater contact between teats and warmed contaminated bedding.

Control measures used in mastitis, while effective against many causative organisms, are not effective against those bacteria such as *E. coli* which are abundant in the environment of intensively managed high-producing dairy cows. To produce mastitis in a healthy cow, bacteria must first be transferred from a reservoir of infection. Bacteria like *Streptococcus* and *Staphylococcus* are present in an infected udder and the spread of infection by these organisms usually occurs at milking. With *E. coli* mastitis, however, the pathogen is in the environment and the organism can therefore be transferred to the udder at any time. There is a correlation between teat end contamination and clinical incidence (Neave and Oliver, 1962). Entry of *E. coli* into the udder is usually followed within 48 h by clinical mastitis and by rapid elimination of the organism by phagocytosis.

There appears to be a spectrum of responses to infection in which the stages of lactation and number of infected quarters are important. At one extreme there is a very mild reaction in the udder and a transient increase in the number of *E. coli* followed by rapid elimination of this organism before clinical signs are evident (Hill *et al.*, 1978). The symptoms include clots, milk discolouration and swelling of the gland, which may become oedematous. In these animals there is little loss of milk production or body condition. At the other extreme there is a very acute reaction, apparently uninhibited multiplication of *E. coli*, diarrhoea, anorexia and acute pyrexia but no early clinical sign in the udder. In such extreme cases, occurring more commonly in very early lactation, no inflammation or swelling of the udder is observed and this constitutes a problem of early diagnosis of mastitis in the field. If there is a complete failure to produce an inflammatory response in the udder, the *E. coli* grow unchecked and produce large amounts of toxin (probably endotoxin) which will not only produce severe damage to the udder, but may also result in general toxaemia when absorbed. Some animals may die within 24 h of becoming ill (Said, 1973).

In coliform mastitis, *E. coli* does not invade (in this respect it resembles isolates from enteric colibacillosis) but is confined to the lumen of the teat canal and lactiferous sinuses. While both forms of colibacillosis in calves, enteric and systemic, are caused by a relatively small number of *E. coli* serotypes, the isolates from bovine *E. coli* mastitis belong to a large number of serological groups. Thus, Linton *et al.* (1979) demonstrated 67 different serological O groups among 279 *E. coli* isolates from individual cows with clinical mastitis. There is a close similarity between strains causing mastitis and those regularly present in the gut of adult cattle, thus indicating that *E. coli* mastitis is an endogenous infection. Most *E. coli* serogroups are equally likely to cause mastitis (Hill and Shears, 1979). Experimental infection with serum-resistant *E. coli* has shown that the variation in response to infection of the mammary gland does not depend upon inoculation size but rather on the animal infected (Hill *et al.*, 1978; Hill and Shears, 1979). Mid-lactation animals respond less severely than newly calved cows but the age of animals does not appear to be important.

It is well known that the dry udder is very resistant to *E. coli* infection. Reiter and Bramley (1975) demonstrated that lactoferrin/citrate/bicarbonate systems in the udder largely account for the resistance of the dry udder to *E. coli* infection. Lactoferrin (LF) in the presence of bicarbonate binds iron and inhibits the multiplication of bacteria which have a high iron requirement, such as *E. coli* (Oram and Reiter, 1966). This effect is countered by citrate, which competes with LF, thereby making iron available to the bacteria (Reiter 1978). The dry gland contains a high concentration of LF and lysozyme (Reiter and Oram, 1967; Reiter and Bramley, 1975). During involution of the gland the citrate is reabsorbed while bicarbonate permeates from the blood into the gland. In this way the conditions for iron-binding and inhibitory activity of LF are optimal. At the time of calving the iron-binding capacity falls and therefore the secretion is no longer inhibitory for *E. coli*. Shortly before parturition, citrate secretion by the gland rapidly increases and the bicarbonate concentration decreases thus permitting multiplication of *E. coli*, culminating in clinical mastitis (Reiter, 1978).

Pathogenesis. Frost *et al.* (1980) examined the pathogenesis of coliform mastitis and found that the main changes in the disease were superficial and confined to the epithelium of the teat sinus, lactiferous sinus and large ducts. Although there were occasional lesions of the basement membrane beneath the damaged epithelium, the damage did not extend beyond this and there was no serious involvement of the secretory tissue. The earliest lesions were seen after 1 h and by 2 h the necrosis and sloughing of epithelial cells were more severe. This was followed by a massive infiltration of neutrophils through the lesions, which were irregularly distributed on the epithelium. These neutrophils often remained attached to the epithelium and formed large mounds of cells. It was suggested that this results in a gross underestimation of the neutrophils in the lumen when this is assessed by counting the neutrophils in the milk. There was no evidence of *E. coli* adhering to the gland despite earlier observations of other bacteria attaching to the ductular epithelium (Frost, 1975; Frost *et al.*, 1977). It was, therefore, assumed that lesions were caused by diffusion of toxins from the organisms. Although endotoxin has been implicated in the disease and the effects of endotoxin on the lactating udder are well known (Carroll *et al.*, 1964; Said, 1973), Frost *et al.* (1977) thought it was unlikely that endotoxin was entirely responsible for the lesions. These workers (Brooker *et al.*, 1981) subsequently presented evidence for the presence of an additional toxin.

Frost *et al.* (1982) have suggested the probable sequence of events in the development of the disease as seen in the field. After infection by the *E. coli*, the organism multiplies rapidly in milk that remains in the teat sinus. This leads to local concentrations of toxins at multiple sites and results in more extensive damage at these sites. These organisms are then distributed first to the lactiferous sinus and then to the large ducts by the normal movement of the cow. If bacterial multiplication is not contained the organisms eventually reach the secreting areas of the gland. As different sites are seeded at different times the degree of tissue damage varies. The multiplication of *E. coli* in the teat canal initiates the inflammatory response early in the infection and the neutrophils then migrate to new areas of epithelial damage. If there is a rapid neutrophil response the secretions become bactericidal and the numbers of *E. coli* decrease. The infection then resolves spontaneously and as the damage is confined to the sinuses without involvement of the secretory tissue the disease has little effect upon the milk yield. However, if the neutrophil response is delayed, the growth of the *E. coli* is unchecked and the high concentrations of toxin cause massive destruction of udder tissue, general toxaemia and perhaps death.

Summary

In the colibacillary diarrhoea form of enteric colibacillosis diarrhoea is caused by enterotoxin. With neonatal diarrhoea the enteropathogenic *E. coli* colonize the small intestine of calves, lambs and piglets within a few days of birth. In piglets colonization is usually facilitated by the fimbrial antigen K88 but *E. coli* that produce K99, 987P or F41 fimbriae occasionally occur. The K99 fimbriae are involved in the colonization of the small intestine of calves and lambs but F41 fimbriae, either alone or in combination with K99, may also be involved. All *E. coli* enteropathogenic for calves, lambs and piglets produce heat-stable entero-toxins, ST. The majority of *E. coli* enteropathogenic for calves and lambs pro-duce STa, whereas most *E. coli* isolated from pigs produce STa and STb. *E. coli* that are enteropathogenic for neonatal and weaned pigs (class 1 *E. coli*) produce different subtypes of ST from those *E. coli* enterotoxigenic only for neonatal pigs (class 2). The difference in response of pig intestine to class 1 and class 2 *E. coli* appears to be a function of both the subtype of ST produced and the age of the pig. In addition to producing ST, most *E. coli* enterotoxigenic for pigs produce heat-labile enterotoxin, LT.

In the colibacillary toxaemic form of enteric colibacillosis, which principally affects pigs at weaning, endotoxin is involved and the toxaemia is manifested as shock in weaner syndrome, haemorrhagic enteritis or possibly oedema disease. In addition, enterotoxigenic strains are capable of causing diarrhoea. The factors that determine whether a pig will be affected by one or other of these conditions have yet to be determined and the processes involved in the pathogenesis of these conditions are unclear. In a given pig the balance between enterotoxin, endotoxin and probably neurotoxin is critical. Thus, if enterotoxin predominates diarrhoea develops and concentrations of the other bacterial products in the lumen remain low. If *E. coli* does not induce diarrhoea, either because it lacks the Ent plasmid or because enterotoxin production is insignificant, then endotoxin and neurotoxin accumulate in the gut. In most pigs the concentrations of these toxins do not clinically affect the animal and it remains healthy. In some animals the con-centration of neurotoxin and/or endotoxin causes oedema disease, while in oth-ers the concentration of endotoxin is such that one of the other forms of toxaemia develops. As in *E. coli* neonatal diarrhoea the bacteria involved in oedema disease are capable of adhering to the mucosal epithelium of the small intestine but the adhesins involved have yet to be determined.

In contrast to the *E. coli* associated with enteric colibacillosis, the organisms that cause systemic colibacillosis are invasive and the disease principally affects poultry, calves and lambs. Colisepticaemia is the acute form of disease and is associated with a generalized infection. Arthritis and meningitis occur in the chronic disease and are associated with a localized infection. In poultry coli-bacillosis is often associated with respiratory conditions involving *Mycoplasma gallisepticum* or infectious bronchitis virus and the *E. coli* may be a secondary invader. In calves colisepticaemia is usually found only in animals that have failed to absorb colostral gammaglobulin. Two independent virulence plasmids have been identified. The ColV plasmid is common in *E. coli* isolated from poultry, calves and lambs affected by systemic colibacillosis and increases the resistance of the pathogen to the host defence mechanisms. The Vir plasmid is less common than ColV and has been found in strains isolated from calves and lambs. The Vir plasmid is associated with the synthesis of a toxin and a surface antigen. Both the ColV and Vir plasmids confer adhesive properties *in vitro* to recombinant *E. coli* and their presence is associated with the production of fimbriae.

Coliform mastitis affects lactating cows and can probably be caused by most serum-resistant *E. coli* from any OK serogroup. In some cows, more frequently in newly calved animals, the neutrophil response is delayed and unable to control the proliferation of *E. coli* in the teat canal. These animals become very ill. This is probably due to toxins that initially cause multiple superficial lesions in the epithelium of the gland and then subsequently enter the blood stream, causing toxaemia. Inflammation of the gland (i.e. mastitis) is not an early feature of this acute infection. In other cows the neutrophils respond rapidly and control the multiplication of *E. coli*. Clinical mastitis occurs soon after infection which resolves spontaneously and the animal quickly recovers. In addition to endotoxin there is evidence for the involvement of a separate toxin. The *E. coli* do not appear to adhere to the ductular or sinus epithelia.

References

Arbuckle, J. B. R. (1970). The location of *Escherichia coli* in the pig intestine. *Journal of Medical Microbiology* **3**, 333–340.

Arbuckle, J. B. R. (1971). Enteropathogenic *Escherichia coli* on the intestinal mucopolysaccharide layer of pigs. *Journal of Pathology and Bacteriology* **104**, 93–98.

Barnum, D. A., Glantz, P. J. and Moon, H. W. (1967). "Colibacillosis." CIBA Veterinary Monograph Series, Summit, New Jersey.

Bertschinger, H. U. and Pohlenz, J. (1980). Bacterial colonization and morphology of the intestine in experimental oedema disease. *Proceedings of the International Pig Veterinary Society 1980 Congress* p.139.

Bertschinger, H. U. and Pohlenz, J. (1983). Bacterial colonization and morphology of the intestine in porcine *Escherichia coli* enterotoxaemia (edema disease). *Veterinary Pathology* **20**, 99–110.

Bertschinger, H. U., Moon, H. W. and Whipp, S. C. (1972). Association of *Escherichia coli* with the small intestinal epithelium. I. Comparison of enteropathogenic and non-enteropathogenic porcine strains in pigs. *Infection and Immunity* **5**, 595–605.

Bijlsma, I. G. W., de Nijs, A. and Frik, J. F. (1981). Adherence of *Escherichia coli* to porcine intestinal brush borders by means of serological variants of the K88 antigen. *Antonie van Leeuwenhoek* **47**, 467–468.

Binns, M. M., Davies, D. L. and Hardy, K. G. (1979). Cloned fragments of the plasmid ColV, I-K94 specifying virulence and serum resistance. *Nature (London)* **279**, 778–781.

Bramley, A. J. and Neave, F. K. (1975). Studies on the control of coliform mastitis in dairy cows. *British Veterinary Journal* **131**, 160–169.

Brooker, B. E., Frost, A. J. and Hill, A. W. (1981). At least two toxins are involved in *Escherichia coli* mastitis. *Experientia* **37**, 290–292.

Burgess, M. N., Bywater, R. J., Cowley, C. M., Mullan, N. A. and Newsome, P. M. (1978). Biological evaluation of a methanol soluble, heat-stable *Escherichia coli* enterotoxin in infant mice, pigs, rabbits and calves. *Infection and Immunity* **21**, 526–531.

Burgess, M. N., Mullan, N. A. and Newsome, P. M. (1980). Heat stable enterotoxins from *Escherichia coli* P-16. *Infection and Immunity* **28**, 1038–1040.

Burrows, M. R., Sellwood, R. and Gibbons, R. A. (1976). Haemagglutination and adhesive properties associated with the K99 antigen of bovine strains of *Escherichia coli*. *Journal of General Microbiology* **96**, 269–275.

Buxton, A. and Thomlinson, J. R. (1961). The detection of tissue sensitizing antibodies to *E. coli* in oedema disease, haemorrhagic gastro-enteritis and in normal pigs. *Research in Veterinary Science* **2**, 73–88.

Carroll, E. J., Schalm, O. and Lasmanis, J. (1964). Experimental coliform mastitis (*Aerobacter aerogenes*) mastitis: Characteristics of endotoxin on its role in pathogenesis. *American Journal of Veterinary Research* **25**, 720–726.

Clancy, J. and Savage, D. C. (1981). Another colicine V phenotype: *In vitro* adhesion of *Escherichia coli* to mouse intestinal epithelium. *Infection and Immunity* **32**, 343–352.

Clugston, R. E., Nielsen, N. O. and Smith, D. L. T. (1973). Experimental edema disease of swine (*E. coli* enterotoxaemia). III. Pathology and pathogenesis. *Canadian Journal of Comparative Medicine* **38**, 34–43.

de Graaf, F. K. and Roorda, I. (1982). Production, purification and characterization of the fimbrial adhesive antigen F41 isolated from the calf enteropathogenic *Escherichia coli* strain B41M. *Infection and Immunity* **36**, 751–753.

de Graaf, F. K., Klemm, P. and Gaastra, W. (1981). Purification, characterization and partial covalent structure of the adhesive antigen K99 of *Escherichia coli*. *Infection and Immunity* **33**, 877–883.

Dress, D. T. and Waxler, G. L. (1970). Enteric colibacillosis in gnotobiotic swine: A fluorescence microscopic study. *American Journal of Veterinary Research* **31**, 1147–1157.

Erskine, R. G., Sojka, W. J. and Lloyd, M. K. (1957). The experimental reproduction of a syndrome indistinguishable from oedema disease. *Veterinary Record* **69**, 301–303.

Fey, H. and Margadant, A. (1962a). Zur Pathogenese der Kälber-Colisepsis. IV. Agammaglobulinämie als disponierender Faktor. *Zentralblatt für Veterinärmedizin* **9**, 653–663.

Fey, H. and Margadant, A. (1962b). Zur Pathogenese der Kälber-Colisepsis V. Versuche zur künstlichen Infektion neugeborener Kälber mit dem Colityp 78:80B. *Zentralblatt für Veterinärmedizin* **9**, 767–778.

Francis, P. G., Sumner, J. and Joyce, D. A. (1980). The influence of the winter environment of the dairy cow on mastitis. *Proceedings of the 11th International Congress on Diseases of Cattle, 1980* Vol. I, pp. 35–43.

Frost, A. J. (1975). Selective adhesion of microorganisms to the ductular epithelium of the bovine mammary gland. *Infection and Immunity* **12**, 1154–1156.

Frost, A. J., Wanasinghe, D. D. and Woolcock, J. B. (1977). Some factors affecting selective adherence of microorganisms in the bovine mammary gland. *Infection and Immunity* **15**, 245–253.

Frost, A. J., Hill, A. W. and Brooker, B. E. (1980). The early pathogenesis of bovine mastitis due to *Escherichia coli*. *Proceedings of the Royal Society of London, Series B* **209**, 431–439.

Frost, A. J., Hill, A. W. and Brooker, B. E. (1982). Pathogenesis of experimental bovine mastitis following a small inoculum of *Escherichia coli*. *Research in Veterinary Science* **33**, 105–112.

Gaastra, W. and de Graaf, F. K. (1982). Host-specific fimbrial adhesins on noninvasive enterotoxigenic *Escherichia coli* strains. *Microbiological Reviews* **46**, 129–161.

Gay, C. C., McKay, K. A. and Barnum, D. A. (1964). Studies on colibacillosis of calves. II. Clinical evaluation of the efficiency of vaccination of the dam as a means of preventing colibacillosis of the calf. *Canadian Veterinary Journal* **5**, 297–308.

Gilka, F. and Salajka, E. (1968). Die Immunfluoreszenz im Studium dersog Colienterotoxämie (oedemkrankheit) der Absetzferkel. *Pathologia Veterinaria* **5**, 475–476.

Gordon, R. F. (1959). Broiler diseases. *Veterinary Record* **71**, 994–1003.

Gross, W. B. (1956). *Escherichia coli* as a complicating factor in chronic respiratory disease of chickens and infectious sinusitis in turkeys. *Poultry Science* **35**, 724–771.

Gross, W. B. (1957). Pathological changes of an *Escherichia coli* infection in chickens and turkeys. *American Journal of Veterinary Research* **18**, 724–730.

Gross, W. B. (1958). Symposium on chronic respiratory diseases of poultry. III. Epizootiology of chronic respiratory diseases in chickens. *American Journal of Veterinary Research* **19**, 448–452.

Gross, W. B. (1972). Colibacillosis. *In* "Diseases of Poultry," (Eds. M. S. Hofstad, B. W. Calnek, C. F. Helmbod, W. M. Reid, and H. W. Yoder), pp. 392–405. Iowa State University Press, Ames.

Guinée, P. A. M. and Jansen, W. H. (1979). Behaviour of *Escherichia coli* K antigens K88ab, K88ac and K88ad in immunoelectrophoresis, double diffusion and haemagglutination. *Infection and Immunity* **23**, 700–705.

Gyles, C. L. (1971). Heat labile and heat stable forms of the enterotoxin from *E. coli* strains enteropathogenic for pigs. *Annals of the New York Academy of Sciences* **176**, 314–322.

Gyles, C. L. (1979). Limitations of the infant mouse test for *Escherichia coli* heat stable enterotoxin. *Canadian Journal of Comparative Medicine* **43**, 371–379.

Gyles, C. L. and Barnum, D. A. (1969). A heat-labile enterotoxin from strains of *Escherichia coli* enteropathogenic for pigs. *Journal of Infectious Diseases* **120**, 419–426.

Harry, E. G. and Hemsley, L. A. (1965). The association between the presence of septicaemia strains of *Escherchia coli* in the respiratory and intestinal tracts of chickens and the occurrence of coli septicaemia. *Veterinary Record* **77**, 35–40.

Hill, A. W. and Shears, A. L. (1979). Recurrent coliform mastitis in the dairy cow. *Veterinary Record* **105**, 299–301.

Hill, A. W., Shears, A. L. and Hibbitt, K. G. (1978). The elimination of serum resistant *Escherichia coli* from experimentally infected single mammary glands of healthy cows. *Research in Veterinary Science* **25**, 89–93.

Isaacson, R. E., Nagy, B. and Moon, H. W. (1977). Colonization of porcine small intestine by *Escherichia coli*: Colonization and adhesion factors of pig enteropathogens that lack K88. *Journal of Infectious Diseases* **135**, 531–539.

Jensen, R. (1974). Colibacillosis in lambs. *In* "Diseases of Sheep," pp. 76–81. Lea & Febiger, Philadelphia.

Jones, G. W. and Rutter, J. M. (1972). Role of the K88 antigen in the pathogenesis of neonatal diarrhoea caused by *Escherichia coli* in piglets. *Infection and Immunity* **6**, 918–927.

Kater, J. C., Davis, E. A., Haughey, K. G. and Hartley, W. J. (1963). *Escherichia coli* infection in lambs. *New Zealand Veterinary Journal* **11**, 32–38.

Kenworthy, R. (1970). Effect of *Escherichia coli* on germ-free and gnotobiotic pigs. I. Light and electron-microscopy of the small intestine. *Journal of Comparative Pathology* **80**, 53–63.

Kohler, E. M. (1971). Enterotoxic activity of whole cell lysates of *Escherichia coli* in young pigs. *American Journal of Veterinary Research* **32**, 731–737.

Kurtz, H. J., Bergeland, M. E. and Barnes, D. M. (1969). Pathologic changes in edema disease of swine. *American Journal of Veterinary Research* **30**, 791–806.

Lariviere, S., Lallier, R. and Morin, M. (1979). Evaluation of various methods for the detection of enteropathogenic *Escherichia coli* in diarrhoeic calves. *American Journal of Veterinary Research* **40**, 130–134.

Linton, A. H., Howe, K., Sojka, W. J. and Wray, C. (1979). A note on the range of *Escherichia coli* O-serotypes causing clinical bovine mastitis and their antibiotic resistance spectra. *Journal of Applied Bacteriology* **46**, 585–590.

Logan, E. F. and Penhale, W. J. (1971). Studies on the immunity of calf to colibacillosis. IV. Prevention of experimental colisepticaemia by the intravenous administration of a bovine serum IgM-rich fraction. *Veterinary Record* **89**, 663–667.

Lopez-Alvarez, J. and Gyles, C. L. (1980). Occurrence of the Vir plasmid among animal and human strains of invasive *Escherichia coli*. *American Journal of Veterinary Research* **41**, 769–774.

Minishew, B. H., Jorgensen, J., Counts, G. W. and Falkov, S. (1978). Association of hemolysin production, haemagglutination of human erythrocytes and virulence for chicken embryos of extra intestinal *Escherichia coli* isolates. *Infection and Immunity* **20**, 50–54.

Moon, H. W. (1965). The association of *Escherichia coli* with diarrhoeal diseases of newborn pig. Ph.D. Thesis, University of Minnesota.

Moon, H. W. (1969). Enteric colibacillosis in the newborn pig: Problems of diagnosis and control. *Journal of the American Veterinary Medical Association* **155**, 1853–1859.

Moon, H. W. and Whipp, S. C. (1970). Development of resistance with age by swine intestine to effects of enteropathogenic *Escherichia coli*. *Journal of Infectious Diseases* **122**, 220–223.

Moon, H. W., Whipp, S. C. and Scartvedt, S. M. (1976). Etiologic diagnosis of diarrhoeal diseases of calves: Frequency and methods for detecting enterotoxin and K99 antigen production by *Escherichia coli*. *American Journal of Veterinary Research* **37**, 1025–1029.

Moon, H. W., Nagy, B., Isaacson, R. E. and Ørskov, I. (1977). Occurrence of K99 antigen on *Escherichia coli* isolated from pigs and colonization of pig ileum by K99+ enterotoxigenic *E. coli* from calves and pigs. *Infection and Immunity* **15**, 614–621.

Moon, H. W., Kohler, E. M., Schneider, R. A. and Whipp, S. C. (1980). Prevalence of pilus antigens, enterotoxin types and enteropathogenicity among K88-negative enterotoxigenic *Escherichia coli* from neonatal piglets. *Infection and Immunity* **27**, 222–230.

Morris, J. A., Stevens, A. E. and Sojka, W. J. (1977). Preliminary characterization of cell-free K99 antigen isolated from *Escherichia coli* B41. *Journal of General Microbiology* **99**, 353–357.

Morris, J. A., Stevens, A. E. and Sojka, W. J. (1978). Anionic and cationic components of the K99 surface antigen from *Escherichia coli* B41. *Journal of General Microbiology* **107**, 173–175.

Morris, J. A., Thorns, C. J. and Sojka, W. J. (1980). Evidence for two adhesive antigens on the K99 reference strain *Escherichia coli* B41. *Journal of General Microbiology* **118**, 107–113.

Morris, J. A., Thorns, C. J., Scott, A. C., Sojka, W. J. and Wells, G. A. W. (1982a). Adhesion in vitro and in vivo associated with an adhesive antigen (F41) produced by a K99− mutant of the reference strain *Escherichia coli* B41. *Infection and Immunity* **36**, 1146–1153.

Morris, J. A., Thorns, C. J., Scott, A. C. and Sojka, W. J. (1982b). Adhesive properties associated with the Vir plasmid: A transmissible pathogenic characteristic associated with strains of invasive *Escherichia coli*. *Journal of General Microbiology* **128**, 2097–2103.

Morris, J. A., Thorns, C. J., Wells, G. A. W. and Sojka, W. J. (1983). The production of F41 fimbriae by piglet strains of enterotoxigenic *Escherichia coli* that lack K88, K99 and 987P fimbriae. *Journal of General Microbiology* **129**, 2753–2759.

Nagy, B., Moon, H. W. and Isaacson, R. E. (1977). Colonization of porcine intestine by enterotoxigenic *Escherichia coli*: Selection of piliated forms *in vivo*, adhesion of piliated forms to epithelial cells *in vitro*, and incidence of a pilus antigen among porcine enteropathogenic *E. coli*. *Infection and Immunity* **16**, 344–352.

Neave, F. K. and Oliver, J. (1962). The relationship between the number of mastitis pathogens placed on the teats of dry cow, their survival, and the amount of intramammary infection caused. *Journal of Dairy Research* **29**, 79–93.

Newsome, P. M. (1980). *E. coli* enterotoxins in colibacillosis: A review. *Proceedings of the British Cattle Veterinary Association 1979–1980*, pp. 91–100.

Nielsen, N. O. (1963). Studies of edema disease of swine. Ph.D. Thesis, University of Minnesota.

Nielsen, N. O. (1981). Oedema disease. *In* "Diseases of Swine," (Eds. A. D. Leman, R. D. Glock, W. L. Mengeking, R. A. C. Penny and F. Scholl), pp. 478–490. Iowa State University Press, Ames.

Nielsen, N. O. and Clugston, R. E. (1971). Comparison of *E. coli* endotoxin shock and acute experimental edema disease in young pigs. *Annals of the New York Academy of Sciences* **176**, 176–189.

Nielsen, N. O., Moon, H. W. and Roe, W. E. (1969). Enteric colibacillosis in swine. *Journal of the American Veterinary Medical Association* **153,** 1590–1606.

Oram, J. D. and Reiter, B. (1966). Inhibitory substances present in milk and secretion of the dry udder. *Report on the National Institute in Dairying* p. 93.

Ørskov, F. and Ørskov, I. (1978). Serotypes of *Escherichia coli* from diseased animals. *Abstracts of the 12th International Association of Microbiological Societies,* p. 29.

Ørskov, I. and Ørskov, F. (1966). Episome carried surface antigen K88 of *Escherichia coli*. I. Transmission of the determinant of the K88 antigen and the influence on the transfer of chromosomal markers. *Journal of Bacteriology* **91,** 69–75.

Ørskov, I., Ørskov, F., Sojka, W. J. and Leach, J. M. (1961). Simultaneous occurrence of *E. coli* B and L antigens in strains from diseased swine. *Acta Pathologica et Microbiologica Scandinavica, Section B* **53,** 404–422.

Ørskov, I., Ørskov F., Sojka, W. J. and Wittig, W. (1964). K antigens K88ab (L) and K88ac (L) in *E. coli*. A new O antigen: O147 and a new K antigen: K89(B). *Acta Pathologica et Microbiologica Scandinavica, Section B* **62,** 439–447.

Ørskov, I., Ørskov, F., Smith, H. W. and Sojka, W. J. (1975). The establishement of K99, a thermolabile, transmissible *Escherichia coli* K antigen, previously called "Kco", possessed by calf and lamb enteropathogenic strains. *Acta Pathologica et Microbiologica Scandinavica, Section B* **83,** 31–36.

Penhale, W. J. and Logan, E. F. (1971). Studies on the immunity of the calf to colibacillosis. II. Preparation of an IgM-rich fraction from bovine serum and its prophylactic use in experimental colisepticaemia. *Veterinary Record* **89,** 623–628.

Penhale, W. J., McEwan, A. D., Selman, I. and Fisher, E. W. (1970). Quantitative studies on bovine immunoglobulins. II. Plasma immunoglobulin levels in market calves and their relationship to neonatal infection. *British Veterinary Journal* **126,** 30–37.

Pesti, L. and Semjen, G. (1973). Studies on enteropathogenicity, loop dilating effect and enterotoxin producing capacity of *Escherichia coli* strains isolated from enteric diseases of swine. *Acta Veterinaria Academiae Scientiarum Hungaricae* **23,** 227–236.

Reiter, B. (1978). Review of the progress of dairy science antimicrobial systems in milk. *Journal of Dairy Science* **45,** 131–147.

Reiter, B., and Bramley, A. J. (1975). Defence mechanisms of the udder and their relevance to mastitis control. *In* "Proceedings of a Seminar on Mastitis Control" (Eds. F. M. Vodd, K. T. Griffin and K. G. Kingwill), pp. 210–222. International Dairy Federation, Reading, England.

Reiter, B. and Oram, J. D. (1967). Bacterial inhibitors in milk and other biological fluids. *Nature (London)* **216,** 328–330.

Renault, K. (1979). La colibacillose du veau diagnostic perspectives nouvelles de prophylaxie medicale. *Bulletin Mensuel de la Société Vétérinaire Pratique de France* **62,** 259–281.

Roberts, D. S. (1957). *Escherichia coli* infection in lambs. *Australian Veterinary Journal* **33,** 43–45.

Roberts, D. S. (1958). Further observation on *E. coli* disease in lambs. *Australian Veterinary Journal* **34,** 152–156.

Runnels, P. L., Moon, H. W. and Schneider, R. A. (1980). Development of resistance with host age to adhesion of K99$^+$ *Escherichia coli* to isolated intestinal epithelial cells. *Infection and Immunity* **28,** 298–300.

Rutter, M. J., Burrows, M. R., Sellwood, R. and Gibbons, R. A. (1975). A genetic basis for resistance to enteric disease caused by *E. coli*. *Nature (London)* **242,** 531–533.

Said, A. H. (1973). Experimental *Escherichia coli* mastitis in cattle, its pathogenesis and treatment. *Tijdschrift voor Diergeneeskunde* **98,** 387–396.

Saunders, C. N., Stevens, A. J., Spence, J. B. and Sojka, W. J. (1960). *Escherichia coli* infection in piglets. *Research in Veterinary Science* **1,** 28–35.

Schimmelpfennig, H. (1970). "Untersuchungen zur Aetiologie der oedem Krankheit des Schweines." Parey, Berlin.

Schimmelpfennig, H. and Weber, R. (1978). Studies on the oedema disease producing toxin of *Escherichia coli* (*E. coli* neurotoxin). *Abstracts of the 12th International Congress of Microbiology, 1978* S21–25.

Schneider, R. A. and To, C. S. M. (1982). Enterotoxigenic *Escherichia coli* strains that express K88 and 987P pilus antigen. *Infection and Immunity* **36**, 417–418.

Schoenaers, F. and Kaeckenbeeck, A. (1958). Etudes sur la colibacillose du veau. Réalization de la maladie. *Annales de Médicine Vétérinaire* **102**, 211–221.

Schultz, L. C., Brass, W. and Nussel, M. (1961). Experimentelle Untersuchungen zur Pathogenese schockartiger und rheumatoider Krankheiten des schweines. I. Schockartige Erkrankungen und die Beteiligung des Zentralen Nervensystem. *Deutsche Tierärztliche Wochenschrift* **68**, 289–296.

Sellwood, R., Gibbons, R. A., Jones, G. W. and Rutter, J. M. (1975). Adhesion of enteropathogenic *Escherichia coli* to pig intestinal brush borders: The existence of two pig phenotypes. *Journal of Medical Microbiology* **8**, 405–411.

Shreeve, B. J. and Thomlinson, J. R. (1971). *Escherichia coli* disease in the piglet. A pathological investigation. *British Veterinary Journal* **126**, 444–451.

Shreeve, B. J. and Thomlinson, J. R. (1972). Absorption of *Escherichia coli* endotoxin by the neonatal pig. *Journal of Medical Microbiology* **5**, 55–59.

Sivaswamy, G. and Gyles, C. L. (1976). The prevalence of enterotoxigenic *Escherichia coli* in the faeces of calves with dirrhoea. *Canadian Journal of Comparative Medicine* **40**, 241 246.

Smith, H. W. (1974). A search for transmissible pathogenic characters in invasive strains of *Escherichia coli*: The discovery of a plasmid-controlled toxin and a plasmid-controlled lethal character closely associated, or identical, with colicine V. *Journal of General Microbiology* **83**, 95–111.

Smith, H. W. and Gyles, C. L. (1970). The relationship between two apparently different enterotoxins produced by enteropathogenic strains of *Escherichia coli* of porcine origin. *Journal of Medical Microbiology* **3**, 387–401.

Smith, H. W. and Halls, S. (1967a). Observations by the ligated intestinal segment and oral inoculation methods in *Escherichia coli* infections in pigs, calves, lambs and rabbits. *Journal of Pathology and Bacteriology* **93**, 499–529.

Smith, H. W. and Halls, S. (1967b). Studies on *Escherichia coli* enterotoxin. *Journal of Pathology and Bacteriology* **93**, 531–543.

Smith, H. W. and Halls, S. (1968). The transmissible nature of the genetic factor in *Escherichia coli* that controls enterotoxin production. *Journal of General Microbiology* **52**, 319–334.

Smith, H. W. and Huggins, M. B. (1976). Further observations on the association of the colicine V plasmid on *Escherichia coli* with pathogenicity and with survival in the alimentary tract. *Journal of General Microbiology* **92**, 335–350.

Smith, H. W. and Huggins, M. B. (1978). The influence of plasmid-determined and other characteristics of enteropathogenic *Escherichia coli* on their ability to proliferate in the alimentary tracts of piglets, calves and lambs. *Journal of Medical Microbiology* **11**, 471–492.

Smith, H. W. and Jones, J. E. T. (1963). Observation on the alimentary tract and its bacterial flora in healthy and diseased pigs. *Journal of Pathology and Bacteriology* **86**, 387–412.

Smith, H. W. and Linggood, M. A. (1971). Observations on the pathogenic properties of the K88, Hly and Ent plasmids of *Escherichia coli* with particular reference to porcine diarrhoea. *Journal of Medical Microbiology* **4**, 467–485.

Smith, H. W. and Linggood, M. A. (1972). Further observations on *Escherichia coli* enterotoxins with particular regard to those produced by atypical piglet strains and by calf and lamb strains: The transmissible nature of these enterotoxins and of a K antigen possessed by calf and lamb strains. *Journal of Medical Microbiology* **5**, 243–250.

Smith, H. W. and Parsell, Z. (1975). Transmissible substrate utilizing ability in enterobacteria. *Journal of General Microbiology* **87**, 129–140.

Sojka, W. J. (1965). "*Escherichia coli* in Domestic Animals and Poultry." Commonwealth Agricultural Bureaux, Farnham Royal.

Sojka, W. J. (1970). *E. coli*—Infektionen beim schweinen, kalbern, Lammern und Geflugel. *Wiener Tierärztliche Monatschrift* **57**, 361–370.

Sojka, W. J. (1971). Enteric diseases in newborn piglets, calves and lambs due to *Escherichia coli* infection. *Veterinary Bulletin (London)* **41**, 509–522.

Sojka, W. J. and Carnaghan, R. B. A. (1961). *Escherichia coli* infection in poultry. *Research in Veterinary Science* **2**, 340–352.

Sojka, W. J., Erskine, R. G. and Lloyd, M. K. (1957). Haemolytic *Escherichia coli* and oedema disease of pigs. *Veterinary Record* **69**, 293–301.

Stevens, A. J. (1963a). Symposium: Enteritis in pigs. 1. Coliform infections in the young pig and practical approach to the control of enteritis. *Veterinary Record* **75**, 1241–1245.

Stevens, A. J. (1963b). Enteritis in pigs—a working hypothesis. *British Veterinary Journal* **119**, 520–526.

Stirm, S., Ørskov, F., Ørskov, I. and Mansa, B. (1967a). Episome-carried surface antigen K88 of *Escherichia coli*. II. Isolation and chemical analysis. *Journal of Bacteriology* **93**, 731–739.

Stirm, S., Ørskov, F., Ørskov, I. and Birch-Andersen, A. (1967b). Episome-carried surface antigen K88 of *Escherichia coli*. III. Morphology. *Journal of Bacteriology* **93**, 740–748.

Tennant, B., Ward, D. E., Braun, R. R., Hunt, E. L. and Baldwin, B. H. (1978). Clinical management and control of neonatal enteric infections of calves. *Journal of the American Veterinary Medical Association* **173**, 654–661.

Terlecki, S. and Shaw, W. B. (1959). *Escherichia coli* infection in lambs. *Veterinary Record* **71**, 181–182.

Terlecki, S. and Sojka, W. J. (1965). The pathogenicity for lambs of *Escherichia coli* of certain serotypes. *British Veterinary Journal* **121**, 462–470.

Thomlinson, J. R. (1963). Observation on the pathogenicity of gastro-enteritis associated with *E. coli*. *Veterinary Record* **75**, 1246–1256.

Thomlinson, J. R. and Buxton, A. (1962). A comparison of experimental anaphylactic shock in guinea pigs with naturally-occurring oedema disease and haemorrhagic gastro-enteritis in pigs. *Research in Veterinary Science* **3**, 186–202.

Truscott, R. B. (1973). Studies on the chicken—lethal toxin of *Escherichia coli*. *Canadian Journal of Comparative Medicine* **37**, 375–381.

Truscott, R. B., Lopez-Alvarez, J. and Pettit, J. R. (1974). Studies on *Escherichia coli* infection in chickens. *Canadian Journal of Comparative Medicine* **38**, 160–167.

Truszczynski, M. and Ciosek, D. (1972). Antigenic properties of enterotoxins and water extracts obtained from *Escherichia coli* pathogenic for pigs. *Research in Veterinary Science* **13**, 205–211.

Whipp, S. C., Moon, H. W. and Argenzio, R. A. (1981). Comparison of enterotoxic activities of heat stable enterotoxins from class 1 and class 2 *Escherichia coli* of swine origin. *Infection and Immunity* **31**, 245–251.

Wilson, M. R. and Hohmann, A. W. (1974). Immunity to *Escherichia coli* in pigs: Adhesion of enteropathogenic *Escherichia coli* to isolated intestinal epithelial cells. *Infection and Immunity* **10**, 776–782.

Wray, C. and Thomlinson, J. R. (1969). *Escherichia coli* products in enteritis. *Veterinary Record* **84**, 645.

4

Adhesins and Colonization Factors of *Escherichia coli*

S. H. PARRY* AND DIANA M. ROOKE

Department of Microbiology, Medical School, University of Newcastle upon Tyne, Newcastle upon Tyne, UK

Introduction

The adhesive properties of *Escherichia coli* were first recognized by Guyot (1908) who observed that some strains possessed the ability to agglutinate red blood cells (RBC) from a number of animal species. Although similar observations were made by subsequent investigators (Rosenthal, 1943; Kauffmann, 1948), it was not until the classic studies of Duguid *et al.* (1955) that haemagglutination (HA) was recognized as an adhesive property of *E. coli* which, in a number of instances, correlated with the presence of fimbriae observed under the electron microscope. The adhesive affinity of *E. coli* for many cell types other than RBC has since been recognized, indicating the potential role of adhesins as colonization factors which allow organisms to establish themselves at body sites in eukaryotic hosts. In recent years the importance of microbial adhesion factors in relation to microbial pathogenicity has emerged and it is clear that certain adhesins of *E. coli* play a significant role in infections at mucosal surfaces and thus constitute true virulence factors.

This article will review the structure and function of adhesins and colonization factors of *E. coli* with particular reference to those that play a role in the pathogenesis of intestinal and urinary tract infection. The genetic control of adhesins is reviewed elsewhere in this volume (Chapter 8) and will not be covered in detail.

*Present address: Department of Immunology, Unilever Research, Colworth Laboratory, Colworth House, Sharnbrook, Bedford MK44 1LQ, U.K.

79

Terminology

The term adhesin, as used here, is defined as a microbial surface component that mediates specific attachment to a eukaryotic cell membrane. The term does not presuppose any particular structure and encompasses the well-defined fimbrial adhesins as well as others that may be poorly characterized and have as yet undefined structures. Other terms are often used to denote adhesins with specific functional attributes such as haemagglutination (haemagglutinin). While the term adhesin does not necessarily suggest a specific functional role in infections in a eurkaryotic host, an adhesin that promotes the colonization of host tissues by an organism may be considered to be a colonization factor.

Thus the presence of the well-defined fimbrial adhesins of enterotoxigenic *E. coli* (e.g. K88, K99 and CFA/I) is strongly associated with the ability to adhere to and proliferate in the intestine (Gaastra and de Graaf, 1982) and these antigens are clearly colonization factors. While type 1 fimbriae are well-characterized adhesins of *E. coli* and mediate attachment to many types of cells (Duguid and Old, 1980), it is not clear whether they are colonization factors since their role in infections is generally poorly understood.

The term 'fimbriae', first used by Duguid (Duguid *et al.,* 1955) to describe non-flagellar proteinaceous filamentous appendages of *E. coli,* is used here in preference to 'pili', following the recommendation by previous authors (Ottow, 1975; Jones, 1977; Ørskov *et al.,* 1977; Pearce and Buchanan, 1980). 'Fimbriae' has temporal priority over the term 'pili' and despite the suggestion by Brinton (1959) that 'pili' (Latin for hairs or fur) is linguistically more correct, the term 'fimbriae' (plural of Latin for thread, fibre or fringe) would seem more appropriate. It was suggested that the name pili be reserved for sex pili, the filamentous appendages involved in the conjugative transfer of DNA (Ottow, 1975; Jones, 1977). Certain adhesins, such as the K88 and K99 antigens, which were originally observed as an amorphous fibrillar mass on the bacterial surface without a definite structure, have been referred to as 'fibrillae' (Jones, 1977). More recently the K88, K99 antigens have been shown to consist of fine filaments (2–3 nm) with a fimbria-like structure and are now considered to be fimbriae (Gaastra and de Graaf, 1982).

Classification and Identification

The functional properties of adhesins provide a useful means of identification, and haemagglutination and adhesion to host cells have proved invaluable for screening isolates for adhesins. In particular, the use of haemagglutination in the presence and absence of D-mannose or its analogues provides a basic distinction between mannose-sensitive (MS) adhesins, which are inhibited by mannose, and mannose-resistant (MR) adhesins, which are not. However, the use of HA alone for identification of adhesins is often insufficient since it is becoming increasingly apparent that many strains of *E. coli* possess multiple adhesins. In addition, some adhesins do not haemagglutinate commonly used RBC. Consequently, identification methods that rely on functional tests require careful interpretation. As the specificity of adhesins lies in their ability to recognize distinct membrane receptors (Jones, 1977) the use of receptor analogues would appear to be a powerful tool for classification and identification. However, characterization of the receptor structure of MR adhesins is, so far, restricted to the P fimbriae of uropathogenic *E. coli*.

MR adhesins have been described in various types of *E. coli*: in facultatively enteropathogenic *E. coli* (FEEC) which colonize the intestine as commensals, giving rise to sporadic extra-intestinal diseases [urinary tract infection (UTI), septicaemia, meningitis]; in enterotoxigenic strains (ETEC) which cause diarrhoea, are non-invasive but are substantial colonizers of the small intestine and produce enterotoxins; in enteropathogenic *E. coli* (EPEC) which are non-toxigenic and non-invasive and in enteroinvasive *E. coli* (EIEC), which cause a dysentery-like syndrome (enterocolitis). *E. coli* adhesins may be classified and identified according to morphology, function or antigenic properties. Morphology was first used in the classification of fimbrial adhesins of *E. coli* and other members of the Enterobacteriacae by Duguid and later by Brinton (see review by Pearce and Buchanan, 1980); however, this scheme does not include the more recently described colonization fimbriae of ETEC strains or the mannose-resistant adhesins of uropathogenic and enteropathogenic *E. coli*. In addition, morphological criteria are of little use for the characterization of non-fimbrial adhesins.

Table 1. *Antigens of fimbrial adhesins of E. coli*

Fimbrial adhesin	Antigen number[a]	Disease association	Species
Type 1	F1	?	—
CFA/I	F2	Diarrhoea	Man
CFA/II	F3	Diarrhoea	Man
K88	F4	Diarrhoea	Pig
K99	F5	Diarrhoea	Sheep, calf, pig
987P	F6	Diarrhoea	Pig
—	F7	UTI	Man
—	F8	UTI	Man
—	F9	UTI	Man
—	F10	UTI	Man
—	F11	UTI	Man
—	F12	UTI	Man

[a]F. Ørskov and Ørskov (1984).

At present the antigenic properties of adhesins are most widely used for identification and classification. Serological methods are particularly useful for the identification of the fimbrial colonization factors of ETEC strains, CFA/I, CFA/II, K88, K99, 987P and F41, which are antigenically distinct from each other and from type 1 fimbriae (Gaastra and de Graaf, 1982). However the MR adhesins of uropathogenic *E. coli* generally appear to be antigenically heterogeneous (Parry *et al.*, 1982; Ørskov *et al.*, 1982), making identification by serological methods difficult. Even P fimbriae of different strains, despite their common receptor specificity, appear to differ antigenically (Svanborg Edén *et al.*, 1983b). F. Ørskov and Ørskov (1984) have recently proposed a typing scheme for fimbrial antigens which at present includes type 1 fimbriae, the well-characterized colonization fimbriae of ETEC strains and a number of fimbrial types from uropathogenic *E. coli* (Table 1).

Purification, Characterization and Antigenic Properties of Adhesins

The characteristics and antigenic properties of type 1 fimbriae have been discussed in a number of reviews (Brinton, 1965; Ottow, 1975; Duguid and Old 1980; Pearce and Buchanan, 1980; I. Ørskov and Ørskov, 1983) and will not be considered in detail.

The major properties of the best characterized MR fimbrial adhesins of ETEC strains and uropathogenic *E. coli* are summarized in Table 2. They will be discussed in detail below together with the less well-characterized MR adhesins of intestinal and extra-intestinal isolates.

Table 2. *Major characteristics of fimbrial adhesins of E. coli*

Fimbrial adhesin	Diameter (nm)	Molecular weight of subunit	pI	Antigenic variation	Location of genetic determinants
Type 1	7	17,000	5.0	Yes	Chromosome
K88	2.1	27,540(ab)[a] 25,000(ac) 26,000(ad)	4.2	Yes	Plasmid
987P	7	18,900	3.7	Not demonstrated	Chromosome?
K99	5	18,500–19,500	9.5	Not demonstrated	Plasmid
F41	3.2	29,500	4.6	Not demonstrated	?
CFA/I	3.2	15,058[b]	4.8	Not demonstrated	Plasmid
CFA/II	3.2	13,000	?	Possibly	Plasmid
P fimbriae	7	17,000–22,000	?	Yes	Chromosome

[a]Apparent molecular weight estimated from SDS–PAGE gels.
[b]Denotes actual molecular weight from sequencing studies.

Animal ETEC Strains

K88 adhesin. Ørskov and co-workers (1961) first described a new L antigen in strains of *E. coli* isolated from pigs with enteritis. The antigen, which was not expressed in cultures grown at 18°C, was numbered K88(L). Further examination of strains isolated from piglets in different countries indicated that the K88 antigen was composed of different antigenic components in different combinations (Ørskov *et al.*, 1964). Cross-absorption studies demonstrated that it existed in at least two forms; each with a common and a distinct determinant, symbolized as K88ab and K88ac. The antigen was found in strains of several different O serotypes and the K88ab antigen could be transferred to another strain and was encoded by a conjugative plasmid (Ørskov *et al.*, 1961; Stirm *et al.*, 1966).

A further antigenic variant, K88ad, has been described although slight serological variations were seen in different strains (Guinée and Jansen, 1979). Similar variations have been observed among the K88ac antigens of different strains (Ørskov *et al.*, 1964). It was suggested that these variants might, in part, result from immunological pressure due to the large-scale vaccination with K88 antigen-containing vaccines. The appearance of the new K88ad variant may also be an attempt by the pathogen to escape such pressure. It has also been suggested (Gaastra and de Graaf, 1982) that the serological variants of K88 antigen may have arisen by selection due to the occurrence of pigs resistant to a particular type of K88 adhesin. This relates to the findings of Bijlsma *et al.* (1981, 1982), who showed that these variants had different adhesive specificities for different pigs, suggesting differences in their intestinal epithelial cell receptors. It is interesting that in recent years strains expressing the K88ab antigen seem largely to have disappeared and K88ac and K88ad-bearing *E. coli* strains are being isolated from infected pigs (Wilson and Hohmann, 1974; Guinée and Jansen, 1979).

K88 antigen was first isolated by heat treatment and homogenization of the bacterial suspension followed by isoelectric precipitation at pH 5.0 (Stirm *et al.*, 1966, 1967a). The resulting material was insoluble at pH 3.5–5.5 and had a high sedimentation rate. The antigen was shown to be a protein containing all the common amino acids except cysteine. On examination by electron microscopy, K88$^+$ strains were surrounded by a mass of fine fibrillar structures and the purified antigen had a similar structure but its morphology was unlike that described for type 1 fimbriae, which were much thicker and straighter (Stirm *et al.*, 1967b). It was suggested that the name fibrillae should be reserved for fine filamentous structures such as the K88 antigen where typical fimbrial morphology cannot be seen (Jones, 1977); however, it seems that the K88 antigen consists of extremely fine fimbria-like structures with a diameter of 2.1 nm (Wadström *et al.*, 1979) and is now generally described as fimbrial.

More recently, the three K88 antigenic variants have been purified by shearing the bacteria for 30 min at 4°C and the resulting crude fimbrial preparations precipitated with ammonium sulphate and dialysed against a buffer containing 2 *M* urea. Final purification was performed by gel filtration on Sephadex CL-4B (Mooi and de Graaf, 1979). Sodium dodecyl sulphate–polyacrylamide gel electrophoresis (SDS–PAGE) demonstrated that this method yielded essentially homogeneous material with apparent molecular weights of 26,000 for K88ad, 23,500 for K88ab and 25,000 for K88ac. It was concluded that the K88 antigen was a pure protein and that the serological variants differed in amino acid composition. Cysteine residues were found only in the K88ad(e) subvariant. Pooled material containing all three antigens had a pI of 4.2, which agrees closely with the pI of 4.1 determined for the K88ab antigen by Parry and Porter (1978).

The complete primary structure of the K88ab antigen has recently been determined by two groups by amino acid sequencing (Klemm, 1981) and nucleotide sequencing (Gaastra *et al.*, 1981). Both methods showed good agreement and it was shown that the K88ab gene encodes a precursor protein of 285 amino acid residues containing a signal peptide of 21 amino acid residues at the N-terminal end. These data confirm those of Mooi *et al.* (1979), who found that the K88ab subunit accumulated in a precursor form when proteolytic processing was inhibited. The signal peptide appears to be removed during translocation of the subunit from the ribosome through the cytoplasmic membrane. The nucleotide sequencing data also indicate that the gene encoding the subunit may be preceded by its own promoter (Gaastra *et al.*, 1981). The K88ab antigen has a true molecular weight of 27,540.

The N-terminal and C-terminal amino acid sequences of the different K88 variants and the N-terminal sequences of cyanogen bromide fragments were determined by Gaastra *et al.* (1979). No difference was observed between K88 variants in the first 23 N-terminal or the last 24 C-terminal amino acids. The N-terminal sequence of some of the cyanogen bromide fragments was different for the different variants. Comparison of amino acid sequences of K88ab and K88ad have shown differences between residues 150 and 220 (Gaastra *et al.*, 1981; Gaastra and de Graaf, 1982). These were dispersed along the polypeptide chain and primarily involved charged amino acid residues which are most likely to be present at the surface of the protein and, therefore, influence the antigenicity of the subunit. This was supported with the use of computer prediction techniques which utilized algorithms incorporating factors such as hydrophilicity and secondary structural potential (Klemm and Mikkelsen, 1982). This study predicted three major antigenic determinants between residues 150 and 220 which showed good correlation with the known sequence differences between the K88ab and K88ad antigenic variants. It has yet to be demonstrated whether these potential determinants relate to observed immunological behaviour. There appears to be no homology between the primary structures of K88 and CFA/I or the N-terminal sequences of type 1, K99 and F41 fimbriae (de Graaf *et al.*, 1980b; Klemm, 1982; Gaastra and de Graaf, 1982).

987P adhesin. The observation that some ETEC strains that lacked the K88 antigen were able to colonize the porcine small intestine led to the description of a distinct adhesin which was designated 987P (Nagy *et al.*, 1976, 1977; Isaacson *et al.*, 1977). The antigen was initially purified by the method of Brinton (1965) and was later observed to have a fimbrial structure with a diameter of 7 nm. It was composed of subunits with an apparent molecular weight of 18,900. Isaacson and Richter (1981) purified 987P by homogenization and five cycles of precipitation with $MgCl_2$. The material, which was homogeneous by SDS–PAGE and electron microscopy, was confirmed to be fimbrial in nature and consisted of 7-nm strands, morphologically indistinguishable from type 1 fimbriae, with an apparent axial hole. The molecular weight of subunits on SDS–PAGE varied depending on the conditions used but was estimated to be about 20,000. Chemically, 987P is composed primarily of protein although there was evidence of an unidentified amino sugar. The amino terminal amino acid was alanine and the pI was 3.7. The 987P antigen has not been shown to be encoded by plasmid DNA and it is likely that it is a chromosomal gene product (Gaastra and de Graaf, 1982).

K99 adhesin. Smith and Linggood (1972) first described a common K antigen produced by calf and lamb ETEC strains which was encoded by a conjugative plasmid designated Kco. The antigen has since been named K99 (Ørskov *et al.*, 1975). K99 was isolated by Isaacson (1977) by salt extraction, ammonium sulphate precipitation and column chromatography on DEAE–Sephadex. The antigen was observed under the electron microscope as rod-shaped fimbriae with a diameter of 8.4 nm which had a strong tendency to aggregate. This value is almost twice that reported by de Graaf *et al.* (1980b) in a later study and may be anomolous because of aggregation. The pI was reported to be greater than 10. Purified K99 no longer haemagglutinated guinea-pig RBC and HA activity was recovered separately from DEAE–Sephadex after application of a salt gradient. This haemagglutinating fraction did not contain K99 antigen. This is in contrast to the findings of Morris *et al.* (1977), who purified K99 from *E. coli* reference strain B41 by a method similar to that described for the K88 antigen (Stirm *et al.*, 1967a), which consisted of five cycles of isoelectric precipitation followed by gel filtration on Sepharose 4B. Fractionation of this material on DEAE–Sephadex resulted in joint elution of HA activity and K99 antigen.

On SDS–PAGE gels K99 was initially reported to consist of two subunits, a minor component with an apparent molecular weight of 29,500 and a major component with a molecular weight of 22,500 (Isaacson, 1977). In a later study Isaacson *et al.* (1981) purified K99 in a similar fashion apart from the initial extraction, which was carried out in urea rather than 1 M NaCl, and it was concluded that only a single subunit species existed with a likely molecular weight of 19,500. Examination of the effect of mercaptoethanol (MetOH) indicated that denaturation in the presence of MetOH could give rise to two mo-

lecular species associated with the breakage and reformation of intrachain disulphide bonds. It was suggested that there was a single intrachain bond which
rapidly reforms after cleavage with and removal of MetOH; the subunits may
then assume one of two conformations depending on whether the disulphide
bond reforms.

Studies by Morris *et al.* (1978b) on purified surface antigen from *E. coli* B41
demonstrated material with a pI of 4.2 which gave MR-HA of sheep RBC. This
MR-HA activity was lost after absorption with antisera containing K99 activity.
On immunoelectrophoresis an anionic and a cationic component were observed
which could be separated by ion-exchange chromatography on DEAE–cellulose
(Morris *et al.*, 1978a). The anionic component consistently exhibited a strong
HA reaction, stable for at least 6 wk at 4°C, while the HA activity of the cationic
component was lost after 3 wk under identical conditions. It was concluded from
this and double diffusion tests that the cell-free K99 from B41 was composed of
two antigenically distinct components, a slowly diffusing anionic component
with stable HA and a faster diffusing cationic component with a labile HA
reaction. The anionic and cationic components were examined further (Morris *et
al.*, 1980) and evidence was provided for the existence of two adhesins on strain
B41. A standard antigenic preparation of K99 from B41 was used to demonstrate
precipitins in antisera to K99 *E. coli* strains. Precipitins to the cationic antigen
were restricted to antisera against K99$^+$ strains from the O9 and O101
serogroups. Neither the anionic nor the cationic antigen was expressed at 18°C.
K99 extracts agglutinated sheep RBC but the activity was lost quickly. It was,
however, only inhibited by antisera to the O9 and O101 serogroups and this
inhibitory antibody could only be removed by absorption with K99$^+$ *E. coli*
from these two serogroups grown at 37 but not those grown at 18°C. Isolation of
the anionic component (ion exchange) confirmed its specificity for sheep cells
and the cationic component was detected by its strong reaction with horse RBC.
It was concluded that the cationic antigen was K99 and the anionic component
represented a distinct adhesin produced only by certain K99$^+$ bacteria. This was
confirmed by isolation of a K99$^-$ mutant of the B41 strain which produced only
the anionic antigen and which was designated F41 (Morris *et al.*, 1982b).

De Graaf *et al.* (1980b) purified K99 from a number of strains by heat
treatment, ammonium sulphate precipitation and gel filtration on Sepharose 4B
followed by deoxycholate treatment. The purified antigen was shown to be
homogeneous on SDS–PAGE gels and it was confirmed that K99 corresponded
to the cationic antigenic component with a pI of 9.5. The antigen had a helical
fimbrial structure with a diameter of approximately 5 nm and it was composed of
subunits with an apparent molecular weight of 18,500. Omission of MetOH did
not affect the molecular weight determination and examination of the same strain
as used by Isaacson (1977) gave a similar value of 18,500. It was suggested that
the material with a molecular weight of 29,500 reported by Isaacson (1977) was

a minor component that co-purified with K99. Amino acid analysis showed the presence of cysteine and a preponderance of amino acids with apolar side chains. The primary structure of K99 has yet to be fully elucidated; however, its N-terminal sequence bears no resemblance to the N-terminal primary structure of the K88 and CFA/I adhesins (Gaastra and de Graaf, 1982).

Chanter (1982) also confirmed that K99 was the cationic adhesive component of strain B41 which had a subunit molecular weight of 19,000. An anionic component consisting of a haemagglutinin with a molecular weight of 34,000 was also described but its relationship to F41 was not examined or discussed. The situation appears to be further complicated by a recent report (Chanter, 1983) providing evidence for two non-K99 MR haemagglutinins on a K99$^-$ mutant of B41. One was an antigen which adhered to calf intestinal brush borders and was composed of protein subunits with a molecular weight of 34,000. This material did not have a regular fimbrial appearance but contained some fine fibrillar structures and from its HA pattern it was suggested to be similar to the F41 antigen described by Morris *et al.* (1980). The second haemagglutinin had a definite fimbrial structure and was composed of subunits with molecular weights of 49,500 and 48,000. This was heat-resistant (100°C for 30 min), antigenically distinct from the anionic adhesin and did not adhere to calf intestinal brush borders. All K99$^+$ strains tested and an *E. coli* K-12 strain produced this MR haemagglutinin when grown at both 37 and 18°C, whereas a K88 strain and a 987P strain did not. The identity of this haemagglutinin remains to be determined although the possibility that it represents type 1 fimbriae or a non-fimbrial haemagglutinin associated with type 1 fimbriae has not been completely discounted.

F41 adhesin. F41, first demonstrated on strain B41, which also expresses K99, is a fimbrial antigen with a diameter of approximately 3 nm which is antigenically distinct from K99 (Morris *et al.*, 1982b). The antigen has been isolated from the K99$^-$ mutant strain B41M by mechanical detachment followed by ammonium sulphate precipitation, gel filtration on Sepharose 4B and treatment with deoxycholate (de Graaf and Roorda, 1982). Its anodic migration and filamentous structure with a diameter of 3.2 nm were confirmed. SDS–PAGE gels gave a single band corresponding to subunits with an apparent molecular weight of 29,500 and its isoelectric point was 4.6. Antiserum against the F41 preparation was absorbed with the live B41M strain grown at 18°C. This reacted in immunoelectrophoresis with purified F41 and no line against K99 antigen was observed nor did a K99 specific antiserum react with F41 antigen (de Graaf and Roorda, 1982).

A potential fimbrial adhesin has been proposed to occur on porcine strains of *E. coli* lacking the K88, K99 and 987P antigens (Moon *et al.*, 1980). It has recently been demonstrated that these isolates express F41 fimbriae (Morris *et al.*, 1983). It is not yet known whether the F41 antigen is encoded by plasmid or chromosomal DNA.

Human ETEC

CFA adhesins. Evans *et al.* (1975) were the first to describe a surface-associated, plasmid-encoded colonization factor designated CFA/I on a strain of *E. coli* isolated from a case of cholera-like diarrhoea. Strains possessing this factor were rapidly able to colonize the intestinal tract of infant rabbits while a laboratory-passed derivative lacking the antigen failed to attain large populations in the intestine. Specific anti-colonization factor antiserum, produced by absorbing hyperimmune serum against the adhesive organism with the colonization factor-negative mutant, protected infant rabbits from challenge, although the serum did not have bactericidal activity. It was demonstrated by electron microscopy that the strain possessing the colonization factor expressed fimbriae which were aggregated by the specific antiserum. The strain lacking CFA was non-fimbriate. No cross-reaction could be observed with a porcine strain of *E. coli* bearing the K88 antigen.

CFA/I purified by ammonium sulphate precipitation and ion-exchange chromatography and examined under the electron microscope was observed as 7-nm strands with a molecular weight of 1.6×10^6 determined by sedimentation equilibrium centrifugation (D. G. Evans *et al.*, 1979). SDS–PAGE of the CFA/I subunit indicated an apparent molecular weight of 23,800. Klemm (1982) prepared CFA/I by heat release at 60°C followed by shearing. The sheared fimbriae were purified by gel filtration on Sepharose 2B and the resulting material gave a single line on SDS–PAGE. The fimbrial subunit had an apparent molecular weight of 14,500. Amino acid analysis revealed a high content of hydrophobic residues (36%) and no cysteine (Klemm, 1979, 1982).

The full primary structure of the CFA/I antigen has recently been elucidated by Klemm (1982). The subunit consists of 147 amino acid residues giving a true molecular weight of 15,058, considerably less than that estimated by D. G. Evans *et al.* (1979). It is likely that this value may be due to dimer formation since a faint 24,000 band was seen together with a major 12,000 band on SDS–PAGE gels in a study by Wevers *et al.* (1980). CFA/I is not a glycoprotein although it has four potential *N*-glycosylation sites (Klemm, 1982). The fimbriae are composed of approximately 100 subunits and their structure is similar to K88 and K99 with a diameter of about 3 nm (Brinton, 1978; Gaastra and de Graaf, 1982). The larger diameter originally described by D. G. Evans *et al.* (1979) was suggested to be due to aggregation (Gaastra and de Graaf, 1982). CFA/I has a pI of 4.8 (Freer *et al.*, 1978). Examination of the amino acid sequence revealed internal homology neither within the CFA/I molecule nor with the amino acid sequence of the K88 subunit (Klemm, 1981) or any other completed or partially completed fimbrial primary structures (Hermodson *et al.*, 1978; de Graaf *et al.*, 1980b; de Graaf and Roorda, 1982). CFA/I has been demonstrated on serogroups O78, O25, O63, O15 and O128 (Evans and Evans, 1978).

Evans and Evans (1978) subsequently described a new colonization factor antigen, designated CFA/II, produced by enterotoxigenic *E. coli* serogroups O6 and O8, which could be distinguished from CFA/I by its inability to agglutinate human erythrocytes. Strains bearing CFA/II colonized the intestine of infant rabbits and this was inhibited by CFA/II-specific antiserum produced by absorption of antiserum against whole organisms with a CFA/II⁻ mutant. Immunodiffusion tests with antisera specific for CFA/I and CFA/II against a mixture of CFA/I and CFA/II demonstrated a line of non-identity indicating that these two antigens represent antigenically distinct fimbriae. In another study antisera against CFA/I antigen did not agglutinate strains bearing CFA/II, K88, K99 or type 1 fimbriae (Wevers *et al.*, 1980).

CFA/II is less well characterized than CFA/I but appears to be of similar fimbrial structure with a diameter of 3.2 nm (Evans and Evans, 1978). These authors purified the antigen by the procedure of Salit and Gotschlich (1977a), originally described for type 1 fimbriae, which consisted of shearing, isoelectric precipitation and three successive precipitations in 10% ammonium sulphate. After final purification by isopycnic centrifugation in caesium chloride the subunits were shown to have an apparent molecular weight of 13,000. As yet no published data are available on the primary structure of CFA/II although its lack of antigenic homology with CFA/I (Evans and Evans, 1978) would suggest considerable differences.

It has recently been reported that putative CFA/II antisera raised against ETEC of the O6:K15:H16 or H⁻ serotypes of different biotypes possess antibodies to three immunologically distinct CS (coli surface-associated) antigens (Smyth, 1982). Both CS1 and CS2 were hamagglutinins (bovine RBC) and were shown to have subunits with molecular weights of 16,300, and 15,300, respectively; the CS3 antigen was non-haemagglutinating and had a 14,800-dalton subunit. Two strains expressed CS1 and CS3, four strains CS2 and CS3 and a further two strains CS2 only. CS3 appeared to be independent of biotype but CS1 was associated with strains of biotype A and CS2 with strains of biotypes B, C and F. As no comparison was made with the original type strain PB-176 (Evans and Evans, 1978) it was not possible to determine which of the CS antigens may correspond to CFA/II. From biotype information the most likely candidates are CS1 or CS3. Further studies are required to clarify the antigenic structure of CFA/II and associated adhesins and it is clear that screening studies carried out with putative CFA/II sera must be interpreted with some caution.

Other distinct fimbrial adhesins have been demonstrated on ETEC strains bearing neither CFA/I nor CFA/II. In a study of the adhesins of ETEC strains Deneke *et al.* (1979) examined fimbriae from a number of different isolates.

Purification of fimbriae from strains bearing adhesins giving MR-HA of guinea-pig RBC was carried out by homogenization followed by absorption with guinea-pig cells at 0°C in the presence of mannose to prevent attachment of type 1 fimbriae. The MR fimbriae, subsequently eluted from RBC at 37, 42 and 50°C, were short and needle-like with a diameter of 5–10 nm. One strain (334) was found to possess two polypeptide chains by SDS–PAGE with subunits of 12,500 and 13,100 and fimbriae with identical molecular weights were also found on other *E. coli* strains. A further study by these authors (Deneke *et al.*, 1981) has examined the antigenic properties of these fimbriae. Three distinct antigenic types were identified and antisera against these three fimbrial types reacted with 60 of 106 (56%) ETEC strains isolated from humans with diarrhoea but not with non-toxigenic, normal human faecal isolates nor with ETEC strains isolated from animals. It has been pointed out by Gaastra and de Graaf (1982) that antiserum against the fimbriae isolated by Deneke *et al.* (1981) reacted with both H10407, the CFA/I type strain, and H10407-P, a CFA/I$^-$ variant that still expresses type 1 fimbriae. It was suggested that the antisera prepared by Deneke *et al.* (1981) may, therefore, contain antibodies to type 1 fimbriae which may account for their observed cross-reactivity.

Thomas *et al.* (1982) described an MR adhesin on a strain of *E. coli* isolated from a patient with diarrhoea. The strain did not show slide agglutination or immunodiffusion precipitin lines with antiserum specific for CFA/I or CFA/II. Examination by electron microscopy revealed fimbriae which were probably not type 1 fimbriae, since the strain failed to cause MS-HA of guinea-pig RBC. Antibodies specific for the fimbriae, prepared by absorption of an antiserum against E8775 with an MR-HA-negative variant, did not react with CFA/I or CFA/II strains. In a survey of 742 ETEC strains, 139 strains caused MR-HA of human and bovine RBC and of these 91 strains were shown to produce CFA/I while the remaining 48 reacted with the E8775 fimbrial antiserum.

Human EPEC, Non-ETEC and FEEC

Williams *et al.* (1977) reported that a human EPEC strain, serotype **O**26:**K**60:**H**11, isolated from a baby with diarrhoea, adhered to the mucosa of the human foetal small intestine in the presence of mannose. Adherence could be transferred to *E. coli* K-12 and correlated with the presence of a conjugative 56-megadalton plasmid (Williams *et al.*, 1978). The adhesin appears to be distinct from the CFA antigens but its nature is unknown.

Fimbriae with a diameter of 5 nm from an MR-HA strain of *E. coli* (**O**18ac:**H**$^-$) isolated from a patient with diarrhoea were purified and charac-terized by Wevers *et al.* (1980). The strain did not produce ST or LT enterotox-ins. SDS–PAGE indicated a subunit molecular weight of 21,000 and on iso-electric focusing two bands were observed with pI values of 5.1 and 5.6. The fimbriae were antigenically distinct from CFA/I and CFA/II adhesins and differed in amino acid composition from CFA/I, K88 and K99 and type 1 fimbriae. It is interesting that antiserum against the purified fimbriae also agglutinated a number of *E. coli* strains isolated from UTI. An apparently novel MR fimbrial adhesin has also been demonstrated in isolates from children after an outbreak of acute diarrhoea in a paediatric intensive care unit (Candy *et al.*, 1982). Strains bearing the fimbriae were antigenically distinct from CFA/I and CFA/II but were not characterized further.

Extra-Intestinal Isolates

Purification and characterization. Svanborg Edén and Hansson (1978) first reported that fimbriae were possible mediators of attachment to human urinary tract epithelial cells based on the strong correlation between the presence of fimbriae and the ability of urinary *E. coli* isolates to adhere to uroepithelial cells (UEC). In their study 11 of 12 strains expressed MS adhesins but one strain had an MR adhesin and lacked MS-HA activity.

Korhonen *et al.* (1980b) described a method for purifying fimbriae by shearing followed by ammonium sulphate precipitation (50% saturation), deoxycholate treatment to solubilize outer membrane fragments and gel filtration in 6 *M* urea to separate dissociated flagella. Binding of purified radio-labelled fimbriae from a pyelonephritogenic *E. coli* strain, having both MS and MR adhesins, to human urinary tract epithelial cells in the presence of D-mannose demonstrated the possible role of MR adhesins in attachment (Korhonen *et al.*, 1980a). A further study demonstrated that this strain and three other isolates from a patient with pyelonephritis recognized the P blood group antigen receptor (Korhonen *et al.*, 1982). When grown on CFA agar these strains were abundantly fimbriated but lacked type 1 fimbriae. The strains were considered to carry P fimbriae which were seen under the electron microscope as 1- to 2-μm filaments with a diameter of 5–7 nm which were morphologically similar to type 1 fimbriae. SDS–PAGE analysis of the various P fimbriae preparations were compared with type 1 fimbriae. In contrast to type 1 fimbrial preparations, all four P fimbriae preparations each showed two or three bands with varied molecular weights ranging from 17,000 to 22,000. Three of the strains had a band with a molecular weight around 17,000, which was similar to the size of the subunit of type 1 fimbriae from *E. coli*.

P fimbriae preparations from two strains were isolated by the same method (Korhonen *et al.*, 1980b) but they were absorbed on columns containing D-mannose to remove type 1 fimbriae (Svanborg Edén *et al.*, 1983b). The fimbriae, which bound to globoseries glycolipid receptors, were characterized more fully with respect to molecular characteristics, amino acid composition and sequence analysis. Protein subunit molecular weights were estimated by SDS–PAGE and found to be 19,500, compared to 17,500 for a preparation of type 1 fimbriae. One of the two strains bearing P fimbriae had, in addition to the 19,500 subunit, a 17,500-dalton protein which was specifically bound by the mannose affinity column.

The two fimbrial proteins consisting of 19,500 subunits and type 1 fimbriae were subjected to amino acid and sequence analysis. The amino acid compositions were similar but not identical and they each contained a high proportion of non-polar residues and two cysteine residues per subunit which were involved in the formation of an intrachain disulphide bond. The N-terminal sequences of the two P fimbriae subunits were found to be highly homologous with each other and with the published N-terminal sequence of type 1 fimbriae (Hermodson et al., 1978). Alignment of sequences showed that the sequence of the subunit of type 1 fimbriae and one of the P fimbriae were two and three residues shorter than the subunit of P fimbriae derived from the second strain. One of the P fimbrial subunits was suggested to be a product of two or more structural genes as the amino terminus and residues 4 and 12 were found to be composed of equimolar amounts of two amino acids which could be accounted for by single nucleotide substitutions (Svanborg Edén et al., 1983b).

Studies on the F7 fimbrial antigen first described by Ørskov et al. (1980b) have shown it to have a subunit molecular weight of 22,000 (Klemm et al., 1982). Van Die et al. (1983) recently cloned the chromosomal genes coding for the MR adhesin of the F7 strain into a fim⁻ strain of E. coli K-12. SDS–PAGE gels of a preparation of fimbriae from the F7 wild type indicated the presence of three bands; however, the E. coli strain carrying the cloned gene produced a single fimbrial subunit with a molecular weight of 17,000, the same as one of the subunits of the parent strain but in conflict with the molecular weight of 22,000 reported for the F7 MR fimbrial subunit by Klemm et al. (1982). Unlike the N-terminal sequence determined for P fimbriae from two isolates (Svanborg Edén et al., 1983b), which was almost identical with that of type 1 fimbriae, greater differences were seen between type 1 and F7 fimbriae (Klemm et al., 1982; F. Ørskov and Ørskov, 1984) although partial sequence homology including identical first three residues was observed. Similar sequence studies of F9 revealed a partial sequence homology with type 1 fimbriae (F. Ørskov and Ørskov, 1984). Furthermore, preliminary studies on the primary structures of F11 and F12 fimbriae indicated that these fimbriae are strongly related to F7 (F. Ørskov and Ørskov, 1984). Klemm et al. (1982) suggested that these studies indicate a possible evolutionary kinship between type 1 fimbriae and MR-HA fimbriae of extra-intestinal E. coli. These fimbriae, unlike K88, K99, CFA/I and CFA/II associated with ETEC, were morphologically indistinguishable from type 1 fimbriae and like type 1 fimbriae they appeared to be encoded by chromosomal, not plasmid, DNA (Hull et al., 1981; Clegg, 1982b; van Die et al., 1983).

Antigenic studies. In enzyme-linked immunosorbent assays (ELISA) antisera raised in rabbits to P fimbriae from each of four strains reacted to different extents with both homologous and heterologous P fimbriae preparations but not with type 1 fimbriae of *E. coli.* It was concluded that the considerable but variable cross-reactivity observed between different P fimbriae indicated a high degree of structural homology (Korhonen *et al.,* 1982). In contrast to these findings, and despite the close N-terminal homology demonstrated between P fimbriae isolated from two pyelonephritogenic isolates, little cross-reactivity could be demonstrated by direct ELISA testing of each fimbrial protein against the heterologous and homologous hyperimmune sera (Svanborg Edén *et al.,* 1983b). These anomalous results may arise from contamination of fimbrial preparations with type 1 fimbriae in the study of Korhonen *et al.* (1982) since in the latter study mannose-binding fimbriae were removed by affinity chromatography. Although Korhonen *et al.* (1982) grew strains on CFA agar to inhibit type 1 fimbriae production, fimbrial preparations showed multiple bands and three preparations had protein subunits with molecular weights around 17,000–17,500, the size of the type 1 fimbrial subunit. Mannose-binding and globotetraose-binding fimbriae showed less than 10% cross-reactivity in competitive ELISA tests (Svanborg Edén *et al.,* 1983b). In functional adhesion tests Fab fragments of antifimbrial antibody to a P fimbriae preparation and type 1 fimbriae significantly reduced the attachment of the homologous strain to buccal epithelial cells (BEC) but had little effect on heterologous strains (Svanborg Edén *et al.,* 1983b).

Ørskov and co-workers carried out a number of serological studies on the fimbrial antigens of urinary tract isolates and, to date, six antigenic types (F7 to F12) have been described. The first, F7, was present on a strain isolated from a patient with pyelonephritis (Ørskov *et al.,* 1980b). This MR adhesin was shown to be fimbrial in nature, it bore no antigenic relationship to K88, K99, CFA/I and 987P and it mediated adhesion to uroepithelial cells. Further studies on the F7 strain (Klemm *et al.,* 1982) by crossed-immunoelectrophoresis indicated that the F7 antigen was composed of more than one fimbrial species and in addition to the MR adhesin two further fimbrial antigens termed 1A and 1C were detected. Type 1A caused MS-HA similar to type 1 fimbriae but the functional properties of type 1C were unknown. Sequence analysis of the 1C antigen indicated close homology with the N-terminal sequence of the type 1 fimbrial subunit and was named psuedo type 1 (I. Ørskov and Ørskov, 1983). Absorption of antiserum to remove anti-psuedo type 1 antibodies provided evidence for two further fimbrial antigens termed $F7_1$ and $F7_2$ but their chemical separation has not yet been achieved. The P and X specificities of the adhesins of F7 were not reported.

Other fimbrial antigens have been demonstrated on urinary isolates (Ørskov *et al.,* 1982a). F8 was particularly associated with **O18:K5** strains although it was first demonstrated in an **O75:K-:H5** strain (I. Ørskov and Ørskov, 1983), F10 was associated with **O7:K1**, F11 with **O1:K1:H7** and **O16:K1** and F12 with **O16:K1**. The **O1:K1** strains bearing the F11 fimbrial serotype had two antigenically different types of fimbriae but only one was numbered. The relationship between these antigens and P- and X-specific adhesins remains to be determined.

Studies in this laboratory have demonstrated a high degree of antigenic heterogeneity of MR-HA fimbriae associated with uropathogenic *E. coli* (Parry *et al.*, 1982). The serological cross-reactions of MR-HA fimbriae of seven *E. coli* isolates, all of which lacked MS adhesins, were examined by slide agglutination with fimbria-specific antisera. The organisms have subsequently been typed with the methods of Väisänen *et al.* (1981) to determine the presence of P-specific and X-specific adhesins and the results are presented in Table 3. Four strains possessed both P- and X-specific fimbriae and the remaining three expressed only P fimbriae. Two of the antisera gave identical agglutination patterns but the other antisera each showed quite different reaction patterns. Thus, despite the presence of P fimbriae on all strains, there was in a number of cases a lack of demonstrable cross-reactivity between strains. It is interesting to note that some one-way cross-reactions occurred. Thus, while anti SP201 cross-reacted with the two other P-fimbriate strains SP133 and SP144 and antisera against the P fimbriae of these two strains reacted with SP201, no reaction was observed between SP133 and SP144. Although further cross-reactions may have been detected with a more sensitive technique such as ELISA it is clear that a considerable degree of antigenic heterogeneity exists, as indicated by other studies (Korhonen *et al.*, 1982; Clegg, 1982a; Svanborg Edén *et al.*, 1983b).

The antigenic structure of the seven MR-fimbriate strains was examined further by absorbing fimbria-specific antisera with cross-reacting heterologous organisms (Parry *et al.*, 1982). Residual activity after absorption was determined by slide agglutination and, if present, was considered to be due to a distinct antigenic grouping. Analysis of such cross-reactions demonstrated seven antigens, desig-

Table 3. *Antigenic cross-reactions between MR adhesins of E. coli strains from lower urinary tract infection*

Strain no.	Receptor specificity of adhesins	Agglutination with antiserum specific for MR adhesins from:						
		SP7	SP57	SP88	SP101	SP133	SP144	SP201
SP7	P and X	+	−	−	+	+	+	−
SP57	P and X	−	+	+	+	−	−	−
SP88	P and X	−	+	+	+	−	−	−
SP101	P and X	+	+	+	+	+	−	+
SP133	P	+	−	−	+	+	−	+
SP144	P	+	−	−	−	−	+	+
SP201	P	−	−	−	+	+	+	+

Table 4. *Antigenic determinants of MR adhesins of E. coli strains*
from lower urinary tract infection

Strain no.	Receptor specificity of adhesins	Antigenic determinants identified
SP7	P and X	c,d,f
SP57	P and X	a,b
SP88	P and X	a,b
SP101	P and X	b,c,e
SP133	P	c,e
SP144	P	f,g
SP201	P	g,e

nated a to g, which occurred in combinations of two or three on an individual strain (Table 4). Antibodies specific for determinants a, b, c and e have so far been confirmed to bind to fimbriae by immune-electron microscopy (Fig. 1). It is not yet clear which determinants relate to P- or X-specific adhesins. It is of interest that strain SP88, which expresses both P- and X-specific adhesins, was examined by Dr. F. Ørskov (Copenhagen) and it did not appear to fall into any of the F antigen groups so far described (F. Ørskov, unpublished data). This strain was originally isolated from a young girl with cystitis and has the serotype **O**157:**K**-:**H**45, which is not commonly associated with UTI.

It is clear from the above studies that the fimbrial antigens of extra-intestinal *E. coli* present an extraordinarily complex picture with strains bearing multiple adhesins and multiple antigenic types. At present it is difficult to draw any firm conclusions because different studies have examined the problem in different ways and it is not clear how antigens determined in one study relate to those defined by others. There is a pressing need to exchange strains between laboratories to rationalize the many antigenic types described. The fimbrial typing scheme proposed by F. Ørskov and Ørskov (1984) (Table 1) goes some way to standardizing fimbrial antigens, although the reference strains for the F antigens of extra-intestinal *E. coli* strains often possess more than one fimbrial type. It is particularly interesting, however, that a number of non-ETEC and non-EPEC isolates from infants with diarrhoea had the fimbrial antigens F7 and F8 also found on urinary tract isolates (Czirók *et al.*, 1982). This raises the question as to whether the MR fimbriae of extra-intestinal *E. coli* form a group of adhesins which are not only important in extra-intestinal infection but are also involved in colonization of the intestine. The relationship between adhesins of extra-intestinal and EPEC strains will also be of considerable interest. Finally, it is important that the relationship between antigenic structure and receptor specificity is determined since there is strong evidence, discussed above, that antigenic heterogeneity exists between P fimbriae derived from different strains. The functional

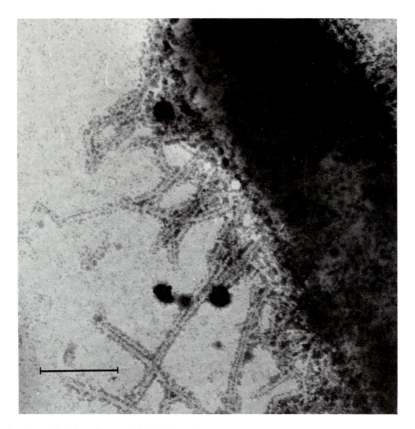

Fig. 1. MR fimbriae of *E. coli*: (a) labelled with homologous rabbit anti-fimbrial antiserum, made specific for a single determinant by cross-absorption, (b) unlabelled control, after treatment with pre-immune rabbit antiserum. The antiserum dilution was 1:20 in phosphate-buffered saline, pH 7.2, and the preparation was negatively stained with 5% phosphotungstic acid, pH 7.2. Magnification ×43,200; bar marker, = 0.5 μm.

heterogeneity likely to exist in X-specific adhesins discussed later in this chapter makes antigenic heterogeneity in this group of adhesins a strong possibility.

Haemagglutination Reactions

Haemagglutination provides a means for detecting and classifying specific adhesins (see reviews by Ottow, 1975; Duguid and Old, 1980; Pearce and Buchanan, 1980; Gaastra and de Graaf, 1982). In this account the relationship between specific HA reactions and the origin of the strains is discussed.

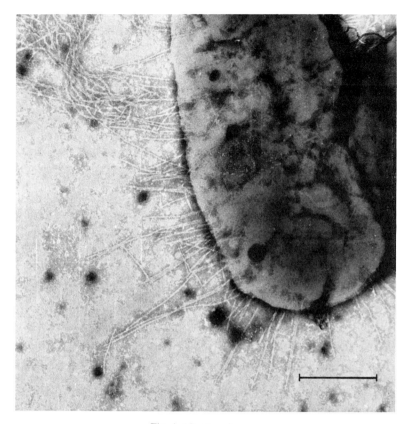

Fig. 1. (*Continued*)

Mannose-Sensitive Haemagglutination

E. coli isolates from a wide variety of sources frequently agglutinate a wide
range of animal RBC. Guinea-pig, fowl, horse, pig and monkey RBC are usually
agglutinated most strongly and HA is inhibited by low concentrations (0.01–
0.5% w/v) of D-mannose, mannan, methyl-α-D-mannoside or other mannose
analogues (Duguid and Old, 1980). The haemagglutinin was termed common,
type 1 or mannose-sensitive (Duguid *et al.*, 1955; Ottow, 1975) and is detected
by agglutination of guinea-pig RBC which is reversibly inhibited by mannose.
The association of mannose-sensitive haemagglutination (MS-HA) with MS
fimbriae was established by electron microscopic observation of fimbriae on
strains producing MS-HA of guinea-pig RBC, and whose presence on the bacteria was affected by the same environmental influences as HA (Duguid and Old,
1980) and phase variation (Brinton, 1965, 1978). Agglutination of RBC by

Table 5. *Mannose-resistant HA of intestinal E. coli*

Isolate	Adhesin	Major serogroup	Erythrocyte specificity[a]									
			Hu	Gp	C	B	P	H	S	R	G	M
ETEC												
Animal	K88ab		+	+	+[c]							
	K88ac			+								
	K88ad			+								
	987P	O9, O20, O141, O8			(+)							
		O101, O149										
	K99	O101						+	+			
	F41	O9, O101	+	+				(+)	(+)			
	Non-anionic (from K99⁻, B41)	O101							+[d]			
Human	CFA/I	O15, O25, O63, O78	+		+	+						
	CFA/II	O6, O8			+	+						
		O6:K15:H⁻:H16										
		CS1				+						
		CS2				+						
	E8775	O25, O11, O167	+[c]			+[c]						

EPEC[b]

Animal	RDEC-1	O7, O15, O25, O115	+		+	
Human		O15	+	+	+	
		O15, **K?, H**$^-$	+			
		O55, O86, O111				
		O119				
		O9			+[c]	

FEEC (stool)

| | O6, O18, O2 | +[e] | + | +[c] | +[e] |
| | O18ac | + | | | |

EIEC

Animal	Vir$^+$			(+)[f]	
Human	O15:**K**$^+$:**H21**				+[e]
	O28ac, O112ac, O124,				
	O136, O143, O144,				
	O152, O164				

[a] Erythrocytes: Hu, human; GP, guinea-pig; C, chicken; B, bovine; P, pig; H, horse; S, sheep; R, rabbit; G, goat; M, monkey. +, Positive MR-HA; (+), Weak positive MR-HA.
[b] Excluding known ETEC and EIEC.
[c] Non-eluting HA.
[d] Heat, formaldehyde, trypsin resistant.
[e] Monospecific HA of RBC indicated.
[f] Formaldehyde resistant.

99

isolated fimbriae (Brinton, 1959; Salit and Gotschlich, 1977a) and inhibition of HA by anti-fimbrial antibodies (Duguid *et al.*, 1955) also testified to the fimbrial identity of these haemagglutinins.

Strains with MS adhesins alone are present in the bowel (Evans *et al.*, 1980) and appear to be associated with asymptomatic bacteriuria or non-significant bacteriuria (Parry *et al.*, 1983). MS adhesins are also co-expressed with MR adhesins in a variety of intestinal and extra-intestinal *E. coli* infections, as discussed below.

Mannose-Resistant Haemagglutination

Many isolates of *E. coli* from a variety of pathogenic situations express haemag-glutinins differing from type 1 fimbriae in their resistance to inhibition by man-nose or mannose analogues, and are termed mannose-resistant. MR-HA pro-duced by whole organisms is reversed by raising the temperature to 50°C. This results from thermal elution of the bacteria from the RBC and probably reflects their weak affinity for red cell receptors (Duguid and Old, 1980). The threshold elution temperature differs between bacterial strains and RBC species; fowl RBC in particular resist elution at high temperature (Duguid *et al.*, 1955). Duguid (1964) described these as mannose-resistant and eluting (MRE) haemagglutinins. Adhesins designated as MR by other authors are probably more correctly referred to as MRE but this property is often not specifically investigated (Duguid and Old, 1980).

There is evidence for an increasing variety of adhesins associated with enteric isolates causing disease in animals and man (Gaastra and de Graaf, 1982). Their general HA patterns are presented in Table 5 and are discussed below. Haemag-glutinins of FEEC isolates involved in extra-intestinal infections are considered in more detail in the section on extra-intestinal isolates.

Animal ETEC

K88 adhesin. The K88 surface antigen produced in serogroups O8, O45, O138, O141, O147, O149 and O157 (Gaastra and de Graaf, 1982) is a fimbrial MR haemagglutinin of guinea-pig and chicken RBC (Duguid *et al.*, 1955; Jones and Rutter, 1974). Although some K88[+] strains also produce MR-HA of sheep, rabbit, cattle, goat and human RBC, they do so much less consistently (Jones and Rutter, 1974; Awad-Masalmeh *et al.*, 1982). Further study has revealed that the strong MR-HA of chicken RBC which is stable at room temperature is associated with the K88ab[+] variant, while guinea-pig HA which elutes at room temperature occurs with K88ab[+], K88ac[+] or K88ad[+] bacteria (Parry and Porter, 1978; Guinée and Jansen, 1979). Haemagglutination inhibition (HAI) with purified fimbriae and specific whole anti-K88ab and anti-K88ac antisera showed that K88ab HA is inhibited by homologous antiserum but only weakly by anti-K88ac. Cross-absorbed antisera with monospecific anti-K88a, K88b, K88c activity showed similar selectivity for homologous fimbriate strains and indicated that these antisera do not cross-react. Antiserum specific for the K88b determinant effectively inhibits chicken MR-HA, suggesting that chicken erythrocyte HA may involve this specific determinant (Parry and Porter, 1978).

987P adhesin. The fimbrial adhesin, termed 987P, may be expressed by porcine ETEC isolates of certain serogroups (Gaastra and de Graaf, 1982) in conjunction with K88 (Guinée and Jansen, 1979; Schneider and To, 1982) or independently of K88 and K99 (Nagy *et al.*, 1977; Isaacson *et al.*, 1977; Moon *et al.*, 1980). Intact cells expressing 987P and purified 987P fimbriae fail to show MR-HA of horse, guinea-pig, rabbit, sheep, pig or bovine RBC (Isaacson *et al.*, 1977) but weak MR-HA of chicken RBC has been demonstrated (Awad-Masalmeh *et al.*, 1982).

K99 adhesin. The K99 fimbrial adhesin which occurs particularly frequently in serogroups **O**8, **O**9, **O**20 and **O**101 (Gaastra and de Graaf, 1982) mediated MR-HA of horse and sheep RBC (Tixier and Gouet, 1975; Burrows *et al.*, 1976; de Graaf and Roorda, 1982). The identity of the fimbriae with K99 antigen was verified by growth at 18°C which prevents fimbrial expression, demonstration of MR-HA activity by isolated K99 fimbriae (de Graaf and Roorda, 1982) and after transfer of the K99 plasmid into a K99$^-$ recipient (Ørskov *et al.*, 1975).

F41 adhesin. The anionic fimbrial adhesin F41, detected on the K99$^+$ reference strain B41 (Morris *et al.*, 1982b) and other ETEC strains of serogroups **O**9 and **O**101 (Gaastra and de Graaf, 1982), has been implicated as responsible for discrepancies in MR-HA patterns among K99$^+$ isolates. HA of a spontaneous K99$^-$ F41$^+$ mutant revealed MR-HA activity for sheep RBC at 4°C that was inhibited by antiserum to K99$^+$ F41$^+$ strains but not K99$^+$ F41$^-$ strains. Further characterization of purified F41 preparations by de Graaf and Roorda (1982) established that the adhesin produces strong MR-HA of human and guinea-pig RBC and weak reactions with sheep and horse RBC. Evidence that F41 is fimbrial comes from absence of MR-HA in cultures grown at 18°C, electron microscopic observations and MR-HA of sheep RBC by isolated cell-free fimbriae (Morris *et al.*, 1982b).

A recent investigation has identified a third-non-anionic MRE haemagglutinin on strain B41 (K99$^-$ F41$^+$), which is specific for sheep RBC and is resistant to heat treatment (100°C for 30 min) but is inactivated by formaldehyde and trypsin. This is in contrast to the heat-labile F41 and K99 haemagglutinins, which are sensitive to formaldehyde and trypsin (Chanter, 1983). The specificity of this novel adhesin for sheep RBC which are also weakly agglutinated by F41 was established by HAI tests with antiserum to the partially purified adhesin.

Human ETEC

ETEC strains, a widespread cause of diarrhoea in adults and children, often possess the well-characterized adhesins CFA/I and CFA/II (Evans and Evans, 1978), which are detected by MR-HA at 4°C of human, bovine and chicken or chicken and bovine RBC, respectively (D. G. Evans *et al.*, 1977, 1979; D. J. Evans, Jr. *et al.*, 1980). CFA/I is found in O15, O25, O63 and O78 strains and CFA/II in O6 and O8 strains (Gaastra and de Graaf, 1982). CFA/I was recognized as fimbrial by abolition of MR-HA after growth at 18°C, heating at 60°C for 1 h, electron microscopy and by HAI with anti-CFA/I antiserum (Evans *et al.*, 1977). Purified CFA/I fimbriae haemagglutinated poorly but MR-HA activity characteristic of whole cells was produced by fimbriae attached to latex particles (D. G. Evans *et al.*, 1979). CFA/I recognizes human RBC in an X-specific manner, as shown by MR-HA of human p̄ RBC and absence of MR-HA of globoside coated RBC (Jann *et al.*, 1981; Källenius *et al.*, 1980b; Leffler and Svanborg Edén, 1981; Parry *et al.*, 1984). This has been verified for CFA/I and CFA/II fimbriae by the particle agglutination method of Svenson *et al.* (1982).

HA typing of these colonization factors is not reliable since HA patterns are not always consistent with the presence of the colonization factor antigens. Detection of CFA/I and CFA/II with specific antisera often discriminates ETEC strains that possess neither CFA/I nor CFA/II yet cause characteristic MR-HA of human and bovine RBC (Cravioto *et al.*, 1982). These may possess an additional fimbrial adhesin, termed E8775, recently described by Thomas *et al.* (1982), which occurs in CFA/I/II-negative ETEC strains of serogroups O25, O115 and O167 and causes MR-HA of bovine and human RBC at room temperature and 4°C. MR-HA was associated with fimbriae, as determined by electron microscopy and production at 37°C but not at 22°C. Furthermore, a mutant lacking fimbriae (E8775B) failed to haemagglutinate (Thomas *et al.*, 1982). However, the colonization potential of this fimbrial adhesin remains to be determined. A minority of ETEC strains possess CFA/I but exhibit an HA profile associated with CFA/II and vice versa (Cravioto *et al.*, 1982). Apparently some ETEC strains produce MR-HA of guinea-pig, bovine and human RBC which cannot be reconciled with the characteristic CFA/I, CFA/II pattern (Deneke *et al.*, 1979; Thorne *et al.*, 1979) and there is a suggestion that other colonization factor antigens may be present. Deneke *et al.* (1979) subsequently determined their fimbrial nature by MR attachment of isolated fimbriae to guinea-pig RBC as a method of concentrating the haemagglutinin. Further study revealed that fimbriae could be detected in fewer strains by HA than with anti-fimbrial antiserum. Whether this results from poor fimbrial expression or the non-haemagglutinating nature of these adhesins remains to be determined. In haemagglutinating isolates, there was no association between HA and fimbrial serogroup specificity (Deneke *et al.*, 1981). Cravioto *et al.* (1982) also reported strains with CFA/I and CFA/II which failed to haemagglutinate indicator red cells and, in particular, they noted differences in the strength of bovine MR-HA attributable to the existence of three antigenic components in CFA/II fimbriae with different affinities for the erythrocyte membrane or varying expression between strains. With putative anti-CFA/II antiserum, Smyth (1982) identified three surface antigens on ETEC strains of serotype **O**6:**K**15:**H**16 or **H**-, of which two, CS1 and CS2, mediated MR-HA of bovine RBC; CS2 also exhibited strong MR-HA of hen RBC (White Leghorn breed). Antigen CS3, which is common to the majority of **O**6 strains, appeared to be non-haemagglutinating. CS1 or CS3 may correspond to CFA/II.

Animal EPEC, Non-ETEC

An adhesin of the rabbit enteropathogen RDEC-1, responsible for colonization of the intestine resulting in diarrhoea (Cantey and Blake, 1977), did not haemagglutinate human, guinea-pig or bovine RBC (Thorne *et al.*, 1979; Cheney and Boedeker, 1983) although by other cultural criteria it possesses type 1 fimbriae (O'Hanley and Cantey, 1978). In view of its MR attachment to isolated rabbit intestinal brush borders (Cheney *et al.*, 1979), it would be of interest to test a wider variety of RBC species to establish whether this adhesin has haemagglutinating properties.

Human EPEC, Non-ETEC and FEEC

D. J. Evans, Jr. *et al.* (1979, 1980) classified human EPEC isolates on the basis of their HA pattern (Table 6). Many (38–42%) isolates of classical EPEC serogroups **O**55, **O**86, **O**111 and **O**119 showed HA type III, and a small subgroup expressed additional MR-HA of human RBC (D. J. Evans, Jr. *et al.*, 1979, 1980; Rothbaum *et al.*, 1982) or no HA (42%). The proportion of EPEC strains giving MR-HA of human and/or bovine RBC varied between 6 and 13% (Evans *et al.*, 1980; Cravioto *et al.*, 1982). In addition, there have been reports of novel haemagglutinins produced by diarrhoeagenic, non-ETEC, non-EIEC isolates (Candy *et al.*, 1982; Rothbaum *et al.*, 1982). Non-eluting MR-HA of bovine RBC was detected in **O**9 strains associated with an acute outbreak of diarrhoea in children. These adhesins were fimbrial, MR-HA and fimbriae being lost together during growth at 18°C. Although the isolates were serologically negative for CFA/II antigen, their fimbriae were interpreted as CFA/II-like on the basis of HA (Candy *et al.*, 1982).

FEEC, comprising a very wide range of serogroups from the normal flora,

Table 6. *HA reactions of human E. coli isolates (after Evans et al., 1980)*

RBC	HA type[a]						
	I	II	III	IV	V[b]	VI	VII
Human	MR	MS or −	—	MS or −	MR[c]	MR	MR
Monkey[d]	MS or −	MS or −	MS	MS or −	MR	MR	MS or −
Guinea-pig	MS or −	MS or −	MS	MS or −	MR	MR or MS or −	MS or −
Chicken	MR	MR	MS	MS or −	MR	MR or MS or −	MS or −
Bovine	MR	MR	—	—	MR	—	MR

[a]MR, mannose-resistant; MS, mannose-sensitive; −, negative.
[b]'Monospecific' HA, i.e. HA with only one of the RBC species listed.
[c]Type V-A in Evans et al. (1980).
[d]African green monkey.

sporadic cases of infantile enteritis and extra-intestinal infection (Ørskov and Ørskov, 1975; Czirók *et al.*, 1976; Ørskov *et al.*, 1977; D. J. Evans *et al.*, 1979), have been studied with respect to their HA patterns. Major FEEC serogroups (O1, O2, O4, O6, O7, O18) isolated from blood of bacteraemic patients in particular, showed HA type VI (see below under extra-intestinal isolates). Those from stool specimens exhibited type VI (59%) with a lesser incidence of type IV (18%) or type V (14%). In general, all FEEC serogroups isolated from blood or stools showed a greater incidence of type VI, especially when compared with non-FEEC serogroups from the same site (17%), the majority of which were non-haemagglutinating (Evans *et al.*, 1980), although a few (21%) gave MR-HA of human and/or bovine cells (Cravioto *et al.*, 1982). The fimbrial nature of the haemagglutinins has been confirmed in selected bacteraemia isolates (Clegg *et al.*, 1982) and those of HA type III, but not in isolates with type V (Evans *et al.*, 1980).

A MR fimbrial haemagglutinin of human RBC observed by Wevers *et al.* (1980) in an O18ac:H⁻ strain associated with diarrhoea could be classified as type V-A, which constitutes a minor HA group of FEEC (Evans *et al.*, 1980). Jann *et al.* (1981) showed that this strain agglutinated human RBC in X-specific fashion.

Animal and Human EIEC

EIEC comprise a subgroup of EPEC that cause a *Shigella* dysentery-like diarrhoeal syndrome in young children, characterized by bacterial penetration and destruction of mucosal tissue (Rowe, 1979). EIEC are often serologically cross-reactive with *Shigella* spp.

Among invasive isolates causing calf and lamb neonatal diarrhoea, the surface antigen associated with a virulence (Vir) plasmid (Smith, 1974; Lopez-Alvarez

and Gyles, 1980) produced weak MR-HA of bovine RBC in fie
after plasmid transfer to K-12, suggesting decreased expressic
environment (Morris *et al.*, 1982a). Evidence for a fimbrial ad
on adhesion tests with isolated brush borders but its HA properti
determined.

The limited data available suggest that the majority of human s ... non-
haemagglutinating with human, bovine, chicken and guinea-pig RBC and presumably lack fimbriae (O'Hanley and Cantey, 1978), or otherwise show predominantly MS-HA of chicken and guinea-pig RBC (D. J. Evans *et al.*, 1979).

Finally, MRE non-fimbrial haemagglutinins were discovered in a few strains
from a wide variety of sources, which with one exception were characterized by
a restricted range of HA comprising MR-HA of ox or sheep RBC or monospecific MR-HA of ox, human or chicken RBC (Ip *et al.*, 1981).

Extra-Intestinal Isolates

Surveys of HA by human extra-intestinal isolates of *E. coli* revealed MR-HA of
human RBC (59–92% incidence, Minshew *et al.*, 1978a,b). The prevalence of
MR haemagglutinins varied extensively between individual studies. In isolates
from cases of meningitis and septicaemia, 57–86% and 14–57% (Cravioto *et al.*,
1979, 1982; Ljungh *et al.*, 1979; Evans *et al.*, 1981), respectively, showed MR-
HA. The incidence of MR-HA strains was 23–61% in unselected cases of urinary tract infection (UTI) (Cravioto *et al.*, 1979, 1982; Ljungh *et al.*, 1979;
Evans *et al.*, 1981; Sussman *et al.*, 1982). Strains from unselected UTI bearing
MR adhesins alone (12–14%) were less common than those with both MR and
MS adhesins (24–49%) (Duguid *et al.*, 1979; Sussman *et al.*, 1982; Parry *et al.*,
1983). Isolates from unselected UTI, especially those from hospital patients,
would be expected to show a lower prevalence of MR adhesins because of the
inclusion of patients with obstruction of the urinary tract or catheter-provoked
infection, in which these adhesins are less likely to play a role.

Investigation of HA of *E. coli* from specific types of UTI suggested that the
prevalence of MR adhesin-bearing strains, with or without MS adhesins, was
highest in those isolated from pyelonephritis (69%) rather than acute symptomatic (34%) and asymptomatic (15%) bacteriuria (Hagberg *et al.*, 1981).
However, studies by Sussman *et al.* (1982) and Parry *et al.* (1982, 1983) showed
a much higher incidence of MR haemagglutinins in isolates from acute cystitis
(75–97%) and suggested that previous surveys underestimated the importance of
these isolates because of inappropriate methods of bacterial cultivation and testing. In the majority of studies CFA agar was used (Cravioto *et al.*, 1979, 1982;
Ljungh *et al.*, 1979; Evans *et al.*, 1980, 1981) which is suited to the expression
of CFA of enterotoxigenic *E. coli* (Evans *et al.*, 1977) but not MR adhesins of
urinary isolates (Parry *et al.*, 1982). Others have used broth media (Svanborg

Éden and Hansson, 1978; Ljungh *et al.*, 1979; Hagberg *et al.*, 1981) which favour the expression of MS adhesins. The effect of cultural conditions on haemagglutinin expression is further considered below and the importance of test conditions for detecting haemagglutination is reviewed in Chapter 10.

Nevertheless, even with the use of appropriate techniques, many non-HA strains have been isolated from unselected UTI (Parry *et al.*, 1982; Sussman *et al.*, 1982) and a smaller but significant number from acute cystitis (Parry *et al.*, 1983). The prevalence of non-HA strains is higher in isolates from non-signifi-cant bacteriuria and unselected UTI, particularly in hospital cases, and highest in urethral or introital contaminants from uninfected urine of healthy individuals (Duguid *et al.*, 1979; Sussman *et al.*, 1982; Parry *et al.*, 1983). The percentage values of this group in other surveys (Evans *et al.*, 1981; Hagberg *et al.*, 1981) were probably over-estimated owing to the test methods used.

In many studies HA testing is restricted to guinea-pig and human RBC in the presence or absence of mannose; relatively few data are available for HA of other animal RBC by extra-intestinal isolates. Comprehensive accounts by Duguid (1964), Duguid *et al.* (1979), and Duguid and Old (1980), in which HA reactions were observed for RBC of 12 animal species other than guinea-pigs, have re-corded a large variety of HA patterns. HA profiles of strains from bacteriuria or UTI for a limited range of animal RBC showed similar heterogeneity (Evans *et al.*, 1980, 1981; Parry *et al.*, 1982). However, the apparent association of a particular HA pattern with specific adhesins is often contradictory. For example, Parry *et al.* (1982) reported an association between MS-HA of guinea-pig cells with MS-HA of chicken RBC in strains possessing only MS adhesins, an asso-ciation not recorded by Evans *et al.* (1980) for similar organisms. In addition, isolates with MR adhesins alone also agglutinated chicken and guinea-pig RBC in MR fashion (Evans *et al.*, 1980), whereas a similar pattern occurred only in strains with both human MR and guinea-pig MS haemagglutinins in the studies of Parry *et al.* (1982). Such differences may reflect the large numbers of strains used by Evans *et al.* (1980) and, more importantly, the different methods of cultivation of test bacteria.

Bacteriuria isolates giving MR-HA of human RBC were classified as type VI (41%), type V (11%) and type VII (8%) in the scheme of Evans *et al.* (1980) (Table 6). Septicaemia isolates were similarly grouped as type VI (45%), type V (9%) and type VII (1%) and the majority (80%) of septicaemia isolates of serogroups O1, O2, O4, O6, O7 and O18 grouped in type VI, indicating a possible relationship between O serogroup and HA type (Evans *et al.*, 1980). Type VI was characteristic of 86% of cerebrospinal fluid (CSF) isolates. It was concluded that extra-intestinal isolates including those from sources as diverse as CSF, bacteraemia and unselected UTI had similar HA patterns (Evans *et al.*, 1981).

Such surveys have illustrated the diversity of MR-HA profiles of *E. coli* but their interpretation has been hampered by an inadequate understanding of the receptor specificities of MR adhesins. However, recent recognition of a glycolipid receptor on human RBC containing an active α-D-Gal$p$-(1→4)-β-D-Gal$p$ residue present in the p^k (trihexosyl ceramide), P (globoside) or P$_1$ blood group antigen (see section on receptors below) has made possible differentiation of MR haemagglutinins of uropathogenic isolates as P-specific (binding to the P receptor) or X-specific (binding to other receptors) (Källenius *et al.*, 1980a,b, 1981a; Leffler and Svanborg Edén, 1981; Väisänen *et al.*, 1981; Korhonen *et al.*, 1982). P adhesins are demonstrated by MR-HA of human P$_1$ RBC (expressing P, P$_1$ and P^k) or P$_2$ phenotype (expressing P and P^k), while X adhesins are recognized by MR-HA of human p̄ RBC which lack P antigens and which P-fimbriate organisms do not agglutinate (Källenius *et al.*, 1980a,b, 1981a; Leffler and Svanborg Edén, 1981; Väisänen *et al.*, 1981).

In practice identification of a P-specific adhesin is not straightforward. To fulfil the identification criteria organisms should show MR-HA of globoside- or trihexosyl ceramide-coated guinea-pig or horse RBC, which otherwise lack the necessary receptor, but not control RBC (Källenius *et al.*, 1981a; Leffler and Svanborg Edén, 1980, 1981; Väisänen *et al*, 1981). However, a number of isolates agglutinate control guinea-pig, bovine or horse RBC directly in an X-specific manner, as indicated by MR-HA of p̄ RBC (Källenius *et al.*, 1980a,b, 1981a; Leffler and Svanborg Edén, 1981; Väisänen *et al.*, 1981). In such strains demonstration of P fimbriae by HA is dependent on partial inhibition of ag-glutination of coated RBC or human RBC with P antigens by the synthetic *O*-Me-galactosyl analogue of the receptor (Källenius *et al.*, 1981a; Väisänen *et al.*, 1981) or an alternative source of receptor material such as hydatid cyst fluid (Parry *et al.*, 1984) (see section on receptors below).

The suggestion that P-specific adhesins of pyelonephritis-associated strains are fimbrial (Källenius and Möllby, 1979; Källenius *et al.*, 1980b) has been verified by the binding of cell-free fimbriae to human neuraminidase-treated human P$_1$ and p^k$_2$ RBC in the presence of α-methyl-D-maunoside (αMM), but not p̄ cells unless these were coated with globoside (Korhonen *et al.*, 1982). The specificity of fimbrial attachment to the P antigen of the erythrocyte was demonstrated by HAI of P$_1$ and P^k$_2$ RBC by the synthetic glycoside P receptor analogue, α-D-Gal$p$-(1→4)-β-D-Gal$p$-1-*O*-Me (Svenson *et al.*, 1982). HA by isolated P-fimbriae occurred only with neuraminidase-treated RBC or with fimbriae bound to latex particles, which overcome charge–charge interactions that would pre-vent HA (Korhonen *et al.*, 1982). These apparently also prevent MR-HA reac-tions of isolated CFA/I fimbriae (D. G. Evans *et al.*, 1979).

The discrimination of two MR adhesin types by their receptor specificities should provide a means of resolving the diverse HA patterns obtained with RBC of different animals. The RBC of pig, sheep, pigeon, fowl, goat and dog, for example, contain the globoside receptor (Leffler and Svanborg Edén, 1980) and complete inhibition of HA by synthetic or naturally occurring receptor would confirm the presence of fimbriae with P specificity. P-specific HA of sheep, pigeon and chicken RBC by HAI with hydatid cyst fluid has been demonstrated (Parry *et al.*, 1984). By present criteria, bacteria with P fimbriae lack MR-HA of bovine, guinea-pig and horse RBC (Leffler and Svanborg Edén, 1980). It might be inferred that these RBC may be agglutinated only by adhesins with X specificity and strong X-specific MR-HA of these RBC has been observed (Svenson *et al.*, 1982; Parry *et al.*, 1984). It is also apparent that fowl and sheep RBC may be agglutinated by bacteria possessing P or X adhesins. The coexistence of both adhesins on the same strain (Väisänen *et al.*, 1981; Svenson *et al.*, 1982) is the basis of the varied range of MR-HA patterns recorded with different animal RBC (Duguid, 1964; Duguid *et al.*, 1979; Duguid and Old, 1980; Evans *et al.*, 1980, 1981; Parry *et al.*, 1982; Sussman *et al.*, 1982) and the use of the synthetic disaccharide P receptor analogue or hydatid fluid allows appraisal of this diversity in relation to P and X adhesin specificity. Further evidence is emerging of variation in HA pattern with adhesins of X specificity. Svenson *et al.* (1982) defined X-specific strains as those with MR-HA of human p̄ RBC but not strains with X-specific guinea-pig and bovine MR-HA. Subdivision of X-specific adhesins must await further characterization of their HA properties. Strains with X adhesins may agglutinate either a wide or a restricted range of animal RBC, including red cells with known P antigens, suggesting the existence of a variety of functional X adhesins (D. M. Rooke and S. H. Parry, unpublished). Detailed examination of these HA patterns, coupled with analysis of the glycosphingolipid constituents of the RBC membranes of various animal species, may provide some evidence for the nature of the red cell receptors in each HA group, although great care will be required to minimize variation in receptor specificity of the RBC from individual animals, as discussed elsewhere (Old, this volume, Chapter 10).

Effect of Cultural Conditions on Expression of HA

Agglutination of RBC by bacteria, which is used as an indicator of their fimbriation, is determined by a complex series of physicochemical interactions described in detail elsewhere (Old, this volume, Chapter 10). In particular, the number of bacteria and their degree of fimbriation determine the intensity of HA of RBC with homologous receptor specificity. Consequently, environmental factors that affect the extent of fimbriation have corresponding effects on adhesion.

Environmental conditions determine the proportion of fimbriate organisms (Eisenstein and Dodd, 1982). This phase variation is a phenotypic control mechanism that switches between the fimbrial and non-fimbrial state in individual bacteria (Brinton, 1965, 1978). It occurs in type 1, CFA/I (Brinton, 1978) and 987P fimbriae (Nagy et al., 1977).

Medium. Serial subculture in static broth enhances expression of type 1 fimbriae at the expense of MR haemagglutinins, while the latter are favoured by growth on solid media (Duguid et al., 1979; Duguid and Old, 1980). Static broth culture is assumed to enrich the population with type 1 fimbriate bacteria because they aggregate into a pellicle, by virtue of their fimbriae, at the medium–air interface, where rapid multiplication is favoured by the abundant oxygen supply. However, MR haemagglutinin production may not be inhibited immediately on primary subculture or by the adverse nature of the medium. For example, Davis et al. (1982) demonstrated retention of MR-HA of human RBC in a urinary isolate possessing MR and MS fimbriae after one subculture in brain heart infusion (BHI) and CFA broth.

Type 1 fimbriae were thought to be subject to catabolite repression by carbohydrates (Ofek et al., 1981) but it has been shown that the degree of fimbriation is determined by alteration of the rate of phase variation, rather than by catabolite repression (Eisenstein and Dodd, 1982).

Some MR adhesins have specific nutritional requirements for maximum expression. Thus, CFA/I and CFA/II are well expressed on a special semi-synthetic agar medium (Evans et al., 1977). Strong expression of K99 requires growth on media with low concentrations of amino acids and carbohydrate (Guinée et al., 1976, 1977) or a glucose-supplemented minimal medium (Gaastra and de Graaf, 1982). The amino acid component of complex media, especially alanine, specifically represses K99 synthesis (de Graaf et al., 1980a), as do glucose, pyruvate and arabinose (Guinée et al., 1976; Isaacson, 1980). Glucose-dependent repression of K99 is reversed by addition of cyclic AMP, demonstrating that it is subject to cAMP-dependent catabolite repression (Isaacson, 1980). Organisms with 987P fimbriae are heavily fimbriate in porcine intestine but become non-fimbriate when grown on medium used for K99 production (Nagy et al., 1977; Moon et al., 1980).

BHI agar or phosphate-buffered agar seem the most suitable for MR-HA by uropathogenic isolates (Parry et al., 1982). P fimbriae from uropathogenic isolates are expressed well on BHI agar (Parry et al., 1984) and CFA agar (Väisänen et al., 1981), media which also support the production of X-specific haemagglutinins (Källenius et al., 1980b; Väisänen et al., 1981; Parry et al., 1984).

Some attention has been addressed to the expression of haemagglutinins in infected urine or after subculture in filter-sterilized human urine. Various haemagglutinins were detected in up to 29% of isolates from the urine of catheterized patients (Ofek et al., 1981) but not in isolates from uncatheterized patients with UTI (Harber et al., 1982). The prevalence of HA$^+$ cultures in the latter increased after subculture in sterile, pooled human urine and in nutrient broth (Harber *et al.*, 1982). Absence of HA activity in the original samples could not be attributed to coating of bacteria from urine with antibody, mannose-containing Tamm-Horsfall glycoprotein in urinary mucus or capsules (Ofek *et al.*, 1981). It has been suggested that in the bladder and in urine expression of MS and MR adhesins is repressed (Ofek *et al.*, 1981; Harber *et al.*, 1982). However, concentrating bacteria from urine by filtration, before examination (Harber *et al.*, 1982), probably excludes the majority of fimbriate bacteria, which are adherent to urinary mucus and exfoliated uroepithelial cells, thus greatly reducing the chances of detecting fimbriate organisms. However, MR and MS haemagglutinins have been detected in artificially inoculated sterile human urine (Davis *et al.*, 1982) and in infected urine by direct HA testing (J. M. Salter, personal communication).

Green and Thomas (1981) suggested that the absorption of blood-reactive substances from the medium onto the surface of bacteria may give a false impression of HA activity. This may be difficult to confirm by comparison of different media since these may affect the degree of fimbrial expression in other ways. Comparison of the IIA activity of washed and unwashed bacteria from the same medium may be no better because of the reduction in fimbriation after mechanical agitation (Svanborg Edén and Hansson, 1978; Avots-Avotins *et al.*, 1981). However, Green and Thomas (1981) were unable to show differences in the HA activity of isolates cultured on a complex and a chemically defined medium.

Aeration. As we have seen, oxygen increases multiplication of type 1 fimbriate organisms. Isaacson (1980) showed that aeration increases K99 production about six-fold during exponential growth but it declines in the stationary phase, thus confirming the observations of Guinée *et al.* (1977). This is in contrast to type 1 fimbriate organisms in which maximum production occurs in the stationary phase (Duguid *et al.*, 1979).

Temperature. Although type 1 fimbriae production is temperature-independent, MR haemagglutinins are not produced at 18°C (Duguid and Old, 1980; Gaastra and de Graaf, 1982). Maximum expression of K88 occurs at 35–40°C and of K99 at 37°C. The optimum temperature and cultural conditions for F41 production are the same as for K99, suggesting that their regulation and biosynthesis are comparable (de Graaf and Roorda, 1982).

Antibiotics. Growth of bacteria in subinhibitory concentrations of various antibiotics reduces HA and fimbriation in strains with type 1 fimbriae (Eisenstein *et al.*, 1980) and uropathogenic strains with MR haemagglutinins (Vosbeck *et al.*, 1982) including P fimbriae of pyelonephritis-associated strains (Väisänen *et al.*, 1982b). The antibiotic effects may result from direct inhibition of fimbrial synthesis or assembly (Eisenstein *et al.*, 1980; Väisänen *et al.*, 1982b).

Adhesion Properties

MS Adhesins

MS fimbriae allow adhesion of organisms to a wide range of cellular targets (Duguid and Old, 1980). Since mannose is ubiquitous in mammalian cell membranes, MS, adhesion to mammalian host cells by type 1 fimbriate bacteria is widespread. MS fimbriae mediate attachment to isolated porcine enterocytes *in vitro* (Isaacson *et al.*, 1978a) but although many ETEC and non-ETEC isolates produce type 1 fimbriae, there is little evidence for their role in intestinal colonization, since it is not known whether type 1 fimbriae are produced in the intestine or adhere to intestinal epithelium *in vivo* (Moon *et al.*, 1979). A CFA/I$^-$ MS$^+$ variant of a human ETEC strain (H-10407, CFA/I$^+$, MS$^+$) which normally colonizes human and rabbit intestine failed to do so in the absence of CFA/I, suggesting that type 1 fimbriae alone are insufficient for colonization (Evans *et al.*, 1978b). This was confirmed by the observation that the CFA/I strain H-10407, enriched for MS fimbriae by broth cultivation, did not adhere to isolated human ileal brush borders (Cheney and Boedeker, 1983). MS fimbriae occurring together with K88, K99, 987P or other MR adhesins are probably of minor importance in adhesion since cell-free MS fimbriae do not inhibit attachment of the bacteria to isolated intestinal cells (Isaacson *et al.*, 1978a; Awad-Masalmeh *et al.*, 1982). MS$^+$ strains attach to human BEC (Ofek *et al.*, 1977; Ofek and Beachey, 1978) but attachment to exfoliated UEC is disputed. Adherence of an MS$^+$ urinary tract isolate to UEC *in vitro* was determined with a radio-labelled counting technique (Schaeffer *et al.*, 1979, 1980) but was lower than that demonstrated by visual counting in surveys of UTI strains (Svanborg Edén and Hansson, 1978; Hagberg *et al.*, 1981). Attachment to particular cell types might account for these discrepancies since UEC derived from urine comprise approximately 90% squamous cells from the bladder trigone, urethra and vaginal vestibule, and the remainder are transitional cells derived from the renal pelvis, ureters and bladder but there is no evidence for this suggestion (Svanborg Edén *et al.*, 1980). Since MS$^+$ organisms attach to urinary mucus, differences in attachment may be due to mucus adhering to UEC (Chick *et al.*, 1981). MS adhesion to human, monkey, pig and rat UEC and BEC is inhibited by anti-type 1 fimbrial antiserum (Parry *et al.*, 1982).

There is some indication that bacterial attachment to UEC, but not vaginal epithelial cells, varies during the menstrual cycle, adherence being greater in the early phase and diminishing just before ovulation (Schaeffer *et al.*, 1979, 1981). However, according to Svanborg Edén *et al.* (1980) this is not the case. Sugarman and Epps (1982) have shown that *E. coli* attaches in larger numbers to HeLa monolayers pre-treated for 18 h with oestrogens, though the concentrations were higher than the serum levels in non-pregnant women. They suggested that adhesion was due to modification of the cell surface but the significance of the hormonal status of the host on cellular attachment by bacteria has yet to be determined.

Difficulties in obtaining consistent results with cells from human donors have led to the use of cultured cell lines for *in vitro* adhesion testing. Thus Jann *et al.* (1981) demonstrated that, with one exception, MS fimbriae, regardless of the presence of MR fimbriae, adhered to a range of epithelial or epithelioid monolayers. The direct adherence of ^{125}I-labelled type 1 fimbriae to a Vero monkey kidney cell line (Salit and Gotschlich, 1977b) supports these findings. One strain apparently exhibited MS attachment to guinea-pig and human RBC, *S. cerevisiae* and cultured tissue cells by an MS wall adhesin (Eshdat *et al.*, 1981; Jann *et al.*, 1981). Although MS fimbriae provide a means of attachment to a wide diversity of cell lines, Chabanon *et al.* (1982) observed that a uropathogenic isolate with MS fimbriae alone failed to adhere to a line of human foetal intestinal cells. However, the demonstration of adhesion to cultured cell lines should be interpreted with caution because such lines may be contaminated (Nelson-Rees *et al.*, 1981).

MS fimbriae also mediate attachment to leucocytes via mannose receptors and this stimulates subsequent phagocytosis (Bar-Shavit *et al.*, 1977, 1980; Mangan and Snyder, 1979a,b; Rottini *et al.*, 1979; Silverblatt *et al.*, 1979; Sussman *et al.*, 1982). This has been confirmed with radio-labelled bacteria by chemiluminescence measurement of the stimulated oxidative phagocyte metabolism (Björksten and Wadström, 1982; Blumenstock and Jann, 1982). However, other bacterial surface components, such as K antigens or lipopolysaccharide, may affect phagocytosis (Öhman *et al.*, 1982b). Non-haemagglutinating and hence non-fimbriate bacteria also adhere to leucocytes (Sussman *et al.*, 1982). A mannose-sensitive non-fimbrial adhesin implicated in leucocyte attachment (Mangan and Snyder, 1979a; Blumenstock and Jann, 1982) is similar to a previously described non-fimbrial MS adhesin for human epithelial cells and yeast cells (Eshdat *et al.*, 1978). The adhesin in the strain used by Blumenstock and Jann (1982) has been identified as a 17,000-dalton protein component of the outer membrane fraction representing either unpolymerized pilin or a separate protein with the same mannose specificity as pilin (Eshdat *et al.*, 1981).

MS fimbriae have a strong affinity for urinary mucus (Ørskov *et al.*, 1980a; Parry *et al.*, 1982) present on the surface of UEC lining the bladder (Ørskov *et al.*, 1980a; Chick *et al.*, 1981). Aqueous extracts of urinary mucus, but not of salivary, gastric or intestinal mucus, agglutinate bacteria with MS fimbriae in a mannose-specific fashion (Parry *et al.*, 1982). Ørskov *et al.* (1980b) noted that type 1 fimbriae are also agglutinated by respiratory mucus. The adhesive specificity resides in the mannose-rich, Tamm-Horsfall glycoprotein (Ørskov *et al.*, 1980a) present in urinary mucus. The uromucoid fraction obtained from urinary mucus by salt precipitation and caesium chloride density gradient centrifugation, the major component of which is the Tamm-Horsfall glycoprotein, has an affinity for MS-fimbriate bacteria (M. J. S. Lee and S. H. Parry, unpublished results). The significance of bacterial attachment to urinary mucus for their persistance in the urinary tract remains to be determined.

MR Adhesins

MR adhesins show pronounced host, individual, and cell type specificity (Table 7). Adhesion may occur *in vitro* to cell types or host species not encountered during natural pathogenic events, or with cultured cell lines.

Animal ETEC

K88 adhesin. K88[+] bacteria have been shown to adhere to the basal portion of the villi in the anterior intestine and to some extent throughout the length of porcine intestine *in vivo* (Hohmann and Wilson, 1975). Other studies (Arbuckle, 1970; Bertschinger *et al.*, 1972) show that K88ab[+] and K88ac[+] strains attach to the brush border of the villous surface of the ileum but not to the intestinal crypts where there is no brush border. It is now established that K88[+] ETEC colonize the entire intestinal tract (Moon *et al.*, 1979). The specificity of adherence by K88 fimbriae has been demonstrated *in vitro* by attachment of K88ab[+] and K88ac[+] bacteria to the brush border of isolated intestinal epithelial cells (IEC) (Wilson and Hohmann, 1974; Parry and Porter, 1978) and MR attachment of cell-free K88 antigen to isolated brush-border membranes (Anderson *et al.*, 1980; Sellwood, 1980). Passive protection of piglets by colostrum antibody raised to partially purified K88ab or K88ac antigen reduced the adherence of challenge K88ab[+] or K88ac[+] bacteria to the intestinal wall. (Rutter and Jones, 1973; Walker and Nagy, 1980). However, the results of antibody inhibition of K88 attachment to IEC *in vitro* are conflicting, Wilson and Hohmann (1974) concluded that the b and c antigenic determinants of the K88 adhesin had binding specificity, as indicated by the ability of homologous monospecific antiserum to inhibit attachment of K88[+] bacteria, whereas K88a antiserum had little or no inhibitory effect on attachment. Parry and Porter (1978) reported that K88a, b and c antisera were inhibitory, indications that all determinants were involved in adhesion and that the conflicting results may be due to poor fimbrial expression in the strain used by Hohmann and Wilson (1975) to raise the K88a antiserum.

The porcine K88 adhesin is host-specific, producing experimental enteric coli-

Table 7. *Mannose-resistant adhesion of intestinal E. coli to isolated epithelial cells*

Isolate	Adhesin[a]	Epithelial cell attachment[b]									
		Intestinal							Buccal	Oesophageal	Tracheal
		P	C	L	AR	IR	AH	FH	H	C	C
ETEC											
Animal	K88ab	+cd	+d	+d							
	ac	+cd	+d	+d							
	ad	+cd									
	987P	+			+						
	K99	+	+	+	+	+					
	F41	+	+								
Human	CFA/I										
	CFA/II						+				
	O6:K15H−/H16										
	CS1										
	CS2										
	CS3										

	O26:K60:H11	Various O serotypes			
EPEC[f]					
Animal	RDEC-1[g]		+		
Human	O9[g]			+	
	Other[g]	+		+	
EIEC					
Animal	Vir[+]	+[e]			+[e]
Human					+[e]

[a] For major serogroups see Table 5.
[b] P, pig; L, lamb; IR, infant rabbit; FH, foetal human; C, calf; AR, adult rabbit; AH, adult human; H, human.
[c] Adherence (+) to specific phenotypes (see receptor section).
[d] Host-specific adherence only.
[e] Resistant to formaldehyde.
[f] Excluding known ETEC and EIEC.
[g] Strains with unspecified adhesins.

115

bacillosis in pigs but not in calves or lambs, nor do calf or lamb strains produce infection in pigs (Gaastra and de Graaf, 1982). In heterologous hosts colonization of the anterior intestinal epithelium does not take place, as shown by the failure of heterologous animal strains to bind to the brush borders of isolated epithelial cells (Wilson and Hohmann, 1974), but non-specific adhesion to basal portions of isolated cells does occur. The specificity of K88 attachment to porcine intestine is determined by the host phenotype and the serological identity of the K88 adhesin (Rutter *et al.*, 1975; Sellwood *et al.*, 1975; Bijlsma *et al.*, 1981), thus demonstrating the complex interplay between serologically heterogeneous adhesins and host cell surface receptor specificity (Gaastra and de Graaf, 1982). Five pig phenotypes have been distinguished which vary in their expression of receptors for the ab, ac and ad variants of K88 (Bijlsma *et al.*, 1981, 1982).

987P adhesin. The fimbrial adhesin 987P attaches to ileal epithelium of porcine hosts *in vivo* over the entire villus surface but not in the crypts (Nagy *et al.*, 1976; Isaacson *et al.*, 1977). In natural infection, colonization is confined to the ileum and large intestine, while ligated gut experiments also showed attachment to jejunum (Nagy *et al.*, 1976; Moon *et al.*, 1979). Supporting evidence *in vivo* for the adhesive properties of 987P includes the inability of non-capsulate, non-fimbriate mutants to colonize pig intestine (Isaacson *et al.*, 1977) and the acquisition of passive immunity to 987P$^+$ bacteria in piglets born to sows immunized during pregnancy with purified 987P (Moon *et al.*, 1979). Demonstration of adhesion *in vitro* to isolated IEC was hampered by the poorly fimbriate state of the bacteria under these conditions but this could be overcome by selection of a heavily fimbriate variant 987P strain which adhered equally *in vitro* to brush borders (Nagy *et al.*, 1977) of ileal and jejunal cells (Isaacson *et al.*, 1978a) whereas *in vivo* adhesion is only to the ileum. 987P$^+$ isolates also attached to rabbit ileal epithelial cells but not to similar cells from infant rabbits, suggesting that the necessary membrane receptor is not expressed in the neonate (Dean and Isaacson, 1982). The specificity of cellular attachment by 987P fimbriae in the rabbit system and to porcine cells was demonstrated by competitive inhibition of *in vitro* adhesion by purified homologous fimbriae and also the Fab fragment of homologous fimbria-specific IgG (Isaacson *et al.*, 1978a; Dean and Isaacson, 1982). Failure of a non-capsulate 987P$^+$ strain to adhere *in vivo* suggested the capsule (K103) as a possible adherence factor *in vivo* (Isaacson *et al.*, 1977). However, more recently Isaacson *et al.* (1978a) have shown that both capsulate and non-capsulate 987P$^+$ mutants adhere equally well to isolated epithelial cells *in vitro* and show similar degrees of competitive inhibition by purified 987P fimbriae.

K99 adhesin. Examination of ileal sections from calves infected with strain B41 (K99$^+$ F41$^+$) by immunofluorescence with specific K99 antiserum illustrated the adherence of bacteria to the mucosa, which is similar to the colonization of the ileum by K88[+] and 987P[+] cells (Isaacson et al., 1978b; Chan et al., 1982). K99[+] ETEC bovine strains and a K-12 transconjugant strain adhered to brush borders of IEC in vitro (Burrows et al., 1976). The K99 adhesin has a wider host range than K88 or 987P adhesins. Some K99[+] strains originally thought of as calf enteropathogens were shown by K99-specific immunofluorescence to produce the adhesin in the pig ileum in close association with the intestinal villous epithelium and to induce diarrhoea in new-born piglets (Moon et al., 1977). K99 adhered equally well in vitro to porcine IEC derived from jejunum or ileum (Isaacson et al., 1978a), indicating that preferential attachment to the ileum of calves, lambs and pigs in vivo (Isaacson et al., 1978b; Moon et al., 1979) depends on factors other than receptor differences. K99 also attaches to infant and adult rabbit ileal epithelial cells and to brush borders (Dean and Isaacson, 1982), which suggests a non-specific binding mechanism. Attachment of K99[+] bacteria to porcine ileal epithelial cells was not inhibited by K88 strains, purified 987P or type 1 fimbriae. Competitive inhibition by homologous fimbriae could not be demonstrated because of aggregation of purified K99 but the Fab fragment of K99-specific IgG inhibited the attachment (Isaacson et al., 1978a). A recent electron microscopic study suggested that a thick bacterial glycocalyx, consisting of an anionic D-glucuronic acid, D-mannose and D-galactose polymer through which the fimbriae protrude, assists the maintenance of a close association between K99[+] organisms and intestinal brush borders in vivo (Chan et al., 1982). These results support the observations that the capsule, in addition to K99 antigen, may be adhesive, since acapsular K99[+] mutants survived in the intestinal lumen but did not adhere to or colonize epithelial brush borders (Smith and Huggins, 1978). Chan et al. (1982), therefore, proposed that K99 provides the initial mechanism of attachment, which is then supported by adherence of the glycocalyx. A similar glycocalyx may be associated with the K88, 987P and F41 adhesins.

F41 adhesin. There is evidence to suggest that the F41 fimbrial adhesin is a determinant in the colonization of animal intestine since E. coli B41M, which expresses only F41, causes diarrhoea and adheres to the microvillous surface of intestinal epithelium in experimentally infected neonatal piglets. Specific indirect immunofluorescence showed the presence of F41 antigen at the villus border (Morris et al., 1982b). However, unlike the bacterial layers observed on the intestinal epithelial cell surface in E. coli possessing K88, K99 and 987P adhesins, scanning electron microscopy (SEM) and fluorescent antibody studies of in vivo adhesion by B41M showed focal distribution of small groups of adherent organisms. This correlated with the relatively low adherence of this strain to isolated calf enterocytes compared with the parental strain (K99[+] F41[+]). F41[+] cells after growth at the permissive temperature of 37°C attached to calf and porcine enterocyte brush borders incubated with mannose but not after growth at the MR fimbrial restrictive temperature of 18°C (Awad-Masalmeh *et al.*, 1982; de Graaf and Roorda, 1982; Morris *et al.*, 1982b; Chanter, 1983) or after pre-treatment with heat or formaldehyde after growth at 37°C (Morris *et al.*, 1982b). Attachment of cell-free F41 to calf enterocytes in the presence of mannose was demonstrated by Morris *et al.* (1982b) and prevented attachment of whole F41 [+] cells by competitive inhibition, indicating that attachment is specific (de Graaf and Roorda, 1982). F41 [+] cells attached to phenyl Sepharose gel, indicating their hydrophobic nature, which may be an additional mechanism of adhesion.

Human ETEC

CFA/I [+] and CFA/II [+] ETEC adhere strongly in an MR fashion to human intestinal mucosa (see reviews by Duguid and Old, 1980; Pearce and Buchanan, 1980; Gaastra and de Graaf, 1982) and MR attachment of CFA/I [+] organisms to isolated human ileal brush borders has been demonstrated (Cheney and Boedeker, 1983). However, there was no correlation between colonization of the infant rabbit intestine, which depends on the presence of the plasmid encoding CFA/I or CFA/II (Moon *et al.*, 1979; Duguid and Old, 1980; Gaastra and de Graaf, 1982), and *in vitro* adhesion to epithelial cells or brush borders from infant or adult rabbits (Dean and Isaacson, 1982). Adherence to human intestinal epithelium correlated with the colonization of human intestine and production of diarrhoea in human volunteers, providing evidence for the involvement of these adhesins in the pathogenesis of *E. coli* diarrhoea. CFA/I and CFA/II attach in relatively low numbers to human blood monocytes, indicating a scarcity of suitable binding sites on the monocyte surface. The attachment appears to be host-dependent, since monocytes obtained from some individuals do not bind CFA/I [+] or CFA/II [+] bacteria (Avril *et al.*, 1981). CFA/I and CFA/II differ markedly in adhesion to cultured cell lines. CFA/I attach in MR fashion to human intestine (407), Vero, feline embryo and porcine kidney cells (Jann *et al.*, 1981) but not to porcine intestine *in vivo* (Moon *et al.*, 1977), which may be due to the tissue specificity of this adhesin or its poor expression in pig intestine. CFA/II does not adhere (Nagy *et al.*, 1977; Jann *et al.*, 1981). Although CFA/II ETEC strains would be expected to bind to human intestinal cell lines, intestine 407 may lack the appropriate receptor because of its foetal origin.

   Toxigenic strains implicated in human diarrhoeal disease, which show differences in HA patterns from CFA/I and CFA/II and are serologically CFA antigen-negative, attach to human BEC. Thorne *et al.* (1979) demonstrated *in vitro* MR adherence to BEC in 52% of toxigenic strains, while control, non-toxigenic human isolates and toxigenic animal pathogens were not adhesive, suggesting the homologous specificity of adhesins of human ETEC isolates. Adherence to BEC presumably reflects the expression of similar cell surface receptors on BEC and IEC. Indeed, evidence exists for the colonization of the oral cavity in infants with diarrhoea (Thorne *et al.*, 1979). Purified fimbriae from several buccal-adherent ETEC strains attached strongly to BEC *in vitro* in MR fashion, indicating their marked avidity for BEC receptors, and attachment of whole bacteria to buccal cells was inhibited specifically by the Fab fragrant of IgG raised against purified fimbriae of a selected strain of the group (Deneke *et al.*, 1979). ETEC isolates of human origin showed striking specificity for host IEC (McNeish *et al.*, 1975) and failed to adhere to isolated porcine epithelial cells *in vitro* (Wilson and Hohmann, 1974) or to cultured epithelial cell lines (Jann *et al.*, 1981). Using human foetal intestinal tissue, Williams *et al.* (1978) identified a plasmid-mediated, MR, host-specific adhesin in a strain of serotype O26;K60:H11 which differs from CFA/I and CFA/II and requires further characterization.

Animal EPEC, Non-ETEC

E. coli RDEC-1, a rabbit anteropathogen that causes diarrhoea by a non-entero-toxigenic, non-invasive mechanism, attaches to the degenerate microvillous borders of the ileal, caecal and colonic mucosa (Cantey and Blake, 1977; Takeuchi *et al.*, 1978). RDEC-1 is strongly species-specific for rabbit intestinal cells *in vivo* (Cantey and Blake, 1977; Takeuchi *et al.*, 1978; Cheney *et al.*, 1980) and displays *in vitro* specificity for ileal rather than jejunal brush borders from rabbits and subcellular specificity for the apical region of villus epithelial cells (Cheney *et al.*, 1980). Recently the initial site of attachment of RDEC-1 to rabbit intestine was identified as the membranous cells of the lymphoid follicles of the ileal Peyer's patches (Cantey and Inman, 1981; Inman and Cantey, 1983) from which the bacteria are continuously shed into the intestinal lumen and subsequently colonize the small and large intestine. In contrast to ETEC colonization of animal intestine, RDEC-1 was observed to be associated with the degenerate micro-villous epithelial borders (Takeuchi *et al.*, 1978). The mechanism of attachment of this strain is not understood but may involve the glycocalyx of epithelial cells, as suggested by EM studies (Takeuchi *et al.*, 1978; Inman and Contey, 1983). The specificity for Peyer's patches in the initial stages of gut infection may represent adherence to suitable receptors at these sites, or areas with a thin glycocalyx facilitating attachment (Cantey and Inman, 1981; Inman and Cantey, 1983). It has been suggested (Cheney *et al.*, 1980) that the receptors are glycoli-pid or glycoprotein but the chemical nature of the receptor is, as yet, uniden-tified. *In vitro* attachment to isolated rabbit brush borders was not inhibited by mannose or its analogues, fucose and a range of sugars and acetylated sugars (Cheney *et al.*, 1979). An understanding of the *in vivo* pathogenesis of diarrhoea in the rabbit by RDEC-1 is likely to be of importance since the destruction of microvilli with bacterial adherence to the damaged luminal intestinal surface in the absence of bacterial invasion or production of ST or LT has recently been reported in some cases of human diarrhoea caused by EPEC (Ulshen and Rollo, 1980; Clausen and Christie, 1982; Rothbaum *et al.*, 1982). The possible involvement of the shiga-like Vero toxin cannot be excluded.

Human EPEC, Non-ETEC and FEEC

EPEC, non-ETEC isolated from children with diarrhoea and expressing monospecific HA of bovine RBC (Candy *et al.*, 1982), showed MR adhesion to buccal cells from healthy adults, which was reduced under cultural conditions that inhibit fimbrial production. EPEC from other intestinal infections that adhere as bacterial clumps *in vivo* only to damaged surfaces of jejunum, duodenum, ileum or rectum lacking microvilli (Ulshen and Rollo, 1980; Clausen and Christie, 1982; Rothbaum *et al.*, 1982) also attach to HEp-2 cells *in vitro* in characteristic clumps (Clausen and Christie, 1982). The nature of these colonization factors remains to be established.

The adhesion characteristics of MR adhesins of FEEC causing extra-intestinal infections are discussed below. An O18ac:H⁻ FEEC strain, isolated from a case of human diarrhoea, with MR fimbriae (Wevers *et al.*, 1980) of X specificity (Jann *et al.*, 1981) did not attach to cultured epithelial cell lines (Jann *et al.*, 1981), rat peritoneal macrophages or human polymorphonuclear leucocytes (PMN) (Blumenstock and Jann, 1982). It remains to be established whether the expected receptors are present on human enterocytes, UEC and periurethral cells.

Animal and Human EIEC

Only limited information is available for animal EIEC isolates. Some strains isolated from new-born calves and lambs with diarrhoea carry the Vir plasmid. Plasmid-positive donor and recombinant strains attach to brush borders of calf ileal epithelium, to squamous oesophageal epithelial cells and ciliated tracheal epithelium in MR fashion, whereas plasmid-free isolates do not (Morris *et al.*, 1982a). In common with MR adhesins of EPEC, adherence was abolished by growth at 18°C or heat treatment but was retained after treatment with formaldehyde (Morris *et al.*, 1982a). Furthermore, attachment to calf ileal brush borders by a K-12 Vir⁺ recombinant was inhibited by a variety of sugars, including glucose, mannose and their derivatives (Morris *et al.*, 1982a). Mannose inhibition indicated the presence of type 1 fimbriae but the significance of inhibition by glucose is not clear. Glucosamine and mannosamine and their acetylated derivatives reduced attachment more than mannose. Similarly, in the presence of mannose and wheat germ lectin, which is specific for *N*-acetylglucosamine residues, attachment was reduced by 90% (Morris *et al.*, 1982a).

Diarrhoea in humans due to EIEC is characterized by a non-febrile syndrome similar to shigellosis, with predominant colonization of the colon and destruction of the mucosa (Dupont *et al.*, 1971; Tulloch *et al.*, 1973). EIEC may produce a fatal enteritis in guinea-pigs with invasion of the lamina propria. Histological studies of rabbit ileal loops with fluorescein-labelled antibody showed penetration of the intestinal mucosa (Tulloch *et al.*, 1973) and destruction of the villi accompanied by acute inflammation (Dupont *et al.*, 1971).

Extra-Intestinal Isolates

Of strains isolated from extra-intestinal infections, those causing meningitis or bacteraemia have been virtually neglected with regard to their adherence to cells other than RBC. A single bacteraemic strain which displayed MR fimbriae alone adhered poorly to a range of cell lines (Jann *et al.*, 1981). In contrast, numerous reports exist of the cellular adherence characteristics of adhesins of uropathogenic strains with UEC, vaginal and periurethral cells. However, many workers have failed to determine the MS or MR adherence specificities of the adhesins concerned and their results are confusing because of inconsistency in cultural conditions used for HA and cellular adhesion tests. In many cases the cultural conditions have been suboptimal for expression of certain adhesin types. For example, Svanborg Edén and Hansson (1978) claimed that several uropathogenic strains adhered to UEC *in vitro* in MR fashion while their HA type indicated the presence of MS fimbriae. Had the authors tested MR-HA with human RBC, rather than with guinea-pig RBC alone, these strains might have been shown to coexpress MR and MS fimbriae, as in a large proportion of isolates from UTI sources (Hagberg *et al.*, 1981; Parry *et al.*, 1982, 1983; Sussman *et al.*, 1982). Moreover, detection of these adhesins would be enhanced by using BHI agar (Parry *et al.*, 1982) in place of nutrient broth (Svanborg Edén and Hansson, 1978).

MR-fimbriate isolates, whether from unselected UTI or cases of pyelonephritis, cystitis or asymptomatic infection, exhibit a restricted range of cellular adherence but show a strong affinity for human or monkey UEC *in vitro* (van den Bosch *et al.*, 1980; Hagberg *et al.*, 1981; Korhonen *et al.*, 1981; Parry *et al.*, 1982). This was confirmed by binding of ^{125}I-labelled fimbriae to UEC (Korhonen *et al.*, 1980b). Non-haemagglutinating strains from various UTI were non-adhesive *in vitro* (Hagberg *et al.*, 1981; Sussman *et al.*, 1982; Parry *et al.*, 1982). Affinity for human UEC seems much greater among isolates of pyelonephritic origin and this is a reflection of the high percentage of isolates producing MR adhesins (Hagberg *et al.*, 1981). Strains bearing only MR fimbriae do not attach to rat UEC (Korhonen *et al.*, 1981; Parry *et al.*, 1982) although there is a contradictory report of a reputedly MS-only strain displaying weak MR attachment to rat UEC (Korhonen *et al.*, 1981). This may result from poor expression of an MRE adhesin, resulting from unsuitable cultural conditions, detected by *in*

vitro epithelial cell adherence but not by HA tests at room temperature. Therefore, the potential for MR fimbrial expression under suitable environmental
conditions should not be excluded. If this is the case, the conflicting evidence for
MR adhesion to rat UEC may be due to differences in genotypic receptor identity
and/or expression. Spurious attachment to UEC due to urinary mucus associated
with the surface of these cells is contraindicated by the MR attachment of this
strain to BEC (Korhonen *et al.*, 1981).

Pyelonephritis isolates bearing MR adhesins alone also showed a much greater
attachment to periurethral epithelial cells than MS-fimbriate bacteria. The adhesins were confirmed as fimbriae by abolition of adhesion with heat treatment
of the bacteria (Källenius *et al.*, 1980c). In general, these strains adhered in
greater numbers to periurethral cells from UTI-prone than from healthy individuals although there was considerable intra-individual variation in periurethral
cell colonization (Källenius and Winberg, 1978; Källenius *et al.*, 1980c). This
was interpreted as variation in bacterial fimbriation rather than host-specific
hormonal effects (Källenius *et al.*, 1980c).

Since MR adhesins from uropathogenic *E. coli* show variation in their receptor
specificity, MR attachment might be expected to exhibit species and tissue
specificity. P fimbriae, which are prevalent among isolates from children with
pyelonephritis (Källenius *et al.*, 1981a; Väisänen *et al.*, 1981), mediate greater
attachment *in vitro* to UEC from individuals expressing the appropriate receptor
for P fimbriae on the UEC membrane (Källenius *et al.*, 1980a,b; Leffler and
Svanborg Edén, 1980) than to those lacking the receptor (Källenius *et al.*, 1981b;
Svenson *et al.*, 1983). P-specific attachment to UEC has been demonstrated by
numerous methods, including inhibition by pre-incubation of bacteria with
globoside and other reactive glycosphingolipids (Leffler and Svanborg Edén,
1980, 1981) or the synthetic α-D-Gal$p$-(1→4)-β-D-Gal$p$-1-OMe glycoside analogue of the disaccharide α-D-Gal$p$-(1→4)-β-D-Gal$p$ binding moiety of P
glycosphingolipids (Svenson *et al.*, 1982, 1983). Svenson *et al.* (1983) suggested that strains expressing P fimbriae have a predilection for squamous UEC,
whereas X-specific adhesin-bearing bacteria adhere to transitional UEC. However, fluorescence-activated cell sorting (FACS) analysis and fluorescence microscopy of the attachment of a single FITC-labelled pyelonephritis isolate to squamous and transitional UEC from healthy male and female human donors
indicated that adherence to each cell type was similar and that UEC subpopula-
tions did not vary in their density of P receptors (Svenson and Källenius, 1983).
Comparison of UEC from healthy and UTI-prone females indicated an increased
density of P receptors on UEC of the latter, suggesting that the extent of P
receptor expression on cells of the urinary tract may be involved in UTI.

Further evidence for variation of adhesive specificities of different MR adhesins was provided by a study of adherence of urinary tract isolates to a range of
animal tissue culture cells (Jann *et al.*, 1981) including kidney cell lines, ex-

pected to express the P receptor. From the HA data provided, it was possible to deduce whether particular strains expressed P- or X-specific adhesins separately, but not whether they were coexpressed. The data presented indicated that two strains possessing P adhesins either failed to attach to any cell line used or exhibited MR adherence to feline embryo cells and kidney cell lines of simian and porcine origin. This may reflect low expression of P fimbriae by one of the strains or may be a consequence of its clinical origin (bacteraemia). Two strains possessing X adhesins attached to the human intestinal cell line 407 or feline embryo and porcine kidney cells (Jann *et al.*, 1981). Since it is recognized that the X-specific adhesins show heterogeneous receptor specificities, differences in cell specificity were not unexpected. However, attachment to feline embryo and porcine kidney cells might also indicate the presence of P fimbriae in one strain.

In contrast to the adhesive affinity for epithelial cells, MR fimbriae attach poorly to leucocytes *in vitro* in the absence of opsonic antibody (Blumenstock and Jann, 1982; Sussman *et al.*, 1982). Sussman *et al.* (1982) noted that 50–60% of leucocytes incubated with MR-fimbriate isolates remained free of attached bacteria and chemiluminescence by MR-HA strains was reduced (Björksten and Wadström, 1982; Blumenstock and Jann, 1982). The coexpression of MS and MR fimbriae appeared to inhibit the usual association of MS^+ organisms with the leucocytes but did not affect subsequent phagocytosis which depends on attachment to the leucocyte cell surface as a triggering mechanism (Blumenstock and Jann, 1982; Sussman *et al.*, 1982).

Hydrophobicity

Hydrophobic interactions may be involved in the adhesion of bacteria to mucous surfaces, by participating in the specific, lectin-like interaction between fimbriae and cell membrane receptors in hydrophobic areas adjacent to the receptor site. Alternatively, masking of the polysaccharide and lipopolysaccharide by hydrophobic fimbriae may allow larger areas of interaction between the bacteria and cell surfaces and possible formation of hydrophobic bonds.

Bacteria with type 1 fimbriae, which contain a preponderance of non-polar amino acids (Brinton, 1965) have a reduced negative surface charge and increased hydrophobicity compared with non-fimbriate strains. The hydrophobic nature of MS-fimbriate organisms has been confirmed by hydrophobic interaction chromatography (HIC) (Jann *et al.*, 1981; Öhman *et al.*, 1982a); some of the MR fimbriae of animal and human ETEC also appear to be hydrophobic.

Smyth *et al.* (1978) demonstrated the hydrophobicity of $K88ab^+$ and $K88ac^+$ organisms by HIC and this correlated with phenotypic expression of the fimbriae and with MR-HA of guinea-pig RBC but was independent of O and K antigens. These authors also commented that if type 1 fimbriae were present on these strains, they did not contribute to hydrophobicity. Presence of F41, K99 (calf and lamb strains), CFA/I or CFA/II resulted in increased hydrophobicity (Morris *et al.*, 1982b). Honda *et al.* (1983) showed, by a salting-out test, that CFA/I$^+$ or CFA/II$^+$ strains were strongly hydrophobic, as were human ETEC strains lacking CFA antigens. The presence of type 1 fimbriae was not established, but the heat stability of the hydrophobic character in some strains suggested that they might possess MS adhesins, though the cultural conditions employed should suppress MS fimbriation and decrease hydrophobicity. Lindahl *et al.* (1981) compared the MR adhesins of ETEC and ranked them in decreasing order of hydrophobicity as CFA/I, CFA/II, K88, K99 and type 1. In contrast, presence of MR adhesins on urinary tract isolates did not correlate with hydrophobicity since strains with P- or X-specific haemagglutinins gave low percentage HIC values if they were not genotypically MS$^+$ or if they were grown on agar to suppress the production of MS adhesins (Jann *et al.*, 1981). Clearly, the potential importance of this character in adhesion of organisms to mucous surfaces requires further investigation.

Relationship between Haemagglutination and Adhesion

The recognition that HA and adhesion were associated originated from observations that *E. coli* isolated from clinical disease of the intestine or urinary tract were haemagglutinating and adhesive for a variety of cell types; a variable proportion (30–50%) of strains of the normal intestinal flora were non-haemagglutinating and variably adhesive *in vitro* (Evans *et al.*, 1980; Hagberg *et al.*, 1981; Sussman *et al.*, 1982). Whether these form part of the extensive population of commensal *E. coli* adherent *in vivo* to brush border cells of the human large intestine and to the mucus gel normally covering the intestinal epithelium (Hartley *et al.*, 1979) is not established. It is possible that expression of potential adhesins is suppressed *in vitro*, as in the case of the 987P adhesin (Nagy *et al.*, 1977).

Direct association between HA and adhesion is supported by observations on MS adhesins in which characteristic MS-HA of guinea-pig RBC correlates with adhesion to a wide range of cell types. However, such an association for MR haemagglutinins is not universal. The cell attachment characteristics have yet to be determined for MR haemagglutinins such as E8775 (Thomas *et al.*, 1982), subclasses of X-specific haemagglutinins of uropathogenic isolates (Källenius *et al.*, 1980b; Väisänen *et al.*, 1981) and MRE non-fimbrial haemagglutinins from a wide range of sources (Ip *et al.*, 1981). Conversely, some MR adhesins, such as that of the rabbit enteropathogen RDEC-1, have not been rigorously tested for HA activity (O'Hanley and Cantey, 1978). A relationship between HA and adhesion is dependent on the expression of suitable membrane receptor molecules on RBC and epithelial cells, but in many cases the identity of these has not been established. An important exception is the receptor for P fimbriae, which is discussed below. Differences in the affinity of adhesins for membrane receptors on RBC and epithelial cells often result in their detection only in HA tests done at low temperature to prevent thermal elution of the bacteria, which will nevertheless adhere to epithelial cells at 37°C. Similarly, poor expression of MR haemagglutinins may result in their demonstration only by attachment to epithelial cells, because of the greater avidity of MR adhesins for such cells. Thus, appropriate conditions for HA testing and the culture of the test bacteria must be provided in order to establish a potential relationship between HA and adhesion.

Intestinal Isolates

Evidence in support of such a relationship for MR adhesins of animal ETEC has been provided by demonstration of the binding to RBC and epithelial cells of purified adhesins such as K88 (Anderson et al., 1980; Sellwood, 1980), K99 (de Graaf and Roorda, 1982) and F41 (de Graaf and Roorda, 1982; Morris et al., 1982b; Chanter, 1983). However, the serological K88 variants display different adhesive affinities for enterocytes of different pig phenotypes (Bijlsma et al., 1981, 1982) despite a common MRE haemagglutinin of guinea-pig RBC (Guinée and Janson, 1979), suggesting functional differences in receptor specificity not detected by HA. K88ab (adhering to pig phenotypes A, B and C and chicken MR-HA[+]) and K88ac (adhering to pig phenotypes A and B and chicken MR-HA[−]) differed in HA and adhesion characteristics, but K88ad, with the same MR-HA as K88ac, had a different but overlapping enterocyte specificity from the former and adhered to pig phenotypes C and D. Such observations may be resolved by identification of the cell receptors involved in the attachment of each variant. Alternatively, other host factors may be involved and these may explain the specificity of the K88 adhesin for epithelial cells of the homologous host (Gaastra and de Graaf, 1982) which is not readily apparent from HA characteristics.

Evidence against an HA–adhesion association is provided by the adhesin of strain RDEC-1 which is yet to be demonstrated as a haemagglutinin and the non-K99, non-F41, heat-resistant sheep haemagglutinin that does not adhere (Chanter, 1983).

The MR-HA pattern of a range of animal RBC has been widely used to screen human ETEC for adherence properties of potential significance in intestinal colonization (D. G. Evans et al., 1977; Ørskov and Ørskov, 1977; Freer et al., 1978; Gross et al., 1978; Cravioto et al., 1979; D. J. Evans, Jr. et al., 1979; Thorne et al., 1979). In many cases, no simple relationship between MR-HA and adhesion was demonstrable. Some strains possessing CFA/I and CFA/II adhesins did not haemagglutinate indicator RBC (Cravioto et al., 1982). Furthermore, the recently described surface antigen CS3 of an O6:K15:H16 ETEC isolate, although non-haemagglutinating with indicator RBC, was reported to

Table 8.   *Classification of ETEC isolates*
(after Thorne et al., 1979)

Group	HA[a]	Mannose-resistant adherence[b]
I	+[c]	+
II	−	+
III	+	−
IV	−	−

[a]Human, guinea-pig RBC.
[b]Human BEC.
[c]+, positive; −, negative.

attach to human IEC (Smyth, 1982). Thorne *et al.* (1979) subdivided ETEC isolates into subgroups (Table 8), depending on their MR-HA of human and guinea-pig RBC and MR adhesive properties for human BEC. Groups I and IV conformed to the general relationship and fimbriae isolated from a group I strain showed MR attachment to guinea-pig RBC and human BEC (Thorne *et al.*, 1979). However, in groups II and III, HA and adhesion were not associated but the use of a limited range of RBC and of inappropriate culture medium may have led to incorrect assumptions about the HA properties of the strains. The contradiction between presence of MR-HA and failure to attach to BEC may reflect poor expression of suitable receptors on buccal cell membranes. Other observations indicate that CFA/I⁻, CFA/II⁻, non-haemagglutinating human ETEC and EPEC nevertheless produce diarrhoea in human volunteers (Dupont *et al.*, 1971; Levine *et al.*, 1980), which would suggest that the organisms are adherent *in vivo* unless other pathogenetic mechanisms are responsible. One of the ETEC isolates also adhered to the villous surface of guinea-pig ileum (Dupont *et al.*, 1971). Additional studies have shown attachment of MR haemagglutinating EPEC strains to human IEC *in vivo* (Rothbaum *et al.*, 1982) and to isolated BEC *in vitro* (Candy *et al.*, 1982). Confirmation of the HA–adhesion association requires further investigation of the functional receptors on RBC and buccal cell membranes and characterization of the adhesins.

Extra-Intestinal Isolates

There are numerous examples in support of an association between HA and *in vitro* adhesion to UEC, periurethral cells and agglutination of human RBC by MR adhesins of isolates associated with disease of the urinary tract (Källenius and Möllby, 1979; Källenius *et al.*, 1980c; van den Bosch *et al.*, 1980; Hagberg *et al.*, 1981; Korhonen *et al.*, 1981; Parry *et al.*, 1982). Non-haemagglutinating strains were generally non-adhesive (Hagberg *et al.*, 1981; Parry *et al.*, 1982;

Sussman *et al.*, 1982) but there are a few instances in which MR adhesion occurred in an apparently non-haemagglutinating strain (van den Bosch *et al.*, 1980). However, the chemical identification of the receptor for P fimbriae of uropathogenic isolates has confirmed the direct association of P-specific HA by isolated P fimbriae (Korhonen *et al.*, 1982) and adhesion to the same receptor on human UEC, periurethral cells (Leffler and Svanborg Edén, 1980, 1981; Svenson *et al.*, 1983; Svenson and Källenius, 1983) and monkey kidney tissue (Svenson *et al.*, 1983). Jann *et al.* (1981) failed to show a direct correlation between HA pattern and adhesion of extra-intestinal isolates to cultured cell lines. A P-fimbriate bacteraemia isolate was non-adherent, whereas a P-fimbriate UTI isolate adhered to pig and monkey tissue, presumably indicating recognition of P receptors. Other isolates from urine or faeces haemagglutinated but did not adhere. X-specific adhesins which gave MR-HA of p̄ RBC seemed to be correlated with adherence to the human intestinal cell line 407. Organisms with X-specific MR-HA activity were reported with higher frequency from cystitis than from pyelonephritis (Väisänen *et al.*, 1981) and since these exhibit heterogeneous RBC receptor specificities it will be important to establish whether they are related to attachment properties for urinary tract cells.

Receptors

Unlike the extensive studies on binding of toxins and hormones to cell membranes there is relatively little information on the nature of the receptors for the adhesins of *E. coli*. The receptor for type 1 fimbriae clearly appears to contain mannose residues (Salit and Gotschlich, 1977a,b; Ofek *et al.*, 1977; Ofek and Beachey, 1978; Öhman *et al.*, 1982b). Although the nature of the carrier molecule or molecules has not yet been determined, one study has indicated, at least in the rat bladder, that the receptors involved in MS adherence are glycolipids (Davis *et al.*, 1981). The cell membrane receptors for K88 and to a lesser extent K99 and CFA/I have received some attention but the receptor structures have not been characterized with certainty. The only clearly defined receptor structure for an MR adhesin is that for P fimbriae of uropathogenic *E. coli*, which has recently been shown to be a digalactose component common to the glycosphingolipids of the human P blood group antigens (Källenius *et al.*, 1980a,b; Leffler and Svanborg Edén, 1980).

Studies on membrane receptors for MR adhesins fall into four main categories: (1) inhibition of adhesin binding by modification of receptor on the target cell by enzymatic or chemical treatment, (2) inhibition of adhesin binding with potential receptor analogues, sometimes with modification by enzyme or chemical treatment, (3) agglutination of normally unreactive cells or latex coated with potential receptor analogues and (4) comparison of receptive target cells with similar cells genotypically or phenotypically deficient in the receptor. Such studies have utilized both host epithelial cells and RBC as target cells, sometimes leading to conflicting results.

Animal and Human ETEC

K88 adhesin. Jones and Rutter (1974) found that the contents of the small intestine of germ-free, colostrum-deprived piglets contained an HA inhibitor which had the properties of either a glycoprotein or polysaccharide which was heat-stable, resistant to proteolytic enzymes and susceptible to oxidation with periodate. This indicated that the K88 receptor in the cell membrane may be a glycoprotein and the possibility that the determinant structure was at or near the terminal non-reducing ends of heterosaccharide residues was suggested. Using inhibition of HA of guinea-pig RBC as indicator, Gibbons *et al.* (1975) tested a number of potential receptor analogues including intact and chemically modified mucous glycoproteins, simple sugars and some of their glycosides and two glycosaminoglycans. Haemagglutination was inhibited by all epithelial glyco-proteins examined but not by structurally different serum glycoproteins, glycosaminoglycans or monosaccharides. Mucus glycoprotein from the pig in-testine was inhibitory in the HA test, indicating that the substance contains the same structure as the intestinal receptor. Chemical modification of the heterosac-charides of the mucous glycoproteins by mild hydrolysis, periodate oxidation or the Smith degradation procedure showed that the unsubstituted β-D-galactosyl residue appeared to be an important feature of the structural requirements of the binding site of the K88 adhesin. Exposure of a terminal β-D-galactosyl residue on human A substance increased its inhibitory activity and similar treatment of the serum glycoprotein fetuin, which is inactive in its native form, rendered it inhibitory. Removal of the terminal α-L-fucosyl-1→2-β-D-galactosyl residue reduced HA inhibition of human H substance. One puzzling feature of this work was that high inhibitory activity was exhibited by some submaxilliary glycopro-teins in which the β-D-galactosyl residue was absent or not prominent. The authors concluded that in some cases inhibition of haemagglutination was non-specific. In this study the lack of inhibition by monosaccharides or by their α- or β-methyl glycosides was taken as an indication that the inhibitor site is larger than a single sugar residue. However, intact heterosaccharide side chains cleaved from one inhibitory glycoprotein (pig submaxillary mucin) were also inactive, suggesting that a minimum molecular size and possibly other stereochemical requirements may be necessary.

In support of these studies, Sellwood (1980) demonstrated a reduction in binding of purified ^{125}I-labelled K88 to porcine brush borders with stachyose (α-D-galactosyl-α-D-galactosyl-α-D-glucosyl-β-D-fructose) and galactan (a poly-mer of D- and L-galactose). However, the use of D-galactose-specific lectins concanavalin A, *Ricinus communis* 120 and soybean agglutinin resulted in poor inhibition, while a fucose-binding protein lectin inhibited binding quite substantially.

Experiments by Anderson *et al.* (1980), who used a similar system to test the inhibition of binding of K88 by sugars and glycoproteins, suggested a role for *N*-acetylhexosamines in the binding process. *N*-Acetylglucosamine, *N*-acetylgalactosamine and *N*-acetylmannosamine exerted the greatest influence although this was substantially less than some of the glycoproteins tested. In contrast to the study of Gibbons *et al.* (1975) ovine submaxillary glycoprotein (OSM) did not inhibit binding. Two types of porcine submaxillary glycoprotein (PSM) were tested, A⁻ PSM, which does not inhibit agglutination of human blood group A RBC by anti-A serum, and A⁺ PSM, which is inhibitory. Neither native A⁻ PSM nor its desialylated derivative inhibited K88 binding; however, desialylated OSM, which has exposed terminal *N*-acetylgalactosamine residues, was inhibitory. Native A⁺ PSM has similar terminal residues but caused aggregation of K88 preparations, preventing its use in assays; however, the desialylated material inhibited K88 attachment. α_1-Acid glycoprotein was also inhibitory but interestingly its asialoagalactosyl derivative was a more potent inhibitor, indicating that *N*-acetylglucosamine residues may also be involved. The degree of inhibition observed for amino sugars is in contrast to the studies of Gibbons *et al.* (1975) and Sellwood (1980). It should be noted that Sellwood (1980) demonstrated a low inhibitory effect by galactosamine, glucosamine and mannosamine and stated that a similar effect could be demonstrated with other compounds possessing free amino groups, e.g. ethanolamine, tris and glycine. One of the authors (S.H.P.) has observed a similar inhibitory effect in haemagglutination tests carried out with K88⁺ organisms in the presence of a tris buffer (S. H. Parry, unpublished observation). Furthermore, studies on the receptor for mannose-sensitive adhesins in rat bladder epithelial cells has provided evidence that *N*-acetylneuraminic acid acts as a non-specific inhibitor of bacterial adherence (Davis *et al.,* 1981).

Also using an ¹²⁵I-labelled K88 antigen preparation, Kearns and Gibbons (1979) demonstrated that plasma membranes prepared from brush borders of receptor-positive pigs had lost 78% of their receptor activity. The K88 binding activity of such membrane preparations was enhanced by incubation with supernatant fractions obtained in the preparation of these membranes. Binding to membrane preparations from pigs lacking the K88 receptor was similarly enhanced. Similar experiments with supernatants from receptor-negative pigs did not enhance uptake by adhesive or non-adhesive membrane preparations. Analysis of membrane fractions revealed consistent differences in glycolipid profiles which corresponded with genotypic differences. Thus this was suggested to be an indication of the glycolipid nature of the K88 receptor.

One possibility that should not be ruled out is that more than one receptor is involved. Evidence in support of this comes from two studies. First, K88ab organisms, in addition to agglutination of guinea-pig RBC, show specific HA of chicken RBC, a reaction not observed with K88ac organisms (Parry and Porter, 1978). Inhibition studies with determinant-specific antisera showed that while antisera specific for the 'a' determinant inhibited adhesion of K88ab organisms to pig enterocyte brush borders they were ineffective in inhibiting the agglutination of chicken RBC. Similar experiments with 'b' specific sera gave inhibition in both tests. This suggests the possibility that selectivity of the K88ab antigen for the chicken RBC involves adhesion via a distinct receptor by the 'b' determinant.

The second line of evidence comes from recent studies on the adhesion of the three antigenic variants of K88 to enterocytes derived from a large number of randomly selected pigs (Bijlsma et al., 1981, 1982). Five different porcine phenotypes were distinguished, each of which showed a different receptive pattern. One phenotype was susceptible to all three variants, three phenotypes were susceptible to only one or two variants and one phenotype was resistant. Receptor-blocking tests carried out by incubation of brush borders of the different phenotypes with partially purified fimbriae of the different variants prior to adhesion tests indicated that the receptor sites for the K88ab and K88ac adhesins have a closer similarity than the receptor site for the K88ad adhesin. Although this suggests separate receptors for the three variants, the authors favoured the hypothesis that a single receptor site existed which may undergo modification. It is clear that characterization of the K88 receptor must await further studies and the most promising avenue appears to be the examination of brush-border membranes from the different adhesive and non-adhesive pig phenotypes.

In summary, it is clear that the binding of K88 to the pig gut is not just a non-specific phenomenon but requires a highly specific interaction with receptors that are present in many but not all animals. Although the exact nature of the receptor has not been defined, there is strong evidence that glycoside residues are involved on glycolipids and/or glycoproteins. Some of the conflicting data may result from the use of different test systems which utilize both RBC and enterocytes and different antigenic types of K88. The source and type of K88 antigen were not reported in many studies.

987P adhesin. The attachment of 987P-fimbriate *E. coli* to small intestinal villous epithelial cells and brush borders of adult female but not infant rabbits has been described (Dean and Isaacson, 1982). A receptor-containing fraction that caused aggregation of strain 987 was released into solution when brush borders were stored at 4°C. The sensitivity of receptor activity to periodate oxidation and pronase digestion was consistent with a glycoprotein-containing molecule. The finding that K99 + strains adhered to both infant and adult epithelium and brush borders indicated that receptors for K99 and 987P are distinct.

K99 and CFA adhesin. There have been no reports of the epithelial cell receptors for K99 and CFA/I and CFA/II adhesins, although one report has examined the erythrocyte receptors for these antigens (Faris *et al.,* 1980). HAI tests with crude ganglioside preparations showed that both monosialoganglio-sides (type III) and disialogangliosides (type II), but not cerebrosides, inhibited HA by both K99 and CFA/I strains. Strains possessing CFA/II and an MR-HA urinary tract isolate were not inhibited by crude gangliosides or cerebrosides. Experiments with highly purified gangliosides showed that only GM_2 inhibited MR-HA of K99 and CFA/I strains and only then at high levels (10 mg/ml), above the critical micellar concentration. It was postulated that both glycolipids and glycoproteins may play a part in the binding of K99 and CFA/I to RBC since trypsin treatment did not affect HA by either CFA/I or K99 antigen but HA of human erythrocytes by CFA/I was competely eliminated after pronase treatment. Treatment of RBC with neuraminidase inhibited the HA of human RBC by CFA/I strains but had no effect on the HA of sheep RBC by K99 strains. In contrast to the studies of D. G. Evans *et al.* (1979), Faris *et al.* (1980) were unable to demonstrate HAI of human erythrocytes by CFA/I and CFA/II strains with *N*-acetylneuraminic acid. Although the latter studies provided evidence for a possible receptor for CFA/I and K99 on RBC, it is possible that different results may be obtained with enterocytes derived from the host species, as was observed in studies with K88. Furthermore, with respect to the host cell specific-ity generally demonstrated by MR adhesins, it is unlikely that CFA/I-bearing human pathogens will share the same receptor as K99-bearing strains causing enteritis in sheep and calves.

Extra-Intestinal Isolates

As discussed earlier, a high proportion of urinary tract isolates of *E. coli* give MR-HA of human RBC. Källenius *et al.* (1980b) first described the agglutina-tion of different phenotypes of human erythrocytes by MR-HA strains isolated from patients with pyelonephritis. No decrease was observed in agglutinability of erythrocytes from cord blood of neonates, in which antigen I is poorly expressed. Furthermore, no significant differences were observed between agglutination of groups O, B and A1 RBC, in which antigen H is less well expressed. However, all but one of the test strains failed to agglutinate RBC from individuals of the rare p̄ phenotype which lack the three P blood group antigens P, P_1 and P^k (Table 9). The different P blood group antigens are correlated to glycosphingolipids of known structures (Table 10). They consist of a lipid moiety (ceramide) anchored in the cell membrane and a carbohydrate residue exposed on the surface.

The terminal non-reducing sugars of the P^k, P_1 and P antigens (α-D-galactose, α-D-galactose and *N*-acetyl-D-galactosamine, respectively) failed to inhibit HA

Table 9. *Human P blood group system*

Phenotype	Antigens	Prevalence
P_1	P, P^k, P_1	75%
P_2	P, P^k	25%
P^k_1	P^k, P_1	Rare
P^k_2	P^k	Rare
$\bar{p}$	—	Rare

of human RBC by pyelonephritogenic strains. Among the saccharides tested only the reduced oligosaccharide isolated from trihexosyl ceramide was inhibitory; the oligosaccharide from globotetraosyl ceramide was not inhibitory at the concentrations tested. Further studies (Källenius *et al.*, 1980a) demonstrated that $P_2{}^k$ RBC which contain only the P^k antigen were agglutinated to the same extent as cells of the P_1 and P_2 phenotypes and it was suggested that the P^k glycosphingolipid was the receptor for pyelonephritic *E. coli*. In the same study the nature of the receptor was investigated by HAI tests with the synthesized disaccharide α-D-Gal$p$-$(1\rightarrow4)$-β-D-Gal$p$-1-O-Ø-NO$_2$, which was found to be inhibitory at concentrations of less than 1 mM. These findings were extended in further studies in which binding of pyelonephritogenic *E. coli* to UEC of the P phenotype was blocked by the synthetic disaccharide (Källenius *et al.*, 1981b). It was suggested that the minimal receptor structure recognized by pyelonephritic strains of *E. coli* is the α-D-Gal$p$-$(1\rightarrow4)$-β-D-Gal$p$ structure. It was also pointed out that this terminal disaccharide occurred on the P_1 antigen.

Independent studies carried out at the same time (Leffler and Svanborg Edén, 1980, 1981) provided three lines of evidence that glycolipids in the globoseries may act as receptors for bacteria attaching to epithelial cells and agglutinating RBC: (1) the adhesion of uropathogenic strains to UEC was inhibited by globoseries glycolipids, (2) binding of bacteria to epithelial or RBC target cells was related to their content of globoseries glycolipids and (3) binding of bacteria to previously unreactive target cells could be induced by coating with globoseries glycolipids. Thus a non-acid glycosphingolipid fraction from human urinary sediment UEC, small intestinal cells and a number of glycolipids from human RBC was tested for ability to inhibit attachment of *E. coli* to UEC. The glycolipids from UEC but not from the small intestinal mucosa were inhibitory. In contrast to the HAI studies of Källenius *et al.* (1980b) the most effective inhibitor of bacterial attachment was found to be globotetraosyl ceramide, although globotriaosyl ceramide also had some activity. It was also demonstrated that glycolipid fractions from erythrocytes of the P_1 and P_2 phenotypes but not fractions from the $\bar{p}$ phenotype were inhibitory in attachment studies. Guinea-pig RBC which were not normally agglutinated by the test strain became susceptible to bacterial HA after

Table 10. *Receptors and potential receptors for P fimbriae*

Receptor or potential receptor	Nature of receptor	Source	Structure[a]
P^k antigen (trihexosyl ceramide)	Glycolipid	Human RBC	Gal-α-(1→4)-Gal-β-(1→4)-Glc-β-(1→1)-Cer
P antigen (globoside)	Glycolipid	Human RBC	Gal-NAc-β-(1→3)-Gal-α-(1→4)-Gal-β-(1→4)-Glc-β-(1→1)-Cer
P$_1$ antigen	Glycolipid	Human RBC	Gal-α-(1→4)-Gal-β-(1→4)-Glc-NAc-β-(1→3)-Gal-β-(1→4)-Glc-β-(1→1)-Cer
Forssman antigen	Glycolipid	Sheep RBC	Gal-NAc-α-(1→3)-Gal-NAc-β-(1→3)-Gal-α-(1→4)-Gal-β-(1→4)-Glc-β-(1→1)-Cer
P$_1$ substance	Glycoprotein	Hydatid cyst fluid	Gal-α-(1→4)-Gal-β-(1→4)-Glc-NAc-[b]

[a]Common receptor structures are underlined.
[b]Terminal trisaccharide.

coating with globoside and there was a rough correlation between agglutinability of different RBC and their globoside content (Leffler and Svanborg Edén, 1980, 1981). It was suggested that the receptor was the α-D-Gal$p$-(1→4)-β-D-Gal$p$ component, a common feature of the globoseries glycolipids recognized by urinary isolates (Kallenius $et\,al.$, 1980b; Leffler and Svanborg Edén, 1980, 1981). Confirmation of the receptor structure recognized by P-specific adhesins comes from a study in which the receptor-specific glycoside containing α-D-Gal$p$-(1→4)-β-D-Gal$p$ was attached to latex particles (Svenson $et\,al.$, 1982). Strains of $E.\,coli$ possessing the P-specific adhesin gave a strong specific agglutination of coated but not uncoated particles.

Strains of $E.\,coli$ recognizing the globoseries glycolipids were shown to possess fimbriae (Kallenius $et\,al.$, 1980b) (termed P fimbriae; Kallenius $et\,al.$, 1981a; Väisänen $et\,al.$, 1981) which were morphologically similar to type 1 fimbriae (Korhonen $et\,al.$, 1982). The criteria used for identification of P fimbriae are (Källenius $et\,al.$, 1981a; Väisänen $et\,al.$, 1981) (1) ability to agglutinate human P$^+$ RBC and inability to agglutinate bovine and guinea-pig RBC, (2) inability to agglutinate p̄ RBC, (3) ability to agglutinate horse or guinea-pig RBC coated with trihexosyl ceramide or globoside but not control uncoated cells of these species and (4) total inhibition of haemagglutination by synthetic α-D-Gal$p$-(1→4)-β-D-Gal$p$.

In studies on $E.\,coli$ isolated from children with pyelonephritis, Väisänen $et\,al.$ (1981) found some strains causing HA of human P$_1$ RBC which also agglutinated p̄ RBC but not globoside-coated guinea-pig cells. These strains, it was suggested, possess a distinct adhesin which recognized a non-P receptor (termed X-specific). Some strains agglutinated both p̄ RBC and globoside-coated RBC and were suggested to have both P- and X-specific adhesins. However, the nature of the X-specific adhesin or adhesins has yet to be determined. A glycoprotein receptor, the M blood antigen, has also been reported as a non-P receptor for adhesins of a single pyelonephritis isolate (Väisänen $et\,al.$, 1982a).

In recent studies in this laboratory we have used a naturally occurring glycoprotein for the identification of P- and X-specific adhesins (Parry $et\,al.$, 1984). The human P$_1$ blood group substance was first described in hydatid cyst fluid by Cameron and Stavely (1957), who demonstrated its ability to inhibit the agglutination of RBC by anti-P$_1$ sera. Subsequently, the P$_1$ substance was purified and characterized and was shown to be a glycoprotein (Morgan and Watkins, 1964) with glycosidic chains containing a terminal trisaccharide sequence α-D-Gal$p$(1→4)-β-D-Gal$p$-β-D-GlcNAc- identical to that of the P$_1$ blood group antigen (Table 10) (Cory $et\,al.$, 1974). Crude hydatid cyst fluid thus provides a convenient source of receptor material for use in the identification and characterization of P fimbriae. Haemagglutination inhibition tests with hydatid cyst fluid have proved to be valuable adjuncts to conventional P-typing methods and have allowed identification of strains bearing X-specific adhesins (no inhibition), X- and P-specific adhesins (partial inhibition) and P-specific adhesins (complete inhibition). Use of this test with a number of uropathogenic isolates and a large panel of RBC from different species has demonstrated that strains with X-specific adhesins give a number of distinct haemagglutination patterns in which different combinations of RBC are reactive (D. M. Rooke and S. H. Parry, unpublished work). This indicates that the X-specific adhesins are probably a heterogeneous group which may recognize a number of different receptors.

P receptors on host tissues. Numerous studies have demonstrated the ability of MR- and P-fimbriate *E. coli* to attach to UEC. Furthermore, a high proportion of *E. coli* isolated from children with pyelonephritis and a smaller but significant proportion of isolates from women with cystitis bear P fimbriae, indicating the potential role of this adhesin in the pathogenesis of UTI (Väisänen *et al.,* 1981; Källenius *et al.,* 1981a). That P receptors are involved in the attachment of organisms is indicated by two lines of evidence. First, there is the ability of the synthetic α-D-Gal$p$-(1→4)-β-D-Gal$p$- disaccharide receptor structure to inhibit the attachment of P-fimbriate organisms to UEC and the identification of globo-series glycolipids in the membrane of UEC (Leffler and Svanborg Edén, 1980, 1981). More recently the presence of P receptors on uroepithelial cells has been confirmed by fluorescence-activated cell sorting analysis in which specific binding of FITC-labelled P-fimbriate *E. coli* was demonstrated (Svenson and Källenius, 1983). Second, globoseries glycolipids have also been identified in human kidney (Mårtensson, 1963, 1966; Marcus and Janis, 1970). Ceramide tetrahexosides (globoside), ceramide trihexosides (trihexosyl ceramide), ceramide dihexosides (lactosyl and digalactosyl ceramide) and monohexosides have all been shown to be present in kidney tissue in a descending order of concentration (see Table 10) (Mårtensson, 1966). It is interesting to note that digalactosyl ceramide, which is not a P antigen, has the α-D-Gal$p$-(1→4)-β-D-Gal$p$ structure. It was also reported that the levels of glycolipids are much higher in juveniles than adults.

Globoside is found at higher concentrations in the kidney than other glycolipids; it is present in the plasma membrane of epithelial cells of the proximal, but not the distal, convoluted tubule (Marcus and Janis, 1970). Recent studies have examined the binding of FITC-labelled P-fimbriate organisms to cryostat sections of female baboon kidney (Svenson and Källenius, 1983). Specific binding of organisms was observed and revealed a high P receptor density on the inner leaflet of blood vessels and on the tubular basement membrane. Adherence of organisms to baboon UEC was also inhibited by synthetic P receptor. It appears, therefore, that receptors are expressed at appropriate sites for colonization of the kidney and the lower urinary tract.

Role of Adhesins in the Pathogenesis of Mucosal Infections

Animal ETEC

K88 adhesin. In neonatal diarrhoea in pigs caused by enterotoxigenic *E. coli* adhesion to the mucosal surface of the intestinal tract was first recognized as an important feature in the pathogenesis of infection. Smith and Linggood (1971) studied strains containing different combinations of conjugative plasmids encoding α-haemolysin (Hly), enterotoxin (Ent) and K88 antigen which were examined for their ability to produce diarrhoea when given orally to naturally reared and recently weaned pigs. Removal of the K88 plasmid from one strain was accompanied by the loss of its diarrhoea-producing capacity while removal of the Hly plasmid had no effect. Introduction of an Ent plasmid into a K88$^+$ strain enhanced its ability to produce diarrhoea but after removal of the K88 plasmid it became non-enteropathogenic. A non-pathogenic wild strain was also rendered enteropathogenic by the introduction of K88 and Ent plasmids and, while strains bearing both plasmids produced the most severe diarrhoea, K88$^+$ Ent$^-$ forms produced milder diarrhoea and K88$^-$ Ent$^+$ and the K88$^-$ Ent$^-$ strains produced no observable effect. Bacteriological examination of the pigs revealed that the K88 antigen enabled the organisms to proliferate high in the small intestine, where the gut flora is minimal and the intestine is particularly sensitive to the effects of enterotoxin (Smith and Halls, 1967; Hohmann and Wilson, 1975). Later studies (Smith and Huggins, 1978) also demonstrated that a non-pathogenic strain (**O9:K36:H**19) could be rendered enteropathogenic for pigs by introduction of K88 and Ent. However, when these plasmids were introduced into *E. coli* K-12 it remained non-pathogenic and was unable to colonize the small intestine, indicating that other factors such as O and K antigens may also play an important role in pathogenesis.

The role of the K88 antigen in pathogenesis of neonatal enteritis has been confirmed in a number of further studies. Jones and Rutter (1972) examined the ability of K88$^+$ and K88$^-$ strains to colonize the intestinal mucosa *in vivo* by immunofluorescence examination of tissue sections from infected pigs. It was shown that K88$^+$ bacteria adhered to the mucosa of the anterior small intestine whereas K88$^-$ organisms were distributed throughout the lumen. Similar findings were reported by Hohmann and Wilson (1975), using immunofluorescence and transmission and scanning electron microscopy. In infection studies K88$^+$ strains killed 50% of conventionally reared pigs compared to only 3% mortality in a group infected with a K88$^-$ strain (Jones and Rutter, 1972). In contrast, in gnotobiotic animals both strains were equally virulent, indicating that in some circumstances enterotoxic activity is sufficient to produce infection. However, gnotobiotic animals lack a normal gut flora which provides competition for pathogens and such animals remain in an enclosed environment heavily contaminated with the infecting strain, resulting in unnatural infection conditions. Such experiments should therefore be interpreted with some caution.

987P adhesin. While strains of *E. coli* bearing the K88 antigen are commonly isolated from cases of porcine diarrhoea, strains that lack the K88 antigen are also common. Several K88$^-$ strains have been shown to colonize the ileum of the pig and layers of adherent bacteria could be observed by immunofluorescence and electron microscopy studies (Bertschinger *et al.*, 1972; Hohmann and Wilson, 1975; Nagy *et al.*, 1976; Moon *et al.*, 1977). In one study a strain of *E. coli* (987) which was able to colonize the intestine was shown by negative staining to produce structures resembling type 1 fimbriae, although these could not readily be observed *in vivo* (Moon *et al.*, 1977). This was confirmed by Isaacson *et al.* (1977), who compared the colonizing and adhesive attributes of enterotoxigenic, acapsular and/or non-fimbriate mutants from a K88$^-$ enteropathogenic strain with the fimbriate parents. Acapsular, non-fimbriate mutants from three different colonizing strains lost their ability to colonize the ileum of new-born piglets. Acapsular, fimbriate and capsular, non-fimbriate mutants derived from 987 both lacked the ability to colonize the ileum. It was concluded that strain 987 may require both capsule and fimbriae although other strains require only fimbriae. The *in vivo* production of fimbriae by *E. coli* 987 was also demonstrated.

In infection studies with pathogenic and non-pathogenic forms of *E. coli*, Smith and Huggins (1978) demonstrated that the 987P fimbrial antigen was a potent factor in the colonization of the small intestine of piglets but not calves or lambs. They showed that unlike K88$^+$ forms, which proliferate in the anterior small intestine, strains with 987P colonization factor proliferated only in the posterior small intestine or to a much greater degree in the posterior than the anterior small intestine. Wild *E. coli* strains lacking 987P, K88 or K99 did not proliferate at all.

K99 adhesin. In studies on enterotoxins of *E. coli* derived from neonatal diarrhoea in lambs and calves, Smith and Linggood (1972) first described a common K antigen produced by calf and lamb enterotoxigenic strains which was encoded by a plasmid designated Kco. Strains lacking this K antigen failed to proliferate in the small intestine and it was suggested that this was a colonization factor responsible for adhesion to the intestine. The antigen now termed K99 (Ørskov *et al.*, 1975) was confirmed to be a colonization factor and strains bearing this antigen were able to proliferate in the posterior small intestine and less well in the anterior small intestine (Smith and Huggins, 1978). A non-pathogenic strain was rendered enteropathogenic for calves and lambs by introduction of K99 and Ent, although, as shown for the K88 antigen, introduction of adhesin and Ent into *E. coli* K-12 failed to induce enteropathogenicity.

Moon *et al.* (1977) reported that several strains of *E. coli* isolated from pigs possessed the K99 antigen previously reported only in strains of lamb and calf origin. The strains caused profuse diarrhoea in new-born piglets and were shown by immunofluorescence to adhere to the villous epithelium of the ileum. However, a K99$^+$ strain from a calf was unable to colonize the pig ileum. This is in contrast to the findings of Smith and Huggins (1978), who produced experimental diarrhoea in pigs with a K99$^+$ *E. coli* isolated from a calf with diarrhoea.

These apparent differences in the ability of K99$^+$ organisms to produce infection in different hosts suggest that other factors may be involved and it is likely that these inconsistencies may reflect the presence of multiple adhesins with different host and cell specificities.

F41 adhesin. The F41 antigen, which appears to be restricted to the O101 and O9 serogroups, is usually found in association with the K99 antigen, although it may occur in the absence of K99 in some isolates from pigs, calves and lambs (Chapter 3). Morris *et al.* (1982b) examined the infectivity of a mutant from strain B41, which lacked the K99 antigen, in germ-free pigs. The piglets developed diarrhoea within 16 h and scanning electron microscopy showed groups of bacteria adherent to the microvilli of villous enterocytes. The *in vivo* production of the F41 antigen was demonstrated by immunofluorescence. It was indicated that in routine monitoring of field isolates of *E. coli,* F41 has been detected only on strains isolated from outbreaks of neonatal diarrhoea, although its precise role in the pathogenesis of neonatal diarrhoea remains to be fully established.

Human ETEC

CFA adhesins. The role of CFA/I and CFA/II in mediating intestinal adhesion and colonization has been demonstrated with an intact infant rabbit diarrhoea model in which adhesion to the mucosal surface was shown by immunofluorescence and the importance of the adhesin in pathogenesis was indicated by the ability of an antiserum specific for CFA to prevent diarrhoea (Evans *et al.*, 1975, 1978a; Evans and Evans, 1978). Because of the host specificity of MR adhesins it is important to use the host species in experiments to determine the role of adhesins in the infectious process. Although this has hampered investigation of the role of CFA in humans, Evans *et al.* (1978b) reported studies in which groups of volunteers ingested a CFA/I$^+$ strain of *E. coli* or a CFA/I$^-$ mutant. The volunteers were examined for clinical signs and excretion pattern; six of seven individuals who received a dose of 10^8 CFA/I$^+$ bacteria had overt clinical symptoms and all persistently excreted organisms for the 8 days of the study. In contrast, none of the group receiving the CFA/I⁻ mutant showed clinical signs and the organism ceased to be excreted after 4 days.

Urinary Tract Infection

There has been substantial interest in recent years in establishing the importance of adhesins in the pathogenesis of UTI. This problem has been examined in a variety of ways and the areas of study fall into three main types: (1) demonstration of adhesins specific for host cells on organisms isolated from UTI, (2) examination of the association of adhesins with particular types of UTI and (3) use of infection models to establish the role of adhesins *in vivo*.

Assessment of the importance of adhesins in UTI is complex because of the nature of the disease. The term UTI encompasses a number of distinct diseases which have two features in common: they arise by the ascending route from organisms originating in the bowel and a central phase in the infection is significant bacteriuria which may or may not give rise to pyelonephritis (Hanson *et al.*, 1977). The infection may, therefore, progress through a number of stages involving colonization of different surfaces and during which different virulence properties of the organism may assume importance. Adhesion may play a role in several phases of colonization. First, attachment may be important in the initiation of urinary infection by allowing colonization of introital and periurethral areas which then provides a reservoir of infective organisms. Ascent of bacteria into the bladder against the flushing action of urine flow and possibly the maintenance of bacterial numbers, once infection has been established in the bladder, may also require this property. Finally, in pyelonephritis, ascent of the ureter and colonization of the kidney are also likely to involve bacterial attachment to mucosal surfaces.

There is substantial evidence that isolates bearing MR adhesins can attach to squamous cells derived from the introitus, periurethral area and urethra and to transitional cells from the bladder. Furthermore, UEC derived from urine possess the globoseries glycolipids which act as receptors for P fimbriae (Leffler and Svanborg Edén, 1981; Svenson *et al.*, 1983). Studies with FITC-labelled P-fimbriate *E. coli* have also demonstrated the affinity of P fimbriae for baboon kidney tissue, indicating the presence of P receptors. Our own studies have also shown that most organisms bearing only X-specific adhesins attach to squamous UEC (unpublished data). Organisms bearing type 1 fimbriae have a specific affinity for urinary slime containing the mannose-rich Tamm-Horsfall glycoprotein (Ørskov *et al.*, 1980b; Parry *et al.*, 1982). It has been argued that this may enhance removal of organisms bearing type 1 fimbriae by acting as a trapping mechanism by which organisms are voided with the mucus (Ørskov *et al.*, 1980a).

Production of an additional MR adhesin with a strong affinity for epithelium would appear to offer significant advantages to an organism and a two-phase attachment process may occur (Ørskov et al., 1980b). Certainly, a high proportion of isolates from lower UTI that express MR adhesins also produce an MS adhesin (Parry et al., 1983). Furthermore, in vivo studies in a mouse model have shown that for bacterial persistence in the kidney adhesins specific for globo-series glycolipids alone were sufficient, whereas strains with both MS and P adhesins remained in higher numbers in the bladder than those with either adhesin alone (Hagberg et al., 1983a; Svanborg Edén et al., 1983a). However, it is also possible that attachment to urinary mucus by MS adhesins alone might maintain organisms in the bladder in the absence of MR adhesins. This is supported by the in vivo mouse study in which a strain bearing an MS adhesin was found in significantly higher numbers in the bladder than a strain bearing only the P adhesin (Hagberg et al., 1983a; Svanborg Edén et al., 1983a). In addition, MS adhesin has been shown to be present on a high proportion of organisms isolated from patients with asymptomatic bacteriuria compared with isolates from healthy individuals, indicating that these strains can colonize the lower urinary tract but may not possess other virulence characteristics necessary to produce a symptomatic infection (Parry et al., 1983).

A strong correlation between adhesive capacity and severity of infection has been reported by Svanborg Edén et al. (1983a), although it appears to be the type of adhesin that is important. A high proportion of MR-HA strains has been reported in E. coli strains isolated from symptomatic infections of the lower urinary tract compared to asymptomatic isolates (Svanborg Edén et al., 1981; Parry et al., 1983) and normal faecal isolates (Svanborg Edén et al., 1981). However, the prevalence of strains with MR-HA activity was considerably higher in one study (Parry et al., 1983). The incidence of P fimbriae has been studied intensively in strains implicated in acute pyelonephritis in children. There seems to be a strong correlation between possession of P fimbriae (81–91%) compared with acute cystitis (19%), asymptomatic bacteriuria (14%) and faecal isolates from healthy children (7%) (Väisänen et al., 1981; Källenius et al., 1981a). However, the percentages reported by Leffler and Svanborg Edén (1981) (92% acute pyelonephritis, 80% acute cystitis, 64% asymptomatic UTI, 50% faecal) for similar groups of bacteria suggest that the adhesin may also play a role in lower UTI. Adhesins with X specificity have been demonstrated independently and in conjunction with P fimbriae and occur at a much lower frequency in strains isolated from pyelonephritis (19%, 6%) (Väisänen et al., 1981; Korhonen et al., 1982) but predominate in faecal strains isolated from healthy patients.

When examining the significance of adhesion in UTI it is important to consider other factors that may influence any observed association. In a study on the adhesins of strains isolated from pregnant women with asymptomatic bacteriuria it was recently reported that there was a significant association between infection with strains bearing MR adhesins and a past history of infection (Parry *et al.*, 1983). Thus in pregnant women with an infection and a past history of UTI there is a four-fold greater chance that this infection is due to an MR-HA-bearing organism than in pregnant women without such a history. The basis for this apparent association is not clear but it may be related to the presence of receptors or greater density of receptors in susceptible individuals. Support for this comes from recent studies which showed that 97% of children without reflux and with recurrent pyelonephritis belonged to the P_1 blood group phenotype compared to 75% of healthy children (Svanborg Edén *et al.*, 1983a). Persons with the P_1 phenotype have a higher density of receptor glycolipids in their erythrocyte membrane than individuals of the P_2 phenotype, although the density of receptors on UEC of individuals with different P phenotypes has yet to be demonstrated. Examination of UEC from one woman prone to UTI and one healthy woman by fluorescence-activated cell sorting assays for binding of P-fimbriate *E. coli* has recently been reported (Svenson and Källenius, 1983). The cells from the infection-prone woman had a much higher tendency to bind P-fimbriate bacteria; however, the P type of the individuals was not reported. With regard to the relationship between past history of lower urinary tract infection and P phenotype, little difference between the number of P_1-positive children with recurrent cystitis (78%) and healthy children (75%) has been observed (Svanborg Edén *et al.*, 1983a). However, the possibility that the situation is different in adult females and that non-P receptors may be of some significance in cystitis has yet to be investigated.

It is also important to consider host susceptibility when assessing the importance of adhesins as it has long been known that susceptibility to UTI is increased by host defects in urinary flow such as obstructive malformations and vesicoureteric reflux. The effect of host factors is clearly demonstrated in patients with recurrent pyelonephritis without reflux, where 74% of strains had adhesive capacity as indicated by HA, compared with only 38% in patients with reflux (Svanborg Edén *et al.*, 1983a). It was concluded that, while attachment to the mucosal lining of the urinary tract is a prerequisite for infection of the kidney in a urinary tract with normal voiding, it does not significantly contribute to infectivity when abnormalities in urine flow impair voiding and aid bacterial persistence. Källenius *et al.* (1983) recently developed an experimental pyelonephritis model in monkeys (*Macaca fascicularis*) which appears closely to parallel the human situation. The model indicated that minor vesicoureteric reflux, combined with attachment of P-fimbriate bacteria, may constitute an alternative mechanism of induction of urethritis and pyelonephritis in the absence of obstruction and gross reflux. In addition, the observed inhibition of pyelonephritis and renal scarring by P receptor blockade with a synthetic disaccharide receptor analogue in this animal model demonstrates the importance of P fimbriae as virulence determinants and also highlights novel potential methods of prophylaxis.

Despite the strong association between possession of adhesins and ability to colonize the normal urinary tract, no study on a specific group of patients has demonstrated adhesins on all isolates. Although it is possible that as yet undefined adhesins are present on some strains, it is likely that adhesion is not an absolute requirement for colonization and other factors are also likely to be involved. It has been shown that other virulence determinants are likely to be required since pyelonephritis and cystitis strains appear to possess more virulence factors than asymptomatic bacteriuria strains (Svanborg Edén et al., 1981). This has been highlighted by in vivo studies of a mouse model in which strains bearing adhesins persisted in significantly higher numbers in kidney and bladder, if associated with serum resistance (Hagberg et al., 1983b; Svanborg Edén et al., 1983a). Furthermore, a laboratory strain of K-12 containing cloned chromosomal genes coding for an MR adhesin (Hull et al., 1981) did not persist in the mouse urinary tract.

Finally, despite the evidence from in vitro and in vivo studies there is still some controversy concerning the possibility that adhesins which have been demonstrated on organisms after isolation are a laboratory artifact and bear no relevance in vivo. Harber et al. (1982) reported that organisms isolated from urinary tract infections possessed few adhesins on testing for haemagglutination after subculture and they found no evidence for fimbrial production by strains isolated directly from urine on examination under the electron microscope. This may be explained in part by the use of inappropriate culture conditions for expression of adhesins and by the selection of patients with urinary tract defects, a group recently shown to have a low percentage of adhesive isolates (Svanborg Edén et al., 1983a). Furthermore, it is unlikely that many organisms free in urine will bear fimbriae since these will be attached to epithelial cells and it is notoriously difficult to demonstrate fimbrial structures on adhering bacteria. It is now clearly evident that MR fimbrial adhesins are expressed in vivo in the urinary tract. Studies with P receptor-coated latex particles, which provide a sensitive and specific means for detection of P fimbriae, (Möllby et al., 1983) showed that in 5 of 10 patients with UTI P-fimbriate E. coli could be detected by direct testing of their urine (Svenson et al., 1982). Our own studies have demonstrated the presence of MR adhesins on bacteria adhering to UEC of patients with cystitis by the use of immunofluorescence with fimbria-specific antisera (Fig. 2). Moreover, centrifuged urine sediment from patients with cystitis can be shown to cause MR-HA of human erythrocytes and this correlates with the in vitro demonstration of MR-HA activity after subculture (J. M. Salter and S. H. Parry, unpublished data).

In conclusion, there is ample evidence that adhesion plays an important role in UTI by contributing to the establishment of bacteria in the normal urinary tract.

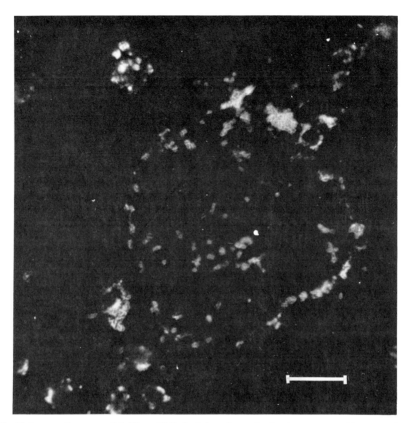

Fig. 2. Immunofluorescent labelling of MR fimbriae of *E. coli* adhering to a uroepithelial cell from the urine of a patient with acute symptomatic bacteriuria. Bacteria were labelled with rabbit anti-MR fimbrial antiserum and FITC-labelled sheep anti-rabbit globulin. Magnification ×329; bar marker, 50 μm.

This is shown by (1) strong association of the adhesive property with urinary isolates compared with *E. coli* from the healthy bowel, (2) demonstration of adhesion to host cells *in vitro*, (3) demonstration of the production of MR adhesins *in vivo*, (4) presence of appropriate receptors on target host tissues and (5) demonstration of enhanced infectivity *in vivo* by strains bearing adhesins compared to those lacking adhesins.

References

Anderson, M. J., Whitehead, J. S. and Kim, Y. S. (1980). Interaction of *Escherichia coli* K88 antigen with porcine intestinal brush border membranes. *Infection and Immunity* **29**, 897–901.

Arbuckle, J. B. R. (1970). The location of *Escherichia coli* in the pig intestine. *Journal of Medical Microbiology* **3**, 333–340.

Avots-Avotins, A. E., Fader, R. C. and Davis, C. P. (1981). Environmental alteration and two distinct mechanisms of *E. coli* adherence to bladder epithelial cells. *Investigative Urology* **18**, 364–370.

Avril, J. L., Fauchère, J. L. and Veron, M. (1981). Adhérence des *Escherichia coli* entéropathogènes pour l'homme aux cellules mononuclées du sang humain. *Annales de Microbiologie (Paris)* **132A**, 141–148.

Awad-Masalmeh, M., Moon, H. W., Runnels, P. L. and Schneider, R. A. (1982). Pilus production, haemagglutination and adhesion by porcine strains of enterotoxigenic *Escherichia coli* lacking K88, K99 and 987P antigens. *Infection and Immunity* **35**, 305–313.

Bar-Shavit, Z., Ofek, I., Goldman, R., Mirelman, D. and Sharon, N. (1977). Mannose residues on phagocytes as receptors for the attachment of *Escherichia coli* and *Salmonella typhi*. *Biochemical and Biophysical Research Communications* **78**, 455–460.

Bar-Shavit, Z., Goldman, R., Ofek, I., Sharon, N. and Mirelman, D. (1980). Mannose binding activity of *Escherichia coli*: A determinant of attachment and ingestion of bacteria by macrophages. *Infection and Immunity* **29**, 417–424.

Bertschinger, H. U., Moon, H. W. and Whipp, S. C. (1972). Association of *Escherichia coli* with the small intestinal epithelium. I. Comparison of enteropathogenic and non-enteropathogenic porcine strains in pigs. *Infection and Immunity* **5**, 595–605.

Bijlsma, I. G. W., de Nijs, A. and Frik, J. F. (1981). Adherence of *Escherichia coli* to porcine intestinal brush borders by means of serological variants of the K88 antigen. *Antonie van Leeuwenhoek* **47**, 467–468.

Bijlsma, I. G. W., de Nijs, A., van der Meer, C. and Frik, J. F. (1982). Different pig phenotypes affect adherence of *Escherichia coli* to jejunal brush borders by K88ab, K88ac, or K88ad antigen. *Infection and Immunity* **37**, 891–894.

Björksten, B. and Wadström, T. (1982). Interaction of *Escherichia coli* with different fimbriae and polymorphonuclear leucocytes. *Infection and Immunity* **38**, 298–305.

Blumenstock, E. and Jann, K. (1982). Adhesion of piliated *Escherichia coli* strains to phagocytes: Differences between bacteria with mannose-sensitive pili and those with mannose-resistant pili. *Infection and Immunity* **35**, 264–269.

Brinton, C. C. (1959). Non-flagellar appendages of bacteria. *Nature (London)* **183**, 782–786.

Brinton, C. C. (1965). The structure, function, synthesis and genetic control of bacterial pili and a molecular model for DNA and RNA transport in gram negative bacteria. *Transactions of the New York Academy of Sciences* **27**, 1003–1054.

Brinton, C. C. (1978). Studies on the piliation of a human enteropathogen, *Escherichia coli*, H-10407. *Proceedings of the 13th Joint U.S.-Japan Conference on Cholera* Publ. No. (NIH) 78-1590, pp. 34–70.

Burrows, M. R., Sellwood, R. and Gibbons, R. A. (1976). Haemagglutinating and adhesive properties associated with the K99 antigen of bovine strains of *Escherichia coli*. *Journal of General Microbiology* **96**, 269–275.

Cameron, G. L. and Staveley, J. M. (1957). Blood group P substance in hydatid cyst fluids. *Nature (London)* **179**, 147–148.

Candy, D. C. A., Leung, T. S. M., Mak, R. H. K., Harries, J. T., Marshall, W. C., Phillips, A. D., Robins-Browne, R., Chadwick, M. V. and Levine, M. M. (1982). A fimbrial antigen mediating haemagglutination of *Escherichia coli* from children. *FEMS Microbiology Letters* **15**, 325–329.

Cantey, J. R. and Blake, R. K. (1977). Diarrhoea due to *Escherichia coli* in the rabbit: A novel mechanism. *Journal of Infectious Diseases* **135**, 454–462.

Cantey, J. R. and Inman, L. R. (1981). Diarrhoea due to *Escherichia coli* strain RDEC-1 in the rabbit: The Peyer's patch as the initial site of attachment and colonization. *Journal of Infectious Diseases* **143**, 440–446.

Chabanon, G., Hartley, C. L. and Richmond, M. H. (1982). *In vitro* attachment to a human cell line of *Escherichia coli* strains causing urinary-tract infection: Occurrence of fimbriae (pili) in adhesive and non-adhesive strains. *Annales de Microbiologie (Paris)* **133A**, 357–369.

Chan, R., Acres, S. D. and Costerton, J. W. (1982). Use of specific antibody to demonstrate glycocalyx, K99 pili and the spatial relationships of K99+ enterotoxigenic *Escherichia coli* in the ileum of colostrum-fed calves. *Infection and Immunity* **37**, 1170–1180.

Chanter, N. (1982). Structural and functional differences of the anionic and cationic antigens in K99 extracts of *Escherichia coli* B41. *Journal of General Microbiology* **128**, 1585–1589.

Chanter, N. (1983). Partial purification and characterization of two non K99 mannose-resistant haemagglutinins of *Escherichia coli* B41. *Journal of General Microbiology* **129**, 235–243.

Cheney, C. P. and Boedeker, E. C. (1983). Adherence of an enterotoxigenic *Escherichia coli* strain, serotype O78:H11, to purified human intestinal brush borders. *Infection and Immunity* **39**, 1280–1284.

Cheney, C. P., Boedeker, E. C. and Formal, S. B. (1979). Quantitation of the adherence of an enteropathogenic *Escherichia coli* to isolated rabbit intestinal brush borders. *Infection and Immunity* **26**, 736–743.

Cheney, C. P., Schad, P. A., Formal, S. B. and Boedeker, E. C. (1980). Species specificity of *in vitro Escherichia coli* adherence to host intestinal cell membranes and its correlation with *in vivo* colonization and infectivity. *Infection and Immunity* **28**, 1019–1027.

Chick, S., Harber, M. J., Mackenzie, R. and Asscher, A. W. (1981). Modified method for studying bacterial adhesion to isolated uroepithelial cells and uromucoid. *Infection and Immunity* **34**, 256–261.

Clausen, C. R. and Christie, D. L. (1982). Chronic diarrhoea in infants caused by adherent enteropathogenic *Escherichia coli*. *Journal of Pediatrics (St. Louis)* **100**, 358–361.

Clegg, S. (1982a). Serological heterogeneity among fimbrial antigens causing mannose-resistant haemagglutination by uropathogenic *Escherichia coli*. *Infection and Immunity* **35**, 745–748.

Clegg, S. (1982b). Cloning of genes determining the production of mannose-resistant fimbriae in a uropathogenic strain of *Escherichia coli* belonging to serogroup O6. *Infection and Immunity* **38**, 739–744.

Clegg, S., Evans, D. J., Jr., and Evans, D. G. (1982). Antigenic heterogeneity of haemagglutination type VI fimbriae produced by *Escherichia coli* isolated from patients with bacteraemia. *Journal of Clinical Microbiology* **16**, 174–180.

Cory, H. T., Yates, A. D., Donald, A. S. R., Watkins, W. M. and Morgan, W. T. J. (1974). The nature of the human blood group P_1 determinant. *Biochemical and Biophysical Research Communications* **61**, 1289–1296.

Cravioto, A., Gross, R. J., Scotland, S. M. and Rowe, B. (1979). Mannose-resistant haemagglutination of human erythrocytes by strains of *Escherichia coli* from extra-intestinal sources: lack of correlation with colonization factor antigen (CFA/I). *FEMS Microbiology Letters* **6**, 41–44.

Cravioto, A., Scotland, S. M. and Rowe, B. (1982). Haemagglutination activity and colonization factor antigens I and II in enterotoxigenic and non-enterotoxigenic strains of *Escherichia coli* isolated from humans. *Infection and Immunity* **36**, 189–197.

Czirók, É., Madár, J., Budai, J., Stverteczky, Z., Kertész, A., Dombi, I. and Gyengési, L. (1976). The role in sporadic enteritis of facultatively enteropathogenic *Escherichia coli* serogroups. *Acta Microbiologica Academiae Scientiarum Hungaricae* **23**, 359–369.

Czirók, É., Ørskov, I. and Ørskov, F. (1982). O:K:H:F serotypes of fimbriated *Escherichia coli* strains isolated from infants with diarrhea. *Infection and Immunity* **37**, 519–525.

Davis, C. P., Avots-Avotins, A. E. and Fader, R. C. (1981). Evidence for a bladder cell glycolipid receptor for *Escherichia coli* and the effect of neuraminic acid and colominic acid on adherence. *Infection and Immunity* **34**, 944–948.

Davis, C. P., Fader, R. C., Avots-Avotins, A. E. and Gratzfeld, S. (1982). Expression of haemagglutination is dependent on environmental factors and bacterial concentration of *Escherichia coli* and *Klebsiella pneumoniae*. *Current Microbiology* **7**, 161–164.

Dean, E. A. and Isaacson, R. E. (1982). *In vitro* adhesion of piliated *Escherichia coli* to small intestinal villous epithelial cells from rabbits and the identification of a soluble 987P pilus receptor-containing fraction. *Infection and Immunity* **36**, 1192–1198.

de Graaf, F. K. and Roorda, I. (1982). Production, purification and characterization of the fimbrial adhesive antigen F41 isolated from the calf enteropathogenic *Escherichia coli* strain B41M. *Infection and Immunity* **36**, 751–758.

de Graaf, F. K., Klaasen-Boor, P. and van Hess, J. E. (1980a). Biosynthesis of the K99 surface antigen is repressed by alanine. *Infection and Immunity* **30**, 125–128.

de Graaf, F. K., Klemm, P. and Gaastra, W. (1980b). Purification, characterization and partial covalent structure of the adhesive antigen K99 of *Escherichia coli*. *Infection and Immunity* **33**, 877–883.

Deneke, C. F., Thorne, G. M. and Gorbach, S. L. (1979). Attachment pili from enterotoxigenic *Escherichia coli* pathogenic for humans. *Infection and Immunity* **26**, 362–368.

Deneke, C. F., Thorne, G. M. and Gorbach, S. L. (1981). Serotypes of attachment pili of enterotoxigenic *Escherichia coli* isolated from humans. *Infection and Immunity* **32**, 1254–1260.

Duguid, J. P. (1964). Functional anatomy of *Escherichia coli* with special reference to entero-pathogenic *E. coli*. *Revista Latinoamericana de Microbiologie* **7**, Supplement 13–14, 1–16.

Duguid, J. P., and Old, D. C. (1980). Adhesive properties of *Enterobacteriacae*. *In* ''Bacterial Adherence'' (Ed. E. H. Beachey), Receptors and Recognition Series B, Vol. 6, pp. 185–217. Chapman and Hall, London.

Duguid, J. P., Smith, I. W., Dempster, G. and Edmunds, P. N. (1955). Non-flagellar filamentous appendages ('fimbriae') and haemagglutinating activity in *Bacterium coli*. *Journal of Pathology and Bacteriology* **70**, 335–348.

Duguid, J. P., Clegg, S. and Wilson, M. I. (1979) The fimbrial and non-fimbrial haemagglutinins of *Escherichia coli*. *Journal of Medical Microbiology* **12**, 213–227.

Dupont, H. L., Formal, S. B., Hornick, R. B., Snyder, M. J., Libonati, J. P., Sheahan, D. G., Labrec, E. H. and Kalas, J. P. (1971). Pathogenesis of *Escherichia coli* diarrhoea. *New England Journal of Medicine* **285**, 1–9.

Eisenstein, B. I. and Dodd, D. C. (1982). Pseudocatabolite repression of type 1 fimbriae of *Escherichia coli*. *Journal of Bacteriology* **151**, 1560–1567.

Eisenstein, B. I., Beachey, E. H. and Ofek, I. (1980). Influence of sublethal concentrations of antibiotics on the expression of the mannose-specific ligand of *Escherichia coli*. *Infection and Immunity* **28**, 154–159.

Eshdat, Y., Ofek, I., Yashouv-Gan, Y., Sharon, N. and Mirelman, D. (1978). Isolation of a mannose-specific lectin from *Escherichia coli* and its role in the adherence of the bacteria to epithelial cells. *Biochemical and Biophysical Research Communications* **85**, 1551–1559.

Eshdat, Y., Speth, V. and Jann, K. (1981). Participation of pili and cell wall adhesin in the yeast agglutination activity of *Escherichia coli*. *Infection and Immunity* **34**, 980–986.

Evans, D. G. and Evans, D. J., Jr. (1978). New surface-associated heat-labile colonization factor (CFA/II) produced by enterotoxigenic *Escherichia coli* of serogroups O6 and O8. *Infection and Immunity* **21**, 638–647.

Evans, D. G., Silver, R. P., Evans, D. J., Jr., Chase, D. G. and Gorbach, S. L. (1975). Plasmid-controlled colonization factor associated with virulence in *Escherichia coli* enterotoxigenic for humans. *Infection and Immunity* **12**, 656–667.

Evans, D. G., Evans, D. J., Jr. and Tjoa, W. (1977). Haemagglutination of human group A erythrocytes by enterotoxigenic *Escherichia coli* isolated from adults with diarrhoea: Correlation with colonization factor. *Infection and Immunity* **18**, 330–337.

Evans, D. G., Evans, D. J., Jr., Tjoa, W. S. and DuPont, H. L. (1978a). Detection and characteriza-tion of colonization factor of enterotoxigenic *Escherichia coli* isolated from adults with diarrhea. *Infection and Immunity* **19**, 727–736.

Evans, D. G., Satterwhite, T. K., Evans, D. J., Jr. and DuPont, H. L. (1978b). Differences in serological responses and excretion patterns of volunteers challenged with enterotoxigenic *Escherichia coli* with and without the colonization factor antigen. *Infection and Immunity* **19**, 883–888.

Evans, D. G., Evans, D. J., Jr., Clegg, S. and Pauley, J. A. (1979). Purification and characterization of the CFA/I antigen of enterotoxigenic *Escherichia coli*. *Infection and Immunity* **25**, 738–748.

Evans, D. J., Jr., Evans, D. G. and Dupont, H. L. (1979). Haemagglutination patterns of enterotoxigenic and enteropathogenic *Escherichia coli* determined with human, bovine, chicken and guinea-pig erythrocytes in the presence and absence of mannose. *Infection and Immunity* **23**, 336–346.

Evans, D. J., Jr., Evans, D. G., Young, L. S. and Pitt, J. (1980). Haemagglutination typing of *Escherichia coli*: Definition of seven haemagglutination types. *Journal of Clinical Microbiology* **12**, 235–242.

Evans, D. J., Jr., Evans, D. G., Höhne, C., Noble, M. A., Haldane, E. V., Lior, H. and Young, L. S. (1981). Haemolysin and K antigens in relation to serotype and haemagglutination type of *Escherichia coli* isolated from extra-intestinal infections. *Journal of Clinical Microbiology* **13**, 171–178.

Faris, A., Lindahl, M. and Wadström, T. (1980). GM2-like glycoconjugate as possible erythrocyte receptor for the CFA/I and K99 haemagglutinins of enterotoxigenic *Escherichia coli*. *FEMS Microbiology Letters* **7**, 265–269.

Freer, J. H., Ellis, A., Wadström, T. and Smyth, C. J. (1978). Occurrence of fimbriae among enterotoxigenic intestinal bacteria isolated from cases of human infantile diarrhoea. *FEMS Microbiology Letters* **3**, 277–281.

Gaastra, W. and de Graaf, F. K. (1982). Host-specific fimbrial adhesins of noninvasive enterotoxigenic *Escherichia coli* strains. *Microbiological Reviews* **46**, 129–161.

Gaastra, W., Klemm, P., Walker, J. M. and de Graaf, F. K. (1979). K88 fimbrial protein: Amino and carboxyl terminal sequences of intact proteins and cyanogen bromide fragments. *FEMS Microbiology Letters* **6**, 15–18.

Gaastra, W., Mooi, F. R., Stuitje, A. R. and de Graaf, F. K. (1981). The nucleotide sequence of the gene encoding the K88ab protein subunit of porcine enterotoxigenic *Escherichia coli*. *FEMS Microbiology Letters* **12**, 41–46.

Gibbons, R. A., Jones, G. W., and Sellwood, R. (1975). An attempt to identify the intestinal receptor for the K88 adhesin by means of haemagglutination inhibition test using glycoproteins and fractions from sow colostrum. *Journal of General Microbiology* **86**, 228–240.

Green, C. P. and Thomas, V. L. (1981). Haemagglutination of human type O erythrocytes, haemolysin production, and serogrouping of *Escherichia coli* isolates from patients with acute pyelonephritis, cystitis and asymptomatic bacteriuria. *Infection and Immunity* **31**, 309–315.

Gross, R. J., Cravioto, A., Scotland, S. M., Cheasty, T. and Rowe, B. (1978). The occurrence of colonization factor (CF) in enterotoxigenic *Escherichia coli*. *FEMS Microbiology Letters* **3**, 231–233.

Guinée, P. A. M. and Jansen, W. H. (1979). Behaviour of *Escherichia coli* K antigens K88ab, K88ac and K88ad in immunoelectrophoresis, double diffusion, and haemagglutination. *Infection and Immunity* **23**, 700–705.

Guinée, P. A. M., Jansen, W. H. and Agterberg, C. M. (1976). Detection of the K99 antigen by means of agglutination and immunoelectrophoresis in *Escherichia coli* isolates from calves and its correlation with enterotoxigenicity. *Infection and Immunity* **13**, 1369–1377.

Guinée, P. A. M., Veltkamp, J. and Jansen, W. H. (1977). Improved Minca medium for the detection of K99 antigen in calf enterotoxigenic strains of *Escherichia coli*. *Infection and Immunity* **15**, 676–678.

Guyot, G. (1908). Ueber die bakterielle Häemagglutination (Bakterio-Haemo-agglutination). *Zentralblatt für Bakteriologie, Parasitenkunde, Infektionskrankheiten und Hygiene, Abteilung 1: Originale* **47**, 640–653.

Hagberg, L., Jodal, U., Korhonen, T. K., Lindin-Janson, G., Lindberg, U. and Svanborg Edén, C.

(1981). Adhesion, haemagglutination and virulence of *Escherichia coli* causing urinary tract infections. *Infection and Immunity* **31**, 564–570.

Hagberg, L., Hull, R., Falkow, S., Freter, R. and Svanborg Edén, C. (1983a). Contribution of adhesion to bacterial persistence in the mouse urinary tract. *Infection and Immunity* **40**, 265–272.

Hagberg, L., Engberg, I., Freter, R., Lam, J., Olling, S. and Svanborg Edén, C. (1983b). Ascending, unobstructed urinary tract infection in mice caused by pyelonephritogenic *Escherichia coli* of human origin. *Infection and Immunity* **40**, 273–283.

Hanson, L. Å., Ahlstedt, S., Fasth, A., Jodal, U., Kaijser, B., Larsson, P., Lindberg, U., Olling, S., Sohl-Åkerlund, A. and Svanborg Edén, C. (1977). Antigens *Escherichia coli,* human immune response and the pathogenesis of urinary tract infections. *Journal of Infectious Diseases* **136**, Supplement, S144–S149.

Harber, M. J., Chick, S., Mackenzie, R. and Asscher, A. W. (1982). Lack of adherence to epithelial cells by freshly isolated urinary pathogens. *Lancet* **1**, 586–588.

Hartley, C. L., Neumann, C. S. and Richmond, M. H. (1979). Adhesion of commensal bacteria to the large intestine wall in humans. *Infection and Immunity* **23**, 128–132.

Hermodson, M. A., Chen, K. C. S. and Buchanan, T. M. (1978). *Neisseria* pili proteins: Amino-terminal amino acid sequences and identification of an unusual amino acid. *Biochemistry* **17**, 442–445.

Hohmann, A. and Wilson, M. R. (1975). Adherence of enteropathogenic *Escherichia coli* to intestinal epithelium *in vivo*. *Infection and Immunity* **12**, 866–880.

Honda, T., Khan, M. M. A., Takeda, Y. and Miwatani, T. (1983). Grouping of enterotoxigenic *Escherichia coli* by hydrophobicity and its relation to haemagglutination and enterotoxin production. *FEMS Microbiology Letters* **17**, 273–276.

Hull, R. A., Gill, R. E., Hsu, P., Minshew, B. H. and Falkow, S. (1981). Construction and expression of recombinant plasmids encoding type 1, or D-mannose-resistant pili from a urinary tract infection *Escherichia coli* isolate. *Infection and Immunity* **33**, 933–938.

Inman, L. R. and Cantey, J. R. (1983). Specific adherence of *Escherichia coli* strain (RDEC-1) to membranous (M) cells of the Peyer's patch in *E. coli* diarrhoea in the rabbit. *Journal of Clinical Investigation* **71**, 1–9.

Ip, S. M., Crichton, P. B., Old, D. C. and Duguid, J. P. (1981). Mannose-resistant and eluting haemagglutinins and fimbriae in *Escherichia coli*. *Journal of Medical Microbiology* **14**, 223–226.

Isaacson, R. E. (1977). K99 surface antigen of *Escherichia coli*: Purification and partial characterization. *Infection and Immunity* **15**, 272–279.

Isaacson, R. E. (1980). Factors affecting expression of the *Escherichia coli* pilus K99. *Infection and Immunity* **28**, 190–194.

Isaacson, R. E. and Richter, P. (1981). *Escherichia coli* 987P pilus: Purification and partial characterization. *Journal of Bacteriology* **146**, 784–789.

Isaacson, R. E., Nagy, B. and Moon, H. W. (1977). Colonization of porcine small intestine by *Escherichia coli*: Colonization and adhesion factors of pig enteropathogens that lack K88. *Journal of Infectious Diseases* **135**, 531–539.

Isaacson, R. E., Fusco, P. C., Brinton, C. C. and Moon, H. W. (1978a). *In vitro* adhesion of *Escherichia coli* to porcine small intestinal epithelial cells: Pili as adhesive factors. *Infection and Immunity* **21**, 392–397.

Isaacson, R. E., Moon, H. W. and Schneider, R. A. (1978b). Distribution and virulence of *Escherichia coli* in the small intestine of calves with and without diarrhoea. *American Journal of Veterinary Research* **39**, 1750–1755.

Isaacson, R. E., Colmenero, J. and Richter, P. (1981). *Escherichia coli* K99 pili are composed of one subunit species. *FEMS Microbiology Letters* **12**, 229–232.

Jann, K., Schmidt, G., Blumenstock, E. and Vosbeck, K. (1981). *Escherichia coli* adhesion to *Saccharomyces cerevisiae* and mammalian cells: Role of piliation and surface hydrophobicity. *Infection and Immunity* **32**, 484–489.

Jones, G. W. (1977). The attachment of bacteria to the surfaces of animal cells. *In* ''Microbial Interactions'' (Ed. J. L. Reissig), Receptors and Recognition Series B, Vol. 3, pp. 139–176. Chapman and Hall, London.

Jones, G. W. and Rutter, J. M. (1972). Role of the K88 antigen in the pathogenesis of neonatal diarrhoea caused by *Escherichia coli* in piglets. *Infection and Immunity* **6**, 918–927.

Jones, G. W. and Rutter, J. M. (1974). The association of K88 antigen with haemagglutinating activity in porcine strains of *Escherichia coli*. *Journal of General Microbiology* **84**, 135–144.

Källenius, G. and Möllby, R. (1979). Adhesion of *Escherichia coli* to human periurethral cells correlated to mannose-resistant agglutination of human erythrocytes. *FEMS Microbiology Letters* **5**, 295–299.

Källenius, G. and Winberg, J. (1978). Bacterial adherence to periurethral epithelial cells in girls prone to urinary tract infections. *Lancet* **2**, 540–543.

Källenius, G., Möllby, R., Svenson, S. B., Winberg, J. and Hultberg, H. (1980a). Identification of a carbohydrate receptor recognized by uropathogenic *Escherichia coli*. *Infection* **8**, Suppl. 3, S288–S293.

Källenius, G., Möllby, R., Svenson, S. B., Winberg, J., Lundblad, A., Svensson, S. and Cedergren, B. (1980b). The P^k antigen as receptor for the haemagglutinin of pyelonephritic *Escherichia coli*. *FEMS Microbiology Letters* **7**, 297–302.

Källenius, G., Möllby, R. and Winberg, J. (1980c). *In vitro* adhesion of uropathogenic *Escherichia coli* to human periurethral cells. *Infection and Immunity* **28**, 972–980.

Källenius, G., Möllby, R., Svenson, S. B., Helin, I., Hultberg, H., Cedergren, B. and Winberg, J. (1981a). Occurrence of P-fimbriated *Escherichia coli* in urinary tract infections. *Lancet* **2**, 1369–1372.

Källenius, G., Svenson, S. B., Möllby, R., Cedergren, B., Hultberg, H. and Winberg, J. (1981b). Structure of carbohydrate part of receptor on human uroepithelial cells for pyelonephritogenic *Escherichia coli*. *Lancet* **2**, 604–606.

Källenius, G., Hultberg, H., Möllby, R., Roberts, J., Svenson, S. B. and Winberg, J. (1983). P-fimbriae of pyelonephritogenic *Escherichia coli*: Significance for reflux and renal scarring—a hypothysis. *Infection* **11**, 73–76.

Kauffmann, F. (1948). On haemagglutination by *Escherichia coli*. *Acta Pathologica et Microbiologica Scandinavica* **25**, 502–506.

Kearns, M. J., and Gibbons, R. A. (1979). The possible nature of the pig intestinal receptor for the K88 antigen of *Escherichia coli*. *FEMS Microbiology Letters* **6**, 165–168.

Klemm, P. (1979). Fimbrial colonization factor CFA/I protein from human enteropathogenic *Escherichia coli* strains: Purification, characterization and N-terminal sequence. *FEBS Letters* **108**, 107–110.

Klemm, P. (1981). The complete amino acid sequence of the K88 antigen, a fimbrial protein from *Escherichia coli*. *European Journal of Biochemistry* **117**, 617–627.

Klemm, P. (1982). Primary structure of the CFA/1 fimbrial protein from human enterotoxigenic *Escherichia coli* strains. *European Journal of Biochemistry* **124**, 339–348.

Klemm, P. and Mikkelsen, L. (1982). Prediction of antigenic determinants and secondary structures of the K88 and CFA/I fimbrial proteins from enteropathogenic *Escherichia coli*. *Infection and Immunity* **38**, 41–45.

Klemm, P., Ørskov, I. and Ørskov, F. (1982). F7 and type 1-like fimbriae from three *Escherichia coli* strains isolated from urinary tract infections: Protein, chemical and immunological aspects. *Infection and Immunity* **36**, 462–468.

Korhonen, T. K., Edén, S. and Svanborg Edén, C. (1980a). Binding of purified *E. coli* pili to human urinary tract epithelial cells. *FEMS Microbiology Letters* **7**, 237–240.

Korhonen, T. K., Nurmiaho, E.-L., Ranta, H. and Svanborg Edén, C. (1980b). New method for isolation of immunologically pure pili from *Escherichia coli*. *Infection and Immunity* **27**, 569–575.

Korhonen, T. K., Leffler, H. and Svanborg Edén, C. (1981). Binding specificity of piliated strains of *Escherichia coli* and *Salmonella typhimurium* to epithelial cells, *Saccharomyces cerevisiae* cells and erythrocytes. *Infection and Immunity* **32**, 796–804.

Korhonen, T. K., Väisänen, V., Saxén, H., Hultberg, H. and Svenson, S.B. (1982). P-antigen-recognizing fimbriae from human uropathogenic *Escherichia coli* strains. *Infection and Immunity* **37**, 286–291.

Leffler, H. and Svanborg Edén, C. (1980). Chemical identification of a glycosphingolipid receptor for *Escherichia coli* attaching to human urinary tract epithelial cells and agglutinating human erythrocytes. *FEMS Microbiology Letters* **8**, 127–134.

Leffler, H. and Svanborg Edén, C. (1981). Glycolipid receptors for uropathogenic *Escherichia coli* on human erythrocytes and uroepithelial cells. *Infection and Immunity* **34**, 920–929.

Levine, M. M., Rennels, M. B., Daya, V. and Hughes, T. P. (1980). Haemagglutination and colonization factors in enterotoxigenic and enteropathogenic *Escherichia coli* that cause diarrhoea. *Journal of Infectious Diseases* **141**, 733–737.

Lindahl, M., Faris, A., Wadström, T. and Hjerten, S. (1981). A new test based on "salting out' to measure relative surface hydrophobicity of bacterial cells. *Biochimica et Biophysica Acta* **677**, 471–476.

Ljungh, Å., Faris, A. and Wadström, T. (1979). Haemagglutination by *Escherichia coli* in septicaemia and urinary tract infections. *Journal of Clinical Microbiology* **10**, 477–481.

Lopez-Alvarez, J. and Gyles, C. L. (1980). Occurrence of the Vir plasmid among animal and human strains of invasive *Escherichia coli*. *American Journal of Veterinary Research* **41**, 769–774.

McNeish, A. S., Turner, P., Fleming, J. and Evans, N. (1975). Mucosal adherence of human enteropathogenic *Escherichia coli*. *Lancet* **2**, 946–948.

Mangan, D. F. and Snyder, I. S. (1979a). Mannose-sensitive interaction of *Escherichia coli* with human peripheral leucocytes *in vitro*. *Infection and Immunity* **26**, 520–527.

Mangan, D. F. and Snyder, I. S. (1979b). Mannose-sensitive stimulation of human leucocyte chemiluminescence by *Escherichia coli*. *Infection and Immunity* **26**, 1014–1019.

Marcus, D. M. and Janis, R. (1970). Localization of glycosphingolipids in human tissues by immunofluorescence, *Journal of Immunology* **104**, 1530–1539.

Mårtensson, E. (1963). On the neutral glycolipids of human kidney. *Acta Chemica Scandinavica* **17**, 2356–2358.

Mårtensson, E. (1966). Neutral glycolipids of human kidney. Isolation, identification and fatty acid composition. *Biochimica et Biophysica Acta* **116**, 296–308.

Minshew, B. H., Jorgensen, J., Counts, G. W. and Falkow, S. (1978a). Association of haemolysin production, haemagglutination of human erythrocytes, and virulence for chicken embryos of extra-intestinal *Escherichia coli* isolates. *Infection and Immunity* **20**, 50–54.

Minshew, B. H., Jorgensen, J., Swanstrum, M., Grootes-Reuvecamp, G. A. and Falkow, S. (1978b). Some characteristics of *Escherichia coli* strains isolated from extra-intestinal infections of humans. *Journal of Infectious Diseases* **137**, 648–653.

Möllby, R., Källenius, G., Korhonen, T. K., Winberg, J. and Svenson, S. B. (1983). P-fimbriae of pyelonephritogenic *Escherichia coli*: Detection in clinical material by a rapid receptor-specific agglutination test. *Infection* **11**, 68–72.

Mooi, F. R., and de Graaf, F. K. (1979). Isolation and characterization of K88 antigens. *FEMS Microbiology Letters* **5**, 17–20.

Mooi, F. R., de Graaf, F. K. and van Embden, J. D. A. (1979). Cloning, mapping and expression of the genetic determinant that encodes for the K88ab antigen. *Nucleic Acids Research* **6**, 849–865.

Moon, H. W., Nagy, B., Isaacson, R. E. and Ørskov, I. (1977). Occurrence of K99 antigen on *Escherichia coli* isolated from pigs and colonization of pig ileum by K99[+] enterotoxigenic, *Escherichia coli* from calves and pigs. *Infection and Immunity* **15**, 614–620.

Moon, H. W., Isaacson, R. E. and Pohlenz, J. (1979). Mechanisms of association of entero-pathogenic *Escherichia coli* with intestinal epithelium. *American Journal of Clinical Nutrition* **32**, 119–127.

Moon, H. W., Kohler, E. M., Schneider, R. A. and Whipp, S. C. (1980). Prevalence of pilus antigens, enterotoxin types and enteropathogenicity among K88-negative enterotoxigenic *Escherichia coli* from neonatal pigs. *Infection and Immunity* **27**, 222–230.

Morgan, W. T. J. and Watkins, W. M. (1964). Blood group P₁ substance. I. Chemical properties. *Proceedings of the 9th Congress of the International Society for Blood Transfusions, 1962* pp. 225–229.

Morris, J. A., Stevens, A. E. and Sojka, W. J. (1977). Preliminary characterization of cell-free K99 antigen isolated from *Escherichia coli* B41. *Journal of General Microbiology* **99**, 353–357.

Morris, J. A., Stevens, A. E., and Sojka, W. J. (1978a). Anionic and cationic components of the K99 surface antigen from *Escherichia coli* B41. *Journal of General Microbiology* **107**, 173–175.

Morris, J. A., Stevens, A. E. and Sojka, W. J. (1978b). Isoelectric point of cell-free K99 antigen exhibiting haemagglutinating properties. *Infection and Immunity* **19**, 1097–1098.

Morris, J. A., Thorns, C. J. and Sojka, W. J. (1980). Evidence for two adhesive antigens on the K99 reference strain *Escherichia coli* B41. *Journal of General Microbiology* **118**, 107–113.

Morris, J. A., Thorns, C. J., Scott, A. C. and Sojka, W. J. (1982a). Adhesive properties associated with the Vir plasmid: A transmissible pathogenic characteristic associated with strains of invasive *Escherichia coli*. *Journal of General Microbiology* **128**, 2097–2103.

Morris, J. A., Thorns, C., Scott, A. C., Sojka, W. J. and Wells, G. A. (1982b). Adhesion *in vitro* and *in vivo* associated with an adhesive antigen (F41) produced by a K99-mutant of the reference strain *Escherichia coli* B41. *Infection and Immunity* **36**, 1146–1153.

Morris, J. A., Thorns, C. J., Wells, G. A. W. and Sojka, W. J. (1983). The production of F41 fimbriae by piglet strains of enterotoxigenic *Escherichia coli* that lack K88, K99 and 987P fimbriae. *Journal of General Microbiology* **129**, 2753–2759.

Nagy, B., Moon, H. W. and Isaacson, R. E. (1976). Colonization of porcine small intestine by *Escherichia coli*: Ileal colonization and adhesion by pig enteropathogens that lack K88 antigen and by some acapsular mutants. *Infection and Immunity* **13**, 1214–1220.

Nagy, B., Moon, H. W. and Isaacson, R. E. (1977). Colonization of porcine intestine by enterotox-igenic *Escherichia coli*: Selection of piliated forms *in vivo*, adhesion of piliated forms to epithelial cells *in vitro*, and incidence of a pilus antigen among porcine enteropathogenic *E. coli*. *Infection and Immunity* **16**, 344–352.

Nelson-Rees, W. A., Daniels, D. W. and Flandermeyer, R. R. (1981). Cross-contamination of cells in culture. *Science* **212**, 446–452.

Ofek, I. and Beachey, E. H. (1978). Mannose-binding and epithelial cell adherence of *Escherichia coli*. *Infection and Immunity* **22**, 247–254.

Ofek, I., Mirelman, D. and Sharon, N. (1977). Adherence of *Escherichia coli* to human mucosal cells mediated by mannose receptors. *Nature (London)* **265**, 623–625.

Ofek, I., Mosek, A. and Sharon, N. (1981). Mannose-specific adherence of *Escherichia coli* freshly excreted in the urine of patients with urinary infections, and of isolates subcultured from the infected urine. *Infection and Immunity* **34**, 708–711.

O'Hanley, P. D. and Cantey, J. R. (1978). Surface structures of *Escherichia coli* that produce diarrhoea by a variety of enteropathogenic mechanisms. *Infection and Immunity* **21**, 874–878.

Öhman, L., Hed, L. and Stendahl, O. (1982a). Interaction between human polymorphonuclear leucocytes and 2 different strains of type 1 fimbriae-bearing *Escherichia coli*. *Journal of Infectious Diseases* **146**, 751–758.

Öhman, L., Magnusson, K.-E. and Stendahl, O. (1982b). The mannose-specific lectin activity of *Escherichia coli* type 1 fimbriae assayed by agglutination of glycolipid-containing liposomes, erythrocytes and yeast cells and hydrophobic interaction chromatography. *FEMS Microbiology Letters* **14**, 149–153.

Ørskov, F. and Ørskov, I. (1975). *Escherichia coli* O:H serotypes isolated from human blood. *Acta Pathologica et Microbiologica Scandinavica, Section B* **83B**, 505–600.

Ørskov, F. and Ørskov, I. (1984). Serotyping of *Escherichia coli*. *In* "Methods in Microbiology." (Eds. T. Bergan and J. R. Norris), Vol. 14. Academic Press, London, (in press).

Ørskov, I. and Ørskov, F. (1977). Special O:K:H serotypes among enterotoxigenic *Escherichia coli* strains from diarrhoea in adults and children. Occurrence of the CF (colonization factor) antigen and of haemagglutinating abilities. *Medical Microbiology and Immunology* **163**, 99–110.

Ørskov, I. and Ørskov, F. (1983). Serology of *Escherichia coli* fimbriae. *Progress in Allergy* **33**, 80–105.

Ørskov, I., Ørskov, F., Sojka, W. J. and Leach, J. M. (1961). Simultaneous occurrence of *E. coli* B and L antigens in strains from diseased swine. *Acta Pathologica et Microbiologica Scandinavica* **53**, 404–422.

Ørskov, I., Ørskov, F., Sojka, W. J. and Wittig, W. (1964). K antigens K88ab(L) and K88ac(L) in *E. coli*. A new O antigen O147 and a new K antigen: K89(B). *Acta Pathologica et Microbiologica Scandinavica* **62**, 439–447.

Ørskov, I., Ørskov, F., Smith, H. W. and Sojka, W. J. (1975). The establishment of K99, a thermolabile, transmissible *Escherichia coli* K antigen previously called "Kco", possessed by calf and lamb enteropathogenic strains. *Acta Pathologica et Microbiologica Scandanavica, Section B* **83**, 31–36.

Ørskov, I., Ørskov, F., Jann, B. and Jann, K. (1977). Serology, chemistry and genetics of O and K antigens of *Escherichia coli*. *Bacteriological Reviews* **41**, 667–710.

Ørskov, I., Ferencz, A. and Ørskov, F. (1980a). Tamm-Horsfall protein or uromucoid is the normal urinary slime that traps type 1 fimbriated *Escherichia coli*. *Lancet* **1**, 887.

Ørskov, I., Ørskov, F. and Birch-Andersen, A. (1980b). Comparison of *Escherichia coli* fimbrial antigen F7 with type 1 fimbriae. *Infection and Immunity* **27**, 657–666.

Ørskov, I., Ørskov, F., Birch-Anderson, A., Kanamori, M. and Svanborg Edén, C. (1982). O, K, H and fimbrial antigens in *Escherichia coli* serotypes associated with pyelonephritis and cystitis. *Scandinavian Journal of Infectious Diseases, Supplementum* **33**, 18–25.

Ottow, J. C. G (1975). Ecology, physiology, and genetics of fimbriae and pili. *Annual Review of Microbiology* **29**, 79–108.

Parry, S. H. and Porter, P. (1978). Immunological aspects of cell membrane adhesion demonstrated by porcine enteropathogenic *Escherichia coli*. *Immunology* **34**, 41–49.

Parry, S. H., Abraham, S. N. and Sussman, M. (1982). The biological and serological properties of adhesion determinants of *Escherichi coli* isolated from urinary tract infections. *In* "Clinical, Bacteriological and Immunological Aspects of Urinary Tract Infections in Children" (Ed. H. Schulte-Wissermann), pp. 113–126. Thieme, Stuttgart.

Parry, S. H., Boonchai, S., Abraham, S. N., Salter, J. M., Rooke, D. M., Simpson, J. M., Bint, A. J. and Sussman, M. (1983). A comparative study of the mannose-resistant and mannose-sensitive haemagglutinins of *Escherichia coli* isolated from urinary tract infections. *Infection* **11**, 123–128.

Parry, S. H., Rooke, D. M. and Sussman, M. (1984). Analysis of mannose-resistant adhesins of *Escherichia coli* by a naturally occurring glycoprotein receptor analogue. *Journal of Microbiological Methods* **2**, 323–331.

Pearce, W. A. and Buchanan, T. M. (1980). Structure and cell membrane-binding properties of bacterial fimbriae. *In* "Bacterial Adherence" (Ed. E. H. Beachey), Receptors and Recognition Series B, Vol. 6, pp. 289–344. Chapman and Hall, London.

Rosenthal, L. (1943). Agglutinating properties of *Escherichia coli*. Agglutination of erythrocytes, leucocytes, thrombocytes, spermatozoa, spores of molds and pollen by strains of *E. coli*. *Journal of Bacteriology* **45**, 545–550.

Rothbaum, R., McAdams, A. J., Giannella, R. and Partin, J. C. (1982). A clinicopathologic study of enterocyte-adherent *Escherichia coli*: A cause of protracted diarrhoea in infants. *Gastroenterology* **83**, 441–454.

Rottini, G., Cian, F. F., Soranzo, M. R., Albirgo, R. and Patriarca, P. (1979). Evidence for the involvement of human polymorphonuclear leucocyte mannoside receptors in the phagocytosis of *Escherichia coli*. *FEBS Letters* **105**, 307–312.

Rowe, B. (1979). The role of *Escherichia coli* in gastroenteritis. *Clinical Gastroenterology* **8**, 625–644.

Rutter, J. M. and Jones, G. W. (1973). Protection against enteric disease caused by *Escherichia coli*—a model for vaccination with a virulence determinant? *Nature (London)* **242**, 531–532.

Rutter, J. M., Burrows, M. R., Sellwood, R. and Gibbons, R. A. (1975). A genetic basis for resistance to enteric disease caused by *Escherichia coli*. *Nature (London)* **257**, 135–136.

Salit, I. E. and Gotschlich, E. C. (1977a). Haemagglutination by purified type I *Escherichia coli* pili. *Journal of Experimental Medicine* **146**, 1169–1181.

Salit, I. E. and Gotschlich, E. C. (1977b). Type 1 *Escherichia coli* pili: Characterization of binding to monkey kidney cells. *Journal of Experimental Medicine* **146**, 1182–1193.

Schaeffer, A. J., Amundsen, S. K. and Schmidt, L. N. (1979). Adherence of *Escherichia coli* to human urinary tract epithelial cells. *Infection and Immunity* **24**, 753–759.

Schaeffer, A. J., Amundsen, S. K. and Jones, J. M. (1980). Effect of carbohydrates on adherence of *Escherichia coli* to human urinary tract epithelial cells. *Infection and Immunity* **30**, 531–537.

Schaeffer, A. J., Jones, J. M. and Dunn, J. K. (1981). Association of *in vitro Escherichia coli* adherence to vaginal and buccal epithelial cells with susceptibility of women to recurrent urinary tract infections. *New England Journal of Medicine* **304**, 1062–1066.

Schneider, R. A. and To, S. C. M. (1982). Enterotoxigenic *Escherichia coli* strains that express K88 and 987P pilus antigens. *Infection and Immunity* **36**, 417–418.

Sellwood, R. (1980). The interaction of the K88 antigen with porcine intestinal epithelial cell brush borders. *Biochimica et Biophysica Acta* **632**, 326–335.

Sellwood, R., Gibbons, R. A., Jones, G. W. and Rutter, J. M. (1975). Adhesion of enteropathogenic *Escherichia coli* to pig intestinal brush borders: The existence of two pig phenotypes. *Journal of Medical Microbiology* **8**, 405–411.

Silverblatt, F. J., Dreyer, J. S. and Schauer, S. (1979). Effect of pili on susceptibility of *Escherichia coli* to phagocytosis. *Infection and Immunity* **24**, 218–223.

Smith, H. W. (1974). A search for transmissible pathogenic characters in invasive strains of *Escherichia coli*: The discovery of a plasmid-controlled toxin and a plasmid-controlled lethal character closely associated, or identical, with colicin V. *Journal of General Microbiology* **83**, 95–111.

Smith, H. W. and Halls, S. (1967). Observations by the ligated intestinal segment and oral inoculation methods in *Escherichia coli* infections in pigs, calves, lambs and rabbits. *Journal of Pathology and Bacteriology* **93**, 499–529.

Smith, H. W. and Huggins, M. B. (1978). The influence of plasmid-determined and other characteristics of enteropathogenic *Escherichia coli* on their ability to proliferate in the alimentary tracts of piglets, calves and lambs. *Journal of Medical Microbiology* **11**, 471–492.

Smith, H. W. and Linggood, M. A. (1971). Observations on the pathogenic properties of the K88, Hly and Ent plasmids of *Escherichia coli* with particular reference to porcine diarrhoea. *Journal of Medical Microbiology* **4**, 467–485.

Smith, H. W. and Linggood, M. A. (1972). Further observations on *Escherichia coli* enterotoxins with particular regard to those produced by atypical piglet strains and by calf and lamb strains: The transmissible nature of these enterotoxins and of a K antigen possessed by calf and lamb strains. *Journal of Medical Microbiology* **5**, 243–250.

Smyth, C. J. (1982). Two mannose-resistant haemagglutinins on enterotoxigenic *Escherichia coli* of serotype O6:K15:H16 or H⁻ isolated from travellers' and infantile diarrhoea. *Journal of General Microbiology* **128**, 2081–2096.

Smyth, C. J., Jonsson, P., Olsson, E., Söderlind, O., Rosengren, J., Hjertén, S. and Wadström, T. (1978). Differences in hydrophobic surface characteristics of porcine enteropathogenic *Escherichia coli* with or without K88 antigen as revealed by hydrophobic interaction chromatography. *Infection and Immunity* **22**, 462–472.

Stirm, S., Ørskov, I. and Ørskov, F. (1966). K88, an episome-determined protein antigen of *Escherichia coli. Nature (London)* **209**, 507–508.

Stirm, S., Ørskov, F., Ørskov, I. and Mansa, B. (1967a). Episome-carried surface antigen K88 of *Escherichia coli*. II. Isolation and chemical analysis. *Journal of Bacteriology* **93**, 731–739.

Stirm, S., Ørskov, F., Ørskov, I. and Birch-Andersen, A. (1967b). Episome-carried surface antigen K88 of *Escherichia coli*. III. Morphology. *Journal of Bacteriology* **93**, 740–748.

Sugarman, B. and Epps, L. R. (1982). Effect of oestrogens on bacterial adherence to HeLa cells. *Infection and Immunity* **35**, 633–638.

Sussman, M., Abraham, S. N. and Parry, S. H. (1982). Bacterial adhesion in the host-parasite relationship of urinary tract infection. *In* "Clinical Bacteriological and Immunological Aspects of Urinary Tract Infections in Children" (Ed. H. Schulte-Wissermann), pp. 103–112. Thieme, Stuttgart.

Svanborg Edén, C. and Hansson, H. A. (1978). *Escherichia coli* as possible mediators of attachment to human urinary tract epithelial cells. *Infection and Immunity* **21**, 229–237.

Svanborg Edén, C., Larsson, P. and Lomberg, H. (1980). Attachment of *Proteus mirabilis* to human urinary sediment uroepithelial cells *in vitro* is different from that of *Escherichia coli*. *Infection and Immunity* **27**, 804–807.

Svanborg Edén, C., Hagberg, L., Hanson, L. Å., Korhonen, T., Leffler, H. and Olling, S. (1981). Adhesion of *Escherichia coli* in urinary tract infection. *In* "Adhesion and Microorganism Pathogenicity" (Eds. K. Elliot, M. O'Connor and J. Whelan), Ciba Foundation Symposium 80, pp. 161–187. Pitman Medical, Tunbridge Wells.

Svanborg Edén, C., Hagberg, L., Hanson, L. Å., Hull, S., Hull, R., Jodal, Y., Leffler, H., Lomberg, H. and Straube, E. (1983a). Bacterial adherence—a pathogenetic mechanism in urinary tract infections caused by *Escherichia coli*. *Progress in Allergy* **33**, 175–188.

Svanborg Edén, C., Gotschlich, E. C., Korhonen, T. K., Leffler, H. and Schoolnik, G. (1983b). Aspects on structure and function of pili on uropathogenic *Escherichia coli*. *Progress in Allergy* **33**, 189–202.

Svenson, S. B. and Källenius, G. (1983). Density and localization of P-fimbriae-specific receptors on mammalian cells: Fluorescence-activated cell analysis. *Infection* **11**, 6–12.

Svenson, S. B., Källenius, G., Möllby, R., Hultberg, H. and Winberg, J. (1982). Rapid identification of P-fimbriated *Escherichia coli* by a receptor-specific particle agglutination test. *Infection* **10**, 209–215.

Svenson, S. B., Hultberg, H., Källenius, G., Korhonen, T. K., Möllby, R. and Winberg, J. (1983). P-fimbriae of pyelonephritogenic *Escherichia coli*: Identification and chemical characterization of receptors. *Infection* **11**, 61–67.

Takeuchi, A., Inman, L. R., O'Hanley, P. D., Cantey, J. R. and Lushbaugh, W. B. (1978). Scanning and transmission electron microscopic study of *Escherichia coli* O15 (RDEC-1) enteric infection in rabbits. *Infection and Immunity* **19**, 686–694.

Thomas, L. V., Cravioto, A., Scotland, S. M. and Rowe, B. (1982). New fimbrial antigenic type (E8775) that may represent a colonization factor in enterotoxigenic *Escherichia coli* in humans. *Infection and Immunity* **35**, 1119–1124.

Thorne, G. M., Deneke, C. F. and Gorbach, S. L. (1979). Haemagglutination and adhesiveness of toxigenic *Escherichia coli* isolated from humans. *Infection and Immunity* **23**, 690–699.

Tixier, G. and Gouet, P. (1975). Mise en évidence d'une structure agglutinant les hématies de cheval en présence de mannose et spécifique des souches d'*Escherichia coli* énterotoxiques d'origine bovine. *Comptes Rendus Hebdomadaires des Séances de l'Académie des Sciences Serie D* **281**, 1641–1644.

Tulloch, E. F., Ryan, K. J., Formal, S. B. and Franklin, F. A. (1973). Invasive enteropathic *Escherichia coli* dysentery. An outbreak in 28 adults. *Annals of Internal Medicine* **79**, 13–17.

Ulshen, M. H. and Rollo, J. L. (1980). Pathogenesis of *Escherichia coli* gastroenteritis in man— another mechanism. *New England Journal of Medicine* **302**, 99–101.

Väisänen, V., Elo, J., Tallgren, L. G., Siitonen, A. Mäkelä, P. H., Svanborg Edén, C., Källenius, G., Svenson, S. B., Hultberg, H. and Korhonen, T. (1981). Mannose-resistant haemagglutination and P antigen recognition are characteristic of *Escherichia coli* causing primary pyelonephritis. *Lancet* **2**, 1366–1369.

Väisänen, V., Korhonen, T. K., Jokinen, M., Gahmberg, C. G. and Ehnholm, C. (1982a). Blood group M-specific haemagglutinin in pyelonephritogenic *Escherichia coli*. *Lancet* **1**, 1192.

Väisänen, V., Lountamaa, K. and Korhonen, T. K. (1982b). Effects of sublethal concentrations of antimicrobial agents on the haemagglutination, adhesion and ultrastructure of pyelonephritogenic *Escherichia coli* strains. *Antimicrobial Agents and Chemotherapy* **22**, 120–127.

van den Bosch, J. F., Verboom-Sohmer, U., Postma, P., de Graaf, J. and Maclaren, D. M. (1980). Mannose-sensitive and mannose-resistant adherence to human uroepithelial cells and urinary virulence of *Escherichia coli*. *Infection and Immunity* **29**, 226–233.

van Die, I., van den Hondel, C., Hamstra, H.-J., Hoekstra, W. and Bergmans, H. (1983). Studies on the fimbriae of an *Escherichia coli* O6:K2:H1:F7 strain: Molecular cloning of a DNA fragment encoding a fimbrial antigen responsible for mannose-resistant haemagglutination of human erythrocytes. *FEMS Microbiology Letters* **19**, 77–82.

Vosbeck, K., Mett, H., Huber, U., Bohn, J. and Petignat, M. (1982). Effects of low concentrations of antibiotics on *Escherichia coli* adhesion. *Antimicrobial Agents and Chemotherapy* **21**, 864–869.

Wadström, T., Smyth, C. J., Faris, A., Johnsson, P. and Freer, J. H. (1979). Hydrophobic adsorptive and haemagglutinating properties of enterotoxigenic *Escherichia coli* with different colonizing factors: K88, K99 and colonization factor antigens and adherence factor. *In* "Proceedings of the International Symposium on Neonatal Diarrhoea" (Ed. S. D. Acres), pp. 29–55. Stuart Brandle Publishing Service, Edmonton, Canada.

Walker, P. D. and Nagy, L. K. (1980). Adhesion of organisms to animal tissues. *In* "Microbial Adhesion to Surfaces" (Eds. R. C. W. Berkeley, J. M. Lynch, J. J. Melling, P. R. Rutter and B. Vincent), pp. 473–494. Ellis Horwood Ltd., Chichester, England.

Wevers, P., Picken, R., Schmidt, G., Jann, B., Jann, K., Golecki, J. R. and Kist, M. (1980). Characterization of pili associated with *Escherichia coli* O18ac. *Infection and Immunity* **29**, 685–691.

Williams, P. H., Evans, N., Turner, P., George, R. H. and McNeish, A. S. (1977). Plasmid mediating mucosal adherence in human enteropathogenic *Escherichia coli*. *Lancet* **1**, 1151.

Williams, P. H., Sedgwick, M. I., Evans, N., Turner, P. J., George, R. H. and McNeish, A. S. (1978). Adherence of an enteropathogenic strain of *Escherichia coli* to human intestinal mucosa is mediated by a colicinogenic conjugative plasmid. *Infection and Immunity* **22**, 393–402.

Wilson, M. R. and Hohmann, A. W. (1974). Immunity to *Escherichia coli* in pigs: Adhesion of enteropathogenic *Escherichia coli* to isolated intestinal epithelial cells. *Infection and Immunity* **10**, 776–782.

5

Cell Surface Components and Virulence: *Escherichia coli* O and K Antigens in Relation to Virulence and Pathogenicity

K. JANN AND BARBARA JANN

Max-Planck-Institut für Immunbiologie, Freiburg-Zähringen, Federal Republic of Germany

Introduction

The cell wall of *Escherichia coli* contains as topmost layer the outer membrane, which is a typical feature of Gram-negative bacteria. As shown in Fig. 1, it contains protein, lipid, and lipopolysaccharide. Potentially, all these components are more or less exposed on the cell surface and thus can interact with cells or substances in the environment. Some proteins form pores through which substances can be exchanged across the cell wall. Others take part in such processes as iron scavenging or cell adhesion. With respect to virulence and pathogenicity of *E. coli* the most prominent outer membrane constituents are the cell wall lipopolysaccharides (LPS), which are not only dominant antigens but also mediators of a great many biological activities (Rietschel *et al.*, 1982). Serologically, the LPS are known as O antigens and pharmacologically as endotoxins.

Certain strains of *E. coli* are encapsulated and have an extracellular polysaccharide envelope. These capsular polysaccharides are termed K antigens. One of the properties of O and K antigens is their protective effect on the bacterial cell. Both protect the coli bacteria against the bactericidal action of phagocytes and complement although the capsular K antigens do so much more effectively than the cell wall O antigens. In what follows we will characterize O and K antigens and try to put our chemical knowledge of these surface components in the perspective of *E. coli* pathogenicity (Jann and Jann, 1982; Robbins *et al.*, 1980). However, it is necessary to stress that the complex phenomenon of bacterial pathogenicity, comprising such processes as bacterial adhesion, evasion of host defence, production of exo-compounds such as toxins and enzymes and mainte-

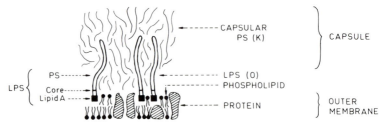

Fig. 1. Diagram of the outer layers of *E. coli.*

nance of bacterial growth under adverse nutritional conditions, can only be touched upon in an analysis of bacterial surface components (Smith, 1977).

Lipid A

Before turning to the polysaccharide antigens and their role in infection, the endotoxic lipid A moiety of the cell wall LPS will be briefly mentioned. It is a rather unusual phosphoglycolipid which has the same basic structure in all enterobacterial Gram-negative LPS (Rietschel *et al.*, 1982; Westphal *et al.*, 1982). As shown in Fig. 2, its backbone is the disaccharide β-glucosaminyl-1,6-glucosamine, in which both amino groups and all but two hydroxyl groups are substituted with saturated and with β-hydroxyl fatty acids. Position 4′ of the nonreducing glucosamine and position 1 of the reducing glucosamine are phosphorylated. These phosphate groups may be further substituted in the LPS of different

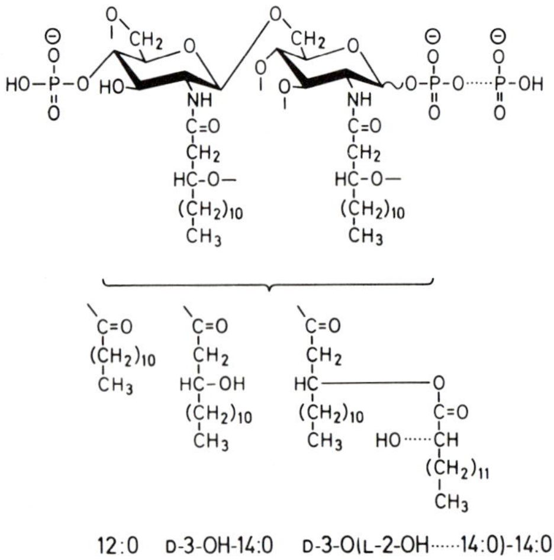

12:0   D-3-OH-14:0   D-3-O(L-2-OH·····14:0)-14:0

Fig. 2. Structure of lipid A from *E. coli.* (Adapted from Rietschel *et al.*, 1982, and Takayama *et al.*, 1983.)

Pyrogenicity
Leucopenia — leucocytosis
Activation   of complement
Activation   of macrophages
Induction   of prostaglandins
Induction   of interferon
Adjuvant activity
Mitogenic activity

Fig. 3. Endotoxin activities that may be relevant to infection.

genera. In *E. coli* there is a phospho-monoester at position 4′ and a pyrophosphoryl group at position 1. The linkage to the carbohydrate part of the LPS is at position 3′. Lipid A is thought to be one of the most potent bacterial non-protein toxins, although strict structure–function relationships are not known. Because of this, and because of the many effects triggered by lipid A, attempts are now being made to synthesize lipid A and its analogues in many laboratories. Some of the effects of lipid A which may be relevant to infection are listed in Fig. 3. They include such different phenomena as pyrogenicity, alteration of the blood cell population, macrophage activation, adjuvanticity and induction of a number of mediators of inflammation. In infections all these effects are probably caused by free LPS, released from the bacteria into the blood stream or into organs.

Lipopolysaccharides and S-R Mutation

For a consideration of the pathogenic effects, the polysaccharide moiety of the cell wall LPS is also important. The general structure of the LPS is shown in Fig. 4. The polysaccharide moiety, which is the outermost part of the molecule, is joined to lipid A through a spacer region, called the common core oligosaccharide. Very many distinct polysaccharides are encountered in *E. coli*, as is the case with Gram-negative bacteria in general. The various polysaccharides differ in the nature of their monosaccharide constituents as well as in their mode of linkage. The chemical structure of these polysaccharides is the molecular basis of their serological specificity, called the bacterial O specificity (Jann and West-

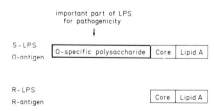

Fig. 4. Lipopolysaccharide structure in relation to the S-R mutation.

phal, 1975). The polysaccharide is therefore called the O-specific polysaccha-ride. Figure 4 also shows the result of a mutation, which may shed some light on the importance of the polysaccharide moiety in the virulence of *E. coli*. Coli bacteria which synthesize the complete LPS are called S forms or wild types. Their LPS are known as O antigens. Through a spontaneous mutation, the S-R mutation, the bacteria lose the ability to synthesize the O-specific polysaccharide moiety of their LPS with concomitant loss of their serological O specificity. The resulting R mutants have R-LPS, which consists of only lipid A and the core oligosaccharide. They are called R antigens and exhibit a previously cryptic specificity, the R specificity. Only a few R structures are known and accordingly the R-antigenic variability is much more restricted than the O-antigenic one. It is important that loss of the O-specific polysaccharide, especially of non-capsulated bacteria, results in loss of virulence and pathogenicity. R mutants are avirulent even when they are derived from virulent S forms. This is evidenced by a faster clearance of R mutants from blood and organs and by the ease with which they are phagocytosed. Thus, the O-specific polysaccharide moiety of the LPS is an important asset of pathogenicity in *E. coli*. This is especially so in non-encapsu-lated strains.

Antigenic Patterns

Structural and immunochemical analyses of the polysaccharide antigens of *E. coli* have revealed that the many different strains exhibit certain antigenic pat-terns (Ørskov *et al.*, 1977). These are combinations of O-antigenic cell wall polysaccharides and K-antigenic capsular polysaccharides, which can easily be demonstrated by immunonelectrophoresis of crude extracts (Ørskov *et al.*, 1971). Figure 5 shows, in a schematic way, the results obtained from the immu-noelectrophoresis of saline extracts of some *E. coli* strains. Similar results were obtained for all known *E. coli* strains with defined O and K antigens, the so-called O and K test strains. The samples were run against O and OK sera. Arcs which appear with both sera were ascribed to O antigens (cell wall LPS) and those appearing in OK but not in O sera were ascribed to K antigens (capsular polysaccharides). Four groups of *E. coli* strains can be differentiated by this method: some strains exhibit neutral O antigens and acidic capsular antigens—the latter having either a high or a low electrophoretic mobility. Other *E. coli* strains exhibit only O antigens, either neutral or acidic. Thus, there are clearly two types of O antigens and two types of K antigens, occurring in restricted combinations. It is striking that the K antigens with low electrophoretic mobility occur only in O groups 8 and 9, whereas those with a high electrophoretic mobility occur in many O groups.

   Before characterizing the various antigens, a rather interesting finding in rela-tion to these results should be discussed. If *E. coli* strains are grouped according

Immunoelectrophoretic pattern	Saline extract from E. coli			
anti-O / anti-OK	O1 : K1 O7 : K7 O15: K14 O23: K22	O3 : K2 O11: K10 O17: K16 O25: K23	O5 : K4 O13: K11 O4 : K12 O16: K54	O4 : K6 O6 : K13 O21: K20 O2 : K56
anti-O / anti-OK	O8 : K8 O8 : K27 O9 : K32	O9 : K9 O9 : K29 O8 : K40	O8 : K25 O9 : K30 O8 : K42	O9 : K26 O9 : K31 O9 : K57
anti-O / anti-OK	O111: K58 O127: K63 O126: K71	O55: K59 O128: K67 O78: K80	O25: K60 O119: K69 O142: K86	O86: K61 O125: K70 O114: K90
anti-O / anti-OK	O112 : K66 O136 : K78 O115	O124 : K72 O137 : K79 O143	O28 : K73 O141 : K85 O144	O113 : K75 (O32): K87

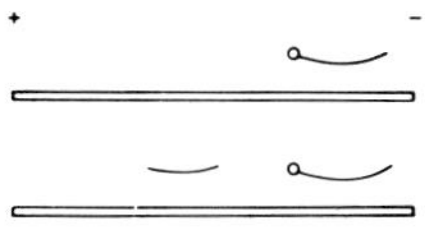

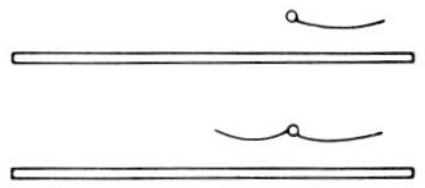

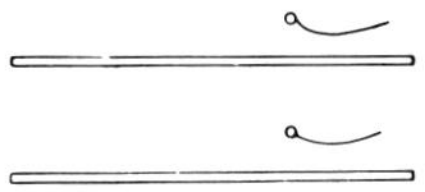

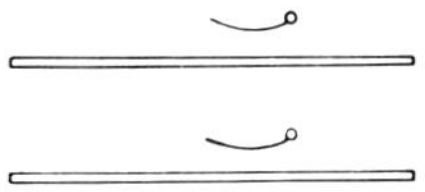

Fig. 5. Immunoelectrophoretic patterns of *E. coli*.

IMMUNOELECTROPHORETIC PATTERN	DISEASES
anti-O / anti-OK	extra-intestinal infections urinary tract infections
anti-O / anti-OK	extra-intestinal infections
anti-O / anti-OK	infantile diarrhoea
anti-O / anti-OK	dysentery - like disease

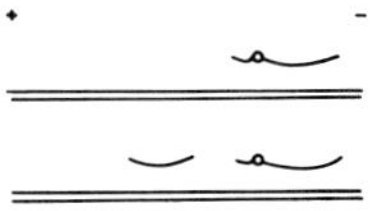

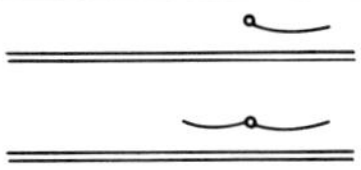

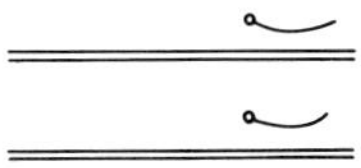

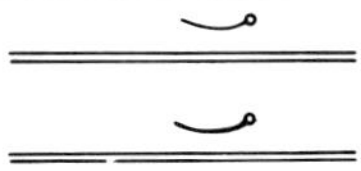

Fig. 6. Immunoelectrophoretic patterns and the pathogenicity of *E. coli*.

to the disease they cause, the same patterns turn up again (Ørskov *et al.*, 1971), as shown in Fig. 6. *E. coli* causing extra-intestinal diseases have neutral O antigens and charged K antigens. K antigens with a high electrophoretic mobility are present in strains causing urinary tract infections and also bacteraemia. Enteropathogenic and enterotoxigenic strains have only neutral O antigens and no capsules (i.e. no K polysaccharides); enteroinvasive strains causing dysentery-like disease exhibit only acidic O antigens and no capsules (i.e. no K polysaccharides). With respect to their surface polysaccharides, some *E. coli* strains resemble *Neisseria*, others resemble *Klebsiella, Salmonella* and perhaps *V. cholerae*, and yet others resemble *Shigella dysenteriae*. These similarities are true not only for the characteristics of their polysaccharide antigens but also for their pathogenicity. Another consistent finding is that invasiveness seems to be correlated with a high surface charge of the *E. coli* cell.

Cell Wall (O) Antigens

Structural studies have revealed that microbial polysaccharides consist of oligosaccharide repeating units (Jann and Westphal, 1975). In structural representations it suffices therefore to formulate the repeating oligosaccharide and the joining linkage. Some repeating units of neutral O-antigenic LPS are shown in Fig. 7. These may be linear homopolysaccharides like that of *E. coli* **O**8 (Reske and Jann, 1972), which is identical with that of *Klebsiella* **O**5. Others, like those of *E. coli* **O**75 and **O**111 (Edström and Heath, 1965; Erbing *et al.*, 1975), are branched heteropolysaccharides. Both these *E. coli* strains are enteropathogenic. Another representative of this group is *E. coli* **O**114. The repeating unit of this O antigen has recently been elaborated (Dmitriev *et al.*, 1983; K. Jann and B. Jann, unpublished) and is also shown in Fig. 7. The repeating units of two acidic O

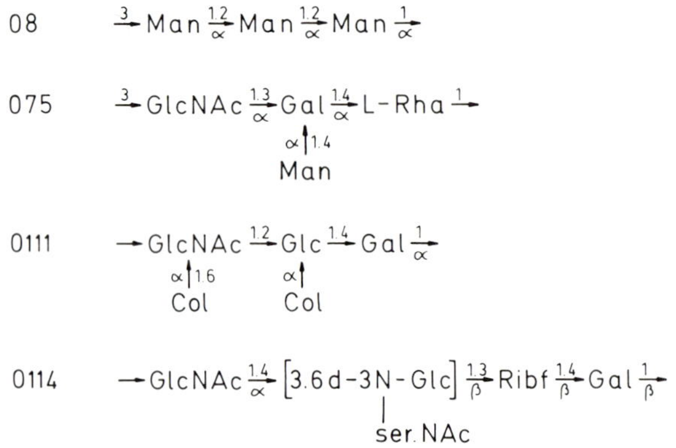

Fig. 7. Repeating units of some neutral O antigens (LPS).

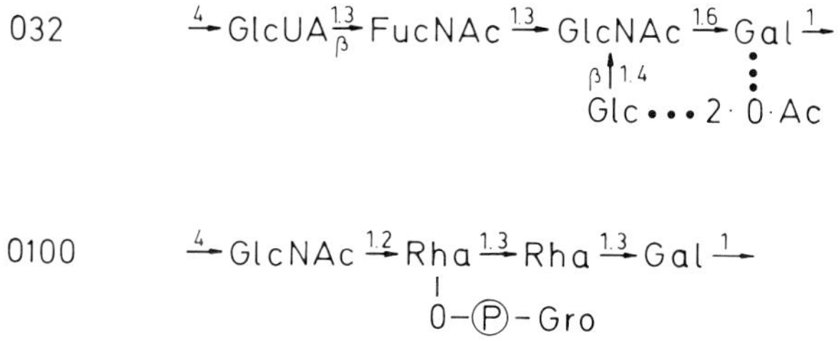

032    $\xrightarrow{4}$ GlcUA $\underset{\beta}{\xrightarrow{13}}$ FucNAc $\xrightarrow{13}$ GlcNAc $\xrightarrow{16}$ Gal $\xrightarrow{1}$

$\qquad\qquad\qquad\qquad\quad \beta{\uparrow}1.4 \qquad \vdots$

$\qquad\qquad\qquad\qquad\quad$ Glc $\bullet\bullet\bullet$ 2·O·Ac

0100 $\xrightarrow{4}$ GlcNAc $\xrightarrow{12}$ Rha $\xrightarrow{13}$ Rha $\xrightarrow{13}$ Gal $\xrightarrow{1}$

$\qquad\qquad\qquad\qquad\quad$ O–(P)–Gro

Fig. 8. Repeating units of two acidic O antigens (LPS).

antigens are shown in Fig. 8. The negative charge may be due to hexuronic acid or to phosphate.

As mentioned above, some *E. coli* strains have acidic O antigens and cause dysentery-like diseases. In this group of bacteria we encounter the interesting phenomenon of the intergeneric identity of O antigens in combination with an intergeneric identity of pathogenic mechanisms. As shown in Fig. 9, the O-specific polysaccharides of *E. coli* **O**58 and **O**124 (Dmitriev *et al.*, 1976; Dmitriev *et al.*, 1977) have the same structures as those of *Shigella dysenteriae* types 3 and 5, respectively. These polysaccharides contain the unusual sugar components rhamnolactylic acid and glucolactylic acid, in which lactic acid is linked to the sugar ring in the same fashion as in muramic acid, the typical constituent of peptidoglycan. The structures of these compounds are compared in Fig. 10. Lactic acid is linked by an ether linkage to C3 of L-rhamnose, to C4 of D-glucose and to C3 of D-*N*-acetylglucosamine. Thus, characteristic structural features of the glycopeptide are also found in components of the outer membrane of *E. coli*. Although the four strains mentioned are all invasive, one cannot directly correlate their invasiveness with the negative surface charge. It was recently found that *E. coli* **O**124 and *S. dysenteriae* type 5 seem to interact with a protein of guinea-pig colonic epthelium (Izhar *et al.*, 1982; D. Mirelman, K. Jann and B. Jann, unpublished). In these cases O-antigenic LPS seem to participate in bacterial adhesion, being recognized by a mammalian lectin.

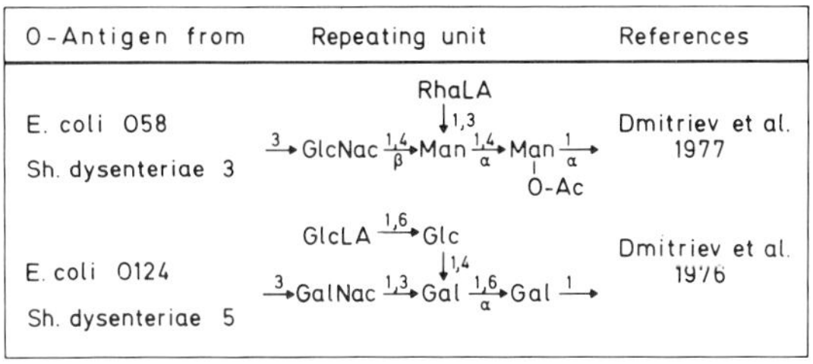

O–Antigen from	Repeating unit	References
E. coli O58 Sh. dysenteriae 3	RhaLA $\downarrow$1,3 $\xrightarrow{3}$ GlcNac $\underset{\beta}{\xrightarrow{1,4}}$ Man $\underset{\alpha}{\xrightarrow{1,4}}$ Man $\underset{\alpha}{\xrightarrow{1}}$ $\qquad\qquad\qquad$ O-Ac	Dmitriev et al. 1977
E. coli O124 Sh. dysenteriae 5	GlcLA $\xrightarrow{1,6}$ Glc $\qquad\qquad\downarrow$1,4 $\xrightarrow{3}$ GalNac $\xrightarrow{1,3}$ Gal $\underset{\alpha}{\xrightarrow{1,6}}$ Gal $\xrightarrow{1}$	Dmitriev et al. 1976

Fig. 9. O antigens (LPS) common to *E. coli* and *S. dysenteriae*. RhaLA, rhamnolactylic acid; GlcLA, glucolactylic acid.

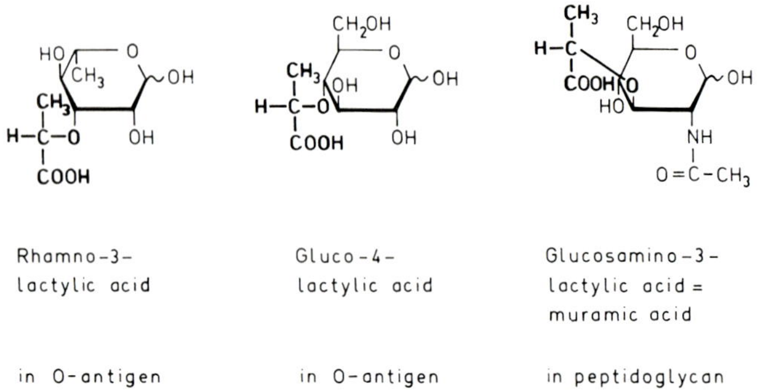

CH$_2$OH

CH$_2$OH

CH$_3$ CH$_2$OH

Rhamno-3-
lactylic acid

Gluco-4-
lactylic acid

Glucosamino-3-
lactylic acid =
muramic acid

in O-antigen

in O-antigen

in peptidoglycan

Fig. 10. O-Lactyl-substituted monosaccharides.

Capsules

Although the interactions of the cell wall LPS with substances or with cells of the infected host are not well understood, their main role in infection seems to be the neutralization of the non-specific host defence (Robbins *et al.,* 1980). This is of special importance for invasive *E. coli.* Most of these strains are encapsulated and it stands to reason that the extra-cellular layer of a capsule provides a good means of overcoming the host defence. In this context we can consider two main effects of capsules as assets of *E. coli* pathogenicity. As is illustrated schematically in Fig. 11, capsules can shield the bacterial cell against the bactericidal action of complement and phagocytes. In this effect, which appears basically to be a physical one, charge seems to play a role. The host defence does not become effective until specific anti-capsular antibodies are formed, which then neutralize the protective effect of the capsule in a process called opsonization. The capsular K-12 antigen, often encountered in urinary tract infections, is a typical example of a shielding capsule. Another means of bacterial self-defence is of an entirely

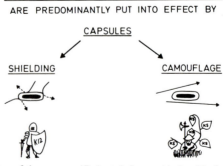

Fig. 11. Evasion of the non-specific host defence with the aid of capsules.

Fig. 12. Electron micrograph of *E. coli* with a thick capsule.

different nature. It is based on structural relations and it may be described as a camouflage of the bacteria with the help of their capsules. Possibly owing to structural similarities with host material, the encapsulated *E. coli* cells cannot be properly recognized as foreign by the immune system of the host and antibodies against them cannot be formed or are only inefficiently formed. In such cases, the non-specific pre-immune phase of the host defence is very much prolonged. Such a camouflage effect, in addition to shielding, is typically exerted by the *E. coli* K1 and K5 capsular polysaccharides.

E. coli capsules can surround the bacterial cell as thick copious layers. As shown in an electron micrograph of an *E. coli* thin section (Fig. 12), the capsules

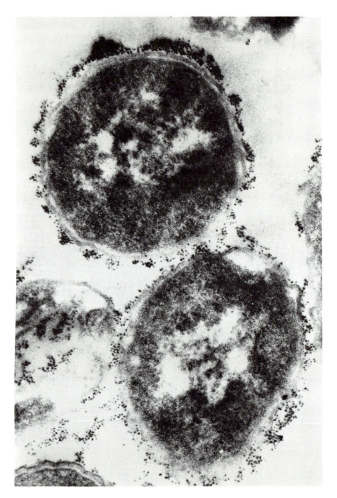

Fig. 13. Electron micrograph of *E. coli* with a thin capsule.

extend far into the environment of the cell. Another capsular form, exhibited by many *E. coli* bacteria, is shown in Fig. 13. Such capsules have quite a thin and patchy appearance. Both forms of capsule are genetically determined and there is no interchange between thick and thin capsules. Although such a differentiation may seem somewhat superficial, there is chemical and physical evidence to support this distinction (Jann and Jann, 1982). Furthermore, to isolate the different capsular antigens the respective coli bacteria must be cultivated under different conditions (Jann and Jann, 1982). Therefore, the distinction between capsular antigens (K polysaccharides) of thick and thin capsules of *E. coli* will be used in the following.

Escherichia coli bacteria may have:	
Thick copious capsules	Thin patchy capsules
K(A) antigens are acidic polysaccharides	K antigens are acidic polysaccharides
Molecular weights above 100,000	Molecular weights below 50,000
Acid components:	Acid components:
hexuronic acids	hexuronic acids, KDO, NANA
Capsules are comparable to:	Capsules are comparable to:
Klebsiella capsules	Neisseria capsules

Fig. 14. Characterization of capsular antigens of *E. coli*.

Capsular (K) Antigens

As is shown in Fig. 14, both capsules are composed of acidic polysaccharides (Jann and Jann, 1982). The polysaccharides forming thick capsules have much higher molecular weights (up to several hundred thousand) than those forming thin capsules. The latter usually have molecular weights ranging between about 5000 and 30,000. There are also characteristic differences in the chemical composition of the two groups of capsular polysaccharides. The acidic components of the polysaccharides forming thick capsules are hexuronic acids, in most cases glucuronic acid. In some of these polysaccharides pyruvate substitution has also been found. In the acidic polysaccharides of thin capsules, hexuronic acids are rarely found. Instead, we encounter *N*-acetylneuraminic acid and frequently 2-keto-3-deoxy-D-manno-octonic acid (KDO). It should be mentioned that high-molecular-weight K polysaccharides from thick capsules never contain amino sugars, a characteristic which they share with the capsular polysaccharides of *Klebsiella*. It is known that the latter bacteria also form thick and copious capsules. The high-molecular-weight polysaccharides from thick capsules have a lower charge density than the low-molecular-weight polysaccharides from thin capsules. Therefore, the high-molecular-weight polysaccharides have a lower electrophoretic mobility than the low-molecular-weight ones. Thus, as in the case of the O-antigenic cell wall LPS, the general physical and chemical character of the capsular polysaccharides is in accordance with the immunoelectrophoretic analysis shown in Figs. 5 and 6 and may also determine the pathogenic properties of the respective *E. coli* strains (Jann and Jann, 1982; Ørskov *et al.*, 1971). The low-molecular-weight polysaccharides from thin capsules are typical of *E. coli* causing urinary tract infections, bacteraemia and neonatal meningitis. Interestingly, bacteraemia and neonatal meningitis can also be caused by certain *Neisseria*, which also have capsules containing *N*-acetylneuraminic acid or KDO.

N-Acetylneuraminic acid and KDO are unusual polysaccharide components. *N*-Acetylneuraminic acid is a constituent of glycoproteins and gangliosides, and

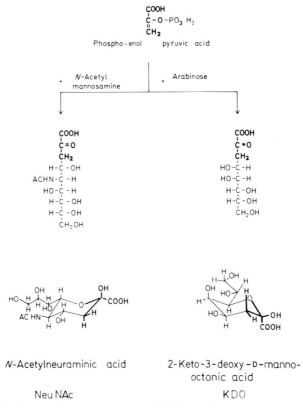

Fig. 15. Common biosynthetic pathway of KDO and N-acetylneuraminic acid.

K antigen	Repeating unit

K27
$$\xrightarrow{6} Glc \xrightarrow{1.3} GlcUA \xrightarrow{1.3} Fuc \xrightarrow{1}$$
$$\Big\downarrow 1.3$$
$$Gal$$

K29
$$\xrightarrow{2} Man \xrightarrow[\beta]{1.3} Glc \xrightarrow{1.3} GlcUA \xrightarrow[\beta]{1.3} Gal \xrightarrow{1}$$
$$\uparrow$$
$$(Py) = Glc \longrightarrow Man$$
$$4,6$$

K30
$$\xrightarrow{2} Man \xrightarrow{1.3} Gal \xrightarrow{1}$$
$$\uparrow 1.3$$
$$GlcUA \xrightarrow[\beta]{1.3} Gal$$

Fig. 16. Repeating units of three polysaccharides from thick capsules.

KDO is characteristically found in enterobacterial LPS, where it forms the linkage region between the lipid and carbohydrate moieties (Jann and Westphal, 1975). These two carbohydrates are biochemically related. As shown in Fig. 15, during their biosynthesis, phosphoenolpyruvate may condense either with *N*-acetylmannosamine to yield *N*-acetylneuraminic acid, or with arabinose to yield KDO.

Using physical and chemical methods, we have characterized a number of capsular polysaccharides of *E. coli*. Figure 16 shows the repeating units of three high-molecular-weight polysaccharides from thick capsules, which bear strong similarity to *Klebsiella* K antigens. The K27 antigen of *E. coli* (Jann *et al.*, 1968) is even a substrate for a *Klebsiella* bacteriophage (Sutherland *et al.*, 1970). *E. coli* strains with the K29 or K30 antigen (Chakraborty *et al.*, 1980; Choy *et al.*, 1975) have been used in model studies of phagocytosis (Horowitz and Silverstein, 1980) and of pathogenicity in the mouse (K. Jann, unpublished).

In the following we shall concentrate on the low-molecular-weight polysac-

K	REPEATING UNIT	REFERENCE
6	$\xrightarrow{3}$ Rib $\xrightarrow[\beta]{1,7}$ KDO $\xrightarrow[\beta]{2}$ $\beta \uparrow 1,2$ Rib	Messmer & Unger 1980 Neszmélyi et al. 1982
	$\xrightarrow{2}$ Rib $\xrightarrow[\beta]{1,2}$ Rib $\xrightarrow[\beta]{1,7}$ KDO $\xrightarrow[\alpha]{2}$	Jennings et al. 1982
12	$\xrightarrow{3}$ Rha $\xrightarrow[\alpha]{1,2}$ Rha $\xrightarrow[\alpha]{1,5}$ KDO $\xrightarrow[\beta]{2}$ $\vdots 7,8$ Oac	Schmidt & Jann 1983
13	$\xrightarrow{3}$ Rib $\xrightarrow[\beta]{1,7}$ KDO $\xrightarrow[\beta]{2}$ $\vdots 5$ Oac	Vann & Jann 1979
14	$\xrightarrow{6}$ GalNac $\xrightarrow[\alpha]{1,5}$ KDO $\xrightarrow[\beta]{2}$ $\vdots 8$ Oac	Jann et al. 1983
15	$\xrightarrow{4}$ GlcNac $\xrightarrow[\alpha]{1,5}$ KDO $\xrightarrow[\beta]{2}$	Vann pers. comm.
20	$\xrightarrow{3}$ Rib $\xrightarrow[\beta]{1,7}$ KDO $\xrightarrow[\beta]{2}$ $\vdots 5$ Oac	Vann pers. comm.
23	$\xrightarrow{3}$ Rib $\xrightarrow[\beta]{1,7}$ KDO $\xrightarrow[\beta]{2}$	Vann & Jann unpubl.

Fig. 17. Repeating units of KDO-containing polysaccharides from thin capsules.

charides from the thin capsules of *E. coli*. The repeating units of KDO-containing polysaccharides are shown in Fig. 17. These capsular antigens are a group of similar polysaccharides, among which only small differences in the nature and linkage of the sugar constituents are found. In all of them KDO, which provides the charge, is present in the β-glycosidic linkage and is closely spaced. Most of the polysaccharides are O-acetylated. This may lead to antigenic variation, as evidenced by a comparison of the K20 and K23 antigens. The capsular polysaccharide of *Neisseria meningitidis* 29-e has the same sugar sequence as the *E. coli* K14 antigen, but different linkages between the sugar units (Bhattacharjee *et al.*, 1978). There is no serological relation between the K14 polysaccharide and the capsular polysaccharide of *N. meningitidis* 29-e. The K12 and K14 capsular polysaccharide, however, cross-react, owing to the same substitution of the KDO unit in both polymers. In comparing native and modified polysaccharides, we found that the immunodominant group is always KDO, whether acetylated or not. For a closer and more reliable serological analysis of these polysaccharides, monoclonal antibodies will have to be used. The construction of anticapsular hybridomas and the characterization of the respective monoclonal anti-K antibodies has begun in our and in other laboratories.

Not all low-molecular-weight capsular polysaccharides of uropathogenic and invasive *E. coli* contain KDO. Some of the non-KDO polysaccharides are shown in Fig. 18. These are the K1, K2, K5 and K51 polysaccharides. The K1 polysaccharide is a poly-o-2.8-*N*-acetylneuraminic acid and has been known as colominic acid for many years (McGuire and Binkley, 1964). Its structure, which is the same as that of the *Neisseria meningitidis* serogroup B capsule, has been verified in two laboratories (Bhattacharjee *et al.*, 1975; Lifely *et al.*, 1981). The K2 antigen (Jann *et al.*, 1980) is reminiscent of certain teichoic acids of *Sta-*

K	REPEATING UNIT	REFERENCE
1Ac+	$\xrightarrow{8}$ NeuNAc $\xrightarrow{2}_{\alpha}$ $|7/9$ OAc	McGuire, Binkley : 1964
1Ac−	$\xrightarrow{8}$ NeuNAc $\xrightarrow{2}_{\alpha}$	Ørskov *et al.* : 1979
2ab	$\left(\xrightarrow{4} \text{Galp} \xrightarrow[\alpha]{1,2} \text{Gro} \xrightarrow{3} \text{P}\right)_{2-4} \left(\xrightarrow{5} \text{Galf} \xrightarrow[\alpha]{1,2} \text{Gro} \xrightarrow{3} \text{P}\right)$ $|2,3$ $|2,3$ OAc OAc	Jann, Schmidt : 1980
2a	$\left(\xrightarrow{4} \text{Galp} \xrightarrow[\alpha]{1,2} \text{Gro} - \text{P}\right)_{2-4} \left(\xrightarrow{5} \text{Galf} \xrightarrow[\alpha]{1,2} \text{Gro} \xrightarrow{3} \text{P}\right)$	Jann, Schmidt : 1980
5	$\xrightarrow{4} \text{GlcUA} \xrightarrow[\beta]{1,4} \text{GlcNAc} \xrightarrow[\alpha]{1}$	Vann *et al.* : 1981
51	$- \text{GlcNAc} - \text{P} -$ $|$ $(\text{OAc})_2$	Jann, Jann unpubl.

Fig. 18. Repeating units of non-KDO polysaccharides from thin capsules.

phylococcus, polymers which have not previously been found in Gram-negative bacteria. It consists of galactose, glycerol and phosphate in a linear sequence and contains galactopyranose and galactofuranose in the same chain. The glycerol phosphate moiety of the K2 antigen has the sn-3 configuration (Fischer *et al.,* 1982), which is the same as that found in teichoic acids and in phospholipids. The K5 antigen has a very simple structure with a repeating sequence of β-glucuronyl-1,4-*N*-acetylglucosamine (Vann *et al.,* 1981). The K51 polysaccharide is a per-O-acetylated poly-*N*-acetylglucosamine-1-phosphate, the linkages of which are not yet known (K. Jann and B. Jann, unpublished).

Like the KDO-containing K polysaccharides, the non-KDO polysaccharides are also subject to antigen modifications by O-acetylation: the K1 antigen and K2 antigen both occur in a non-acetylated and an acetylated form (Jann and Schmidt, 1980; Ørskov *et al.,* 1979). Bacteria with the non-acetylated polymers seem to be more pathogenic.

A common feature of the KDO-containing and non-KDO polysaccharides of invasive *E. coli* is their low molecular weight in combination with very high charge density. Considering these properties, it appears difficult to envisage how these polysaccharides can form a capsule which covers the bacterial cell. In a detailed chemical analysis we found that these low-molecular-weight antigens contain a hydrophobic moiety, as shown in Fig. 19 with the K-12 polysaccharide as an example (Schmidt and Jann, 1982). The lipid part is a phosphatidic acid, linked to the reducing end of the polysaccharides through a phosphodiester bridge. Similar results were obtained with some poly-*N*-acetylneuraminic acid antigens (Gotschlich *et al.,* 1981), although the phosphatidic acid was not isolated in that study. By using radio-reduction, the reducing sugar was found to be KDO. We assume that the polysaccharides are kept in association with the bacterial cell through hydrophobic interactions of their lipid moiety with the cell wall. This results in the formation of a capsule around the *E. coli* cell, exerting its protective effects.

One of these effects, evasion of the non-specific host defence by shielding of the bacteria with their negatively charged capsules, can eventually be overcome by anti-capsular antibodies, which are formed in the immune phase of the host defence. As previously noted, some capsular polysaccharides may escape im-

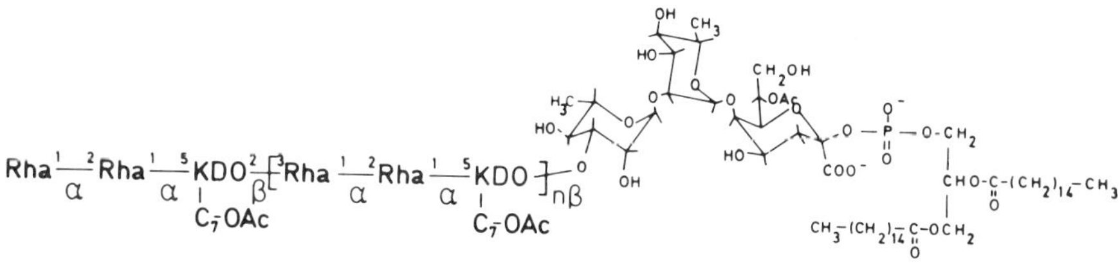

Fig. 19. Structure of the K-12 antigen, consisting of a polysaccharide and a lipid moiety.

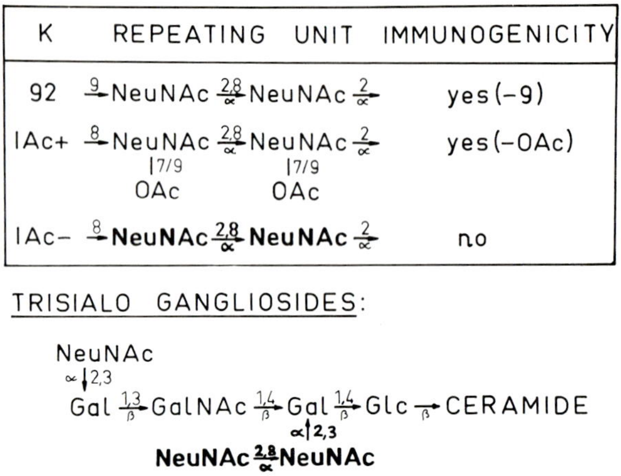

K	REPEATING UNIT	IMMUNOGENICITY
92	$\overset{9}{-}$NeuNAc $\overset{2,8}{\underset{\alpha}{-}}$NeuNAc $\overset{2}{\underset{\alpha}{-}}$	yes (−9)
IAc+	$\overset{8}{-}$NeuNAc $\overset{2,8}{\underset{\alpha}{-}}$NeuNAc $\overset{2}{\underset{\alpha}{-}}$ $\quad$ |7/9 $\qquad$ |7/9 $\quad$ OAc $\qquad\quad$ OAc	yes (−OAc)
IAc−	$\overset{8}{-}$**NeuNAc**$\overset{2,8}{\underset{\alpha}{-}}$**NeuNAc** $\overset{2}{\underset{\alpha}{-}}$	no

TRISIALO GANGLIOSIDES:

NeuNAc
$\quad \alpha\downarrow 2,3$
Gal $\overset{1,3}{\underset{\beta}{-}}$GalNAc $\overset{1,4}{\underset{\beta}{-}}$Gal $\overset{1,4}{\underset{\beta}{-}}$Glc $\overset{}{\underset{\beta}{-}}$CERAMIDE
$\qquad\qquad\qquad \alpha\downarrow 2,3$
$\qquad\qquad$ **NeuNAc**$\overset{2,8}{\underset{\alpha}{-}}$**NeuNAc**

Fig. 20. Possible reason for the non-immunogenicity of K1 polysaccharide.

mune recognition and the onset of the specific host defence, which is governed by immune reactions against the bacteria, may be delayed or even abrogated. Capsular antigens which exert this camouflage effect are the K1 and K5 antigens—and this is probably one of the reasons why *E. coli* bacteria with these capsular antigens are particularly virulent. An explanation of this fact can be attempted by structural comparisons, as shown in Fig. 20 with the K1 antigen. Three very similar α-poly-*N*-actylneuraminic acid antigens of *E. coli* are compared in this picture: the K92 antigen in which α-2.9 and α-2.8 linkages alternate, the acetylated K1 antigen and its non-acetylated form which both contain only α-2.8 linkages (Egan *et al.*, 1977; Ørskov *et al.*, 1979). The K92 antigen gives rise to antibodies which are specific for the 2.9-linked and not for the 2.8-linked α-*N*-acetylneuraminic acid. The acetylated K1ac$^+$ antigen is a weak immunogen and the immunodominant group is the *O*-acetyl substituent. The non-acetylated K1ac$^-$ antigen has a very low immunogenicity in rabbits and is not immunogenic in man (Glode *et al.*, 1977; Kasper *et al.*, 1973). Thus, it is the α-2.8-linked *N*-acetylneuraminic acid which cannot be recognized by the host. As shown in the lower part of Fig. 20, this structure is also present in certain gangliosides. Recognition of the K1 antigen by the immune system may, therefore, be hampered by this structural relation with gangliosides of host cells. The copy number and/or expression of these glycolipids on host cell surfaces may dictate whether the K1 antigen is a weak antigen or none at all.

The capsular antigen of *Neisseria meningitidis* B has the same structure as the K1 antigen of *E. coli* (Bhattacharjee *et al.*, 1975). It is also practically non-immunogenic and the discussion of structural relationships concerning the K1 antigen also applies to the capsular antigen of *N. meningitidis* B. It is interesting to note that the capsular antigens of *N. meningitidis* B and *E. coli* K1 not only have

K5 POLYSACCHARIDE:

$\cdots\overset{4}{\rightarrow}$GlcNAc$\overset{1,4}{\underset{\alpha}{\rightarrow}}$GlcUA$\overset{1,4}{\underset{\beta}{\rightarrow}}$GlcNAc$\overset{1,4}{\underset{\alpha}{\rightarrow}}$GlcUA$----$ⓟ$-$DIGLYCERIDE

INTERMEDIATE OF HEPARIN BIOSYNTHESIS:

$\cdots\overset{4}{\rightarrow}$GlcNAc$\overset{1,4}{\underset{\alpha}{\rightarrow}}$GlcUA$\overset{1,4}{\underset{\beta}{\rightarrow}}$GlcNAc$\overset{1,4}{\underset{\alpha}{\rightarrow}}$GlcUA$\rightarrow$(Gal)$_2$-Xyl-PROTEIN

Fig. 21. Possible reason for the non-immunogenicity of K5 polysaccharide.

the same structure but produce the same disease, neonatal meningitis. This is, therefore, another example of intergeneric identity of surface structure and pathogenicity.

The non-immunogenicity of the *E. coli* K5 antigen may be discussed in a fashion similar to that of the *E. coli* K1 antigen. Figure 21 shows that this capsular polysaccharide is practically identical with a biochemical precursor of heparin. It has, in fact, been shown recently that the capsular K5 polysaccharide is a substrate for the enzyme de-*N*-acetylase, which initiates the processing of the precursor of heparin (Rodén, 1980). This precursor is probably sufficiently exposed in mammals to prevent the K5 polysaccharide from being recognized as foreign by the immune system.

Structural relatedness such as presented here is probably not the only reason for the very low or absent immunogenicity of certain bacterial surface polysaccharides. Such a correlation, however, appears quite suggestive.

Conclusion

Although the involvement of cell wall O antigens and capsular K antigens is difficult to demonstrate directly, there is now ample evidence for their importance in host–parasite interactions. The intergeneric identity of structural expression in a similar pathogenic situation shows that certain surface structures seem to provide optimal means for counteracting the non-specific host defence, or for establishing an advantage in certain micro-ecological situations. For a given pathogenic process there seem to exist preferred combinations of polysaccharides on the bacterial surface and differences in charge seem to dictate differences in pathogenic properties of *E. coli*. The most important polysaccharides of pathogenic bacteria are undoubtedly the capsular ones. In view of the fact that all capsules are charged, it is an open question whether the acidic cell wall LPS can also assume capsular functions in pathogens like the enteroinvasive *Shigella*-like *E. coli*.

A closer examination of the role which the bacterial surface polysaccharides play in *E. coli* infections has become possible by their structural elucidation and by their immunochemical analysis with conventional sera and more recently with monoclonal antibodies. Our present knowledge provides a means not only for a study of host–parasite interactions on a molecular level but also for tackling such practical problems as the production of polysaccharide vaccines for the prevention of bacterial infections.

NOTE ADDED IN PROOF: Recently, a glycoprotein containing an α-2.8-oligo-*N*-acetylneuraminic acid has been found in neuronal cell membranes (Finne, 1982; Hoffman *et al.*, 1982). It was suggested by Edelman (1983) that this glycoprotein is identical with the neuronal cell adhesion molecule (n-CAM). The structural identity of the oligo-*N*-acetylneuraminic acid of this recognition molecule with the *E. coli* K1 antigen has shed a new light on the low immunogenicity of the K1 antigen. The immunogenic blindness to the K1 capsule may possibly be due to transient expression of a complex carbohydrate essential in brain development.

References

Bhattacharjee, A. K., Jennings, H. J., Kenny, C. P., Martin, A. and Smith, I. C. P. (1975). Structural determination of the sialic acid polysaccharide antigens of *Neisseria meningitidis* serogroups B and C with carbon 13 nuclear magnetic resonance. *Journal of Biological Chemistry* **250**, 1926–1932.

Bhattacharjee, A. K., Jennings, H. J. and Kenny, C. P. (1978). Structural elucidation of the 3-deoxy-D-manno-octulosonic acid containing meningococcal 29-e capsular polysaccharide antigen using carbon-13 nuclear magnetic resonance. *Biochemistry* **17**, 645–651.

Chakraborty, A. K., Friebolin, H. and Stirm, S. (1980). Primary structure of the *Escherichia coli* serotype K30 capsular polysaccharide. *Journal of Bacteriology* **141**, 971–972.

Choy, Y. M., Fehmel, F., Frank, N. and Stirm, S. (1975). *Escherichia coli* capsule bacteriophages. VI. Primary structure of the bacteriophage 29 receptor, the *E. coli* serotype 29 capsular polysaccharide. *Journal of Virology* **16**, 581–590.

Dmitriev, B. A., Lvov, V. L., Kochetkov, N. K., Jann, B. and Jann, K. (1976). Cell wall lipopolysaccharide of the "Shigella-like" *Escherichia coli* O124. Structure of the polysaccharide chain. *European Journal of Biochemistry* **64**, 491–498.

Dmitriev, B. A., Knirel, Y. A., Kochetkov, N. K., Jann, B. and Jann, K. (1977). Cell wall lipopolysaccharide of the "Shigella-like" *Escherichia coli* O58. Structure of the polysaccharide chain. *European Journal of Biochemistry* **79**, 111–115.

Dmitriev, B. A., Lvov, V., Tochtamysheva, N. V., Shashkov, A. S., Kochetkov, N. K., Jann, B. and Jann, K. (1983). Cell wall lipopolysaccharide of *Escherichia coli* O114:H4. Structure of the polysaccharide chain. *European Journal of Biochemistry* **134**, 517–521.

Edelman, G. M. (1983). Cell adhesion molecules. *Science* **219**, 450–457.

Edström, R. D. and Heath, E. E. (1965). Isolation of colitose-containing oligosaccharides from the cell wall lipopolysaccharide of *Escherichia coli*. *Biochemical and Biophysical Research Communications* **21**, 638–643.

Egan, W., Liu, T. Y., Dorow, D., Cohen, J. S., Robbins, J. D., Gotschlich, E. C. and Robbins, J. R. (1977). Structural studies on the sialic acid polysaccharide antigen of *Escherichia coli* strain Bos-12. *Biochemistry* **16**, 3687–3692.

Erbing, C., Svensson, S. and Hammarström, S. (1975). Structural studies on the O-specific side-chains of the cell wall lipopolysaccharide from *E. coli* O75. *Carbohydrate Research* **44**, 259–265.

Finne, J. (1982). Occurrence of unique polysialyl carbohydrate units in glycoprotein of developing brain. *Journal of Biological Chemistry* **257**, 11966–11970.

Fischer, W., Schmidt, M. A., Jann, B. and Jann, K. (1982). Structure of the *Escherichia coli* K2 capsular antigen. Stereochemical configuration of the glycerol phosphate and distribution of galactopyranosyl and galactofuranosyl residues. *Biochemistry* **21**, 1279–1284.

Glode, M. P., Robbins, J. B., Liu, T. Y., Gotschlich, E. C., Ørskov, I. and Ørskov, F. (1977).

Cross antigenicity and immunogenicity between capsular polysaccharides of group C *Neisseria meningitidis* and of *Escherichia coli* K92. *Journal of Infectious Diseases* **135**, 94–102.

Gotschlich, E. C., Frazer, B. A., Nishimura, O., Robbins, J. B. and Liu, T. Y. (1981). Lipid on capsular polysaccharides of Gram-negative bacteria. *Journal of Biological Chemistry* **256**, 8915–8921.

Hoffman, S., Sorkin, B., White, P. C., Brackenburg, R., Mailhammer, R., Rutishauser, U., Cunningham, B. A. and Edelman, G. M. (1982). Chemical characterization of a neural cell adhesion molecule purified from embryonic brain membranes. *Journal of Biological Chemistry* **257**, 7720–7729.

Horowitz, M. A. and Silverstein, S. C. (1980). Influence of the *Escherichia coli* capsule on complement fixation and phagocytosis and killing by human phagocytes. *Journal of Clinical Investigation* **65**, 82–94.

Izhar, M., Nuchamowitz, Y. and Mirelman, D. (1982). Adherence of *Shigella flexneri* to guinea pig intestinal cells is mediated by a mucosal adhesin. *Infection and Immunity* **35**, 1110–1118.

Jann, K. and Jann, B. (1982). The K antigens of *Escherichia coli*. *Progress in Allergy* **33**, 53–79.

Jann, K. and Schmidt, M. A. (1980). Comparative chemical analysis of two variants of the *Escherichia coli* K2 antigen. *FEMS Microbiology Letters* **7**, 79–81.

Jann, K. and Westphal, O. (1975). Microbial polysaccharides. *In* "The Antigens," Vol. 3 (Ed. M. Sela), pp. 1–125. Academic Press, New York.

Jann, K., Jann, B., Schneider, K. F., Ørskov, F. and Ørskov, I. (1968). Immunochemistry of K antigens of *Escherichia coli*. 5. The K antigen of *E. coli* O8:K27(A):H⁻. *European Journal of Biochemistry* **5**, 456–465.

Jann, K., Jann, B., Schmidt, M. A. and Vann, W. F. (1980). Structure of the *Escherichia coli* K2 capsular antigen, a teichoic acid like polymer. *Journal of Bacteriology* **143**, 1108–1115.

Jann, K., Hofmann, P. and Jann, B. (1983). Structure of the 2-keto-3-deoxy-D-manno-octonic acid (KDO) containing capsular polysaccharide (K14 antigen) from *Escherichia coli* O6:K14:H. *Carbohydrate Research* **120**, 131–141.

Jennings, H. J., Rosell, K. G. and Johnson, K. G. (1982). Structure of the 3-deoxy-D-manno-octulosonic acid containing polysaccharide (K6 antigen) from *Escherichia coli* LP 1092. *Carbohydrate Research* **105**, 45–56.

Kasper, D. L., Winkelhake, J. L., Zollinger, W. D., Brandt, B. L. and Artenstein, M. S. (1973). Immunochemical similarity between polysaccharide antigens of *Escherichia coli* O7:K1(L):NM and group B *Neisseria meningitidis*. *Journal of Immunology* **110**, 262–268.

Lifely, M. R., Gilbert, A. S. and Moreno, C. (1981). Sialic acid polysaccharide antigens of *Neisseria meningitidis* and *Escherichia coli*: Esterification between adjacent residues. *Carbohydrate Research* **94**, 193–203.

McGuire, E. J. and Binkley, S. B. (1964). The structure and chemistry of colominic acid. *Biochemistry* **3**, 247–251.

Messmer, P. and Unger, F. M. (1980). Structure of the capsular polysaccharide (K6 antigen) from *Escherichia coli* LP1O92. *Biochemical and Biophysical Research Communications* **96**, 1003–1010.

Neszmélyi, A., Jann, K., Messner, P. and Unger, F. (1982). Constitutional and configurational assignments by ¹³C N.M.R. spectroscopy of *Escherichia coli* capsular polysaccharides containing ribose and 3-deoxy-D-mann-2-octulosonic acid (KDO). *Journal of the Chemical Society, Chemical Communications* pp. 1017–1019.

Ørskov, F., Ørskov, I., Jann, B. and Jann, K. (1971). Immunoelectrophoretic patterns of extracts from all *Escherichia coli* O and K antigen test strains. Correlation with pathogenicity. *Acta Pathologica et Microbiologica Scandinavica, Section B: Microbiology and Immunology* **79B**, 142–152.

Ørskov, F., Ørskov, I., Sutton, A., Schneerson, R., Lind W., Egan, W., Hoff, G. E. and Robbins, J. B. (1979). Form variation in *Escherichia coli* K1: Determined by O-acetylation of the capsular polysaccharide. *Journal of Experimental Medicine* **149**, 669–685.

Ørskov, I., Ørskov, F., Jann, B. and Jann, K. (1977). Serology, chemistry and genetics of O and K antigens of *Escherichia coli. Bacteriological Reviews* **41**, 667–710.

Reske, K., and Jann, K. (1972). The O8 antigen of *Escherichia coli.* Structure of the polysaccharide chain. *European Journal of Biochemistry* **31**, 320–328.

Rietschel, E. T., Schade, U., Jensen, M., Wollenweber, H. W., Lüderitz, O. and Greisman, S. G. (1982). Bacterial endotoxins: Chemical structure, biological activity and role in septicaemia. *Scandinavian Journal of Infectious Diseases, Supplementum* **31**, 8–21.

Robbins, J. B., Schneerson, R., Egan, W., Vann, W. and Liu, D. T. (1980). Virulence properties of bacterial capsular polysaccharides—unanswered questions. *In* ''The Molecular Basis of Microbial Pathogenicity,'' (Eds. H. Smith, J. J. Skehel and M. J. Turner), pp. 115–132. Verlag Chemie, Weinheim.

Rodén, L. (1980). ''The Biochemistry of Glycoproteins and Proteoglycans,'' pp. 267–371. Plenum, New York.

Schmidt, M. A. and Jann, K. (1982). Phospholipid substitution of capsular (K) polysacchrides from *Escherichia coli* causing extraintestinal infections. *FEMS Microbiology Letters* **14**, 69–74.

Schmidt, M. A. and Jann, K. (1983). Structure of the 2-keto-3-deoxy-D-manno-octonic acid contain-ing capsular polysaccharide (K12 antigen) of the urinary tract infective *Escherichia coli* O4:K12:H$^-$. *European Journal of Biochemistry* **131**, 509–517.

Smith, H. (1977). Microbial surfaces in relation to pathogenicity. *Bacteriological Reviews* **41**, 475–500.

Sutherland, I. W., Jann, K. and Jann, B. (1970). The isolation of O-acetylated fragments from the K antigen of *Escherichia coli* O8:K27(A):H$^-$. On the action of a phage induced enzyme from *Klebsiella aerogenes. European Journal of Biochemistry* **12**, 285–288.

Takayama, K., Qureshi, N., Mascagni, P., Nashed, M. A., Anderson, L. and Raetz, Ch. Rh. (1983). Fatty acyl derivatives of glucosamine-1-phosphate in *Escherichia coli* and their relation to lipid A. Complete structure of a diacyl GlcN-1-phosphatidylglycerol-deficient mutant. *Journal of Biological Chemistry* **258**, 7379–7385.

Vann, W. F. and Jann, K. (1979). Structure and serological specificity of the K13 antigenic polysac-charide (K13 antigen) of urinary tract infective *Escherichia coli. Infection and Immunity* **25**, 85–92.

Vann, W. F., Schmidt, M. A., Jann, B. and Jann, K. (1981). The structure of the capsular polysaccharide (K5 antigen) of urinary tract infective *Escherichia coli* O10:K5H4. A polymer similar to desulfo heparin. *European Journal of Biochemistry* **116**, 359–364.

Westphal, O., Jann, K., and Himmelspach, K. (1982). Chemistry and immunochemistry of bacterial lipopolysaccharides as cell wall antigens and endotoxins. *Progress in Allergy* **3**, 9–39.

6

Toxins Affecting Intestinal Transport Processes

J. HOLMGREN

Department of Medical Microbiology, University of Göteborg, Göteborg, Sweden

Introduction

It is now well established that *E. coli* bacteria can cause diarrhoeal disease in people and animals by producing one or both of two classes of enterotoxins, one being heat-labile (LT) and the other heat-stable (ST). Smith and Halls (1967) first described that enterotoxigenic *E. coli* (ETEC) isolates from the stool of young domestic animals with severe diarrhoeal disease could produce a heat-labile as well as a heat-stable enterotoxin and further showed that these toxins were both genetically controlled by plasmids. A few years later *E. coli* isolates from people with cholera-like disease were likewise shown to produce LT and/or ST with properties similar, if not identical, to those produced by the animal-pathogenic *E. coli* strains (Sack *et al.*, 1971; Gorbach and Khurana, 1972; Sack, 1975, 1978). Despite their similar effect on intestinal fluid transport processes the LT and ST toxins are clearly very different from each other structurally and functionally (Table 1). The purpose of this account is to summarize current knowledge of those aspects of *E. coli* enterotoxins that are relevant for an understanding of their ability to give rise to diarrhoeal disease and to discuss approaches to the specific counteraction of their effects on intestinal transport processes.

The LT and ST Molecules

E. coli LT is strikingly similar to but not structurally, immunologically and functionally identical with cholera toxin and most of the knowledge about LT and its mode of action clearly has followed directly from earlier studies of cholera toxin (Sack, 1975, 1978; Holmgren and Lönnroth, 1980; Gill and Enomoto, 1980; Holmgren, 1981).

Similar to cholera toxin, *E. coli* LT has one toxic-active A subunit and five receptor-binding B subunits. The B subunits are aggregated in a ring by tight,

THE VIRULENCE OF ESCHERICHIA COLI

178   J. HOLMGREN

Table 1. *Properties of E. coli enterotoxins (modified from Sack, 1978)*

Property	LT	ST
Molecular weight	86,500	~1,500–5,000
Subunits	Five B + one A	None identified
Immunological properties	Closely related to cholera toxin and to other LTs	Not immunogenic in natural state
Mucosal receptors	GM$_1$ ganglioside plus related glycoprotein	None identified
Biochemical action	Activates adenylate cyclase	Activates guanylate cyclase
Physiological action	Prolonged fluid hypersecretion after lag	Rapid-onset, short-lived hyper-secretion
Genetic control in bacteria	Plasmid	Plasmid
Assay methods	Animal models, tissue culture, immuno- or receptor immu-noassays, DNA hybridization	Animal models, immunoassays, DNA hybridization

non-covalent bonds. The A subunit is linked to and partially inserted in the B ring through weaker non-covalent interactions (Dafni and Robbins, 1976; Clements and Finkelstein, 1979; Holmgren and Svennerholm, 1978; Kunkel and Robertson, 1979; Wolk et al., 1980; Gill et al., 1981) (Fig. 1).

Recent studies by Dallas and Falkow (1980) and Spicer and Noble (1982) have clarified the primary structures of the B as well as the A subunits by establishing the DNA nucleotide sequences of the corresponding *eltB* and *eltA* genes. Of the 103 amino acid residues of the LT B subunit only 22 differ from the corresponding residues in cholera toxin, and in 20 of these a single base change in the *eltB* codon could account for the difference (Dallas and Falkow, 1980). The A subunit contains 236 amino acid residues and shows considerable homology with the cholera toxin A subunit (Spicer et al., 1981; Spicer and Noble, 1982). Like the latter, it may be proteolytically 'nicked' between its two cysteine residues to yield the two disulphide-linked fragments A, and A$_2$, probably by cleavage and

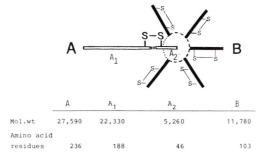

Fig. 1. Proposed structure model of *E. coli* LT. LT is usually an unnicked periplasmic protein in *E. coli* and nicking may be mediated by host intestinal fluid proteases.

removal of two amino acid residues after Arg-188 (Spicer and Noble, 1982). Such nicking is important for the biological–enzymatic activity of the toxin, which resides specifically in the A_1 fragment (Moss and Richardson, 1978; Gill and Richardson, 1980). Essentially all cholera toxin molecules are already nicked when purified, while most *E. coli* LT toxin molecules are not (Gill *et al.*, 1981), probably because the main portion of the latter is obtained from the periplasmic space rather than from the cell supernatant. Consequently, trypsin treatment concomitantly nicks the LT A subunit and increases the biological activity of the LT molecule (Rappaport *et al.*, 1976). In the human or animal intestine *in vivo* gut fluid factors may indeed be mainly responsible for both releasing *E. coli* LT from its periplasmic containment and activating it by proteolytic nicking.

Analysis of the DNA sequences of the genes coding for the toxin suggest that the A and B subunits are initially synthesized as precursors with amino terminal signal regions of 18 and 21 amino acids, respectively (Dallas and Falkow, 1980; Spicer and Noble, 1982). Such hydrophobic signal sequences facilitate translocation of precursors across the cytoplasmic membrane into the periplasmic space or outer membrane. Synthesis of the B subunit precursor and removal of the signal sequence by proteolytic processing has been shown (Palva *et al.*, 1981) and can be strongly predicted for LT A as well. Assembly of the B subunit pentamer and presumably of the holotoxin occurs after maturation of the precursor (Hirst *et al.*, 1983). The fact that the holotoxin has five B subunits and one A subunit suggests that the *eltB* gene is expressed more efficiently than the *eltA* gene, perhaps due to a more efficient ribosome binding site on the *eltB* mRNA.

Immunologically, *E. coli* LT and cholera toxin are closely related but not identical. This was first demonstrated for the holotoxins (Holmgren *et al.*, 1973; Smith and Sack, 1973; Gyles, 1974) and was later extended to each of the composite subunits (Clements and Finkelstein, 1978; Holmgren and Svennerholm, 1979; Kunkel and Robertson, 1979). The recent development of monoclonal antibodies against cholera toxin as well as *E. coli* LT has enabled a closer analysis of the immunological relationship between these toxins and also of the possible heterogeneity of toxins produced by different bacterial strains (Table 2). In each subunit there are fully shared, partially cross-reactive and unique antigen determinants as compared with the corresponding subunit of the other toxin. Honda *et al.* (1981) recently described the immunological non-identity between LT produced by human and porcine *E. coli* strains, indicating interspecies heterogeneity. To date no intraspecies toxin heterogeneity has been identified but the availability of monoclonal antibodies immediately provides much sharper tools to observe discrete changes than those hitherto available.

E. coli ST from human- as well as animal-pathogenic strains has recently been purified and characterized (Alderete and Robertson, 1978; Field *et al.*, 1978; Takeda *et al.*, 1979; Lallier *et al.*, 1980; Staples *et al.*, 1980; Stavric *et al.*, 1981). Different purified ST preparations appear to vary in size from by *ca.* 2000–5000 daltons and correspondingly in their number of amino acid residues.

Table 2. *Immunological relationship between cholera toxin and E. coli LT as determined with monoclonal antibodies (data from Lindholm et al., 1983; Svennerholm et al., 1983)*

Immunological reactivity	Monoclonal antibodies against			
	Cholera toxin		E. coli LT	
	B	A	B	A
CT only	40	4	—	—
CT > LT	10	2	—	—
CT = LT	11	3	11	0
CT < LT	—	—	4	0
LT only	—	—	1	2
Total	70		18	

The mechanism of action and potency of these preparations, which were all active in the infant mouse test, are similar, suggesting that they share a stretch of closely homologous or identical amino acid residues containing the biological activity. Indeed, studies of the primary structures of *E. coli* ST support such a notion. Chan and Gianella (1981) recently described the amino acid sequence of a 2300-dalton purified ST from a human *E. coli* strain and found that its sequence of 18 amino acid residues, except in two positions, was identical with the last 18 amino acids of a *ca.* 5000-dalton porcine ST nucleotide sequence (So and McCarthy 1980). Six cysteine residues that can engage in intrachain disulphide linkages plus only one charged amino acid (Glu) could satisfactorily explain the characteristic heat and pH stabilities as well as the hydrophobicity of the ST molecule. We may envisage the various molecular forms of ST that are active in the infant mouse model as having a common or very similar 14-residue carboxy terminal sequence containing the key biological activity preceded by an NH_2-terminal stretch of amino acids of different length (Chan and Gianella, 1981) (Fig. 2).

In contrast to LT, which is cell-associated and localized in the periplasmic space (Hirst *et al.*, 1984), ST appears to be transported and released into the extra-cellular milieu (Robertson and Alderete, 1978).

Burgess *et al.* (1978) recently described two *E. coli* ST activities (ST_A and ST_B) that were elaborated by single strains of porcine- or bovine-pathogenic *E. coli*. ST_A is methanol-soluble and active in the infant mouse assay, while ST_B is methanol-insoluble and active in the ligated pig or rabbit intestinal loop but negative in the infant mouse assay (Burgess *et al.*, 1978; Kapitany *et al.*, 1979). So far, no *E. coli* strain isolated from humans has been shown to produce ST_B but the search for infant mouse-negative ST has been limited to date.

DNA hybridization tests have provided preliminary evidence for molecular

Human ST (Chan & Gianella, 1981)

          5   10   15
Asn-Thr-Phe-Tyr-Cys-Cys-Glu-Leu-Cys-Cys-Tyr-Pro-Ala-Cys-Ala-Gly-Cys-Asn

Porcine ST (So & McCarthy, 1980)

--- Asn-Thr-Phe-Tyr-Cys-Cys-Glu-Leu-Cys-Cys-<u>Asn</u>-Pro-Ala-Cys-Ala-Gly-Cys-<u>Tyr</u>

Fig. 2. Primary structures of *E. coli* ST (ST_A) from strains pathogenic for man and pig.

heterogeneity among infant mouse-active ST produced by different human strains (Moseley *et al.*, 1980, 1982). A radio-labelled fragment of DNA encoding a porcine ST_A was used as a hybridization probe for homologous DNA sequences in lysed *E. coli* colonies. Using stringent conditions for the hybridization, ST_A-encoding DNA from less than half of the human strains was detected, while with less stringent conditions the previously negative ST_A strains also became positive, suggesting related but not identical DNA sequences (Moseley *et al.*, 1980). This suggestion has been confirmed in a subsequent study by Moseley *et al.* (1982) in which a second probe for genes coding for ST—ST_A-H consisting of a radio-labelled DNA fragment encoding a human ST_A—was used along with the previously constructed porcine ST_A-P probe. All strains of *E. coli* producing an ST that was active in the suckling mouse assay were identified by using both the ST_A-H and ST_A-P probes in stringent hybridization tests; about half of the strains were identified with each probe (Moseley *et al.*, 1982). The observation that some strains of ETEC were detected in both the ST_A-H and ST_A-P probes suggests that these strains possess two genes encoding ST_A production, perhaps residing on different plasmids.

Receptors and Molecular Action of LT and ST

It is now well established that LT from *E. coli* activates adenylate cyclase (Evans *et al.*, 1972; Moss and Richardson, 1978; Gill and Richardson, 1980) and that ST (at least ST_A) stimulates guanylate cyclase activity (Field *et al.*, 1978; Hughes *et al.*, 1978; Guerrant *et al.*, 1980).

The mode of action of *E. coli* LT on adenylate cyclase is in many regards indistinguishable from that of cholera toxin, which has been extensively reviewed (e.g., Gill and Enomoto, 1978; Holmgren and Lönnroth, 1978; Fishman, 1980; Holmgren, 1981). The effect of LT is not apparent until after a distinct lag period of more than 30 min but is then sustained and unaffected by washing or disintegration of the intoxicated cells. The toxin binds to specific receptors on the brush-border membrane via its five B subunits, and this is followed by translocation of the A subunit through the membrane and release of the A_1 fragment into the cytosol. The A_1 fragment enzymatically splits intracellular NAD and, possibly in cooperation with as yet undefined cytosol factors, covalently links the

ADP-ribose moiety of NAD onto the GTP-binding protein of the adenylate cyclase complex at the inner side of the basolateral membrane of the mucosal cells. This ''locks'' adenylate cyclase in an active form by inhibiting an inherent feed-back regulatory mechanism which normally involves the hydrolysis of GTP to GDP and P_i and this results in the accumulation of intracellular cyclic AMP.

Recent studies indicate that while the receptors for cholera toxin in the gut mucosa are GM_1 gangliosides the receptors for *E. coli* LT may consist of GM_1 gangliosides as well as a structurally related glycoprotein (Holmgren *et al.*, 1982; Holmgren, 1981). It was known that GM_1 ganglioside has specific binding affinity for *E. coli* LT similar to that for cholera toxin (Svennerholm and Holmgren, 1978); indeed, diagnostic assays of LT in bacterial culture filtrates or diarrhoeal stools are based upon the receptor-like binding of the toxin to plastic-adsorbed GM_1 ganglioside (Svennerholm and Holmgren, 1978; Czerkinsky and Svennerholm, 1983). Moreover, the incorporation of GM_1 ganglioside into a ganglioside-deficient cell line of mouse fibroblasts sensitized these normally toxin-resistant cells so that they responded with cyclic AMP formation when exposed to either cholera toxin or *E. coli* LT (Moss *et al.*, 1979). These findings are consistent with a receptor function of GM_1 for LT as well as for cholera toxin. On the other hand, recent studies of rabbit intestinal mucosal cells or isolated brush-border membranes showed that:

1. The number of *E. coli* LT binding sites, in contrast to the number of cholera toxin receptors, did not correlate well with the tissue content of GM_1 ganglioside (Holmgren *et al.*, 1982).
2. Delipidization by means of a chloroform–methanol–water mixture, which removes all lipids, including GM_1, but does not extract glycoproteins, did not extract LT binding sites quantitatively although it completely removed cholera toxin binding sites; the non-extracted LT-specific binding sites had properties consistent with a glycoprotein nature (Holmgren *et al.*, 1982).
3. Cholera B subunit did not block the *in vivo* fluid secretogenic action of *E. coli* LT even when used in a 100-fold higher concentration than that completely blocking cholera toxin action; LT B subunit, on the other hand, efficiently blocked the action of *E. coli* LT and cholera toxin (Holmgren *et al.*, 1982).

Preliminary results with isolated mucosal cells and purified brush-border membranes of human small intestine indicate that in the human gut LT can also choose between glycoprotein and GM_1 ganglioside-specific binding sites while cholera toxin selectively binds to GM_1 ganglioside (J. H. Holmgren, unpublished observations). However, whereas the LT glycoprotein receptors seem to exceed the GM_1 receptors about 10-fold in the rabbit intestine they seem to be present in roughly the same number as GM_1 in human gut epithelium.

Much less is known about the detailed mechanism by which ST stimulates guanylate cyclase in the intestinal mucosa. In contrast to the delayed but sustained and irreversible effect of LT on adenylate cyclase and on fluid secretion, the action of ST on guanylate cyclase as well as fluid secretion is almost instantaneous, relatively short-lasting and readily reversed by rinsing of the intestine (Field *et al.*, 1978; Hughes *et al.*, 1978; Guerrant *et al.*, 1980). It is possible that diarrhoea itself may reverse the ST action by 'diluting' the toxin from its mucosal receptors.

Studies of tissue specificity have shown that the action of ST, in contrast to that of LT, is restricted to intestinal cells and is much more pronounced on the mucosa of the small intestine and caecum than the colon (Guerrant *et al.*, 1980). This probably reflects the tissue distribution of ST receptors, the molecular nature of which is unknown. Guanylate cyclase is present in brush-border membranes and, according to the current view, the binding of ST to its receptors on the brush-border membrane probably couples these receptors to guanylate cyclase, thereby stimulating the activity of this enzyme. However, Guerrant *et al.* (1980) observed that indomethacin, a potent inhibitor of prostaglandin synthesis, inhibited ST-induced intestinal guanylate cyclase activation as well as fluid secretion although it did not inhibit the fluid secretion induced by 8-bromo-cyclic GMP. This could suggest a role for prostaglandin or leukotriene synthesis in ST-induced guanylate cyclase stimulation but other explanations are also possible.

Toxin-Induced Intestinal Secretion

Intestinal epithelium can simultaneously absorb and secrete electrolytes and water. The movement of water across the intestinal epithelium is largely due to osmotic gradients created by the active transport of electrolytes from one side of the epithelium to the other. Absorption primarily takes place in the colon and by the villus cells of the small intestine, while small intestinal crypt cells are mainly responsible for electrolyte and fluid secretion. Changes in either or both absorption and secretion may result in diarrhoea. Since the *E. coli* enterotoxins mainly, if not exclusively, affect fluid transport processes of the small intestine, the absorptive and secretory processes of this tissue will be briefly summarized (for a thorough treatment of this subject the reader is referred to Johnson, 1981).

Absorption of water by the small intestinal epithelium is largely determined by the absorption of sodium, which can take place by three transcellular processes: (1) Na diffusion through the brush-border membrane due to the electronegativity and low Na concentration inside the villus cells, (2) Na absorption coupled to the absorption of various organic solutes such as sugars and amino acids and (3) neutral one-for-one coupled NaCl transport. The absorbed, intracellular sodium is then actively extruded through the basolateral membrane by the membrane-associated Na-K-activated ATPase. This builds up an increased concentration of sodium in the lateral intercellular space and on the serosal side, providing an osmotic driving force for water to flow from the lumen to the intercellular space and then to the blood.

The secretory process by crypt cells can be seen as the opposite of absorption. Coupled NaCl entry processes on the basolateral membrane increase the chloride concentration within the crypt cell to a level above electrochemical equilibrium, while the sodium entering with the chloride is recycled back across the basolateral membrane by the Na,K-ATPase, Various secretory stimuli, via intracellular messengers such as cyclic nucleotides and calcium, increase the luminal membrane permeability to chloride, allowing it to be 'secreted'.

Cyclic AMP was the first identified intracellular mediator of intestinal secretion. The current view of the role of cyclic AMP in intestinal secretion is that this nucleotide stimulates active secretion of chloride (and probably also bicarbonate) by crypt cells and inhibits normal absorption of chloride coupled to sodium by villus cells. Since *E. coli* LT increases cyclic AMP in both villus and crypt cells, it probably produces diarrhoea both by inhibiting absorption and by stimulating secretion, using cyclic AMP as an intracellular messenger (Field, 1981; Moss and Vaughan, 1980) (Fig. 3).

Recent findings suggest that calcium might be an important intracellular regulator of intestinal electrolyte transport (Field, 1981). Results of studies with calcium ionophores, calcium-chelating agents, and calcium-blocking drugs have all indicated that an increase in intracellular calcium stimulates secretion, while a decrease enhances absorption. Calcium may even be the ultimate messenger for cyclic AMP-induced secretion and hence for *E. coli* LT and cholera toxin. According to this view, the increased level of cyclic AMP would release calcium from intracellular stores, thus triggering activation of the calcium–calmodulin complex, resulting in increased permeability for chloride from intestinal crypt cells. This suggestion is based on the observation that various phenothiazines and some other substances that are known to bind to, and inactivate, the calcium-binding protein calmodulin reverse both cyclic AMP-mediated secretion induced by cholera toxin or theophylline and non-cyclic AMP-mediated secretion induced by a calcium ionophore. However, the same drugs also have other effects, for instance they inactivate adenylate cyclase, and it is difficult to know if their anti-secretory action is related only to their interaction with calmodulin.

The observation that *E. coli* ST increases the intestinal mucosal cyclic GMP concentration concomitantly with its effect on fluid secretion suggests that cyclic GMP is also an intracellular mediator of intestinal secretion (Field *et al.*, 1978; Hughes *et al.*, 1978; Guerrant *et al.*, 1980). Results from Ussing chamber experiments with rabbit or human intestinal epithelium suggest that ST is as inhibitory as cyclic AMP-inducing secretagogues such as LT on coupled sodium

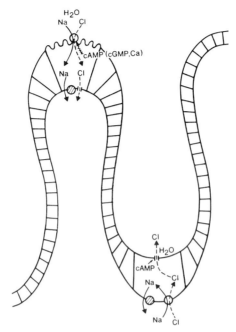

Fig. 3. Intestinal fluid transport processes affected by *E. coli* enterotoxins through intracellular mediator substances. Absorption of sodium, chloride and water by villus cells through and electrically neutral NaCl co-transport across the brush-border membrane are inhibited by both *E. coli* LT and ST, probably via elevated levels of cyclic AMP and cyclic GMP, respectively, and perhaps calcium. In crypt cell, cyclic AMP (and hence LT) instead increases the luminal permeability for chloride, allowing it to be secreted into the lumen; the effect of ST on this process is much weaker.

chloride absorption but is not as effective in stimulating crypt cell chloride secretion (Fig. 3). This difference is probably due to the higher concentration of guanylate cyclase in villus as compared with crypt cells as well as to different responses to the two cyclic nucleotides. The latter possibility is suggested by observations of a weaker chloride secretory response to 8-bromo-cyclic GMP than to an equal concentration of 8-bromo-cyclic AMP (Field, 1981).

Prospects for Anti-Binding and Anti-Secretory Agents

The new knowledge about the specific binding of enterotoxins to intestinal epithelium as an essential step in the pathogenesis of enterotoxic diarrhoeal disease suggests various manoeuvres that might prevent or reverse disease due to these toxins (Fig 4):

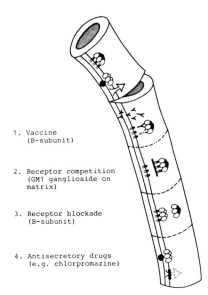

1. Vaccine
   (B-subunit)

2. Receptor competition
   (GM1 ganglioside on
   matrix)

3. Receptor blockade
   (B-subunit)

4. Antisecretory drugs
   (e.g. chlorpromazine)

Fig. 4. Possible approaches to the prevention and treatment of *E. coli* LT pathogenic action.

1. *Immunization-stimulated formation of mucosal antibodies to prevent bind-ing of toxin to epithelium.* In experimental animals such immunization with LT or cholera toxins has been shown to confer protection against *E. coli* LT-induced diarrhoea. Likewise, immunization with the purified LT B or cholera B subunits has conferred protection to challenge with LT or LT-producing *E. coli* organisms. *E. coli* ST is ordinarily a poor immunogen but, coupled to a suitable carrier protein (LT B or cholera B subunit?), it may give rise to antibodies with neutralizing capacity.

2. *Competitive inhibition of toxin binding to epithelium by use of free recep-tors or receptor analogues.* The administration of highly purified receptor GM_1 ganglioside to rabbits has been shown to prevent fluid secretion in response to challenge with either purified LT or live, LT-producing *E. coli*. The same preventive effect was achieved by GM_1 covalently coupled to cellulose powder, a form in which the ganglioside retains its receptor-binding capacity without the risk of being incorporated into the epithelium as a functional receptor (L. Svennerholm, unpublished). No specific recep-tor for *E. coli* ST has as yet been identified but perhaps strongly hydro-phobic compounds or gels could have some non-specific ST-binding ca-pacity *in vivo*.

3. *Receptor blockade by use of non-toxic binding agents.* As mentioned above, purified LT B subunit can completely prevent fluid secretion in ligated rabbit intestine due to *E. coli* LT or LT-toxigenic *E. coli* (Holmgren *et al.*, 1981). As yet, no non-toxic derivative of ST that could be used for

blocking the ST receptors has been developed. Such an agent would be particularly attractive to develop though, since in contrast to *E. coli* LT, the binding of ST to receptors appears to be a reversible process directly linked to its toxic action and, therefore, an efficient blocking structure might be useful for both prevention and treatment of ST-mediated disease.

4. *Reversal of diarrhoea by anti-secretory agents.* These agents could act by either stimulating normal electrolyte and water absorption by intestine or inhibiting fluid secretion. A variety of agents have been identified as having anti-secretory properties in experimental diarrhoea models in animals. These drugs include chlorpromazine and other phenothiazines, nicotinic acids, opiates including loperamide, local anaesthetics, prostaglandin synthesis inhibitors such as indomethacin and aspirin, chloroquine, lithium, berberine, somatostatin, glucocorticoids, amphotericin B, prostacyclin and others (Powell and Field, 1980; Powell *et al.*, 1979; Roberts *et al.*, 1979). The best studied of these agents to date is chlorpromazine. In animal experiments chlorpromazine has shown distinct anti-secretory activity against cyclic AMP-associated diarrhoea including that induced by *E. coli* LT (Holmgren *et al.*, 1978) and also an effect against diarrhoea due to *E. coli* ST (Abbey and Knoop, 1979). Clinical trials with single-dose chlorpromazine treatment of diarrhoea have shown (1) a 50–65% reduction of fluid loss in intravenously hydrated Bangladeshi patients with severe cholera (Rabbani *et al.*, 1979, 1982), (2) a 50% reduction in the failure rate of treatment of Bangladeshi children with severe cholera who were given an oral rehydration solution plus tetracycline (Islam *et al.*, 1982) and (3) a significant reduction of fluid loss in children with non-cholera diarrhoea mainly due to *E. coli* (M. M. Khin, unpublished). The anti-secretory effect of chlorpromazine might be multifactorial; independent or interrelated effects on calmodulin, adenylate cyclase, and protein kinase activities have been observed. The primary effect could well be binding of the phenothiazine drug to intracellular calmodulin, because the calmodulin and calcium system is known to influence the activity of both adenylate and guanylate cyclase as well as of the protein kinases that mediate the biological effects of cyclic AMP or cyclic GMP.

References

Abbey, D. M. and Knoop, F. C. (1979). Effect of chlorpromazine on the secretory activity of *Escherichia coli* heat-stable enterotoxin. *Infection and Immunity* **26**, 1000–1003.

Alderete, J. F. and Robertson, D. C. (1978). Purification and chemical characterization of the heat-stable enterotoxin produced by porcine strains of the enterotoxigenic *Escherichia coli*. *Infection and Immunity* **19**, 1021–1030.

Burgess, M. N., Bywater, R. J., Cowley, C. M., Mullan, N. A. and Newsome, P. M. (1978). Biological evaluation of a methanol-soluble, heat-stable *E. coli* enterotoxin in infant mice, pigs, rabbits, and calves. *Infection and Immunity* **21**, 526–531.

Chan, S.-K. and Gianella, R. A. (1981). Amino acid sequence of heat-stable enterotoxin produced by *Escherichia coli* pathogenic for man. *Journal of Biological Chemistry* **256**, 7744–7746.

Clements, J. D. and Finkelstein, R. A. (1978). Immunological cross-reactivity between the heat-labile enterotoxin(s) of *Escherichia coli* and subunits of *Vibrio cholerae* enterotoxin. *Infection and Immunity* **21**, 1036–1039.

Clements, J. D. and Finkelstein, R. A. (1979). Isolation and characterization of homogenous heat-labile enterotoxins with high specific activity from *Escherichia coli* cultures. *Infection and Immunity* **24**, 760–769.

Czerkinsky, C. C. and Svennerholm, A.-M. (1983). Ganglioside GM_1 enzyme-linked immunospot (ELISPOT) assay for simple identification of heat-labile enterotoxin-producing *Escherichia coli*. *Journal of Clinical Microbiology* **17**, 965–969.

Dafni, Z. and Robbins, J. B. (1976). Purification of heat-labile enterotoxin from *Escherichia coli* O78:H11 by affinity chromatography with antiserum to *Vibrio cholerae* toxin. *Journal of Infectious Diseases* **133**, S138–S141.

Dallas, W. S. and Falkow, S. (1980). Amino acid sequence homology between cholera toxin and *Escherichia coli* heat-labile toxin. *Nature (London)* **288**, 499–501.

Evans, D. J., Jr., Chen, L. C., Curlin, G. T. and Evans, D. G. (1972). Stimulation of adenyl cyclase by *Escherichia coli* enterotoxin. *Nature (London)* **236**, 137–138.

Field, M. (1981). Secretion of the small intestine. *In* "Physiology of the Gastrointestinal Tract, Vol. 2" (Ed. L. R. Johnson), pp. 963–982. Raven Press, New York.

Field, M., Graf, L. H., Jr., Laird, W. J. and Smith, T. L. (1978). Heat-stable enterotoxin of *Escherichia coli: In vitro* effects on adenylate cyclase activity, cyclic GMP concentration, and ion transport in small intestine. *Proceedings of the National Academy of Sciences of the U.S.A.* **75**, 2800–2804.

Fishman, P. H. (1980). Mechanism of action of cholera toxin: Events on the cell surface. *In* "Secretory Diarrhea" (Eds. M. Field, J. S. Fordtran and S. G. Schultz), pp. 85–106. American Physiological Society, Bethesda, Maryland.

Gill, D. M. and Enomoto, K. (1980). Intracellular, enzymic action of enterotoxins: The biochemical basis of cholera. *In* "Cholera and Related Diarrheas" (Eds. Ö. Ouchterlony and J. Holmgren), pp. 104–114. Karger, Basel.

Gill, D. M. and Richardson, S. H. (1980). Adenosinediphosphate-ribosylation of adenylate cyclase catalyzed by heat-labile enterotoxin of *Escherichia coli:* Comparison with cholera toxin. *Journal of Infectious Diseases* **141**, 64–70.

Gill, D. M., Clements, J. D., Robertson, D. C. and Finkelstein, R. A. (1981). Subunit number and arrangement in *Escherichia coli* heat-labile enterotoxin. *Infection and Immunity* **33**, 677–682.

Gorbach, S. L. and Khurana, C. M. (1972). Toxigenic *Escherichia coli*. A cause of infantile diarrhea in Chicago. *New England Journal of Medicine* **287**, 791–795.

Guerrant, R. L., Hughes, J. M., Chang, B., Robertson, D. C. and Murad, F. (1980). Activation of intestinal granulate cyclase by heat-stable enterotoxin of *Escherichia coli:* Studies of tissue specificity, potential receptors, and intermediates. *Journal of Infectious Diseases* **142**, 220–228.

Gyles, C. L. (1974). Immunological study of the heat-labile enterotoxins of *Escherichia coli* and *Vibrio cholerae. Infection and Immunity* **9**, 564–570.

Hirst, T. R., Hardy, S. J. S. and Randall, L. L. (1983). Assembly *in vivo* of enterotoxin from *Escherichia coli:* Formation of the B subunit oligomer. *Journal of Bacteriology* **153**, 21–26.

Hirst, T. R., Randall, L. L. and Hardy, S. J. S. (1984). Cellular location of heat-labile enterotoxin in *Escherichia coli. Journal of Bacteriology* **157**, 637–642.

Holmgren, J. (1981). Actions of cholera toxin and the prevention and treatment of cholera. *Nature (London)* **292**, 413–417.

Holmgren, J. (1981). Actions of cholera toxin and the prevention and treatment of cholera. *Nature (London)* **292**, 413–417.

Holmgren, J. and Lönnroth, I. (1980). Structure and function of enterotoxins and their receptors. *In* "Cholera and Related Diarrheas" (Eds. Ö. Ouchterlony and J. Holmgren), pp. 88–103. Karger, Basel.

Holmgren, J. and Svennerholm, A.-M. (1979). Immunologic cross-reactivity between *Escherichia coli* heat-labile enterotoxin and cholera toxin A and B subunits. *Current Microbiology* **2**, 55–58.

Holmgren, J., Söderlind, O. and Wadström, T. (1973). Cross-reactivity between heat-labile enterotoxins of *Vibrio cholerae* and *Escherichia coli* in neutralization tests in rabbit ileum and skin. *Acta Pathologica et Microbiologica Scandinavica, Section B* **81**, 757–762.

Holmgren, J., Lange, S. and Lönnroth, I. (1978). Reversal of cyclic AMP-mediated intestinal secretion in mice by chlorpromazine. *Gastroenterology* **75**, 1103–1108.

Holmgren, J., Svennerholm, A-M. and Åhrén, C. (1981). Non-immunoglobulin fraction of human milk inhibits bacterial adhesion (hemagglutination) and enterotoxin binding of *Escherichia coli* and *Vibrio cholerae*. *Infection and Immunity* **33**, 136–141.

Holmgren, J., Fredman, P., Lindblad, M., Svennerholm, A. M. and Svennerholm, L. (1982). Rabbit intestinal glycoprotein receptor for *Escherichia coli* heat labile enterotoxin lacking affinity for cholera toxin. *Infection and Immunity* **38**, 424–433.

Holmgren, J., Svennerholm, A. M. and Lindblad, M. (1983). Receptor-like glycocompounds in human milk that inhibit classical and El Tor *Vibrio cholerae* cell adherence (haemagglutination). *Infection and Immunity* **39**, 147–154.

Honda, T., Tsujii, T., Takeda, Y. and Miwatani, T. (1981). Immunological nonidentity of Heat-labile enterotoxins from human and porcine enterotoxigenic *Escherichia coli*. *Infection and Immunity* **34**, 337–340.

Hughes, J. N., Murad, F., Chang, B. and Guerrant, R. L. (1978). Role of cyclic GMP in the action of heat-stable enterotoxin of *Escherichia coli*. *Nature (London)* **271**, 755–756.

Islam, M. R., Sack, D. A., Holmgren, J., Bardhan, P. K. and Rabbani, G. H. (1982). Use of chlorpromazine in the treatment of cholera and other severe acute watery diarrhoeal diseases. *Gastroenterology* **82**, 1335–1340.

Johnson, L. R. (Ed.) (1981). "Physiology of the Gastrointestinal Tract." Vols. 1 and 2. Raven Press, New York.

Kapitany, R. A., Forsythe, G. W., Scoot, A., McKenzie, S. F. and Worthington, R. W. (1979). Isolation and partial characterization of two different heat-stable enterotoxins produced by bovine and porcine strains of enterotoxigenic *E. coli*. *Infection and Immunity* **26**, 173–177.

Kunkel, S. L. and Robertson, D. C. (1979). Purification and chemical characterization of the heat-labile enterotoxin produced by enterotoxigenic *Escherichia coli*. *Infection and Immunity* **25**, 586–596.

Lallier, R., Lariviere, S. and St. Pierre, S. (1980). *Escherichia coli* heat-stable enterotoxin: Rapid method of purification and some characteristics of the toxin. *Infection and Immunity* **28**, 469–474.

Lindholm, L., Holmgren, J., Wikström, M., Karlsson, U., Andersson, K. and Lycke, N. (1983). Monoclonal antibodies to cholera toxin with special reference to cross-reactions with *Escherichia coli* heat-labile enterotoxin. *Infection and Immunity* **40**, 570–576.

Moseley, S. L., Huq, I., Alim, A. R. M. A., So, M., Samadpour-Motalebi, M. and Falkow, S. (1980). Detection of enterotoxigenic *Escherichia coli* by DNA colonihybridization. *Journal of Infectious Diseases* **142**, 892–898.

Moseley, S. L., Echeverria, P., Seriwatana, J., Tirapat, C., Chaicumpa, W., Sakuldaipeara, T. and Falkow, S. (1982). Identification of enterotoxigenic *Escherichia coli* by colonihybridization using free enterotoxin gene. *Journal of Infectious Diseases* **145**, 863–869.

Moss, J. and Richardson, S. H. (1978). Activation of adenylate cyclase by heat-labile *Escherichia coli* enterotoxin. Evidence for ADP-ribosyltransferase activity similar to that of choleragen. *Journal of Clinical Investigation* **62**, 281–285.

Moss, J. and Vaughan, N. (1980). Mechanism of activation of adenylate cyclase by choleragen and *E. coli* heat-labile enterotoxin. *In* "Secretory Diarrhea" (Eds. M. Field, J. S. Fordtran and S. G. Schultz), pp. 107–126. American Physiological Society, Bethesda, Maryland.

Moss, J., Garrison, S., Fishman, P. H. and Richardson, S. H. (1979). Ganglioside sensitizes unresponsive fibroblasts to *E. coli* heat-labile enterotoxin. *Journal of Clinical Investigation* **64**, 381–384.

Palva, E. T., Hirst, T. R., Hardy, S. J. S., Holmgren, J. and Randall, L. (1981). Synthesis of precursors to the B subunit of heat-labile enterotoxin in *Escherichia coli*. *Journal of Bacteriology* **146**, 325–339.

Powell, D. W. and Field, M. (1980). Pharmacological approaches to treatment of secretory diarrhea. *In* "Secretory Diarrhea" (Eds. M. Field, J. S. Fordtran and S. G. Schultz), pp. 187–209. American Physiological Society, Bethesda, Maryland.

Powell, D. W., Tapper, E. J. and Morris, S. M. (1979). Aspirin-stimulated intestinal electrolyte transport. *Gastroenterology* **76**, 1429–1437.

Rabbani, G. H., Greenough, W. B., Holmgren, J. and Lönnroth, I. (1979). Chlorpromazine reduces fluid-loss in cholera. *Lancet* **1**, 410–412.

Rabbani, G. H., Greenough, W. B., Holmgren, J. and Kirkwood, B. (1982). Controlled trial of chlorpromazine as anti-secretory agent in intravenously hydrated cholera patients. *British Medical Journal* **284**, 1361–1364.

Rappaport, R. S., Sagin, J. F., Pierzchala, W. A., Bonde, G., Rubin, B. A. and Tint, H. (1976). Activation of heat-labile *Escherichia coli* enterotoxin by trypsin. *Journal of Infectious Diseases* **133**, S41–S54.

Roberts, A., Hanchar, A. J., Lancaster, C. and Nezamis, J. E. (1979). Prostacyclin inhibits entero-pooling and diarrhea. *In* "Prostacyclin" (Eds. J. R. Vane and S. Bergström), pp. 147–158. Raven Press, New York.

Robertson, D. C. and Alderete, J. F. (1978). Chemistry and biology of the heat-stable *Escherichia coli* enterotoxin. *In* "Cholera and Related Diarrheas" (Eds. Ö. Ouchterlony and J. Holmgren), pp. 115–126. Karger, Basel.

Sack, R. B. (1975). Human diarrheal disease caused by enterotoxigenic *Escherichia coli*. *Annual Review of Microbiology* **29**, 333–353.

Sack, R. B. (1978). Pathogenesis and pathophysiology of diarrheal diseases caused by *Vibrio cholerae* and enterotoxigenic *Escherichia coli*. *In* "Cholera and Related Diarrheas" (Eds. O. Ouchterlony and J. Holmgren), pp. 53–63. Karger, Basel.

Sack, R. B., Gorbach, S. L., Banwell, J. G., Jacobs, B., Chatterjee, B. D. and Mitra, R. C. (1971). Enterotoxigenic *Escherichia coli* isolated from patients with severe cholera-like disease. *Journal of Infectious Diseases* **123**, 378–385.

Smith, H. W. and Halls, S. (1967). Studies on *Escherichia coli* enterotoxin. *Journal of Pathology and Bacteriology* **93**, 531–543.

Smith, N. W. and Sack, R. B. (1973). Immunologic cross-reactions of enterotoxins from *Escherichia coli* and *Vibrio cholerae*. *Journal of Infectious Diseases* **127**, 164–170.

So, M. and McCarthy, B. J. (1980). Nucleotide sequence of the bacterial transposon Tn 1681 encoding a heat-stable (ST) toxin and its identification in enterotoxigenic *Escherichia coli* strains. *Proceedings of the National Academy of Sciences of the U.S.A.* **77**, 4011–4015.

Spicer, E. K. and Noble, J. A. (1982). *Escherichia coli* heat-labile enterotoxin. Nucleotide sequence of the A subunit gene. *Journal of Biological Chemistry* **257**, 5716–5721.

Spicer, E. K., Kavanaugh, W. M., Dallas, W. S., Falkow, S., Konigsberg, W. H. and Schafer, D. E. (1981). Sequence homologies between A subunits of *Escherichia coli* and *Vibrio cholerae* enterotoxins. *Proceedings of the National Academy of Sciences of the U.S.A.* **78**, 50–54.

Staples, S. J., Ascher, S. E. and Gianella, R. A. (1980). Purification and characterization of heat-stable enterotoxin produced by a strain of *E. coli* pathogenic for man. *Journal of Biological Chemistry* **257**, 4716–4721.

Stavric, S., Dickie, N., Gleeson, T. M. and Akhtar, N. (1981). Application of hydrophobic interaction chromatography to purification of *Escherichia coli* heat-stable enterotoxin. *Toxicon* **19,** 743–747.

Svennerholm, A.-M. and Holmgren, J. (1978). Identification of *Escherichia coli* heat-labile enterotoxin by means of a ganglioside-immunosorbent assay (GM1-ELISA) procedure. *Current Microbiology* **1,** 19–23.

Svennerholm, A.-M., Wikström, M., Lindholm, L. and Holmgren, J. (1983). Characterization of monoclonal antibodies to *Escherichia coli* heat-labile enterotoxin. To be published.

Takeda, Y., Takeda, T., Yano, T., Yamamoto, K. and Miwatani, T. (1979). Purification and partial characterization of heat-stable enterotoxin of enterotoxigenic *Escherichia coli*. *Infection and Immunity* **25,** 978–985.

Wolk, W., Svennerholm, A.-M. and Holmgren, J. (1980). Isolation of *Escherichia coli* heat-labile enterotoxin by affinity chromatography: Characterization of subunits. *Current Microbiology* **3,** 339–344.

7

Candidate Virulence Markers

E. GRIFFITHS

National Institute for Biological Standards and Control, Hampstead, London, UK

Introduction

A significant development in our understanding of *Escherichia coli* infections took place when it was discovered that the genetic determinants for certain virulence characteristics, such as enterotoxin production and adhesiveness, could be carried by plasmids (see Smith *et al.,* this volume, Chapter 8; Elwell and Shipley, 1980). However, the presence of a plasmid is not always required for virulence, nor is it the only factor involved. Smith and Linggood (1971) noted that the acquisition of plasmids, which coded for enterotoxin synthesis and for adhesiveness, was not sufficient to convert all *E. coli* strains into highly virulent enteropathogens. We now know that there is undoubtedly some connection between the structures of the complex polysaccharides which make up the O and K antigens and the virulence of these organisms (Ørskov *et al.,* 1977; Merson *et al.,* 1979). For example, it seems that enterotoxigenic *E. coli* belong to only a small number of serotypes and serogroups, perhaps because the virulence plasmids are more easily acquired and retained by *E. coli* with specific surface properties (Ørskov *et al.,* 1977; Merson *et al.,* 1979). There are also *E. coli* strains that belong to the classically recognized enteropathogenic serotypes that are non-toxigenic and non-invasive according to the usual criteria and do not harbour recognized virulence plasmids but which, nevertheless, cause severe diarrhoeal disease (Levine *et al.,* 1978; Scotland *et al.,* 1980; Ulshen and Rollo, 1980; Cantley and Blake, 1977). Clearly, although a great deal is known about *E. coli,* its heat-labile and heat-stable enterotoxins, its adhesive or colonization factors and the part they play in diarrhoeal disease, knowledge of factors involved in all enteric *E. coli* infections is still incomplete. Our understanding of infections caused by *E. coli* that invade the body or produce localized extraintestinal infections is even more elementary. In this chapter a few, perhaps less well-known, virulence markers will be discussed, the importance of which is only now beginning to be appreciated and understood. Some of these are chro-

193

mosomally determined and some plasmid-mediated and they can be conveniently divided into two groups: factors that enable *E. coli* to multiply successfully under the iron-restricted conditions found in host tissues and those that confer on the bacteria an ability to resist the bactericidal action of serum.

Virulence and Iron

One common and essential factor in all infections is the ability of the invading pathogen to multiply successfully in host tissues. This property has long been known to be greatly influenced by the availability of iron. Animals injected with various forms of iron are much more susceptible to infection with a variety of bacteria, including *E. coli,* than are untreated controls (Griffiths, 1981). Iron compounds also abolish the antibacterial effects of body fluids *in vitro* (Bullen *et al.,* 1978; Weinberg, 1978; Griffiths, 1981). In general, iron affects the susceptibility of normal animals to infection by reducing the lethal dose of bacteria. For example, the lethal dose of *E. coli* **O**111 for normal guinea-pigs is about 10^8 cells, whereas in guinea-pigs injected with ferric ammonium citrate or haem compounds this dose may be reduced to about 10^3 cells, a 100,000-fold reduction (Bullen *et al.,* 1968). Only iron has been found to act in this way and the investigation of the phenomenon has not only led to an explanation of the changes in resistance which sometimes accompany clinical alterations in the iron status of the host but has also increased our understanding of the mechanisms involved in microbial pathogenicity.

Iron-Binding Proteins

It is now known that although plenty of iron is present in the body fluids of man and animals, the amount of free iron readily available to bacteria is extremely small. Most of the iron in the body is found intracellularly as ferritin, haemosiderin or haem and that which is extracellular in serum or other body fluids is attached to high-affinity iron-binding proteins, transferrin in blood and lymph and lactoferrin in external secretions and milk; a related protein called ovotransferrin occurs in avian egg white (Bezkorovainy, 1980; Morgan, 1981). Transferrin is synthesized mainly in the liver (Morgan, 1981; Bezkoravainy 1977, 1980) and immunohistochemical techniques indicate that lactoferrin is synthesized locally and secreted by the mucosal lining of the respiratory, gastrointestinal and urogenital tracts (Masson and Heremans, 1966; Masson *et al.,* 1966a,b; Tourville *et al.,* 1969). These glycoproteins have molecular weights of about 80,000 and each is capable of reversibly binding two Fe^{3+} ions with the simultaneous incorporation of two bicarbonate ions (Bezkorovainy, 1980; Morgan, 1981). They have association constants for iron of approximately 10^{36} and are usually only partially saturated with iron. Lactoferrin has the additional property of being able to bind iron under the more acidic conditions often prevailing at sites of inflammation. The iron-binding property of transferrin is diminished below pH 6, whereas lactoferrin holds its iron until the pH is below 4 (McClelland and van Furth, 1976; Morgan, 1981; Ainscough et al., 1980). Thus, although there is normally an abundance of iron in body fluids, the amount of free iron in equilibrium with the iron-binding proteins is of the order of 10^{-18} M, which is far too low to sustain bacterial growth (Bullen et al., 1978). In addition, during infection the host reduces even further the amount of iron bound to serum transferrin (Cartwright et al., 1946). This decrease is called the hypoferraemia of infection and is due to a movement of iron from serum to the iron stores. The effect can be reproduced by injecting small amounts of bacterial endotoxin (Baker and Wilson, 1965).

Acquisition of Iron by E. coli: Enterochelin, Ferrichrome, Citrate

Since all known bacterial pathogens need iron to multiply, and therefore to establish an infection, those that multiply successfully under the iron-restricted conditions found in body fluids must be able to develop a mechanism for assimilating protein-bound iron, or for acquiring it from liberated haem. Many micro-organisms combat the biological unavailability of iron in aerobic environments, where iron exists in the ferric form, by producing low-molecular-weight iron-chelating compounds known as siderophores (Neilands, 1981; Raymond and Carrano, 1979). Production of siderophores can be regarded as an evolutionary response to the appearance of oxygen in the atmosphere and the subsequent oxidation of ferrous to ferric iron. In aqueous media at neutral pH all simple ferric salts are hydrolysed to form highly insoluble ferric hydroxide, which has a solubility product of about 10^{-38} M (Biedermann and Schindler, 1957). With the equilibrium concentration of ferric iron at neutral pH about 10^{-18} M, diffusion-limited transport would be many orders of magnitude too slow to provide the bacterial cell with sufficient amounts of iron. This problem is circumvented by aerobic and facultative anaerobic micro-organisms through the production of siderophores that solubilize ferric iron and transport it back into the bacterial cell. These low-molecular-weight siderophores are part of what is termed the 'high-affinity' iron transport system; a 'low-affinity' iron transport system operates when iron is freely available but little is known about this mechanism. Siderophores are also produced by several pathogenic organisms and these chelators can remove iron from iron-binding proteins (Griffiths, 1983).

E. coli, as well as Salmonella typhimurium and Klebsiella pneumoniae, produces the iron chelator enterochelin (also called enterobactin) under conditions of iron restriction in vitro (Pollack and Neilands, 1970; O'Brien and Gibson, 1970; Rogers et al., 1977). This compound, which is the cyclic triester of 2,3-di-

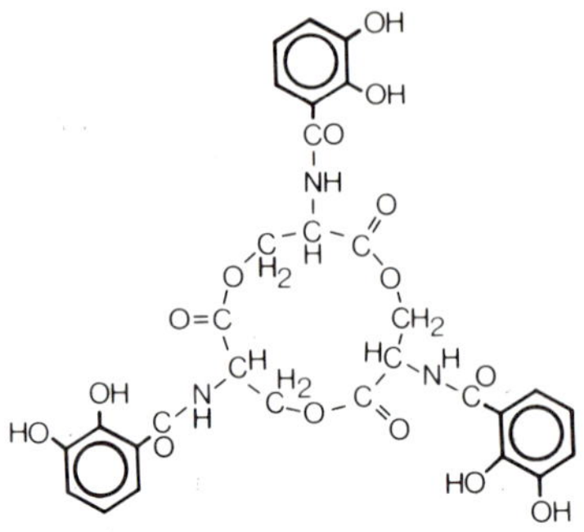

Fig. 1. Structural formula of enterochelin.

hydroxy-*N*-benzoyl-L-serine (Fig. 1), is made only under conditions of iron restriction and it very efficiently removes iron from the iron-binding proteins, promoting bacterial growth at extremely low concentrations (0.01 μ*M*) (Rogers, 1973; Rogers *et al.*, 1977; Miles and Khimji, 1975; Carrano and Raymond, 1979). Enterochelin acts as a hexadentate chelating agent through its three cate-chol groups and, when co-ordinating with ferric ion, it acts as a six-proton acid; therefore, at neutral pH the hexa co-ordinate complex carries three negative charges $[Fe(ent)]^{3-}$ (Harris *et al.*, 1979a). The single most outstanding feature of siderophores is their extremely high affinity for ferric iron; the formation constant for ferric enterochelin at neutral pH is near 10^{52} and is the highest ever found for a ferric ion-binding chelate (Harris *et al.*, 1979a,c). Organisms using enterochelin-mediated iron transport are, therefore, able to compete very effec-tively for complexed iron.

However, enterochelin is used only once for transporting iron into the cell. This is because the cyclic triester linkages of Fe^{3+}- enterochelin are cleaved by a specific esterase when it enters the bacterial cell, apparently in order to release the iron, ultimately producing 2,3-dihydroxybenzoylserine, which is then dis-carded (Rosenberg and Young, 1974; Greenwood and Luke, 1978). It has been suggested that hydrolysis of the ester bonds is necessary because ferric entero-chelin has too low a reduction potential to be physiologically reducible and consequently must be hydrolysed to permit the reduction and subsequent release of iron; cleavage of the ester bonds certainly increases the redox potential of the ferric complex of enterochelin (Harris *et al.*, 1979c; Raymond and Carrano, 1979; Cooper *et al.*, 1978). However, the fact that a carbocyclic analogue of enterochelin, which lacks ester bonds, effectively supplies iron to an entero-chelin-deficient strain of *E. coli* has raised questions about the exact role of the esterase, questions which have not yet been satisfactorily answered (Hollifield and Neilands, 1978). Whatever the role of the esterase, the genetic lesion called *fes* is real; *E. coli fes* mutants lack esterase activity and are unable to use ferric enterochelin; the ferric complex accumulates in the cell with the result that a sedimented pellet of these cells show a distinct pinkish colour (Langman *et al.*, 1972; Rosenberg and Young, 1974).

Although enterochelin production by *E. coli* has mostly been studied *in vitro* with *E. coli* K-12, it is known that pathogenic *E. coli* secrete this siderophore while infecting normal guinea-pigs (Griffiths and Humphreys, 1980). Enterochelin and its degradation products have been detected in the peritoneal washings from guinea-pigs lethally infected with *E. coli* **O**111, a finding which supports the notion that the presence of iron-binding proteins in body fluids creates an iron-restricted environment *in vivo* and affects bacterial metabolism accordingly. Furthermore, loss by mutation of the ability to synthesize enterochelin has been shown to reduce greatly the virulence of *S. typhimurium* for mice and to inhibit the ability of the organism to grow in human serum, clearly demonstrating how essential enterochelin production can be for virulence (Yancey *et al.*, 1979). Hoiseth and Stocker (1981) have made use of this fact in designing a new live oral attenuated *Salmonella* vaccine that confers excellent protection against challenge with a virulent strain. Enterochelin is synthesized in both *Salmonella* and *E. coli* from chorismate, the final product of the common aromatic (aro) biosynthetic pathway (Fig. 2). Using the tetracycline-resistance transposon Tn10 inserted into gene *aro*A, Hoiseth and Stocker (1981) produced a non-reverting aromatic-requiring derivative of virulent *S. typhimurium*. This block in the aromatic pathway made the organism auxotrophic for two compounds that are not available in vertebrate tissues, enterochelin and *p*-aminobenzoic acid, and in doing so, rendered it non-virulent. Such *Salmonella* derivatives conferred excellent protection when used as live vaccines but vaccines from similar derivatives of *E. coli* have not yet been made.

Limiting the availability of iron by reducing the ability of a pathogen to synthesize iron chelators, or by increasing the rate of destruction of these compounds in tissue fluids, by high temperatures might also be expected to restrict the ability of an organism to multiply successfully in the presence of iron-binding

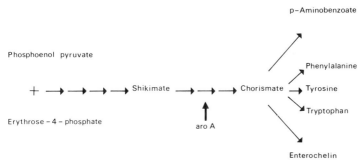

Fig. 2. Relationship of the common aromatic amino acid biosynthetic pathway to the synthesis of enterochelin and to *p*-aminobenzoate.

proteins. Indeed it has been proposed that this is one of the mechanisms accounting for the adaptive role of fever (Garibaldi, 1972; Grieger and Kluger, 1978; Kluger and Rothenburg, 1979). Buck *et al.* (1979) showed that *E. coli* **O**111 produced less enterochelin and related catechols at 42°C than at 37°C when growing *in vitro* in an iron-restricted environment. There was also a small decrease in growth rate at the elevated temperature, but only in the presence of the iron-binding protein. While these results support the idea that increased temperature reduces the ability of certain bacteria to grow in iron-restricted environments by suppressing the production of siderophores, much more work is needed to show that this functions as a protective mechanism *in vivo*.

Other host defensive mechanisms, which depend upon the immune system, may also involve enterochelin. Recent work suggests that normal human serum contains antibodies specific for enterochelin which, when acting in concert with transferrin, limit the growth of certain *E. coli* strains (Moore *et al.*, 1980; Moore and Earhart, 1981). These antibodies appear to block enterochelin-mediated iron uptake but not the uptake of iron from ferrichrome or citrate (see below). Since enterochelin is an aromatic molecule it may well adhere to protein and act as a hapten for generating antibodies. An alternative immune mechanism has been described by Fitzgerald and Rogers (1980), who studied serotype-specific horse serum antibodies and secretory IgA in human milk. These antibodies, which also act in conjunction with transferrin or lactoferrin to inhibit the multiplication of *E. coli,* appear to interfere with enterochelin synthesis or secretion. They recognize colitose, which is the terminal sugar of the O-specific side chain of the lipopolysaccharide of the serotype of *E. coli* used (**O**111) and block the production of enterochelin, thus depriving the bacteria of essential iron (Bullen *et al.*, 1978). Although the mechanisms involved in inhibiting enterochelin production are not known, it is thought that the bacteriostatic action of IgA and lactoferrin plays a part in the protection afforded by breast milk against neonatal enteritis caused by *E. coli* (Bullen *et al.*, 1972; Bullen, 1976, 1981; Honour and Dolby, 1979; Dolby and Honour, 1979; Dolby *et al.*, 1980).

Enterochelin seems to be the main endogenous siderophore made by *E. coli* when growing in media containing low levels of available iron. However, *E. coli* also has the remarkable capability for obtaining iron by means of a variety of exogenous hydroxamate-type siderophores which are not synthesized by the organism itself but produced by other micro-organisms (Leong and Neilands, 1976; Raymond and Carrano, 1979; Konisky, 1979; Neilands, 1981). *Salmonella* and *Klebsiella* can also obtain iron in this way. The best studied of these exogenous chelators is the fungal trihydroxamate siderophore, ferrichrome. Unlike enterochelin, ferrichrome is not destroyed after transporting iron into *E. coli* (Leong and Neilands, 1976) although recent results have shown that it is acetylated on one of the hydroxylamino oxygens after reductive separation of Fe^{3+} (Hartmann and Braun, 1980; Schneider *et al.*, 1981). It is not known whether acetylation is an essential step in the utilization of ferrichrome nor is it known whether ferrichrome-mediated iron transport could play any part in supplying iron to pathogenic *E. coli in vivo* during infection. With the continuing search for improved iron chelators for the treatment of patients with iron overload, and since many of these chelators are based on the microbial siderophores (Raymond *et al.*, 1981), the possibility that such hydroxamate or, indeed, catecholate-type chelators might increase the susceptibility of treated individuals to bacterial infections should not be overlooked. Similarly, it is not known whether iron chelated to citrate *in vivo* can serve as a source of iron for *E. coli*. *E. coli* possesses a citrate-mediated iron transport system which is induced when the bacteria are grown in an iron-limited medium containing citrate (Rosenberg and Young, 1974; Woodrow *et al.*, 1978; Hussein *et al.*, 1981). This is in contrast to *Salmonella typhimurium*, which lacks this particular iron uptake mechanism, although, paradoxically, citrate can be used as a carbon source by *Salmonella* but not normally by *E. coli* (Neilands, 1981; Wagegg and Braun, 1981; Kay and Cameron, 1978). *E. coli* containing H1 plasmids can, however, utilize citrate as sole energy source (Smith *et al.*, 1978).

Adaptation to Iron Restriction

The process of adapting to growth in an iron-restricted environment involves not only the synthesis of enzymes for enterochelin synthesis but also the production of outer-membrane protein receptors and enzymes which are involved in the uptake and release of iron from this and other siderophores. The need for specific receptors is due in part to the fact that the molecular weights of the siderophores exceed the diffusion limit (about 600) of the small, water-filled pores of the outer membrane (Nikaido, 1979). The molecular weight of the iron–citrate complex, however, is probably no greater than 443 [Fe (citrate)$_2$] and the fact that there is also a specific receptor for this iron complex suggests that the iron requirements of the cell can be satisfied only by initial adsorption of the Fe–siderophore to surface receptors where iron can be accumulated relative to the concentration in the growth medium. The protein receptors of the enteric bacteria, which have apparent molecular weights in the range 74,000–83,000 (74K–83K), are essential to transport the iron chelators into the bacterial cell and are not synthesized in any significant amounts by the iron-replete organisms. In *E. coli* the receptor proteins are usually designated according to their molecular weights, 83K, 81K, 78K and 74K proteins (Braun *et al.*, 1976). The 81K protein (Fep A protein) is the receptor for Fe–enterochelin and is the product of the *fep* gene, mapping in the *ent fep fes* gene cluster at approximately 13 min (Hollifield and Neilands, 1978; Konisky, 1979; Braun and Hantke, 1981; Neilands, 1982; Fiss *et al.*, 1982). This gene cluster controls the biosynthesis, uptake and hydrolysis of enterochelin (Rosenberg and Young, 1974; Bachmann and Low, 1980). The 78K protein (Ton A protein) is the receptor for ferrichrome and the product of the *ton*A gene (Konisky, 1979; Braun and Hantke, 1981; Neilands, 1982). The functions of the 83K protein and the 74K protein (Cir protein), which is the product of the *cir* gene (Hancock *et al.*, 1977; Konisky, 1979), are currently unknown. *E. coli* also produces another outer-membrane protein with an apparent molecular weight of about 81K (Fec A protein) when growing in iron-restricted media containing 1 m*M* citrate (Hancock *et al.*, 1976; Wagegg and Braun, 1981). This protein appears to be part of the citrate-mediated iron transport system (Hussein *et al.*, 1981; Wagegg and Braun, 1981).

Most of the work on the iron-regulated outer-membrane proteins of *E. coli* has been carried out with laboratory strains, such as *E. coli* K-12. Recently, however, we have shown that pathogenic strains produce the same new proteins when growing *in vitro* in the presence of iron-binding proteins (Griffiths *et al.*, 1983). It is interesting that the relative abundance of the 83K, 81K, 78K and 74K proteins expressed in different pathogenic strains varied considerably. Also, the pathogens produced more of the iron-regulated proteins than did *E. coli* K-12 when grown under the same conditions. Since surface components play such an important part in host–bacteria interactions these alterations, induced in the bacterial cell envelope by iron-binding proteins, are clearly of interest, especially since it has been shown that the same new outer-membrane proteins are expressed by *E. coli* growing *in vivo* during lethal peritoneal infections (Fig. 3) (Griffiths *et al.*, 1983, and unpublished data). Indeed, the iron-regulated outer-membrane proteins of *E. coli* O111 grown *in vivo* during infection are major components of the outer membrane and are present in amounts equal to, or even slightly greater than, the so-called 'major' outer-membrane proteins with apparent molecular weights in the range 30K–42K (Overbeeke and Lugtenberg, 1980). Of course, it remains to be seen if *E. coli* express these proteins when colonizing body sites other than the peritoneal cavity. However, the possibility that the siderophore receptors are important protective antigens seems worthy of exploration. Some of the iron-regulated proteins are exposed on the surface of the pathogenic strain *E. coli* O111 and can interact with large molecules in solution (Griffiths *et al.*, 1983). It is possible, therefore, that they are exposed to, and can interact with, antibodies. Preliminary work has shown that the iron-regulated outer-membrane proteins are immunogenic and that some normal sera from mice, rabbits, guinea-pigs and man contain anti-siderophore-receptor IgG antibodies (E. Griffiths, P. Stevenson, and R. Thorpe, unpublished data). However, it is not yet known whether these antibodies inhibit *E. coli* multiplication or contribute, together with the antibodies to enterochelin (Moore *et al.*, 1980; Moore and Earhart, 1981) or to the O antigen (Fitzgerald and Rogers, 1980), towards the bacteriostatic effect of sera on *E. coli* (Bullen and Rogers, 1969; Rogers, 1973, 1976). Similarly, it is not known whether individuals with IgG antibodies against the iron-regulated outer-membrane proteins also have IgA

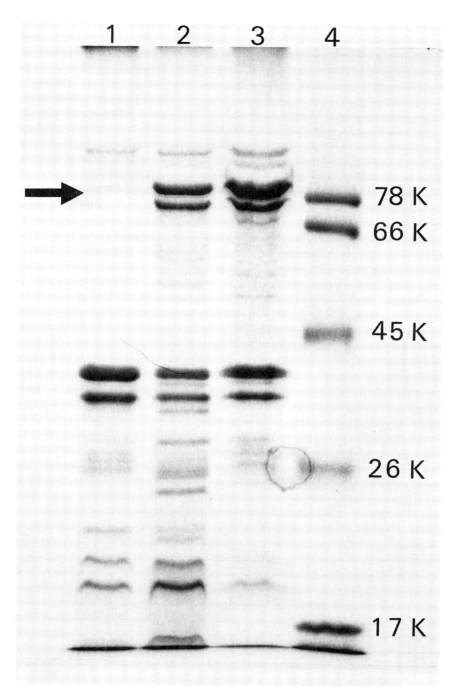

Fig. 3. Polyacrylamide gel electrophoresis of the outer-membrane proteins of *E. coli* O111 grown in (1) trypticase soy broth, (2) trypticase soy broth containing ovotransferrin, (3) the peritoneal cavities of lethally infected guinea-pigs. Lane 4 carried molecular weight markers, the sizes of which are indicated. The positions of the iron-regulated outer-membrane proteins are shown by the arrow. (From Griffiths *et al.*, 1983.)

and/or IgM antibodies against these proteins. It might be expected that the level of secretory IgA would be important for protection on mucosal surfaces, such as in the small intestines.

Although much is now known about enterochelin and about the outer-membrane siderophore receptors, little is known about the way these high-affinity iron-sequestering systems are regulated, except to say that their synthesis is repressed when iron is freely available in the medium (Rosenberg and Young, 1974; Konisky, 1979; Braun and Hantke, 1981; Neilands, 1981, 1982). The synthesis of enterochelin and of some of the siderophore receptors appears to be co-ordinately regulated by iron even though the genes are not all included in the same transcriptional unit (McIntosh and Earhart, 1977; Bachmann and Low, 1980; Klebba *et al.*, 1982). The 83K, 81K and 74K proteins are regulated together but the synthesis of the 78K protein is controlled separately (Plastow *et al.*, 1981; Klebba *et al.*, 1982). Induction of the Fec A protein depends not only on low levels of iron but, in addition, on relatively high concentrations of exogenous citrate (Hussein *et al.*, 1981). However, constitutive mutants of *E. coli* K-12 which express the 83K, 81K, 78K, 74K and Fec A protein in an iron-rich medium have recently been produced (Hantke, 1981). This suggests that a regulatory mechanism is acting which covers all iron-uptake systems and which is superimposed upon the regulation of the individual systems (Hantke, 1981; Braun and Burkhardt, 1982). These constitutive mutants have been called *fur* mutants and are similar to those found in *S. typhimurium* (Ernst *et al.*, 1978). The nature of the regulatory mechanism affected by the *fur* mutation is currently unknown (Hantke, 1982).

In addition to synthesizing enterochelin and the outer-membrane siderophore receptors, other more fundamental changes also take place in *E. coli* as it adapts to growth under the iron-restricted conditions imposed by iron-binding proteins. Growth of *E. coli* in the presence of transferrin, lactoferrin or ovotransferrin results in the generation of specifically altered tRNA molecules which elute earlier on chromatography than do the same tRNAs from the iron-replete cells (Griffiths and Humphreys, 1978; Buck and Griffiths, 1982). The same iron-related tRNA alterations have been shown to occur in *E. coli* O111 growing *in vivo* during infection (Griffiths *et al.*, 1978). These chromatographically altered tRNAs lack the methylthio (ms^2) moiety of 2-methylthio-N^6-(Δ^2-isopentenyl)-adenosine (ms^2i^6A) which occurs adjacent to the 3' end of the anticodon of the tRNAs from iron-replete *E. coli* (Buck and Griffiths, 1982); the change in structure is shown for tRNA^{phe} in Fig. 4. The tRNAs which are altered during iron restriction include tRNA^{phe}, tRNA^{tyr}, tRNA^{trp} and two minor tRNA^{ser} species and recent work suggests a regulatory role for these changes (Griffiths and Humphreys, 1978; Buck and Griffiths, 1981, 1982; M. Buck and E. Griffiths, unpublished data). Variations in the extent of modification of the adenosine located next to the anticodon affect codon–anticodon interactions and

tRNA ^{Phe}

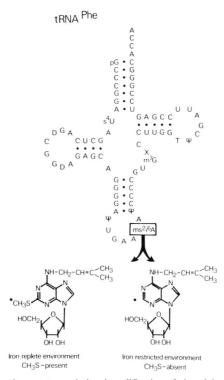

Fig. 4. Effect of iron on the post-transcriptional modification of phenylalanine tRNA from *E. coli*. [From Griffiths, 1981. *In* "Nutritional Factors" (ed. R. F. Beers and E. G. Bassett). Copyright 1981 by Raven Press, New York.]

lead to changes in the translational efficiency of the tRNA. Loss of the ms^2 group reduces the translational efficiency of the molecule (Buck and Griffiths, 1982). Furthermore, the evidence suggests that the ms^2-deficient tRNAs, with their lowered translational efficiencies, relieve transcription termination at the attenuators of certain operons of the aromatic amino acid biosynthetic pathway and thus lead to their increased expression under iron-restricted conditions. Attenuation is also regulated by the level of charged tRNA which, in turn, depends upon the availability of the amino acid (Yanofsky, 1981). The regulation of the phenylalanine and tryptophan operons, and possibly other operons of the aromatic amino acid biosynthetic pathway, may be involved with the adaptation of *E. coli* for growth in iron-restricted environments. Enterochelin is synthesized from chorismic acid by way of a branch of the aromatic amino acid biosynthetic pathway (Fig. 2) (Rosenberg and Young, 1974) and iron itself is an essential component of the first enzyme of the common pathway (McCandliss and Herrmann, 1978). The ability to regulate the expression of various operons in this pathway through the level of charging of the tRNA and by the degree of modification of the molecule, which are in turn dependent upon iron levels, may well provide the cell with useful regulatory flexibility when it makes enterochelin. Methylthiolation of tRNA also influences aromatic amino acid transport in *E. coli* (Buck and Griffiths, 1981) and it only remains to be seen whether there are other metabolic consequences of tRNA alteration. In some operons, such as the His operon, transcription termination at the attenuator is the major controlling element and there is no repressor molecule (Yanofsky, 1981). The absence of the methylthio group in tRNA of *E. coli* is not, however, the major controlling signal that regulates enterochelin synthesis or the synthesis of the iron-regulated outer-membrane proteins in this organism (Griffiths and Buck, 1979, and unpublished data).

The ColV Plasmid Iron Sequestering System

Certain strains of *E. coli* are capable of causing generalized extra-intestinal infections in man and animals. Many of these have been shown to carry a colicin V (ColV) plasmid and to have a higher level of survival in blood and peritoneal fluid when compared with strains lacking ColV plasmid (Smith, 1974; Minshew *et al.*, 1978; Silver *et al.*, 1980; Davies *et al.*, 1981). Indeed, Smith and Huggins (1976, 1980) have shown that the elimination of the ColV plasmid from virulent *E. coli* strains was always accompanied by a decrease in pathogenicity; reintroduction of the ColV plasmid into cured, derivative strains restored pathogenicity to its original level. Although possession of a ColV plasmid markedly enhances virulence, and is clearly a virulence factor in these invasive strains, colicin V activity itself is not essential (Binns *et al.*, 1979; Quakenbush and Falkow, 1979; Williams and Warner, 1980). Williams (1979a) was the first to show that some ColV plasmids encode an iron-sequestering system and suggested that this plasmid-mediated iron transport mechanism might play a key role in the pathogenicity of organisms harbouring such plasmids. Two components of the ColV-plasmid-mediated iron uptake system have now been identified: one is a hydroxamate-type siderophore called aerobactin (Stuart *et al.*, 1980; Williams and Warner, 1980; Braun, 1981; Warner *et al.*, 1981) and the other is an iron-regulated outer-membrane protein which appears to form at least part of the ferric–aerobactin receptor (Grewal *et al.*, 1982; Bindereif *et al.*, 1982).

Aerobactin, a member of the hydroxamic acid–citrate family of siderophores and originally isolated from *Aerobacter aerogenes* (Gibson and Magrath, 1969), is a conjugate of 6-(*N*-acetyl-*N*-hydroxyamino)-2-aminohexanoic acid and citric acid (Fig. 5). It forms an octahedral ferric complex using the two bidentate hydroxamate groups, the central carboxylate and probably the citrate hydroxyl group (Harris *et al.*, 1979b). Although its secretion is thought to play an important part in the virulence of the invasive *E. coli* strains (Williams, 1979a; Stuart *et al.*, 1980; Warner *et al.*, 1981; Williams and Warner, 1980), it is not clear

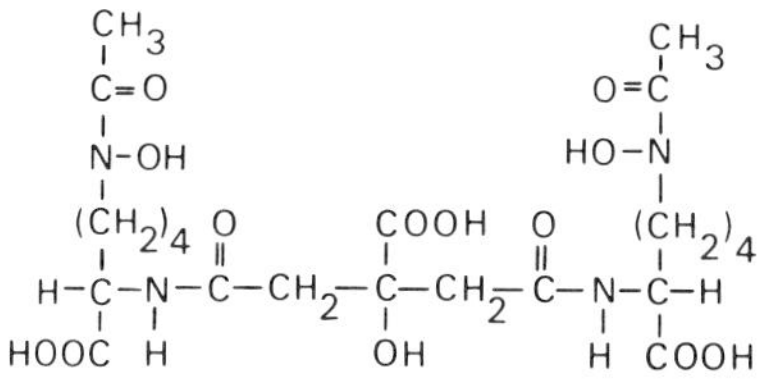

Fig. 5. Structural formula of aerobactin.

how acquisition of the ability to make this siderophore confers a selective advantage on organisms already capable of synthesizing enterochelin. At pH 7.4, the normal physiological pH, enterochelin is by far the most effective siderophore characterized (Harris *et al.,* 1979c). The formation constant for ferric enterochelin (log Kf) is 52, which is roughly 29 orders of magnitude larger than that of aerobactin (log Kf 22.9) (Harris *et al.,* 1979b). Formation constants by themselves, however, say nothing about the relative ability of ligands to compete with one another for ferric ion at a given pH. This is because of variations in the ligand protonation constants and of concentration dependencies. In order to obtain a more direct ranking of ligands under specified conditions, Harris *et al.* (1979a,c) report what they call pM values, which are the negative logarithms of the free ferric ion concentrations under different conditions, calculated from the various equilibrium values involved. However, as can be seen in Table 1, the pM value of enterochelin at pH 7.4 is still more than 10 orders of magnitude higher than that of aerobactin; indeed, aerobactin is marginally less effective than transferrin. Quite apart from these thermodynamic considerations, it appears that hydroxamate-based chelating agents are kinetically inferior to catecholate-type siderophores in removing iron from transferrin (Carrano and Raymond, 1979). Stuart *et al.* (1980) proposed that ColV-plasmid-specified aerobactin is bound to the bacterial cell wall and is, therefore, functionally different and, somehow,

Table 1. *Values of pM for solutions of ferric complexes of biological ligands (taken from the data of Harris et al., 1979b; Raymond and Carrano, 1979)*

Ligand	pM_1[a]	pM_2[b]
Enterochelin	37.6	35.52
Ferrichrome	27.25	25.20
Transferrin	25.62	23.6
Aerobactin	25.37	23.32

[a]Total metal 10^{-6} M: total ligand 10^{-3} M, pH 7.4.

[b]Total metal 10^{-6} M: total ligand 10^{-5} M, pH 7.4.

more efficient than the secreted enterochelin in acquiring iron from iron-binding proteins. However, this conclusion contrasts with the known relative effectiveness of the two compounds as iron chelators and the cell-free mode of action suggested for aerobactin by genetic and other data (Williams and Warner, 1980; Warner *et al.*, 1981). Aerobactin is secreted into culture media of both ColV-plasmid-bearing *E. coli* and other aerobactin producing bacteria, such as *Aerobacter aerogenes,* which incidentally also make enterochelin (Gibson and Magrath, 1969; Payne, 1980; Braun, 1981; Warner *et al.*, 1981; van Tiel-Menkveld *et al.*, 1982). There is, in addition, a specific receptor for ferric aerobactin in the outer membrane of these organisms (see below).

One possibly important difference between the mode of action of enterochelin and the hydroxamate siderophores, such as aerobactin, is that the latter may be recycled. There is as yet no evidence that aerobactin cannot be re-used, whereas enterochelin is known to be used only once for transporting iron into the bacterial cell; enterochelin-mediated iron transport is, therefore, an energetically expensive way of assimilating iron (Raymond and Carrano, 1979). In contrast, iron can be removed from the hydroxamates by simple reduction of Fe^{3+} to Fe^{2+} without destruction of the chelator (Harris *et al.*, 1979b). Harris *et al.* (1979b) consider that the high-affinity, low-capacity and energetically expensive iron transport system using enterochelin might operate under conditions of extreme iron stress and the low-affinity, high-capacity system using aerobactin, under conditions of mild iron stress. Others, however, conclude that enterochelin is produced under conditions of mild iron limitation and that aerobactin production is maximal under iron-deficient conditions (van Tiel-Menkveld *et al.*, 1982). Experimentally it can be seen that both these chelators are produced simultaneously, although acrobactin is synthesized mainly in the late exponential and stationary phases of growth (Gibson and Magrath, 1969; Stuart *et al.*, 1980; van Tiel-Menkveld *et al.*, 1982). The relevance of data obtained with bacteria in iron-deficient media, in which the quantity of iron is limiting and the organisms starved of iron, to the situation *in vivo* during infection is questionable. A clear distinction should be made between the quantity of iron present in a medium and its availability to bacteria. There is plenty of iron present in body fluids although it is not readily available because of the presence of the iron-binding proteins. As far as infection is concerned, it is the relative ability of these chelators to remove iron from transferrin or lactoferrin and to transport it back to *E. coli* that is important. It must not be forgotten, therefore, that these reactions take place in body fluids, such as serum or mucosal secretions, and not in trypticase soy broth or minimal media. The presence of other factors, in serum or body fluids, could well affect the overall effectiveness of different siderophores. It is known that the kinetic barrier to efficient hydroxamate-mediated removal of iron from transferrin can be partially overcome by the addition of other anions to the system, a phenomenon which could well function in body fluids such as serum (Pollack *et al.,* 1976). We might also consider the presence of the anti-enterochelin anti-bodies, the anti-O-antibodies and the anti-receptor antibodies, all of which might restrict the effectiveness of an enterochelin-mediated iron uptake system in serum and favour aerobactin-mediated iron transport.

Although there is no convincing answer as yet to why acquisition of the ability to synthesize an apparently inferior iron chelator gives *E. coli* an advantage during growth *in vivo,* it is clear that mere synthesis of this chelator is insufficient for virulence; a specific receptor for the uptake of ferric aerobactin is also required (Williams and Warner, 1980). Recently the identity of an outer-membrane protein receptor, specified by a ColV plasmid, regulated by iron and involved in aerobactin-mediated iron transport, has been reported (Grewal *et al.,* 1982; Bindereif *et al.,* 1982). This protein has an apparent molecular weight of 74K and is revealed only in *cir* mutants which lack the 74K Cir protein; it serves not only as part of the receptor for ferric aerobactin but also for cloacin, a bacteriocin produced by *Enterobacter cloacae* (Oudega *et al.,* 1979; van Tiel-Menkveld *et al.,* 1982; Bindereif *et al.,* 1982). It is interesting that the iron-controlled expression of this 74K protein is also regulated by the chromosomal *fur* gene and is thus expressed constitutively in *fur* mutants (Braun and Burkhardt, 1982). To complicate matters further, Braun *et al.* (1982) have reported that the presence of the ColV-plasmid-specified 74K protein in the outer membrane of *E. coli* is not by itself sufficient for transport of ferric aerobactin; the 78K (Ton A) protein, which is the receptor for ferrichrome, also seems somehow to be involved. Clearly, at present we know only part of the ColV plasmid story.

α-*Haemolysin and the Use of Haem Iron*

Not all strains of *E. coli* that cause extra-intestinal infections carry a ColV plasmid. Some possess instead the ability to synthesize a haemolysin. Haemolysins are bacterial toxins that mediate the lysis of erythrocytes; colonies of haemolytic *E. coli* can easily be identified by the clear zone of haemolysis that surrounds them on blood agar plates. The association of haemolytic *E. coli* with disease has long been recognized. As early as 1921 Dudgeon *et al.* suggested that haemolytic activity might play a part in the virulence of *E. coli* in the human urinary tract. Since then many studies have shown that a high proportion of *E. coli* isolated from all sites of extra-intestinal infection in man are haemolytic (Fried and Wong, 1970; Cooke and Ewins, 1975; Green and Thomas, 1981; Minshew *et al.,* 1978; De Boy *et al.,* 1980). Although haemolytic *E. coli* are isolated much less frequently from human faeces (Minshew *et al.,* 1978; De Boy *et al.,* 1980), many enteropathogenic *E. coli* of porcine origin produce haemolysins (Smith and Linggood, 1971). In addition, haemolytic strains have been found to be much more virulent for mice and rats than haemolysin-negative strains or mutants (Smith and Linggood, 1971; Emödy *et al.,* 1980; van den

Bosch *et al.*, 1979, 1981, 1982; Welch *et al.*, 1981). Using the chicken embryo model developed by Powell and Finkelstein (1966), Minshew *et al.* (1978) found that about 80% of strains which were virulent for chicken embryos were also haemolytic.

Lovell and Rees (1960) were the first to describe a filterable haemolysin produced by haemolytic *E. coli.* Smith (1963) called this toxin α-haemolysin, to distinguish it from another cell-bound haemolysin which he called β-haemolysin. The α-haemolysins synthesized by *E. coli* from a variety of sources are antigenically cross-reactive and have similar but not identical physical properties (Smith, 1963). In general, the α-haemolysins are heat-labile acidic proteins with molecular weights estimated at 58,000 to 120,000 and they require Ca^{2+} for full activity (Short and Kurtz 1971; Smith, 1963; Rennie and Arbuthnott, 1974; Rennie *et al.*, 1974; Williams, 1979b). The synthesis of α-haemolysin in *E. coli* has, in many cases, been shown to be mediated by plasmids (Smith and Gyles, 1970; Smith and Halls, 1967; Smith and Heller, 1973; Smith and Linggood, 1970; Waalwijk *et al.*, 1982a; Goebel *et al.*, 1974; Noegel *et al.*, 1979). Many such α-haemolysin (Hly) plasmids have been identified but they appear to be genetically diverse, except for the region encoding α-haemolysin production which, in one case, has been shown to consist of three cistrons, one coding for a precursor of the toxin molecule and the other two for mechanisms involved in processing and transporting the toxin out of the bacterial cell (Royer-Pokora and Goebel, 1976; Waalwijk *et al.*, 1982b; Noegel *et al.*, 1979; de la Cruz *et al.*, 1979; Stark and Shuster, 1982). Recent work suggests that genes encoding haemolysin production may be on a transposable DNA segment able to jump from one plasmid into another or even into the chromosome (de la Cruz *et al.*, 1980; Noegel *et al.*, 1981; Waalwijk *et al.*, 1982a; Royer-Pokora and Goebel, 1976). Indeed, there are several instances, especially in haemolytic *E. coli* from human extra-intestinal infections, where the haemolysin determinant is not transferable from strain to strain and is thought to be located either in the chromosome or in a non-mobilized plasmid (Smith and Linggood, 1970; Minshew *et al.*, 1978; Hull *et al.*, 1982). The genetic basis for haemolysin production by one such *E. coli* strain has now been investigated in some detail and the genes for haemolysin production found to be located on the chromosome and to map near the *ilv* gene cluster (Hull *et al.*, 1982).

Although considerable information has accumulated about the nature of haemolysins and the genetic basis for their production, until recently very little was known about their role in pathogenesis. It is now clear, however, that haemolysin activity itself contributes towards virulence (Welch *et al.*, 1981) and that it does so, at least sometimes, by enabling haemolytic strains of *E. coli* to obtain iron for growth from the haem released from lysed erythrocytes (Linggood and Ingram, 1982). In addition to obtaining iron from the iron-binding proteins, some pathogens, including *E. coli,* can obtain iron from cell-free haem (Bullen,

1981; Griffiths, 1981). Haem iron seems to be more readily available to *E. coli* than iron bound to transferrin or lactoferrin but normally there is only a trace of free haem in serum and this is bound to haemopixin or serum albumin (Morgan, 1981). It is necessary, therefore, to liberate haemoglobin from the erythrocytes by haemolysis in order to promote bacterial growth. Since haemolysis resulting from haemolytic diseases such as sickle-cell anaemia, bartonellosis and malaria leads to increased susceptibility to bacterial infections (Barrett-Conner, 1971; Cuadra, 1956, Kaye *et al.*, 1967), probably due in part to the freely available haem, increasing the level of available haem by haemolysin-induced lysis of erythrocytes might be expected to do the same. The experiments of Linggood and Ingram (1982) strongly suggest that this does indeed occur. The transfer of an Hly plasmid encoding haemolysin production to a non-haemolytic strain of *E. coli* increased the virulence of the strain for mice. Injections of non-toxic amounts of α-haemolysin, phenylhydrazine (which causes haemolysis), iron or haemoglobin simulated the effect of the Hly plasmid by stimulating bacterial growth. Active or passive immunization against α-haemolysin protected mice on challenge with a haemolytic *E. coli in vivo;* this protection was overcome by administration of iron salts at the time of challenge. An iron-sequestering mechanism depending upon the utilization of the liberated haem would, of course, be expected to be important only for growth of the bacteria in blood and the ability to make haemolysin should have little effect on bacterial growth on mucosal surfaces such as in the gut, as actually seems to be the case (Smith and Linggood, 1971). It is not clear, however, why haemolysin production should be associated with the virulence of human pyelonephritogenic strains of *E. coli;* perhaps this property is important only during a certain stage of the infection. There seems to be a particular connection between haemolysin production and the ability of the *E. coli* to produce pyelonephritis in mice and rats after intravenous injection, possibly indicating that an ability to multiply effectively in the blood stream is an important factor related to the initiation of infection by this route (Fried *et al.*, 1971; van den Bosch *et al.*, 1979, 1981). It would be interesting to see what effect passive or active immunization against α-haemolysin would have in such experimental kidney infections in mice. Anti-α-haemolysin antibodies have been found in patients with infections caused by haemolytic *E. coli* and also in healthy controls (Smith, 1963; Emödy *et al.*, 1982). However, the titres were very much higher in the infected group, which included patients with cystitis and pyelonephritis (Emödy *et al.*, 1982). It is also interesting to note that the concentration of haptoglobin, a naturally occurring haemoglobin-binding protein, increases dramatically in the plasma of patients with both acute and chronic infections (Owen *et al.*, 1964; Chiancone *et al.*, 1968).

Although α-haemolysin might sometimes act as a virulence factor by enabling haemolytic strains of *E. coli* to obtain iron for growth from the lysed erythrocytes of infected animals, it may in addition function simply as a cytotoxin. The cytotoxic activity of α-haemolysin has been shown with chicken embryo fibro-blasts and mouse fibroblasts (Chaturvedi *et al.*, 1969; Cavalieri and Snyder, 1982a). α-Haemolysin also affects both the activity and viability of human peripheral blood leukocytes *in vitro* (Cavalieri and Snyder, 1982b,c). It is not known whether these events occur *in vivo* but if they do, then together they would result in the host having a greatly reduced capacity to phagocytose and kill the invading bacteria, two events which should increase the survival of *E. coli* during infection. It is possible, of course, that both the haemolytic and cytotoxic mechanisms function *in vivo*, possibly operating independently at different body sites. Nothing, however, is known about possible variations in the ability of different *E. coli* strains actually to utilize the iron once the haemoglobin has been released from the erythrocytes. It is known that haem iron can be used by many *E. coli* strains independently of haemolysin production (Griffiths, 1981). Thus, the lethal dose of non-haemolytic *E. coli* for guinea-pigs can be dramatically reduced by treating the animals with haemoglobin (Bullen *et al.*, 1968). It has been suggested that the increased availability of iron in haem abolishes the need for the *E. coli* to adapt to the otherwise iron-restricted environment found in host tissues and allows them to multiply more rapidly (Griffiths, 1981). *E. coli* has a doubling time of about 35 min when growing *in vitro* in the presence of iron-binding proteins but this is reduced by 10 min when iron is freely available in the medium (Griffiths and Humphreys, 1978). Although this difference is not very large it can have a significant effect on the size of the bacterial population over a matter of hours; such differences in generation times *in vivo* could be crucial in determining the outcome of an infection (Griffiths, 1981). Lysis of red cells is not an uncommon injury and the life-threatening combination of the presence together of *E. coli* and haemoglobin in the peritoneal cavity is well documented clinically (Davis and Yull, 1964; Bornside *et al.*, 1968; Bornside and Cohn, 1968; Bullen *et al.*, 1978). This combination may arise from any form of trauma (Davis and Yull, 1964) and it has been shown that it is the iron in haemoglobin that is responsible for stimulating bacterial growth (Bornside *et al.*, 1968; Lee *et al.*, 1979). It has also been found that the administration of haptoglobin protects against experimental infections promoted by haemoglobin and such work has suggested a possible use for haptoglobin in treating potentially fatal haemoglobin-induced bacterial infections (Eaton *et al.*, 1982). However, the mechanism whereby *E. coli* is able to utilize the iron in haem is still unknown and this aspect is clearly deserving of investigation.

Serum Resistance

While the successful multiplication of *E. coli* in the normally iron-restricted environment *in vivo* is obviously essential for their pathogenicity, ability to resist the bactericidal action of serum can also be considered as a virulence factor. It is well known that blood, or serum, is bactericidal for many Gram-negative organisms and this bactericidal activity is thought to be an important host defence mechanism against infections caused by such bacteria. No attempt will be made here to discuss all that is known about the bactericidal effects of serum and the various and numerous properties of resistant bacteria (see Taylor, this volume, Chapter 25). Rather, those aspects will be emphasized about which we now have clearer insight and which can with some confidence be regarded as virulence factors.

The lethal activity of normal serum results primarily from the action of the complement system on the bacterial surface; complement may be activated by the classical pathway involving antibody or by the alternative pathway (Inoue *et al.*, 1968; Müller-Eberhard, 1975; Frank, 1979). Although lysozyme is present in serum, it is not essential for the bactericidal action of complement (Inoue *et al.*, 1959; Glynn, 1969). Complement alone damages both the outer and inner membranes of *E. coli*, although it seems that it is the damage done to the inner membrane that kills the cell (Wright and Levine, 1981a,b). The role of resistance to this bactericidal system in the virulence of invasive *E. coli* is suggested, on the one hand, by epidemiological evidence, and on the other, by results obtained with experimental invasive infections in animal models (Olling, 1977; Durack and Beeson, 1977; Björksten and Kaijser, 1978; Roantree and Rantz, 1960; McCabe *et al.*, 1978). Two cellular components, an outer-membrane protein and a polysaccharide capsular antigen, have been investigated in some detail and shown to be independently capable of providing *E. coli* with substantial resistance to serum killing.

The K1 Capsular Antigen

The K1 polysaccharide capsular antigen is the polysaccharide most frequently found on *E. coli* strains isolated from human adults and neonates with bacteraemia and urinary tract infections; *E. coli* K1 is also associated with invasive diseases in laboratory and domestic animals (Schiffer *et al.*, 1976; McCabe *et al.*, 1978; Sarff *et al.*, 1975). An even more striking relationship has been shown between the antigenic make-up of *E. coli* and neonatal meningitis; strains of *E. coli* possessing the K1 antigen have been almost the only serological type isolated from such cases of meningitis (Robbins *et al.*, 1974; Sarff *et al.*, 1975). The K1 capsular polysaccharide is a linear homopolymer of $\alpha 2 \rightarrow 8$-linked N-acctylneuraminic acid and is structurally and antigenically identical with the meningococcal group B capsular polysaccharide (Kasper *et al.*, 1973; Robbins *et al.*, 1980). Passive administration of anti-K1 antibodies prevented bacteraemia and meningitis in infant rats fed *E. coli* K1 (Glode *et al.*, 1977) and K1-negative derivatives of these invasive K1 strains failed to produce bacteraemia in infant rats (Silver *et al.*, 1980). The precise role of the K1 antigen in bacterial pathogenicity has, however, been something of a mystery. Some reports show that it exerts an antiphagocytic effect (Bortolussi *et al.*, 1979; Weinstein and Young, 1978) and others that it provides *E. coli* with resistance to the bactericidal effects of serum (Glynn and Howard, 1970; Gemski *et al.*, 1980). K1-positive *E. coli* have been made extremely sensitive to serum killing as a consequence of single-step mutations which produce K1-negative derivatives (Gemski *et al.*, 1980). However, others have suggested that the K antigens provide little if any protection against serum bactericidal activity (Björksten and Kaijser, 1978; McCabe *et al.*, 1978; Pitt, 1978; Taylor and Robinson, 1980). Nevertheless, more recent work clearly shows that the K1 antigen itself can provide rough *E. coli* strains with significant resistance against the bactericidal effects of normal human serum, although another K antigen, the K27 antigen, had no effect (Opal *et al.*, 1982). Molecular cloning of the K1 polysaccharide genes of *E. coli* in *E. coli* K-12 has provided unequivocal evidence that this antigen does provide the bacteria with protection against serum killing (Timmis *et al.*, 1981; Silver *et al.*, 1981). Although *E. coli* K-12 was sensitive to the bactericidal effects of serum, *E. coli* K-12 expressing the K1 antigen was resistant. Protection is probably related to the shielding effects of the K1 polysaccharide capsule on the cell structures capable of activating or interacting with complement (Robbins *et al.*, 1980) (see Jann and Jann, this volume, Chapter 5).

Plasmid-Encoded Outer Membrane Proteins

In recent years several groups have reported that certain plasmids also substantially increase the resistance of *E. coli* to the bactericidal action of serum (Fietta *et al.*, 1977; Reynard and Beck, 1976; Taylor and Hughes, 1978; Binns *et al.*, 1979; Reynard *et al.*, 1978). Some of these are conjugative (self-transmissible) multiple-antibiotic-resistance plasmids and two have been investigated in some detail; they are the large (100 kb) plasmid R6-5 (Silver and Cohen, 1972) and the closely related R100 plasmid, both of which can reduce the serum sensitivity of smooth and rough *E. coli* strains (Reynard and Beck, 1976; Taylor and Hughes, 1978; Reynard *et al.*, 1978; Ogata and Levine, 1980). The effect of these R plasmids on the serum sensitivity of rough K-12 strains has, however, been questioned by Taylor and Hughes (1978), who detected plasmid-dependent serum resistance only with a smooth *E. coli* strain. Such a discrepancy might be attributed to the use of serum from different animals and/or to differing definitions of resistance (Taylor and Hughes, 1978; Binns *et al.*, 1982; Ogata and Levine, 1980). More recent data show, without a doubt, that R plasmids can increase the serum resistance of rough *E. coli* K-12; however, the absolute level of resistance does, indeed, depend upon the intrinsic sensitivity of the host bacterium (Ogata and Levine, 1980). R-plasmid-encoded resistance determinants seem to be fully functional only when superimposed upon a full complement of O-specific lipopolysaccharide chains (Ogata and Levine, 1980; Taylor and Robinson, 1980; Timmis *et al.*, 1981). The R6-5- and R100-encoded serum resistance determinant has been identified as a major outer-membrane protein. Gene cloning and mapping techniques have shown that it is the plasmid *tra*T gene product which is responsible for increasing the resistance of *E. coli* to the bactericidal action of complement (Moll *et al.*, 1980; Binns *et al.*, 1982; Ogata *et al.*, 1982). The *tra*T gene product is an outer-membrane protein with a molecular weight of 25,000 and is one of the major exposed proteins on the surface of *E. coli* containing F-like conjugative plasmids (Manning *et al.*, 1980; Ferraza and Levy, 1980; Timmis *et al.*, 1981). The *tra*T gene lies in a large plasmid operon that encodes most of the functions necessary for conjugation and, together with the *tra*S gene, mediates what is called surface (or entry) exclusion; this is the reduced ability of bacteria carrying plasmids to act as effective recipients in conjugation with closely related donor strains (Achtmann *et al.*, 1977; Manning and Achtmann, 1979).

Fietta *et al.* (1977) reported that a number of different antibiotic-resistance plasmids confer upon host bacteria the ability to resist the lethal action of serum. They suggest that several distinct types of plasmid gene products may be responsible for this property. Because R6-5 shows extensive polynucleotide sequence homology with other Inc FII-group plasmids, like R100 and R1 (Sharp *et al.*, 1973), it is almost certain that serum resistance encoded by this group of plasmids is due to the *tra*T gene product. However, R plasmids are not the only plasmids that increase the serum resistance of *E. coli*. A ColV plasmid, which did not confer a growth advantage on *E. coli* under conditions of iron stress (Williams, 1979a), has been shown by Binns *et al.* (1979) to carry determinants for resistance to the bactericidal effects of serum. This conjugative plasmid, ColV, I-K94, specifies colicin I as well as colicin V production and was originally found in *E. coli* K94, a strain isolated from the faeces of patients suffering from *Salmonella paratyphi* B infection (Fredericq and Joiris, 1950). Although plasmid ColV, I-K94 seems to have undergone some changes during its maintenance in culture collections (Davies *et al.*, 1982), it clearly enhances the virulence of *E. coli* and this has been traced to its increased resistance to complement (Binns *et al.*, 1979). The genetic locus in ColV, I-K94 that is responsible for conferring serum resistance has been called *iss,* for increased survival in serum. Since the *iss* gene maps well outside the transfer genes of this plasmid and does not hybridize with the *tra*T gene, it seems that the two genes are quite distinct; the gene products are also immunologically different (Binns *et al.*, 1982; Ogata *et al.*, 1982). Although a large part of the transfer region of the ColV, I-K94 plasmid shows homology with that of the F plasmid, there is a section in the region of the *tra*T and *tra*S genes which shows no homology; ColV, I-K94 and F are not in the same exclusion group (Sharp *et al.*, 1973;

Willetts and Maule, 1973; Binns *et al.*, 1982). The identity of the *iss* gene product is unknown but it may well be an outer-membrane protein which, like the *tra*T protein, is exposed on the bacterial cell surface (Binns *et al.*, 1982). It seems probable that the *tra*T protein and the *iss* gene product function in a similar way, although at present we know little about the mechanisms involved. Binns *et al.* (1982) found no additive effect on resistance when both genes were present together in *E. coli* and neither gene product prevented the formation of the terminal complement complex. Consumption of the terminal components of complement, C6–C9, was the same whether *E. coli* contained the *tra*T or *iss* genes or not, suggesting that it is the action of the terminal complement complex that is blocked by *tra*T or *iss*, not its formation (Binns *et al.*, 1982; Ogata and Levine, 1980; Wright and Levine, 1981a,b). There is currently insufficient information to allow us to say how this is achieved but it is known that the role of the *tra*T protein is highly specific. Plasmids containing point mutations in the *tra*T gene fail to protect their bacterial hosts from the killing effect of serum, even though the host organism may contain large amounts of *tra*T-like protein in their outer membrane (Moll *et al.*, 1980).

Concluding Remarks

Although the various candidate virulence markers (Table 2) have been discussed separately in this chapter, for the sake of convenience and clarity, it is now clear that the nature and outcome of an *E. coli* infection depends not on one virulence factor acting alone but on a combination of these determinants acting together. Each factor is, as the title of this chapter implies, a candidate or potential virulence marker and is only fully effective when present in *E. coli* with the appropriate genetic make-up. The multifactorial nature of virulence in *E. coli* has been convincingly demonstrated by Smith and Huggins (1980) with strain O18ac:**K**1:**H**7:ColV, an organism isolated from a human baby with meningitis. Forms of this organism lacking either the O18 or K1 antigens or the ColV

Table 2. *Candidate virulence markers*

Marker	Function
Enterochelin/81K outer-membrane protein	Iron uptake
ColV-plasmid-encoded aerobactin/74K outer-membrane protein	Iron uptake
α-Haemolysin	Iron uptake/cytotoxin
colV, I-K94 plasmid *iss* gene product	Serum resistance
R plasmid *tra*T gene product	Serum resistance
K1 antigen	Serum resistance

plasmid were significantly less lethal than the parent strain in experimental infections. Absence of the K1 antigen was associated with a greater loss of virulence than was absence of the O18 antigen or the ColV plasmid. In fact, loss of the ColV plasmid was associated with the least loss of virulence and higher doses of *E. coli* O18ac:**K**1:**H**7:ColV$^-$ bacteria resulted in a disease which, clinically and bacteriologically, resembled that produced by organisms of the ColV$^+$ parent form. The fact that the higher levels of ColV$^-$ bacteria, which lack the aerobactin-mediated iron uptake system (Williams, 1979a; Smith and Huggins, 1980), grew as well as the ColV$^+$ strain in host tissues suggests that the enterochelin-mediated iron uptake mechanism can function efficiently *in vivo*, or that some other iron uptake system is operating. Whether enterochelin itself should be considered a virulence factor is debatable. Rogers (1973) considers it to be a virulence determinant for some *E. coli* strains and Yancey *et al.* (1979) have shown that it is essential for the virulence of *S. typhimurium*. It is also known that it is produced *in vivo* during *E. coli* infection (Griffiths and Humphreys, 1980). However, Miles and Khimji (1975) and Montgomerie *et al.* (1979) found that enterochelin production does not correlate with virulence and is no index of the capacity of a strain to proliferate *in vivo*. They conclude, therefore, that enterochelin cannot usefully be called a virulence factor. Enterochelin is, in fact, produced by many if not most *E. coli* strains, including *E. coli* K-12, so that it is clearly not peculiar to pathogenic *E. coli*. But other so-called 'virulence determinants', such as the K1 antigen, are also found both in *E. coli* strains associated with disease and in those isolated from healthy individuals (Sarff *et al.*, 1975). In addition, it must be remembered that enterochelin production itself is only part of this particular iron-sequestering system. The 81K outer membrane receptor is also an essential component and possible variations in the amount of this protein in the envelopes of different strains might be important. Preliminary work has already shown differences in the relative abundance of each of the iron-regulated outer-membrane proteins in different strains of *E. coli* (Griffiths *et al.*, 1983); furthermore, all of the pathogenic strains examined produced more of the iron-regulated outer-membrane proteins during iron restriction than *E. coli* K-12. It remains to be seen whether these differences are related to virulence. It also remains to be seen whether acrobactin-mediated iron uptake is important only at the start of an infection, when small numbers of bacteria are present, and whether later on, when bacterial numbers have increased, the potentially more efficient siderophore, enterochelin, takes over.

Although ColV$^-$ forms of *E. coli* can be pathogenic, there is no doubt that the presence of the ColV plasmid increases the basic pathogenic potential of certain *E. coli* strains by allowing them to initiate an infection at a lower infective dose (Smith and Huggins, 1980; Cabello, 1979). However, for the ColV plasmid to have this effect on virulence it must be present in an organism which has the correct genetic background. Smith and Huggins (1980) found that the presence of the ColV plasmid hardly affected the virulence of *E. coli* K-12. Indeed, even the creation of an $O18^+$ and $K1^+$ form of *E. coli* K-12, containing the ColV plasmid, resulted in a strain that was only slightly more virulent than the $O18^-$ $K1^-$ $ColV^-$ form, which was very much less virulent than the parent $O18^+$ $K1^+$ $ColV^+$ clinical isolate; the H7 antigen has no effect on virulence. This implies that the original wild-type strain must have at least one more factor which, in co-operation with $O18$, $K1$ and ColV, is responsible for its high degree of virulence. This result is in keeping with the fact that treatment of this parent strain with the mutagen N-methyl-N'-nitro-N-nitrosoguanidine (NTG) resulted in a derivative that was much less virulent but still possessed the $O18$, $K1$ and ColV markers. The nature of the determinant destroyed or altered by NTG is unknown.

Although many of the factors associated with virulence in *E. coli* have been recognized for some time, it is only recently, and after much detailed study at the molecular level, that meaningful progress has been made in understanding their mode of action and complex interrelationships to each other and to virulence. An apparently simple virulence trait such as the possession of a ColV plasmid is now seen to be quite complex and to operate by more than one possible mechanism (Williams, 1979a; Binns *et al.*, 1979; Clancy and Savage, 1981). Even the structure of the K1 antigen is susceptible to subtle changes. The chemical basis for the alternating antigenic change called 'form variation' in *E. coli* K1 capsular polysaccharide has been shown to be the result of random O-acetylation of C7 and C9 carbons of the $\alpha 2 \rightarrow 8$-linked sialic acid polymer (Ørskov *et al.*, 1979). The two K1 form variants have different antigenic, immunogenic and bio-chemical properties and although their relationship to virulence is unclear, 90% cerebrospinal fluid and blood isolates are of the O-acetyl-negative form (Ørskov *et al.*, 1979; Robbins *et al.*, 1980). A better understanding of the environment encountered by pathogenic bacteria *in vivo* has also contributed to our knowledge of virulence determinants. The fact that *E. coli* can undergo considerable phe-notypic change both in its metabolism and in the composition of its outer-membrane proteins when growing in the iron-restricted environment found in host tissues, clearly adds another dimension to what is an already complex situation. Possible connections between these changes, which are associated with the adaptation of the organisms to grow in the iron-restricted environment found *in vivo,* and other bacterial properties, such as serum resistance or the ability to exchange plasmids, now need to be explored. It will be seen, therefore, that although our knowledge of *E. coli* and its virulence markers has expanded remarkably during the past decade, there is still much to learn about this versatile organism. It will also be seen that understanding the mechanisms of *E. coli* pathogenicity at the molecular level is essential if we hope to develop new approaches to the treatment, or preferably, the prevention of the diseases they cause.

References

Achtmann, M., Kennedy, N. and Skurray, R. (1977). Cell–cell interactions in conjugating *Escherichia coli:* Role of *tra*T protein in surface exclusion. *Proceedings of the National Academy of Sciences of the U.S.A.* **74,** 5104–5108.

Ainscough, E. W., Brodie, A. M., Plowman, J. E., Bloor, S. J., Loehr, J. S. and Loehr, T. M. (1980). Studies on human lactoferrin by electron paramagnetic resonance, fluorescence and resonance Raman spectroscopy. *Biochemistry* **19,** 4072–4079.

Bachmann, B. J. and Low, K. B. (1980). Linkage map of *Escherichia coli* K12 Edition 6. *Microbiological Reviews* **44,** 1–56.

Baker, P. J. and Wilson, J. B. (1965). Hypoferremia in mice and its application to the assay of endotoxin. *Journal of Bacteriology* **90,** 903–910.

Barrett-Conner, E. (1971). Bacterial infection and sickle cell anaemia. *Medicine (Baltimore)* **50,** 97–112.

Bezkorovainy, A. (1977). Human milk and colostrum proteins: A review. *Journal of Dairy Science* **60,** 1023–1037.

Bezkorovainy, A. (1980). "Biochemistry of Non-heme Iron." Plenum, New York.

Biedermann, G. and Schindler, L. (1957). On the solubility product of precipitated iron (III) hydroxide. *Acta Chemica Scandinavica* **11,** 731–740.

Bindereif, A., Braun, V. and Hantke, K. (1982). The cloacin receptor of Col V-bearing *Escherichia coli* is part of the Fe^{3+}-aerobactin transport system. *Journal of Bacteriology* **150,** 1472–1475.

Binns, M. M., Davies, D. L. and Hardy, K. G. (1979). Cloned fragments of the plasmid ColV, I-K94 specifying virulence and serum resistance. *Nature (London)* **279,** 778–781.

Binns, M. M., Mayden, J. and Levine, R. P. (1982). Further characterization of complement resistance conferred on *Escherichia coli* by the plasmid genes *tra*T of R100 and *iss* of Col V, I-K94. *Infection and Immunity* **35,** 654–659.

Björksten, B. and Kaijser, B. (1978). Interaction of human serum and neutrophils with *Escherichia coli* strains: Differences between strains isolated from urine of patients with pyelonephritis or asymptomatic bacteruria. *Infection and Immunity* **22,** 308–311.

Bornside, G. H. and Cohn, I. (1968). Hemoglobin as a bacterial virulence-enhancing factor in fluids produced in strangulation intestinal obstruction. *American Surgeon* **34,** 63–67.

Bornside, G. H., Bouis, P. J., Jr. and Cohn, I., Jr. (1968). Hemoglobin and *Escherichia coli*, a lethal intraperitoneal combination. *Journal of Bacteriology* **95,** 1567–1571.

Bortolussi, R., Ferrieri, P., Bjorksten, B. and Quie, P. G. (1979). Capsular K1 polysaccharide of *Escherichia coli:* Relationship to virulence in newborn rats and resistance to phagocytosis. *Infection and Immunity* **25,** 293–298.

Braun, V. (1981). *Escherichia coli* cells containing the plasmid Col V produce the iron ionophore aerobactin. *FEMS Microbiology Letters* **11,** 225–228.

Braun, V. and Burkhardt, R. (1982). Regulation of the Col V plasmid-determined iron (III)–aerobactin transport system in *Escherichia coli*. *Journal of Bacteriology* **152,** 223–231.

Braun, V. and Hantke, K. (1981). Bacterial cell surface receptors. *In* "Organization of Prokaryotic Cell Membranes," Vol. II (Ed. B. K. Ghosh), pp. 1–73. CRC Press, Boca Raton, Florida.

Braun, V., Hancock, R. E. W., Hantke, K. and Hartmann, A. (1976). Functional organization of the outer membrane of *Escherichia coli:* Phage and colicin receptors as components of iron uptake systems *Journal of Supramolecular Structure* **5,** 37–58.

Braun, V., Burkhardt, R., Schneider, R. and Zimmermann, L. (1982). Chromosomal genes for Col V plasmid-determined iron III–aerobactin transport in *Escherichia coli*. *Journal of Bacteriology* **151,** 553–559.

Buck, M. and Griffiths, E. (1981). Regulation of aromatic amino acid transport by tRNA: Role of 2-methylthio-N^6- (Δ^2-isopentenyl)-adenosine. *Nucleic Acids Research* **9,** 401–414.

Buck, M. and Griffiths, E. (1982). Iron mediated methylthiolation of tRNA as a regulator of operon expression in *Escherichia coli*. *Nucleic Acids Research* **10**, 2609–2624.

Buck, M., Humphreys, J. and Griffiths, E. (1979). Iron binding catechols, fever and growth of *Escherichia coli* in lethally infected animals. *Society for General Microbiology, Quarterly* **7**, 9–10.

Bullen, J. J. (1976). Iron-binding proteins and other factors in milk responsible for resistance to *Escherichia coli*. *In* "Acute Diarrhoea in Childhood" (K. Elliott and J. Knight, eds.), CIBA Foundation Symposium, vol. 42, pp. 149–169. Elsevier, Amsterdam.

Bullen, J. J. (1981). The significance of iron in infection. *Reviews of Infectious Diseases* **3**, 1127–1138.

Bullen, J. J. and Rogers, H. J. (1969). Bacterial iron metabolism and immunity to *Pasteurella septica* and *Escherichia coli*. *Nature (London)* **224**, 380–382.

Bullen, J. J., Leigh, L. C. and Rogers, H. J. (1968). The effect of iron compounds on the virulence of *Eschericia coli* for guinea pigs. *Immunology* **15**, 581–588.

Bullen, J. J., Rogers, H. J. and Leigh, L. (1972). Iron-binding proteins in milk and resistance to *Escherichia coli* infections in infants. *British Medical Journal* **1**, 69–75.

Bullen, J. J., Rogers, H. J. and Griffiths, E. (1978). Role of iron in bacterial infection. *Current Topics in Microbiology and Immunology* **80**, 1–35.

Cabello, F. C. (1979). Determinants of pathogenicity of *E. coli* K1. *In* "Plasmids of Medical, Environmental and Commercial Importance" (Eds. K. N. Timmis and A. Pühler), pp. 115–160. Elsevier/North-Holland Biomedical Press, Amsterdam.

Cantley, J. R. and Blake, R. K. (1977). Diarrhea due to *Escherichia coli* in the rabbit, a novel mechanism. *Journal of Infectious Diseases* **135**, 454–462.

Carrano, C. J. and Raymond, K. N. (1979). Ferric iron sequestering agents. 2. Kinetics and mechanism of iron removal from transferrin by enterochelin and synthetic tricatechols. *Journal of the American Chemical Society* **101**, 5401–5404.

Cartwright, G. E., Lauritsen, A., Humphreys, S., Jones, P. J., Merrill, I. M. and Wintrobe, M. M. (1946). The anemia of infection. II. The experimental production of hypoferremia and anemia in dogs. *Journal of Clinical Investigation* **25**, 81–86.

Cavalieri, S. J. and Snyder, I. S. (1982a). Cytotoxic activity of a partially purified *Escherichia coli* alpha-hemolysin. *Journal of Medical Microbiology* **15**, 11–21.

Cavalieri, S. J. and Snyder, I. S. (1982b). Effect of *Escherichia coli* alpha-hemolysin on human peripheral leukocyte viability *in vitro*. *Infection and Immunity* **36**, 455–461.

Cavalieri, S. J. and Snyder, I. S. (1982c). Effect of *Escherichia coli* alpha-hemolysin on human peripheral leukocyte function *in vitro*. *Infection and Immunity* **37**, 966–974.

Chaturvedi, U. C., Mathur, A., Khan, A. M. and Mehrotra, R. M. L. (1969). Cytotoxicity of filtrates of haemolytic *Escherichia coli*. *Journal of Medical Microbiology* **2**, 211–218.

Chiancone, E., Afsen, A., Ioppolo, C., Vecchini, P., Finazzi-Agro, A., Wyman, J. and Antonini, E. (1968). Studies on the reaction of haptoglobin with haemoglobin and haemoglobin chains. I. Stoichiometry and affinity. *Journal of Molecular Biology* **34**, 347–356.

Clancy, J. and Savage, D. C. (1981). Another colicin V phenotype: *In vitro* adhesion of *Escherichia coli* to mouse intestinal epithelium. *Infection and Immunity* **32**, 343–352.

Cooke, E. M. and Ewins, S. P. (1975). Properties of strains of *Escherichia coli* isolated from a variety of sources. *Journal of Medical Microbiology* **8**, 107–111.

Cooper, S. R., McArdle, J. V. and Raymond, K. N. (1978). Siderophore electrochemistry: Relation to intracellular iron release mechanism. *Proceedings of the National Academy of Sciences of the U.S.A.* **75**, 3551–3554.

Cuadra, M. (1956). Salmonellosis complications in human bartonellosis. *Texas Reports on Biology and Medicine* **14**, 97–113.

Davies, D. L., Falkiner, F. R. and Hardy, K. G. (1981). Colicin V production by clinical isolates of *Escherichia coli*. *Infection and Immunity* **31**, 574–579.

Davies, D. L., Binns, M. M. and Hardy, K. G. (1982). Sequence rearrangements in the plasmid Col V, I-K94. *Plasmid* **8**, 55–72.

Davis, J. H. and Yull, A. B. (1964). A toxic factor in abdominal injury. II. The role of the red cell component. *Journal of Trauma* **4**, 84–90.

De Boy, J. M., II, Wachsmuth, K. I. and Davis, B. R. (1980). Hemolytic activity in enterotoxigenic and non-enterotoxigenic strains of *Escherichia coli*. *Journal of Clinical Microbiology* **12**, 193–198.

de la Cruz, F., Zabala, J. C. and Ortiz, J. M. (1979). Incompatability among alpha-hemolytic plasmids studied after inactivation of the hemolysin gene by transposition of TN802. *Plasmid* **2**, 507–519.

de la Cruz, F., Müller, D., Ortiz, J. M. and Goebel, W. (1980). Hemolysis determinant common to *Escherichia coli* hemolytic plasmids of different incompatibility groups. *Journal of Bacteriology* **143**, 825–833.

Dolby, J. M. and Honour, P. (1979). Bacteriostasis of *Escherichia coli* by milk IV. The bacteriostatic antibody of human milk. *Journal of Hygiene* **83**, 255–265.

Dolby, J. M., Honour, P. and Rowland, M. G. M. (1980). Bacteriostasis of *Escherichia coli* by milk. V. The bacteriostatic properties of milk of West African mothers in the Gambia: *in vitro* studies. *Journal of Hygiene* **85**, 347–358.

Dudgeon, L. S., Wordley, E. and Bawtree, F. (1921). On bacillus coli infections of the urinary tract, especially in relation to haemolytic organisms. *Journal of Hygiene* **20**, 137–164.

Durack, D. T. and Beeson, P. B. (1977). Protective role of complement in experimental *Escherichia coli* endocarditis. *Infection and Immunity* **16**, 213–217.

Eaton, J. W., Brandt, P., Mahoney, J. R. and Lee, J. T., Jr. (1982). Haptoglobin: A natural bacteriostat. *Science* **215**, 691–693.

Elwell, L. P. and Shipley, P. L. (1980). Plasmid-mediated factors associated with virulence of bacteria to animals. *Annual Review of Microbiology* **34**, 465–496.

Emödy, L., Pal, T., Safonova, N. V., Kuch, B. and Golutra, N. K. (1980). Alpha-haemolysin: An additive virulence factor in *Escherichia coli*. *Acta Microbiologica Academiae Scientarum Hungaricae* **27**, 333–342.

Emödy, L., Batai, I., Jr., Kerényi, M., Székely, J., Jr. and Polyak, L. (1982). Anti-*Escherichia coli* alpha-haemolysin in control and patient sera. *Lancet* **2**, 986.

Ernst, J. F., Bennett, R. L. and Rothfield, L. I. (1978). Constitutive expression of the iron-enterochelin and ferrichrome uptake systems in a mutant strain of *Salmonella typhimurium*. *Journal of Bacteriology* **135**, 928–934.

Ferrazza, D. and Levy, S. B. (1980). Biochemical and immunological characterization of an R. plasmid-encoded protein with properties resembling those of major cellular outer membrane proteins. *Journal of Bacteriology* **144**, 149–158.

Fietta, A., Romero, E. and Siccordi, A. G. (1977). Effect of some R. factors on the sensitivity of rough Enterobacteriaceae to human serum. *Infection and Immunity* **18**, 278–282.

Fiss, E. H., Stanley-Samuelson, P. and Neilands, J. B. (1982). Properties and proteolysis of ferric enterobactin outer membrane receptor in *Escherichia coli* K12. *Biochemistry* **21**, 4517–4522.

Fitzgerald, S. P. and Rogers, H. J. (1980). Bacteriostatic effect of serum: Role of antibody to lipopolysaccharide. *Infection and Immunity* **27**, 302–308.

Frank, M. F. (1979). The complement system in host defense and inflammation. *Reviews of Infectious Diseases* **1**, 483–501.

Fredericq, P. and Joiris, E. (1950). Distribution des souches productrices de colicine V dans selles normales et pathologiques. *Comptes Rendus des Séances de la Société de Biologie et de Ses Filiales* **144**, 435–437.

Fried, F. A. and Wong, R. J. (1970). Etiology of pyelonephritis: Significance of hemolytic *Escherichia coli*. *Journal of Urology* **103**, 718–721.

Fried, F. A., Vermeulen, C. W., Ginsburg, M. J. and Cone, C. M. (1971). Etiology of pyelonephritis: Further evidence associating the production of experimental pyelonephritis and hemolysis in *Escherichia coli*. *Journal of Urology* **106**, 351–354.

Garibaldi, J. A. (1972). Influence of temperature on the biosynthesis of iron transport compounds by *Salmonella typhimurium*. *Journal of Bacteriology* **110**, 262–265.

Gemski, P., Cross, A. S. and Sadoff, J. C. (1980). K1 antigen-associated resistance to the bactericidal activity of serum. *FEMS Microbiology Letters* **9**, 193–197.

Gibson, F. and Magrath, D. I. (1969). The isolation and characterization of a hydroxamic acid (aerobactin) formed by *Aerobacter aerogenes* 62-1. *Biochimica Biophysica Acta* **192**, 175–184.

Glode, M. P., Sutton, A., Moxon, E. R. and Robbins, J. B. (1977). Pathogenesis of neonatal *Escherichia coli* meningitis: Induction of bacteremia and meningitis in infant rats fed *E. coli* K1. *Infection and Immunity* **16**, 75–80.

Glynn, A. A. (1969). The complement lysozyme sequence in immune bacteriolysis. *Immunology* **16**, 463–471.

Glynn, A. A. and Howard, C. J. (1970). The sensitivity to complement of strains of *Escherichia coli* related to their K antigens. *Immunology* **18**, 331–346.

Goebel, W., Royer-Pokora, B., Lindenmaier, W. and Bujord, H. (1974). Plasmids controlling synthesis of hemolysin in *Escherichia coli:* Molecular properties. *Journal of Bacteriology* **118**, 964–973.

Green, C. P. and Thomas, V. L. (1981). Hemagglutination of human type O erythrocytes, hemolysin production and serogrouping of *Escherichia coli* isolates from patients with acute pyelonephritis, cystitis and asymptomatic bacteriuria. *Infection and Immunity* **31**, 309–315.

Greenwood, K. T. and Luke, R. K. J. (1978). Enzymic hydrolysis of enterochelin and its iron complex in *Escherichia coli* K12. *Biochimica Biophysica Acta* **525**, 209–218.

Grewal, K. K., Warner, P. J. and Williams, P. H. (1982). An inducible outer membrane protein involved in aerobactin-mediated iron transport by Col V strains of *Escherichia coli*. *FEBS Letters* **140**, 27–30.

Grieger, T. A. and Kluger, M. J. (1978). Fever and survival: The role of serum iron. *Journal of Physiology (London)* **279**, 187–196.

Griffiths, E. (1981). Iron and the susceptibility to bacterial infection. *In* "Nutritional Factors: Modulating Effects on Metabolic Processes" (Eds. R. F. Beers and E. G. Bassett), pp. 463–676. Raven Press, New York.

Griffiths, E. (1983). Bacterial adaptation to a low iron environment. *In* "Microbiology—1983" (Ed. D. Schlessinger), pp. 329–333. American Society for Microbiology, Washington, D.C.

Griffiths, E. and Buck, M. (1979). tRNA alterations and adaptation of pathogenic *Escherichia coli* to growth *in vivo*. *Society for General Microbiology, Quarterly* **7**, 10–11.

Griffiths, E. and Humphreys, J. (1978). Alterations in tRNAs containing 2-methylthio-N^6-(Δ^2-isopentenyl)-adenosine during growth of enteropathogenic *Escherichia coli* in the presence of iron binding proteins. *European Journal of Biochemistry* **82**, 503–513.

Griffiths, E. and Humphreys, J. (1980). Isolation of enterochelin from the peritoneal washings of guinea-pigs lethally infected with *Escherichia coli*. *Infection and Immunity* **28**, 286–289.

Griffiths, E., Humphreys, J., Leach, A. and Scanlon, L. (1978). Alterations in the tRNAs of *Escherichia coli* recovered from lethally infected animals: *Infection and Immunity* **22**, 312–317.

Griffiths, E., Stevenson, P. and Joyce, P. (1983). Pathogenic *Escherichia coli* express new outer membrane proteins when growing *in vivo*. *FEMS Microbiology Letters* **16**, 95–99.

Hancock, R. E. W., Hantke, K. and Braun, V. (1976). Iron transport in *Escherichia coli* K12: Involvement of the colicin B receptor and of a citrate-inducible protein. *Journal of Bacteriology* **127**, 1370–1375.

Hancock, R. E. W., Hantke, K., and Braun, V. (1977). Iron transport in *Escherichia coli* K12: 2,3-dihydroxy-benzoate promoted iron uptake. *Archives of Microbiology* **114**, 231–239.

Hantke, K. (1981). Regulation of ferric iron transport in *Escherichia coli* K12: Isolation of a constitutive mutant. *Molecular and General Genetics* **182**, 288–292.

Hantke, K. (1982). Negative control of iron uptake systems in *Escherichia coli*. *FEMS Microbiology Letters* **15**, 83–86.

Harris, W. R., Carrano, C. J. and Raymond, K. N. (1979a). Spectrophotometric determination of the proton-dependent stability constant of ferric enterobactin. *Journal of the American Chemical Society* **101**, 2213–2214.

Harris, W. R., Carrano, C. J. and Raymond, K. N. (1979b). Co-ordination chemistry of microbial iron transport compounds 16: Isolation, characterization and formation constants of ferric aerobactin. *Journal of the American Chemical Society* **101**, 2722–2727.

Harris, W. R., Carrano, C. J., Cooper, S. R., Sofen, S. R., Avdeef, A. E., McArdle, J. V. and Raymond, K. N. (1979c). Co-ordination chemistry of microbial iron transport compounds 19: Stability constants and electrochemical behaviour of ferric enterobactin and model complexes. *Journal of the American Chemical Society* **101**, 6097–6104.

Hartmann, A. and Braun, V. (1980). Iron transport in *Escherichia coli:* Uptake and modification of ferrichrome. *Journal of Bacteriology* **143**, 246–255.

Hoiseth, S. K. and Stocker, B. A. D. (1981). Aromatic-dependent *Salmonella typhimurium* are non-virulent and effective as live vaccines. *Nature (London)* **291**, 238–239.

Hollifield, W. C., Jr. and Neilands, J. B. (1978). Ferric enterobactin transport system in *Escherichia coli* K12. Extraction, assay and specificity of the outer membrane receptor. *Biochemistry* **17**, 1922–1928.

Honour, P. and Dolby, J. M. (1979). Bacteriostasis of *Escherichia coli* by milk. III. The activity and stability of early transitional and mature human milk collected locally. *Journal of Hygiene* **83**, 243–254.

Hull, S. I., Hull, R. A., Minshew, B. H., and Falkow, S. (1982). Genetics of hemolysin of *Escherichia coli*. *Journal of Bacteriology* **151**, 1006–1012.

Hussein, S., Hantke, K. and Braun, V. (1981). Citrate-dependent iron transport system in *Escherichia coli* K12. *European Journal of Biochemistry* **117**, 431–437.

Inoue, K., Tanigowa, T., Takubo, M., Satani, M. and Amano, T. (1959). Quantitative studies on immune bacteriolysis. II. The role of lysozyme in immune bacteriolysis. *Biken Journal* **2**, 1–20.

Inoue, K. K., Yonemasu, A., Takimizawa, A. and Amano, T. (1968). Studies on the immune bacteriolysis. XIV. Requirement of all nine components of complement for immune bacteriolysis. *Biken Journal* **11**, 203–206.

Kasper, D. L., Winkelhake, J. L., Zollinger, W. D., Brandt, B. L. and Artenstein, M. S. (1973). Immunochemical similarity between polysaccharide antigens of *Escherichia coli* O7 K1(L): NM and group B *Neisseria meningitidis*. *Journal of Immunology* **110**, 262–268.

Kay, W. W. and Cameron, M. (1978). Citrate transport in *Salmonella typhimurium*. *Archives of Biochemistry and Biophysics* **190**, 270–280.

Kaye, D., Gill, F. A. and Hook, E. W. (1967). Factors influencing host resistance to salmonella infections. *American Journal of the Medical Sciences* **254**, 205–215.

Klebba, P. E., McIntosh, M. A. and Neilands, J. B. (1982). Kinetics of biosynthesis of iron-regulated membrane proteins in *Escherichia coli*. *Journal of Bacteriology* **149**, 880–888.

Kluger, M. J. and Rothenburg, B. A. (1979). Fever and reduced iron: Their interaction as a host defense response to bacterial infection. *Science* **203**, 374–376.

Konisky, J. (1979). Specific transport systems and receptors for colicins and phages. *In* ''Bacterial Outer Membranes, Biogenesis and Functions'' (Ed. M. Inouye), pp. 319–359. Wiley, New York.

Langman, L., Young, I. G., Frost, G., Rosenberg, H. and Gibson, F. (1972). Enterochelin system of iron transport in *Escherichia coli:* Mutations affecting ferri-enterochelin esterase. *Journal of Bacteriology* **112**, 1142–1149.

Lee, J. T., Jr., Ahrenhoz, D. H., Nelson, R. D. and Simmons, R. L. (1979). Mechanisms of the adjuvant effect of hemoglobin in experimental peritonitis. V. The significance of the co-ordinated iron component. *Surgery* **86**, 41–48.

Leong, J. and Neilands, J. B. (1976). Mechanisms of siderophore iron transport in enteric bacteria. *Journal of Bacteriology* **126**, 823–830.

Levine, M. M., Nalin, D. R., Hornick, R. B., Bergquist, E. J., Waterman, D. H., Young, C. R. and Sotman, S. (1978). *Escherichia coli* strains that cause diarrhoea but do not produce heat-labile or heat stable enterotoxins and are non-invasive. *Lancet* **1**, 1119–1122.

Linggood, M. A. and Ingram, P. L. (1982). The role of alpha haemolysin in the virulence of *Escherichia coli* for mice. *Journal of Medical Microbiology* **15**, 23–30.

Lovell, R. and Rees, T. A. (1960). A filterable haemolysin from *Escherichia coli*. *Nature (London)* **188**, 755–756.

McCabe, W. R., Kaijser, B., Olling, S., Uwaydah, M. and Hanson, L. A. (1978). *Escherichia coli* in bacteraemia: K and O antigens and serum sensitivity of strains from adults and neonates. *Journal of Infectious Diseases* **138**, 33–41.

McCandliss, R. J. and Herrmann, K. M. (1978). Iron, an essential element for biosynthesis of aromatic compounds. *Proceedings of the National Academy of Sciences of the U.S.A.* **75**, 4810–4813.

McClelland, D. D. L. and van Furth, R. (1976). Antimicrobial factors in the exudates of skin windows in human subjects. *Clinical and Experimental Immunology* **25**, 442–448.

McIntosh, M. A. and Earhart, C. F. (1977). Co-ordinate regulation by iron of the synthesis of phenolate compounds and three outer membrane proteins in *Escherichia coli*. *Journal of Bacteriology* **131**, 331–339.

Manning, P. A. and Achtmann, M. (1979). Cell to cell interactions in conjugating *Escherichia coli*: The involvement of the cell envelope. *In* "Bacterial Outer Membranes: Biogenesis and Functions" (Ed. M. Inouye), pp. 409–447. Wiley, New York.

Manning, P. A., Bentin, L. and Achtman, M. (1980). Outer membrane of *Escherichia coli*: Properties of the F sex factor *tra*T protein which is involved in surface exclusion. *Journal of Bacteriology* **142**, 285–294.

Masson, P. L. and Heremans, J. F. (1966). Studies on lactoferrin the iron-binding protein of secretions. *Protides of the Biological Fluids* **14**, 115–124.

Masson, P. L., Heremans, J. F. and Dive, C. (1966a). An iron-binding protein common to many external secretions. *Clinica Chimica Acta* **14**, 735–739.

Masson, P. L., Heremans, J. F., Pignot, J. J. and Wauters, G. (1966b). Immunohistochemical localization and bacteriostatic properties of an iron-binding protein from bronchial mucus. *Thorax* **21**, 538–544.

Merson, M. H., Ørskov, F., Ørskov, I., Sack, R. B., Huq, I. and Koster, F. T. (1979). Relationship between enterotoxin production and serotype in enterotoxigenic *Escherichia coli*. *Infection and Immunity* **23**, 325–329.

Miles, A. A. and Khimji, P. L. (1975). Enterobacterial chelators of iron: Their occurrence, detection and relation to pathogenicity. *Journal of Medical Microbiology* **8**, 477–490.

Minshew, B. H., Jorgensen, J., Counts, G. W., and Falkow, S. (1978). Association of hemolysin production, hemagglutination of human erythrocytes and virulence for chicken embryos of extra-intestinal *Escherichia coli* isolates. *Infection and Immunity* **20**, 50–54.

Moll, A., Manning, P. A. and Timmis, K. N. (1980). Plasmid-determined resistance to serum bactericidal activity: A major outer membrane protein, the *tra*T gene product, is responsible for plasmid-specified serum resistance in *Escherichia coli*. *Infection and Immunity* **28**, 359–367.

Montgomerie, J. Z., Kalmanson, G. M. and Guze, L. B. (1979). Enterobactin and virulence of *Escherichia coli* pyelonephritis. *Journal of Infectious Diseases* **140**, 1013.

Moore, D. G. and Earhart, C. F. (1981). Specific inhibition of *Escherichia coli* ferrienterochelin uptake by a normal human serum immunoglobulin. *Infection and Immunity* **31**, 631–635.

Moore, D. G., Yancey, R. J., Lankford, C. E. and Earhart, C. F. (1980). Bacteriostatic entero-chelin-specific immunoglobulin from normal human serum. *Infection and Immunity* **27**, 418–423.

Morgan, E. H. (1981). Transferrin, biochemistry, physiology and clinical significance. *Molecular Aspects of Medicine* **4**, 1–123.

Müller-Eberhard, H. J. (1975). Complement. *Annual Reviews of Biochemistry* **44**, 697–724.

Neilands, J. B. (1981). Microbial iron compounds. *Annual Reviews of Biochemistry* **50**, 715–731.

Neilands, J. B. (1982). Microbial envelope proteins related to iron. *Annual Reviews of Microbiology* **36**, 285–309.

Nikaido, H. (1979). Nonspecific transport through the outer membrane. *In* "Bacterial Outer Membranes, Biogenesis and Functions" (Ed. M. Inouye), pp. 361–407. Wiley, New York.

Noegel, A., Rdest, U., Springer, W. and Goebel, W. (1979). Plasmid cistrons controlling synthesis and excretion of the exotoxin alpha-hemolysin of *Escherichia coli*. *Molecular and General Genetics* **175**, 343–350.

Noegel, A., Rdest, U. and Goebel, W. (1981). Determination of the functions of hemolytic plasmid pHly 152 of *Escherichia coli*. *Journal of Bacteriology* **145**, 233–247.

O'Brien, I. G. and Gibson, F. (1970). The structure of enterochelin and related 2,3-dihydroxy-*N*-benzoylserine conjugates from *Escherichia coli*. *Biochimica Biophysica Acta* **215**, 393–402.

Ogata, R. T. and Levine, R. P. (1980). Characterization of complement resistance in *Escherichia coli* conferred by the antibiotic resistance plasmid R100. *Journal of Immunology* **125**, 1494–1498.

Ogata, R. T., Wintens, C. and Levine, R. P. (1982). Nucleotide sequence analysis of the complement resistance gene from plasmid R100. *Journal of Bacteriology* **151**, 819–827.

Olling, S. (1977). Sensitivity of gram-negative bacilli to the serum bactericidal activity: A marker of the host–parasite relationship in acute and persisting infections. *Scandinavian Journal of Infectious Diseases, Supplementum* **10**, 1–10.

Opal, S., Cross, A. and Gemski, P. (1982). K antigen and serum sensitivity of rough *Escherichia coli*. *Infection and Immunity* **37**, 956–960.

Ørskov, F., Ørskov, I., Sutton, A., Schneerson, R., Lin, W., Egan, W., Hoff, G. E. and Robbins, J. B. (1979). Form variation in *Escherichia coli* K1: Determined by O-acetylation of the capsular polysaccharide. *Journal of Experimental Medicine* **149**, 669–685.

Ørskov, I., Ørskov, F., Jann, B. and Jann, K. (1977). Serology, chemistry and genetics of O and K antigens of *Escherichia coli*. *Bacteriological Reviews* **41**, 667–710.

Oudega, B., Oldziel-Werner, W. J. M., Klaasen-Bloor, P., Rezee, A., Glas, J. and de Graaf, F. K. (1979). Purification and chacterization of cloacin DF13 receptor from *Enterobacter cloacae* and its interaction with cloacin DF13 *in vitro*. *Journal of Bacteriology* **138**, 7–16.

Overbeeke, N. and Lugtenberg, B. (1980). Major outer membrane proteins of *Escherichia coli* strains of human origin. *Journal of General Microbiology* **121**, 373–380.

Owen, J. A., Smith, R., Padayi, R. and Martin, J. (1964). Serum haptoglobin in disease. *Clinical Science* **26**, 1–6.

Payne, S. (1980). Synthesis and utilization of siderophores by *Shigella flexneri*. *Journal of Bacteriology* **143**, 1420–1424.

Pitt, J. (1978). K1 antigen of *Escherichia coli*: Epidemiology and serum sensitivity of pathogenic strains. *Infection and Immunity* **22**, 219–224.

Plastow, G. S., Pratt, J. M. and Holland, I. B. (1981). The ferrichrome receptor protein (ton A) of *Escherichia coli* is synthesized as a precursor *in vitro*. *FEBS Letters* **131**, 262–264.

Pollack, J. R. and Neilands, J. B. (1970). Enterobactin, an iron transport compound from *Salmonella typhimurium*, *Biochemical and Biophysical Research Communications* **38**, 989–992.

Pollack, S., Aisen, P., Lasky, F. D. and Vanderhoff, G. (1976). Chelate mediated transfer of iron from transferrin to desferrioxamine. *British Journal of Haematology* **34**, 231–235.

Powell, C. J., Jr. and Finkelstein, R. A. (1966). Virulence of *Escherichia coli* strains for chick embryos. *Journal of Bacteriology* **91**, 1410–1417.

Quakenbush, R. L. and Falkow, S. (1979). Relationship between colicin V activity and virulence in *Escherichia coli*. *Infection and Immunity* **24**, 562–564.

Raymond, K. N. and Carrano, C. J. (1979). Co-ordination chemistry of microbial iron transport. *Accounts of Chemical Research* **12**, 183–190.

Raymond, K. N., Pecoraro, V. L. and Weitl, F. L. (1981). Design of new chelating agents. *In* "Development of Iron Chelators for Clinical Use" (Eds. A. E. Martell, W. F. Anderson and D. G. Badman), pp. 165–187. Elsevier/North-Holland Publ., Amsterdam.

Rennie, R. P. and Arbuthnott, J. P. (1974). Partial characterization of *Escherichia coli* haemolysin. *Journal of Medical Microbiology* **7**, 179–188.

Rennie, R. P., Freer, J. H. and Arbuthnott, J. P. (1974). The kinetics of erythrocyte lysis by *Escherichia coli* haemolysin. *Journal of Medical Microbiology* **7**, 189–195.

Reynard, A. M. and Beck, M. E. (1976). Plasmid-mediated resistance to the bactericidal effects of normal rabbit serum. *Infection and Immunity* **14**, 848–850.

Reynard, A. M., Beck, M. E. and Cunningham, R. K. (1978). Effects of antibiotic resistance plasmids on the bactericidal activity of normal rabbit serum. *Infection and Immunity* **19**, 861–866.

Roantree, R. J. and Rantz, L. A. (1960). A study of the relationship of the normal bactericidal activity of human serum to bacterial infection. *Journal of Clinical Investigation* **39**, 72–81.

Robbins, J. B., McCracken, G. H., Gotschlich, E. C., Ørskov, F., Ørskov, I. and Hanson, L. (1974). *Escherichia coli* K1 capsular polysaccharide associated with neonatal meningitis. *New England Journal of Medicine* **290**, 1216–1220.

Robbins, J. B., Schneerson, R., Egan, W. B., Vann, W. and Liu, D. T. (1980). Virulence properties of bacterial capsular polysaccharide—unanswered questions. *In* "The Molecular Basis of Microbial Pathogenicity" (Eds. H. Smith, J. J. Skehel and M. J. Turner), pp. 115–132. Verlag Chemie, Weinheim.

Rogers, H. J. (1973). Iron-binding catechols and virulence in *Escherichia coli*. *Infection and Immunity* **7**, 445–456.

Rogers, H. J. (1976). Ferric iron and the antibacterial effect of horse 7S antibodies to *Escherichia coli* O111. *Immunology* **30**, 425–433.

Rogers, H. J., Synge, C., Kimber, B. and Bayley, P. M. (1977). Production of enterochelin by *Escherichia coli* O111. *Biochimica Biophysica Acta* **497**, 548–557.

Rosenberg, H. and Young, I. G. (1974). Iron transport in the enteric bacteria. *In* "Microbial Iron Metabolism" (Ed. J. B. Neilands), pp. 67–82. Academic Press, New York.

Royer-Pokora, B. and Goebel, W. (1976). Plasmids controlling synthesis of hemolysin in *Escherichia coli*. II. Polynucleotide sequence relationships among hemolytic plasmids. *Molecular and General Genetics* **144**, 177–183.

Sarff, L. D., McCracken, G. H., Jr., Schiffer, M. S., Glode, M. P., Robbins, J. B., Ørskov, I. and Ørskov, F. (1975). Epidemiology of *Escherichia coli* K1 in healthy and diseased newborns. *Lancet* **1**, 1099–1104.

Schiffer, M. S., Oliveira, E., Glode, M. P., McCracken, G. H., Jr., Sarff, M. and Robbins, J. B. (1976). A review: Relation between invasiveness and the K1 capsular polysaccharide of *E. coli*. *Pediatric Research* **10**, 82–89.

Schneider, R., Hartmann, A. and Braun, V. (1981). Transport of the iron ionophore ferrichrome in *Escherichia coli* K12 and *Salmonella typhimurium* LT2. *FEMS Microbiology Letters* **11**, 115–119.

Scotland, S., Day, N. P. and Rowe, B. (1980). Production of a cytotoxin affecting Vero cells by strains of *Escherichia coli* belonging to traditional enteropathogenic serogroups. *FEMS Microbiology Letters* **7**, 15–17.

Sharp, P. A., Cohen, S. N. and Davidson, N. (1973). Electron microscope heteroduplex studies of sequence relations among plasmids of *Escherichia coli*. II. Structure of drug resistance (R) factors and F factors. *Journal of Molecular Biology* **75**, 235–255.

Short, E. C., Jr. and Kurtz, H. J. (1971). Properties of the hemolytic activities of *Escherichia coli*. *Infection and Immunity* **3**, 678–687.

Silver, R. P. and Cohen, S. N. (1972). Nonchromosomal antibiotic resistance in bacteria. V. Isolation and characterization of R factor mutants exhibiting temperature sensitive repression of fertility. *Journal of Bacteriology* **110**, 1082–1088.

Silver, R. P., Aaronson, W., Sutton, A. and Schneerson, R. (1980). Comparative analysis of plasmids and some metabolic characteristics of *E. coli* K1 from diseased and healthy individuals. *Infection and Immunity* **29**, 200–206.

Silver, R. P., Finn, C. W., Vann, W. F., Aaronson, W., Schneerson, R., Kretschmer, P. J. and Garon, C. F. (1981). Molecular cloning of the K1 capsular polysaccharide genes of *E. coli*. *Nature (London)* **289**, 696–698.

Smith, H. W. (1963). The haemolysins of *Escherichia coli*. *Journal of Pathology and Bacteriology* **85**, 197–211.

Smith, H. W. (1974). A search for transmissible pathogenic characters in invasive strains of *Escherichia coli:* The discovery of a plasmid-controlled toxin and a plasmid-controlled lethal character closely associated, or identical, with colicine V. *Journal of General Microbiology* **83**, 95–111.

Smith, H. W. and Gyles, C. L. (1970). The relationship between different plasmids introduced by F into the same strain of *Escherichia coli* K12. *Journal of General Microbiology* **62**, 277–285.

Smith, H. W. and Halls, S. (1967). The transmissible nature of the genetic factor in *Escherichia coli* that controls haemolysin production. *Journal of General Microbiology* **47**, 153–161.

Smith, H. W. and Heller, E. D. (1973). The activity of different transfer factors introduced into the same plasmid-containing strain of *Escherichia coli* K12. *Journal of General Microbiology* **78**, 89–99.

Smith, H. W. and Huggins, M. B. (1976). Further observations on the association of the colicine V plasmid of *Escherichia coli* with pathogenicity and with survival in the alimentary tract. *Journal of General Microbiology* **92**, 335–350.

Smith, H. W. and Huggins, M. B. (1980). The association of the O18, K1 and H7 antigens and the Col V plasmid of a strain of *Escherichia coli* with its virulence and immunogenicity. *Journal of General Microbiology* **121**, 387–400.

Smith, H. W. and Linggood, M. A. (1970). Transfer factors in *Escherichia coli* with particular regard to their incidence in enteropathogenic strains. *Journal of General Microbiology* **62**, 287–299.

Smith, H. W. and Linggood, M. A. (1971). Observations on the pathogenic properties of the K88, Hly and Ent plasmids of *Escherichia coli* with particular reference to porcine diarrhoea. *Journal of Medical Microbiology* **4**, 467–485.

Smith, H. W., Parsell, Z. and Green, P. (1978). Thermosensitive H1 plasmids determining citrate utilization. *Journal of General Microbiology* **109**, 305–311.

Stark, J. M. and Shuster, C. W. (1982). Analysis of hemolytic determinants of plasmid pHly 185 by Tn5 mutagenesis. *Journal of Bacteriology* **152**, 963–967.

Stuart, S. J., Greenwood, K. T. and Luke, R. K. J. (1980). Hydroxamate-mediated transport of iron controlled by Col V plasmids. *Journal of Bacteriology* **143**, 35–42.

Taylor, P. W. and Hughes, C. (1978). Plasmid carriage and the serum sensitivity of enterobacteria. *Infection and Immunity* **22**, 10–17.

Taylor, P. W. and Robinson, M. K. (1980). Determinants that increase the serum resistance of *Escherichia coli*. *Infection and Immunity* **29**, 278–280.

Timmis, K. N., Manning, P. A., Echarti, C., Timmis, J. K. and Moll, A. (1981). Serum resistance in *E. coli*. *In* "The Molecular Biology, Pathogenicity and Ecology of Bacterial Plasmids" (Eds. S. B. Levy, R. C. Clowes and E. L. Koenig) pp. 113–144. Plenum, London.

Tourville, D. R., Adler, R. H., Bienenstock, J. and Tomasi, T. B. (1969). The human secretory immunoglobulin system: Immunohistological localization of γA secretory 'piece' and lactoferrin in normal human tissues. *Journal of Experimental Medicine* **129**, 411–423.

Ulshen, M. H. and Rollo, J. L. (1980). Pathogenesis of *Escherichia coli* gastroenteritis in man—another mechanism. *New England Journal of Medicine* **302**, 99–101.

van den Bosch, J. F., de Graaf, J. and MacLaren, D. M. (1979). Virulence of *Escherichia coli* in experimental hematogenous pyelonephritis in mice. *Infection and Immunity* **25**, 68–74.

van den Bosch, J. F., Postma, P., de Graaf, J. and MacLaren, D. M. (1981). Haemolysis by urinary *Escherichia coli* and virulence in mice. *Journal of Medical Microbiology* **14**, 321–331.

van den Bosch, J. F., Emödy, L. and Ketyi, I. (1982). Virulence of haemolytic strains of *Escherichia coli* in various animal models. *FEMS Microbiology Letters* **13**, 427–430.

van Tiel-Menkveld, G. J., Mentjox-Vervuust, J. M., Oudega, B. and de Graaf, F. K. (1982). Siderophore production by *Enterobacter cloacae* and a common receptor protein for the uptake of aerobactin and cloacin DF13. *Journal of Bacteriology* **150**, 490–497.

Waalwijk, C., van den Bosch, J. F., MacLaren, D. M. and de Graff, J. (1982a). Hemolysin plasmid coding for the virulence of a nephropathogenic *Escherichia coli* strain. *Infection and Immunity* **35**, 32–37.

Waalwijk, C., van den Bosch, J. F., MacLaren, D. M. and de Graaff, J. (1982b). Plasmid content and virulence properties of urinary *Escherichia coli* strains. *FEMS Microbiology Letters* **14**, 171–175.

Wagegg, W. and Braun, V. (1981). Ferric citrate transport in *Escherichia coli* requires outer membrane receptor protein FecA. *Journal of Bacteriology* **145**, 156–163.

Warner, P. J., Williams, P. H., Bindereif, A. and Neilands, J. B. (1981). Col V plasmid-specified aerobactin synthesis by invasive strains of *Escherichia coli*. *Infection and Immunity* **33**, 540–545.

Weinberg, E. D. (1978). Iron and infection. *Microbiological Reviews* **42**, 45–66.

Weinstein, R. and Young, L. S. (1978). Phagocytic resistance of *Escherichia coli* K1 isolates and relationship to virulence. *Journal of Clinical Microbiology* **8**, 748–755.

Welch, R. A., Dellinger, E. P., Minshew, B. and Falkow, S. (1981). Haemolysin contributes to virulence of extra-intestinal *E. coli* infections. *Nature (London)* **294**, 665–667.

Willetts, N. and Maule, J. (1973). Surface exclusion by Col V - K94. *Genetical Research* **21**, 297–299.

Williams, P. H. (1979a). Novel iron uptake system specified by Col V plasmids: An important component in the virulence of invasive strains of *Escherichia coli*. *Infection and Immunity* **26**, 925–932.

Williams, P. H. (1979b). Determination of the molecular weight of *Escherichia coli* α-hemolysin. *FEMS Microbiology Letters* **5**, 21–24.

Williams, P. H. and Warner, P. J. (1980). Col V plasmid-mediated, colicin V-independent iron uptake system of invasive strains of *Escherichia coli*. *Infection and Immunity* **29**, 411–416.

Woodrow, G. C., Langman, L., Young, I. G. and Gibson, F. (1978). Mutations affecting the citrate-dependent iron uptake systems in *Escherichia coli*. *Journal of Bacteriology* **133**, 1524–1526.

Wright, S. D. and Levine, R. P. (1981a). How complement kills *E. coli*. 1. Location of the lethal lesion. *Journal of Immunology* **127**, 1146–1151.

Wright, S. D. and Levine, R. P. (1981b). How complement kills *E. coli*. II. The apparent two-hit nature of the lethal event. *Journal of Immunology* **127**, 1152–1156.

Yancey, R. J., Breeding, S. A. L. and Lankford, C. E. (1979). Enterochelin (enterobactin): Virulence factor for *Salmonella typhimurium*. *Infection and Immunity* **24**, 174–180.

Yanofsky, C. (1981). Attenuation in the control of expression of bacterial operons. *Nature (London)* **289**, 751–758.

8

Genetics of *Escherichia coli* Virulence

H. R. SMITH, SYLVIA M. SCOTLAND AND B. ROWE

Division of Enteric Pathogens, Central Public Health Laboratory, London, UK

Introduction

Certain strains of *Escherichia coli* cause disease in man and animals and this symposium clearly shows how knowledge of the pathogenic mechanisms involved has increased in the past 10 years. The organisms will be discussed under the two main headings of *E. coli* causing diarrhoeal diseases and *E. coli* causing extra-intestinal infections. For strains such as enterotoxigenic *E. coli* (ETEC) the pathogenic mechanisms are reasonably well understood but for other types of *E. coli*, such as those causing urinary tract infection, the mechanisms are less clear and are the subject of controversy. Consequently, genetic studies of the various groups will reflect this disparity but will, it is hoped, indicate how further work could evaluate the importance of various bacterial functions in virulence.

Many of the *E. coli* virulence factors that will be described are plasmid-determined, that is they are carried on extra-chromosomal genetic elements. Plasmids are sometimes unstable and this may make it possible to relate changes in the virulence of the organism with the loss of plasmids. Reintroduction of the plasmid into the wild-type strains lacking the relevant plasmid or transfer of the plasmid to other recipient strains can then confirm that genes coding for a particular character are carried on the plasmid. A plasmid is initially characterized by properties such as size, conjugative ability and incompatibility grouping. Restriction enzyme analysis and DNA hybridization experiments are used to compare the relatedness of different plasmids. Studies of the properties of a plasmid in its entirety are usually followed by more detailed mapping and location of the regions coding for the pathogenic character under investigation.

Cloning techniques have now made it possible to isolate small genetic elements from chromosomal and plasmid DNA and to insert them into plasmid or bacteriophage cloning vectors. After transfer into suitable recipients, the expression of the genes can be determined. The products encoded by the cloned genes and by mutant derivatives may give additional information on the function-

THE VIRULENCE OF ESCHERICHIA COLI

ing of the genes. All these techniques have been used in the study of *E. coli* virulence factors (see Chapter 24).

E. coli Causing Diarrhoeal Disease

Three groups of *E. coli* causing diarrhoeal disease are recognized at the present time: (1) ETEC, in which production of a heat-stable enterotoxin (ST) or a heat-labile enterotoxin (LT) or both toxins has been demonstrated, (2) entero-pathogenic *E. coli* (EPEC), which were the first of the groups to be recognized as causing diarrhoea but whose pathogenic mechanisms are uncertain and future work may require that this group be subdivided should a variety of mechanisms be identified and (3) enteroinvasive *E. coli* (EIEC), which resemble shigellae in their ability to penetrate and multiply in the intestinal mucosa.

Enterotoxigenic E. coli

These diarrhoeagenic organisms have the ability to adhere to and colonize the small intestine of humans and animals and produce enterotoxins. Many studies have shown that ETEC commonly belong to a restricted range of serogroups (Ørskov and Ørskov, 1977; Scotland *et al.,* 1977; Merson *et al.,* 1979; Rowe, 1979) although, as more laboratories have tested for the presence of these organisms, small numbers belonging to many other serogroups have been reported (Table 1). The genes coding for the production of enterotoxins have been located

Table 1.   *Enterotoxin production in E. coli*

Types and combinations of enterotoxins produced	Host	Main O serogroups[a]	Genetic control
ST_A LT ST_A, LT	Man	6, 8, 15, 20, 25 27, 63, 78, 114 115, 128, 148, 153, 159	Plasmid or phage[b]
ST_A ST_B ST_A, ST_B ST_B, LT ST_A, ST_B, LT	Piglet	8, 9, 45, 64 78, 101, 138 139, 141, 147 148, 149, 157	Plasmid
ST_A	Calf, lamb	8, 9, 20, 101	Plasmid

[a]The majority of ETEC belong to the O serogroups shown in the table but small numbers of ETEC belonging to many other serogroups have also been isolated.
[b]In one strain the LT genes were phage-determined (Takeda and Murphy, 1978).

on plasmids (Smith and Halls, 1968; for review, see Elwell and Shipley, 1980) apart from a strain of human origin in which the LT genes were phage-determined (Takeda and Murphy, 1978). However, the ability to produce enterotoxins has only been transferred from a relatively small number of strains and the possibility cannot be excluded that this property may be chromosomally determined in some strains. The enterotoxins and adhesive factors produced by ETEC have been described in detail in other chapters. In this section the different types of enterotoxin or Ent plasmid will be considered. The genetic control of adhesion in ETEC will then be discussed, particularly in relation to the linkage of the genes encoding adhesion and enterotoxin production.

Enterotoxins and enterotoxin plasmids. In the 1950s, a series of investigations identified enterotoxin production by testing bacterial cultures of *E. coli* strains or sterile bacterial preparations in ligated intestinal loops of several animal species (De *et al.,* 1956; Taylor *et al.,* 1961; Smith and Halls, 1967a,b). The enterotoxins identified were of two types: heat-labile enterotoxin, which was very similar in structure and mode of action to cholera toxin, and heat-stable enterotoxin (Smith and Halls, 1967a,b; Gyles and Barnum, 1969). Studies of enterotoxigenic strains isolated from pigs, calves and lambs suggested that there were different types of ST. Many strains of porcine origin produced an enterotoxin which caused fluid accumulation when injected into a ligated gut loop of a weaned pig (Smith and Halls, 1967a). A few 'atypical' strains from pigs produced an enterotoxin that was positive in ligated gut loops of piglets less than 1 wk old but negative in weaned pigs (Moon and Whipp, 1970; Smith and Linggood, 1972). Strains isolated from calves and lambs gave the same result in neonatal piglet loops as the atypical porcine strains and these enterotoxins were also positive in tests with calf and lamb intestinal loops (Smith and Linggood, 1972). Dean *et al.* (1972) used an infant mouse assay to detect ST production in strains of human origin. Subsequently Sivaswamy and Gyles (1976) used this test to examine bovine strains and found complete correlation between the results of the calf gut loop and infant mouse tests. Clearly, two types of ST could be differentiated by these tests in different animals.

Burgess *et al.* (1978) showed that two STs could be separated by methanol extraction; they were termed ST_A and ST_B. ST_A was methanol-soluble, positive in the infant mouse assay and in loops of neonatal piglets and calves, but negative in the weaned pig gut loop test. ST_B was methanol-insoluble and positive in neonatal and weaned pig gut loops but not in the infant mouse test.

Enterotoxin plasmids have been divided into the following groups based on the enterotoxin(s) determined by the plasmids. The types of enterotoxins produced by different strains vary with the hosts from which the strains are isolated (Table 1). Ent plasmids which have been transferred from strains of different origins will be described.

1) ST$_A$ PLASMIDS. Plasmids coding for production of a heat-stable enterotoxin, positive in neonatal piglet gut loop tests but inactive in loops of weaned pigs, were transferred to *E. coli* strain K-12 from two porcine strains; one belonged to O serogroup 101 and the other was untypable (Smith and Linggood, 1972). Later studies showed the two strains were positive in the infant mouse assay (Burgess *et al.*, 1978). ST$_A$ plasmids have also been transferred from porcine strains of O serogroups 138 and 149 (Gyles, 1979; Franklin and Möllby, 1981; Franklin *et al.*, 1981). Further studies with two ST$_A$ plasmids, identified in O149:K91:H1O strains of porcine origin, showed they were conjugative, that is promoted a transfer system and had molecular weights of about 46×10^6; they both belonged to the FII incompatibility group (M. M. McConnell, unpublished). Incompatible plasmids cannot coexist stably in the same cell and, by testing pairs of plasmids, over 20 incompatibility groups have been defined (Jacob *et al.*, 1977). F-like plasmids are related to the F factor of K-12 and strains carrying these plasmids possess fimbriae which are receptors for the F-specific phages (Meynell *et al.*, 1968). These F-like plasmids can be subdivided into a number of incompatibility groups which form the so-called 'F-incompatibility complex'.

An ST$_A$ plasmid, molecular weight 65×10^6, was transferred to K-12 from the bovine strain B41 of serotype O101:K99 (Gyles *et al.*, 1974; So *et al.*, 1976). This plasmid was conjugative and showed between 33 and 41% DNA homology with the F factor and F-like plasmids.

A conjugative ST$_A$ plasmid TP224 was transferred to K-12 from a human strain of serotype O166:H27; it had a molecular weight of 97×10^6 (Scotland *et al.*, 1979). TP224 appeared to belong to a new incompatibility group as it was compatible with members of all known groups (Bradley, 1981; M. M. McConnell, unpublished). Other ST$_A$ plasmids, which also code for the production of adhesive factors, have been identified in strains of human origin (see below).

2) ST$_B$ PLASMIDS. Plasmids coding for the production of ST$_B$ were transferred to K-12 from a number of porcine strains (Smith and Halls, 1968; Smith and Gyles, 1970a; Gyles *et al.*, 1974). K-12 strains carrying ST$_B$ plasmids were not always examined for ST$_A$ production but when this was done the ST$_B$ plasmids did not determine ST$_A$. Gyles *et al.* (1974) showed that some of these ST$_B$ plasmids were heterogeneous in molecular weight, ranging from 21×10^6 to 80×10^6. DNA hybridization studies demonstrated that the ST$_B$ plasmids had no significant homology with Ent plasmids coding for production of other enterotoxins and drug resistance (R) plasmids which belonged to the F-like type (So *et al.*, 1974, 1975).

3) LT PLASMIDS. Plasmids coding for LT production alone have been identified in strains of human origin (Echeverria *et al.*, 1978; Scotland *et al.*, 1979; Echeverria and Murphy, 1980; McConnell *et al.*, 1980). LT plasmids transferred to K-12

from strains of O serogroup 78 of different geographical origins all belonged to the F-incompatibility complex and were fertility inhibition positive (fi^+). However, they could be divided into three groups based on properties such as incompatibility, phage restriction, molecular weight and DNA homology (McConnell *et al.*, 1980; Willshaw *et al.*, 1980). Plasmids belonging to two of the groups were conjugative, F-like and had molecular weights between 51×10^6 and 60×10^6, whereas members of the third group were non-conjugative with molecular weights of about 42×10^6. Within a group the LT plasmids were very similar in their genetic and molecular properties. Members of different groups shared between 40 and 60% homology and LT plasmids belonging to all three groups had between 20 and 40% homology with standard F-like plasmids.

4) ST_A-LT PLASMIDS. The few plasmids of this type that have been identified were conjugative, F-like, had a molecular weight between 55×10^6 and 60×10^6 and were from strains of human origin (So *et al.*, 1975; McConnell *et al.*, 1979; Scotland *et al.*, 1979). Some ST_A-LT plasmids also determine the production of adhesive factors (see below).

5) ST_B-LT PLASMIDS. Plasmids coding for production of these enterotoxins were identified in porcine ETEC and transferred to K-12 (Smith and Gyles, 1970a; Gyles *et al.*, 1974; Franklin and Möllby, 1981). So *et al.* (1975) demonstrated that different ST_B-LT plasmids were closely related to each other and shared about 87% of their sequences. The plasmid P307, from a strain of O serogroup 8, was analysed in detail; it was F-like, belonged to the FI incompatibility group and had a molecular weight of 60×10^6. Hybridization studies showed that P307 had between 21 and 45% DNA homology with F-like plasmids in regions coding for transfer, replication and incompatibility (So *et al.*, 1975; Santos *et al.*, 1975). It appears that ST_B-LT plasmids with properties like those of P307 are very widely distributed in strains of porcine origin.

Some enterotoxigenic strains of both animal and human origin carry more than one plasmid coding for enterotoxin production. Several porcine strains of O serogroup 149, positive for ST_A, ST_B and LT, carried both an ST_A and an ST_B-LT plasmid (Franklin and Möllby, 1981; Franklin *et al.*, 1981). The porcine strain P16 of O serogroup 9 also produced ST_A and ST_B and the genes coding for their production were located on separate plasmids (Gyles *et al.*, 1974; So and McCarthy, 1980). So far, there appear to be no reports of plasmids coding for both ST_A and ST_B. In strains of human origin the genes for ST_A and LT are frequently encoded by separate plasmids. For example, in a study of strains belonging to O serogroup 78, isolated in different parts of Asia and Africa, most of the ST^+ LT^+ strains carried an LT plasmid and a second plasmid coding for the production of ST_A and colonization factor antigen I (CFA/I) (Smith *et al.*, 1979; McConnell *et al.*, 1980, 1981).

Enterotoxin plasmids coding for drug resistance. Enterotoxigenic strains resistant to antimicrobial drugs have been reported. In several studies of drug-resistant ETEC, selection for transfer of drug resistance to K-12 resulted in co-transfer of enterotoxin production and usually the two properties were determined by separate plasmids in the transconjugants (Wachsmuth *et al.*, 1976; Echeverria *et al.*, 1978; McConnell *et al.*, 1979, 1980; Scotland *et al.*, 1979). In a few cases it was demonstrated that a single plasmid coded for enterotoxin production and drug resistance (Gyles *et al.*, 1977; McConnell *et al.*, 1979; Scotland *et al.*, 1979; Echeverria and Murphy, 1980; Stieglitz *et al.*, 1980).

A strain of serotype O159:H34 of human origin carried an ST_A-LT plasmid, a separate non-autotransferring ampicillin resistance plasmid and four other plasmids. Transposition of ampicillin resistance (Ap) onto the Ent plasmid resulted in formation of Ap-Ent plasmids and these recombinants were transferred to K-12 (McConnell *et al.*, 1979; Scotland *et al.*, 1979). In some transconjugants the Ap-Ent plasmids transferred at much higher frequencies than the original ST_A-LT plasmid; this suggested that in these cases there was insertion of the ampicillin resistance transposon *Tn*A into the transfer region of the Ent plasmid.

The plasmid pCG86 which codes for production of ST_B and LT and resistance to streptomycin, sulphonamides, tetracyclines and mercury has been mapped by electron microscopic heteroduplex analysis. pCG86 shared a 48-kilobase (kb) region of complete homology with P307 and this included the ST_B and LT genes. The results suggested that pCG86 may have arisen by recombination between an Ent plasmid such as P307 and an R factor (Mazaitis *et al.*, 1981). Silva *et al.* (1978) used pCG86 for isolation of mutants affecting enterotoxin production. Strains carrying plasmid mutants defective in LT synthesis were obtained and were used to study LT production and assembly. Insertions of the transposon *Tn*3 into pCG86 resulted in increased LT production. These mutants were present in a higher plasmid copy number and the insertions mapped into a region affecting replication control (Picken *et al.*, 1982). Strains with enhanced LT production were also isolated in other experiments but in these cases the mutations mapped on the chromosome (Bramucci *et al.*, 1981). In some mutants the total LT production was higher while in others an increased amount of LT was released from the cells.

So *et al.* (1978b) have used drug resistance transposons such as *Tn*A to mark Ent plasmids *in vitro* so that there is a selectable marker for genetic studies of these plasmids. The transfer of three different Ent plasmids marked with drug resistance was examined with non-toxigenic strains of serogroups to which ETEC commonly belong and also with strains of serogroups in which ETEC are rarely found. Strains of both classes acquired and maintained the three plasmids, suggesting that the prevalence of certain serogroups among ETEC is due to other factors (Scotland *et al.*, 1983a).

Cloning of genes coding for enterotoxin production. Much of the recent work on the genetic control of enterotoxin production has involved cloning and identification of plasmid-specified products.

ST_A. Genes determining ST_A have been cloned from plasmids found in strains of bovine, porcine and human origin. The ST_A gene from the bovine strain B41 was cloned into the vector pBR322 and it was located within a transposable element (*Tn*1681) which was flanked by inverted repeats of the insertion sequence IS1 (So *et al.*, 1979; So and McCarthy, 1980). From strains P288 and P310 of porcine origin, ST_A genes were cloned on fragments that had an identical structure to *Tn*1681 (Lathe *et al.*, 1980). A similar structure was observed in an ST_A gene cloned from strain H10407 of human origin (Yamamoto and Yokota, 1980). Since ST_A was identified on a transposable element it was suggested that this could explain why ST_A genes had been found on several different types of plasmid.

Moseley *et al.* (1980) isolated ETEC from patients with diarrhoea and tested these strains by colony hybridization with an ST_A probe derived from *Tn*1681. This ST_A probe detected only 15 of 43 strains which were positive in the infant mouse assay for ST_A. A second ST_A probe, prepared from a strain of human origin, detected all the ST_A producers which were negative with the first probe (Moseley *et al.*, 1982). The two ST_A genes were designated *est*A1 (present in *Tn*1681) and *est*A2, cloned from the strain of human origin. These two genes did not hybridize with each other under stringent conditions for DNA hybridization, that is in $5\times$ standard saline citrate (SSC), 50% formamide at 37°C for 64 h (Harford *et al.*, 1981). When relaxed conditions were used ($5\times$ SSC, 20% formamide at 37°C for 64 h) there was partial hybridization between *est*A1 and *est*A2.

The *est*A1 gene has been found in both animal and human strains, whereas *est*A2 has been detected only in human strains so far (So *et al.*, 1981; Moseley *et al.*, 1982; N. Harford, unpublished). An ST_A gene has been cloned from the CFA/I-ST plasmid NTP113 of human origin and the distribution of restriction sites within the cloned fragment indicated the ST_A gene was probably identical with *est*A2 (Willshaw *et al.*, 1983; N. Harford, personal communication). The ST genes, *est*A1 and *est*A2, have been sequenced and the gene products, ST_{A1} and ST_{A2}, compared. The genes had 72% homology and the enterotoxins shared 62% of their amino acid sequences with the C-terminal region of 18 amino acids being highly conserved (De Wilde *et al.*, 1981).

A limitation in the use of probes such as *est*A1 or *est*A2 is that they detect gene sequences that may not be expressed. Strain H10407 carries two plasmids associated with enterotoxin production and adhesion. One plasmid, with a molecular weight of 56×10^6, codes for CFA/I and ST_A production while the plasmid of

42×10^6 codes for synthesis of LT but not ST_A (Smith *et al.*, 1979; McConnell *et al.*, 1980, 1981). From the hybridization results with H10407 it was clear that the *est*A2 probe hybridized to the CFA/I-ST plasmid while *est*A1 hybridized to the LT plasmid (N. Harford, unpublished); this suggested that H10407 had an *est*A1 gene that was not expressed. In contrast, an enterotoxin gene may be expressed in one host but not in another. Two different plasmids coding for ST_A production were transferred to several wild-type strains. With a non-motile strain, of O serogroup 88, there was no expression of ST_A in the infant mouse assay; in all the other strains to which the same plasmids were transferred ST_A was produced (Scotland *et al.*, 1983a).

ST_B. Compared with ST_A much less is known about the organization of the ST_B gene(s). The ST_B sequences that were cloned from the ST_B-LT plasmid pCG86 were flanked by inverted repeats, suggesting that they were on a transposable element (Mazaitis *et al.*, 1981). There was no hybridization between the ST_B sequences and *est*A1 or *est*A2 (Maas, 1981).

LT. The LT genes of the plasmid P307 have been cloned in pBR313 on a fragment of 1.2×10^6 (So *et al.*, 1978a; Dallas and Falkow, 1979; Dallas *et al.*, 1979a). The fragment coded for two proteins: LT_A, with a molecular weight of 25,000, which had the ability to activate adenyl cyclase, and LT_B, molecular weight 11,500, which was required for binding of the toxin to the intestinal epithelium. It appeared that the cistrons for LT_A (*elt*A) and LT_B (*elt*B) were transcribed into a single messenger RNA with a promoter located at the N-terminal end of LT_A. Sequencing studies demonstrated that *elt*A and *elt*B coded for a signal peptide at the N-terminal end which was required for transport through the cytoplasmic membrane (Dallas and Falkow, 1980; Spicer *et al.*, 1981; Spicer and Noble, 1982). Yamamoto and Yokota (1980) cloned LT genes from a plasmid in strain H10407 of human origin; the structure of the cloned LT region was very similar to that reported for the genes cloned from P307. The distal end of *elt*A, also termed *tox*A, overlapped with the proximal end of the B subunit gene *elt*B or *tox*B (Yamamoto *et al.*, 1982b). The genes encoding LT_A and LT_B were flanked by repeated DNA sequences, termed β, and the ST_A gene present on the same plasmid was also bounded by inverted repeats, designated α. The ST_A gene was adjacent to the LT genes and within the repeated β sequences (Yamamoto and Yokota, 1981). The LT_A and LT_B genes have also been cloned separately (Yamamoto *et al.*, 1981; Sanchez *et al.*, 1982). An LT_A derivative called LT_{A*} was constructed which lacked the adenyl cyclase-activating function of LT but retained the immunological properties. Strains carrying plasmids coding for LT_B alone still had the binding activity of LT and might be useful for the development of vaccine strains to protect against diarrhoeal disease caused by ETEC.

Subcloning of the LT region showed that a 0.5×10^6 *Hind*III fragment specified LT_B and about 10% of the cistron encoding LT_A. DNA hybridization experiments showed that this fragment was conserved in two Ent plasmids of human origin and two from porcine sources (Dallas *et al.*, 1979b). Willshaw *et al.* (1980) demonstrated a fragment of approximately 0.5×10^6 after *Hind*III treatment of seven LT plasmids and an ST_A-LT factor, all isolated from human strains of diverse origins. The cloned LT genes have been used to prepare a DNA probe for the detection of ETEC (Moseley *et al.*, 1980, 1982; Echeverria *et al.*, 1982). The LT probe hybridized with all the strains shown to produce LT in the Chinese hamster ovary (CHO) cell assay.

The amino acid sequences of LT_B molecules of human and porcine origin have been compared. The regions covered the signal peptide and the amino terminus of the mature peptide and only three amino acids were different (Yamamoto *et al.*, 1982a). Immunodiffusion studies with LTs of porcine and human origin showed that although the two LTs shared major antigenic determinants each LT also had unique antigens (Honda *et al.*, 1981; Geary *et al.*, 1982). There were also differences in the molecular weights of the B subunit oligomers as well as the B subunit monomers.

Plasmids coding for production of adhesive factors. A number of different adhesive factors have been identified in enterotoxigenic strains (Table 2). Most of these factors are plasmid-determined but the genes encoding production of 987P or F41 have not yet been located (for review, see Gaastra and de Graaf, 1982).

κ88. K88 was shown to be plasmid-mediated by Ørskov and Ørskov (1966) and its importance in colonization of the small intestine in piglets was demonstrated

Table 2.   *Adhesive factors found in enterotoxigenic E. coli*

Adhesive factor	Host	O serogroups	Genetic control
K88[a]	Piglet	8, 45, 138, 141 147, 149, 157	Plasmid
K99	Calf, lamb, piglet	8, 9, 20, 64, 101	Plasmid
987P	Piglet	9, 20, 141	?
F41	Calf	9, 101	?
CFA/I	Man	4, 15, 25, 63, 78 90, 110, 114, 126, 128, 153	Plasmid
CFA/II	Man	6, 8, 9, 78, 80, 85, 115, 139, 168	Plasmid
E8775	Man	25, 115, 167	Plasmid

[a]K88 has been found in a number of different antigenic forms, K88ab, K88ac and K88ad.

by the use of K88 plasmid-carrying strains (Smith and Linggood, 1971b; Smith, 1976; Smith and Huggins, 1978). K88 plasmids usually carried the genes determining raffinose utilization (Smith and Parsell, 1975). Most of these K88-Raf plasmids had a molecular weight of approximately 50×10^6, were non-conjugative and had a very high degree of DNA homology with each other (Shipley *et al.*, 1978). In transfer experiments some K88 plasmids recombined with transfer factors to form larger conjugative plasmids. Detailed studies of one K88 plasmid, pSF711, showed that the region with the K88 and *raf* determinants was flanked by direct repeats of IS1 and could be transposed to another replicon (Schmitt *et al.*, 1979).

κ99. Smith and Linggood (1972) first demonstrated that production of K99 was plasmid-determined. In strain B41 the genes were located on a conjugative plasmid with a molecular weight of 52×10^6 which also coded for resistance to streptomycin and tetracyclines (So *et al.*, 1976). Some K-12 transconjugants derived from B41 carried a single plasmid coding for production of K99 and ST_A and also drug resistance (H. R. Smith, unpublished). In other bovine strains it appeared there was linkage of the genes for ST_A and K99 production on a single plasmid (Harnett and Gyles, 1982). K99 plasmids have been used to study the expression of the antigen; certain O serogroups such as 8, 9 and 20 produced considerably less K99 than strains belonging to O serogroup 101. This regulation appeared to be host-dependent because no difference in the level of expression was observed when K99 plasmids from different serogroups were transferred to K-12 (de Graaf *et al.*, 1980).

CFA/I. This adhesive factor has been detected in strains of several serogroups (Table 2) (Evans *et al.*, 1975; Ørskov and Ørskov, 1977; Gross *et al.*, 1978; Cravioto *et al.*, 1979a; Smyth *et al.*, 1979; Scotland *et al.*, 1981; Thomas and Rowe, 1982). Strains possessing CFA/I were shown to colonize the intestine of rabbits and humans (Evans *et al.*, 1975; Satterwhite *et al.*, 1978). In volunteer studies with H10407 (ST^+ LT^+ CFA/I^+) and H10407-P (ST^- LT^+ CFA/I^-) diarrhoea and persistence of the ingested *E. coli* occurred only in subjects challenged with CFA/I^+ organisms.

In contrast to the situation in most animal strains, plasmids coding for adhesive factors in ETEC of human origin usually also determine enterotoxin production. Production of CFA/I is encoded by a plasmid that also carries the gene for the synthesis of ST_A (Evans *et al.*, 1975; Evans and Evans, 1978; Smith *et al.*, 1979). CFA/I-ST plasmids were mobilized into K-12 from wild-type *E. coli* belonging to O serogroups 78 and 128 and the plasmids were shown to possess closely similar genetic and molecular properties (Reis *et al.*, 1980; McConnell *et al.*, 1981; Willshaw *et al.*, 1982a; Murray *et al.*, 1983). The plasmids were non-conjugative but could be mobilized into K-12 efficiently by the R factor R1-19; the molecular weights of the CFA/I-ST plasmids ranged from 52×10^6 to 72×10^6. A plasmid coding for CFA/I, ST and LT was mobilized from a non-motile strain of O serogroup 63 and closely resembled the CFA/I-ST plasmids in its properties (McConnell *et al.*, 1981; Willshaw *et al.*, 1982a). The CFA/I-ST plasmid NTP113, from an O78:H12 strain from South Africa, was mapped by insertion and deletion of drug resistance transposons. Mutations in plasmid derivatives that no longer coded for CFA/I were located in two widely separated regions of the plasmid and complemented to restore production of CFA/I. The ST_A gene mapped very close to CFA/I region 1 and explained the close association of the two properties (Smith *et al.*, 1982).

CFA/I was not expressed in a K-12 strain which had a chromosomal mutation preventing the production of type 1 fimbriae (K-12 *fim*). Production of K88 and K99 was not affected in this K-12 *fim* strain. However, while no expression of CFA/I occurred in some wild-type *fim* strains this was not always the case, suggesting that the chromosomal genes which affect CFA/I synthesis are located close to those controlling type 1 fimbriae but the two are not identical (M. M. McConnell, unpublished).

CFA/II. Production of CFA/II is also plasmid-determined and plasmids encoding the synthesis of ST and LT in addition to CFA/II were identified in wild-type strains belonging to O serogroups 6 and 85 (Penaranda *et al.*, 1980). One of these CFA/II-ST-LT plasmids was introduced into K-12 by co-transformation with a drug resistance plasmid. In studies of CFA/II strains belonging to O serogroups 6 and 8 Evans and Evans (1978) reported that cell extracts heated to 60°C for 30 min gave a single precipitin line in immunodiffusion tests with a specific CFA/II antiserum. However, recent studies showed that in O6:H16 strains three CFA/II antigenic components could be detected with antisera prepared against strains of different biotypes (Cravioto *et al.*, 1982; Smyth, 1982). These antigens were designated components 1, 2 and 3 or coli surface-associated antigens CS1, CS2 and CS3. Strains of O6:H16 biotype A produced CS1 and CS3 whereas, in general, strains of biotypes B, C and F possessed CS2 and CS3. In contrast to the O6:H16 strains, CFA/II$^+$ strains belonging to the other serogroups produced only CS3 (Cravioto *et al.*, 1982; Smyth, 1982) apart from one strain of serotype O139:H28 which had CS1 and CS3 (M. M. McConnell and S. M. Scotland, unpublished). Plasmid studies demonstrated that the production of both antigens (CS1 and CS3 or CS2 and CS3) by the O6:H16 strains was controlled by a single plasmid which also coded for ST and LT production. CFA/II-ST-LT plasmids were transferred to K-12 from O6:H16 strains of biotypes A, B and C (Mullany *et al.*, 1983; Smith *et al.*, 1983) and also from an O168:H16 strain (M. M. McConnell, unpublished). The plasmids were non-conjugative but could be mobilized efficiently by the R factor R100; they had very similar genetic properties but their molecular weights ranged from 62×10^6

to 115×10^6. When the CFA/II-ST-LT plasmids were mobilized into K-12 only the CS3 antigen was expressed. Transfer of the plasmids to O6:H16 strains showed that the expression of CS1 or CS2 was restored but the antigen produced was dependent on the properties of the host. The plasmid coded for production of CS1 and CS3 in strains of biotype A and CS2 and CS3 in strains of biotype B, C and F. Electron microscopy of plasmid-carrying strains showed that CS1 and CS2 were serologically distinct fimbrial antigens while CS3 was a non-fimbrial surface component (Mullany *et al.*, 1983; Smyth, 1984).

E8775. The fimbrial antigen E8775 has been detected in ETEC strains of O serogroups 25, 115 and 167 (Table 2) (Thomas and Rowe, 1982; Thomas *et al.*, 1982). A plasmid with a molecular weight of 85×10^6 was mobilized from a strain of O serogroup 167 into K-12 and coded for production of E8775 fimbriae and also ST. However, in strains of O serogroup 25 the genes for the E8775 antigen appeared to be on a separate plasmid from the one determining enterotoxin production (L. V. Thomas, unpublished).

Cloning of genes coding for production of adhesive factors. The K88, K99 and CFA/I adhesion systems have been examined in more detail by cloning the genetic determinants and analysing the products.

K88. In the case of K88, the determinants for K88ab, K88ac and K88ad have all been cloned (Mooi *et al.*, 1979; Shipley *et al.*, 1979, 1981; Gaastra and de Graaf, 1982). In all three cases the K88 genes were located on a *Hin*dIII fragment of 7×10^6 to 8×10^6 and this was reduced to a fragment of 4.3×10^6 in the case of K88ab without loss of activity. This 4.3×10^6 fragment coded for at least six polypeptides detected in minicells; these small cells which lack chromosomal DNA can be formed at cell division from certain strains of *E. coli*. A polypeptide with a molecular weight of 30,000 was thought to be the precursor of a 27,000 polypeptide. The K88 fimbrial subunit was identified as a polypeptide with a molecular weight of 26,000. All five polypeptides appeared to be transported across the cytoplasmic membrane by means of a signal peptide. The genes determining these five polypeptides were mapped by analysis of deletion derivatives and the possible functions of the gene products were investigated. The 81,000 polypeptide, which was located in the outer membrane, appeared to be necessary for the assembly or anchorage of the fimbrial subunits. The 27,000 polypeptide was found mainly in the periplasm and might be necessary for integration of the fimbrial subunit into the outer membrane. It was suggested that the role of the 17,000 polypeptide, also periplasmically located, involved the post-translational modification of fimbrial subunits (Mooi *et al.*, 1981, 1982). The nucleotide sequences for the structural genes for K88ab and K88ad protein subunits have been determined (Gaastra *et al.*, 1981, 1983). There were 47 differences in the nucleotide sequences, resulting in 34 differences between the amino acid sequences. These changes in amino acid sequence, which accounted for the b and d antigenic variants, were not restricted to a particular part of the molecule but the majority occurred in clusters.

Analysis of the cloned K88ac genes has shown similar findings to those reported for K88ab. The genes were designated *adh*A, *adh*B and *adh*C in operon I and *adh*D in operon II with *adh*D encoding the K88 fimbrial subunit (Kehoe *et al.*, 1981). It was suggested that the product of *adh*C might be a positive regulator required for expression of operon II. A fifth gene, *adh*E, which mapped close to *adh*A, has been identified (Dougan *et al.*, 1983b; Kehoe *et al.*, 1983). These recent studies suggested that a single transcription unit was involved in K88ac expression. The *adh*A polypeptide was located in the outer membrane while the products of *adh*B and *adh*C were mainly in the periplasm (Dougan *et al.*, 1983a).

K99. Van Embden *et al.* (1980) have cloned the K99 genes on a 4.5×10^6 fragment. Deletion analysis located the approximate position of the K99 structural gene and this was determined precisely by nucleotide sequencing (de Graaf and Mooi, 1982). Like K88, the K99 subunit was produced as a precursor containing a signal peptide of 22 amino acids. A fragment carrying the K99 genes cloned in the vector pACYC184 was analysed in minicells. At least five polypeptides were expressed in addition to those specified by the vector; the polypeptide of 22,000 corresponded in size to that of the purified K99 fimbrial subunit (Kehoe *et al.*, 1982). Three of the polypeptides encoded by this K99 fragment appeared to be of similar size to the K88 *adh*A, *adh*C and *adh*D products. The K88 and K99 genes were compared in genetic complementation and DNA hybridization studies. There appeared to be no complementation of K88 mutants by K99 genes but homology was observed when two probes containing regions of the K88 determinant were hybridized to the cloned K99 genes. This indicated a close relationship between some of the DNA sequences involved in expression of the K88 and K99 adhesion systems (Kehoe *et al.*, 1982).

CFA/I. The genes encoding production of CFA/I have been cloned from the CFA/I-ST plasmid NTP113. Since two widely separated regions of the plasmid were required for CFA/I expression it was necessary to clone the regions separately. Strains carrying clones were tested for CFA/I by genetic complementation using recipient strains with plasmids coding for either region 1 or region 2 separately. Region 1 was cloned on a 5×10^6 fragment and region 2 on a fragment of 1.4×10^6. The ST_A gene mapped very close to region 1 but was cloned independently of CFA/I. A strain carrying the two CFA/I regions cloned on compatible vectors produced CFA/I and was non-toxigenic (Willshaw *et al.*, 1983).

Strains carrying plasmids with the cloned regions were also tested for CFA/I production by an enzyme-linked immunosorbent assay. It was concluded that region 1 contained the CFA/I structural gene. The polypeptides produced in *E. coli* minicells carrying the cloned derivatives were also examined. Region 1 directed the production of up to eight polypeptides but it appeared that some of these represented precursor forms detected because of incomplete proteolytic processing in minicells. The polypeptide with a molecular weight of approximately 13,000 was identified as the CFA/I fimbrial subunit with CFA/I antiserum in the presence of staphylococcal protein A. Region 2 appeared to specify three polypeptides (Willshaw *et al.,* 1982b; G. A. Willshaw, H. R. Smith and M. M. McConnell, unpublished). The complete amino acid sequence of the CFA/I fimbrial subunit has been determined. The protein comprised 147 amino acids with a molecular weight of 15,058; it did not show any homology with the K88 subunit or other fimbrial proteins (Klemm, 1982).

Enteropathogenic E. coli

Enteropathogenic strains of *E. coli* are a cause of diarrhoeal disease of infants in both developing and developed countries. They have been recognized by epidemiological studies combined with serotyping. They belong to a restricted range of serogroups which in general differ from those to which ETEC and EIEC belong (Rowe, 1979); however, EPEC serogroups occasionally contain strains which produce heat-labile or heat-stable enterotoxins normally produced by ETEC. Although still isolated frequently from sporadic cases of diarrhoea in infants, the number of outbreaks due to EPEC has declined in the developed countries since the early 1970s.

The mechanisms by which EPEC strains elicit a fluid response in the intestine of the human host are unclear. Unlike ETEC they have not been shown to produce either ST or LT (Goldschmidt and DuPont, 1976; Gross *et al.,* 1976; Gurwith *et al.,* 1977) nor did they possess the genes coding for the production of these enterotoxins (Robins-Browne *et al.,* 1982). There have, however, been several reports of toxin production by EPEC, using assay systems which differed from those commonly used for ST or LT. Klipstein *et al.* (1978) showed that certain EPEC strains isolated from diarrhoeal outbreaks produced an extra-cellular toxin causing net water secretion in a rat perfusion model.

Konowalchuk *et al.* (1977) demonstrated that culture filtrates of certain *E. coli* strains had a cytotonic effect on monolayers of Vero cells which contrasted with the cytotoxic effect of LT (Keusch and Donta, 1975) and they termed this cytotoxin VT. VT was distinct from LT and ST as it had no effect on Y1 or CHO cells commonly used to detect LT, and caused no fluid accumulation in the infant mouse test used to detect ST_A. Partially purified VT did, however, induce some fluid accumulation in the rabbit ileal loop test (Konowalchuk *et al.,* 1978). In their initial study Konowalchuk *et al.* (1977) showed VT production by 7 of 35 EPEC strains isolated from infants with diarrhoea. Three of the seven strains belonged to O serogroup 26 and subsequently an association of VT production with strains of *E. coli* O serogroup 26 has been reported (Wade *et al.*, 1979; Scotland *et al.*, 1980a; Wilson and Bettelheim, 1980). In the study of Scotland *et al.* (1980a) 25 strains were VT$^+$ among 253 EPEC strains isolated from infants with diarrhoea in the United Kingdom and belonging to 11 different serogroups; VT$^+$ strains included 20 strains of 31 tested belonging to serotype **O26:H**11. Johnson *et al.* (1983) demonstrated VT production by strains of O serogroup 157 associated with haemorrhagic colitis.

Smith and Gyles (1970b) used rabbit intestinal loops to examine strains from infant diarrhoea and showed that certain strains of O serogroup 26 gave particularly consistent dilatation. The enterotoxin produced by these strains was destroyed by heating to 65°C for 15 min and was therefore heat-labile (Smith and Gyles, 1970b). However, Smith and Linggood (1971a) showed that with one strain, H19, of serotype **O26:H**11, the effect in the rabbit loop was not neutralized by antiserum raised against strain P307 of porcine origin now known to produce a classical LT which can be detected in tissue culture tests. Later studies showed that H19 did not produce LT as recognized by action on Y1 mouse adrenal or CHO cells.

Smith and Linggood (1971a) examined 26 EPEC strains isolated from infants with diarrhoea and attempted to transfer to K-12 the ability to dilate rabbit intestinal loops, termed Ent. Transfer of Ent was detected from strain H19. They also reported that H19 possessed one plasmid coding for drug resistance and a second coding for colicin production; both plasmids were transferable to K-12. Not all transconjugants which acquired drug resistance or the ability to produce colicin were Ent$^+$ and it was concluded that Ent was not carried by these plasmids.

H19 was one of the strains of O serogroup 26 shown to be VT$^+$ by Konowalchuk *et al.* (1977) and the K-12 Ent$^+$ derivative obtained by Smith and Linggood (1971a) was also shown to be VT$^+$ by Scotland *et al.* (1980b). It is therefore likely that the Ent and VT properties of H19 are the same and that the first transfer of toxin production from human *E. coli* strains by Smith and Linggood was of VT. Scotland *et al.* (1980b) were able to transfer the VT$^+$ property from H19 to K-12, at a frequency of 10^{-2} per recipient cell. Drug resistance and colicin production were transferred at a frequency of 5×10^{-1} and were encoded by plasmids with molecular weights of 52×10^6 and 61×10^6, respectively. Derivatives of strain H19 which had lost both these plasmids were still VT$^+$ and retained one large plasmid and two small plasmids. Examination of K-12 transconjugants failed to show that VT was plasmid-encoded; further work indicated that VT production in these transconjugants was associated with a bacteriophage (Scotland *et al.*, 1983c). Wade *et al.* (1979) were unable to transfer VT production from H19 although transfer of drug resistance and colicin production was obtained; but they, too, reported that loss from strain H19 of the plasmids coding for these properties was not accompanied by loss of VT production. The role of VT in the pathogenicity of EPEC remains to be established. Toxins from EPEC which acted on mouse fibroblasts (Farkas-Himsley *et al.*, 1978) or HeLa cells (O'Brien *et al.*, 1982) have also been reported but there have been no published studies on the genetic control of these toxins.

The ability to colonize the epithelial mucosa of the upper intestine is important in the pathogenicity of EPEC; however, in contrast to ETEC, the mechanism of this colonization is not known. A number of *in vitro* models have been used to investigate these mechanisms. McNeish *et al.* (1975) studied the adhesion of EPEC to the mucosa of pieces of human foetal small intestine. One of the strains possessing the ability to adhere to this tissue, that is Adh$^+$, was H19. A derivative of H19 which had lost the plasmid coding for colicin production (and also the plasmid coding for drug resistance) no longer adhered to intestinal tissue. In mating experiments with H19, 4 of 12 K-12 transconjugants were Adh$^+$ (Williams *et al.*, 1977). Subsequently mucosal adherence was associated with the acquisition of the plasmid with a molecular weight of 56×10^6 coding for production of colicin Ib (Williams *et al.*, 1978).

Another *in vitro* model for adhesiveness tested the ability of bacteria to adhere to HEp-2 cells grown in tissue culture (Cravioto *et al.*, 1979b). Adherence to HEp-2 cells was found in a significantly higher proportion of strains belonging to EPEC serogroups and isolated from outbreaks of infantile enteritis than in strains isolated from healthy humans. The ability to adhere to HEp-2 cells was not common in ETEC, even when they possessed CFA/I or CFA/II. It has recently been shown that adhesion to HEp-2 cells by several EPEC is mediated by non-fimbrial adhesins (Scotland *et al.*, 1983b). Strain H19 was able to adhere to HEp-2 cells (Cravioto *et al.*, 1979b) and derivatives of this strain which had lost the plasmid coding for colicin production (or the plasmid coding for drug resistance) were still HEp-2 cell adhesive. It, therefore, seems unlikely that the bacterial surface structures that enable EPEC strains to adhere to HEp-2 cells are the same as those that facilitate adhesion to foetal intestinal tissue. Baldini *et al.* (1983) recently reported the transfer of a plasmid from another EPEC strain of O serogroup 127 to K-12, which resulted in the transconjugant becoming HEp-2 cell adhesive.

A number of other *in vitro* tests have been used to study the adhesion of EPEC. These include adhesion to buccal epithelial cells (Candy *et al.*, 1978), to brush borders of human intestinal epithelial cells obtained from autopsies (Lafeuille *et al.*, 1981) and to HeLa cells grown in tissue culture (Chiarini and Giammanco, 1978). EPEC strains belonging to many different serogroups and isolated from sporadic cases of diarrhoea were able to agglutinate erythrocytes from several species, including man and calf in the presence of mannose (Evans *et al.*, 1979;

Cravioto *et al.*, 1982). However, there have been no genetic studies of the factors involved in these interactions.

Cantey and Blake (1977) described a diarrhoeal disease of rabbits caused by a strain of O serogroup 15 termed RDEC-1. This organism resembled EPEC in failing to produce ST or LT and in being non-invasive. It adhered in large numbers to the epithelial cells of the ileum, caecum and colon and resulted in destruction of the microvilli. Although there had been previous reports (Cantey *et al.*, 1981) of a possible role of a bacterial capsule in adhesion of strain RDEC-1, Boedeker (1983) reported that adhesion to rabbit epithelial cells was due to specific RDEC-1 fimbriae immunologically distinct from type 1 fimbriae. Transfer of a plasmid with a molecular weight of 80×10^6 from RDEC-1 to recipients was accompanied by the ability to adhere *in vitro* to intestinal brush borders of the rabbit but not of rat, guinea-pig or man.

Enteroinvasive E. coli

The third class of *E. coli* causing intestinal disease has been termed enteroinvasive. They resemble shigellae in their ability to penetrate epithelial cells of the intestinal mucosa, particularly of the colon, but rarely to deeper levels. The intracellular multiplication of the bacteria and destruction of mucosal tissue results in bacillary dysentery. In the laboratory, invasiveness of EIEC and of shigellae is indicated by a positive Serény test in which keratoconjunctivitis develops after application of the bacteria to the conjunctiva of the guinea pig eye (Serény, 1957) or alternatively by the invasion of cells grown in tissue culture (La Brec *et al.*, 1964). Neither of these tests has been used widely in epidemiological studies and the incidence of disease due to EIEC is difficult to evaluate, although many outbreaks have been reported. EIEC belong to a restricted range of serogroups, which differ from those to which EPEC and ETEC belong (Rowe, 1979).

In addition to the similarity of the infections caused by EIEC and shigellae, these two groups of organisms are related in other properties (Edwards and Ewing, 1972). They may share or have identical somatic antigens and EIEC are often non-motile. EIEC are also less active biochemically than typical *E. coli* strains and in particular may fail to produce gas from glucose or to decarboxylate lysine.

Recent work on the genetics of the virulence of shigellae has been followed by studies of EIEC because of the similarity of their pathogenic mechanisms. In early studies of virulence in shigellae (for reviews, see Formal and Hornick, 1978; Petrovskaya and Licheva, 1982) hybrids of virulent *Shigella* strains and avirulent strains of *E. coli* were obtained by conjugation; a number of chromosomal sites were mapped in different regions which affected both penetration of and multiplication in epithelial cells. Other genes determining lipopolysaccharide synthesis also affected virulence. It was concluded that a multiplicity of chromosomal genes were involved in pathogenicity. More recently plasmids carried by virulent *Shigella* strains have also been shown to be necessary for virulence. Loss of a large plasmid with a molecular weight of 120×10^6 from strains of *Sh. sonnei* was associated with loss of the form I surface antigen and of the ability to cause keratoconjunctivitis in the Serény test (Kopecko *et al.*, 1980; Sansonetti *et al.*, 1980). The large plasmid was marked with drug resistance and introduced by mobilization with plasmid R386 into form II variants resulting in the restoration of form I antigen production and virulence (Sansonetti *et al.*, 1981b). Large plasmids, with molecular weights of $120–140 \times 10^6$, associated with virulence have also been reported in strains of *Sh. flexneri, Sh. dysenteriae* and *Sh. boydii* (Sansonetti *et al.*, 1981a, 1982b; Silva *et al.*, 1982a).

A search for large plasmids involved in pathogenicity has now been made in EIEC and all strains which were positive in the Serény test harboured a plasmid with a molecular weight of 140×10^6 (Harris *et al.*, 1982; Sansonetti *et al.*, 1982a; Silva *et al.*, 1982b); these included strains belonging to O serogroups 28, 29, 124, 136, 143, 144 and 152. All three studies noted that certain strains which had spontaneously become avirulent in the guinea-pig eye test no longer had the large plasmid. Sansonetti *et al.* (1982b) also noted the presence of a new, smaller plasmid in certain avirulent mutants and suggested that a deletion had occurred in the large plasmid.

Although all invasive strains possessed a plasmid of molecular weight 140×10^6, such plasmids were also present in non-invasive strains studied by Harris *et al.* (1982) and Silva *et al.* (1982b). These strains belonged to serogroups considered to be characteristic of EIEC and some of the strains had been positive in Serény tests. It is possible that mutations had occurred either in chromosomal or plasmid genes which altered the invasive ability of such strains or that small deletions had occurred which were not detected by agarose gel electrophoresis. Sansonetti *et al.* (1982a,b) showed that some invasive strains harboured more than one large plasmid and some of these large plasmids may be unrelated to invasiveness.

Using the large plasmid derived from *Sh. flexneri* 5, which had been marked with drug resistance and co-integrated with a mobilizing plasmid, Sansonetti *et al.* (1982a) were able to restore the virulence in tissue culture and Serény tests of *E. coli* strains of O serogroups 124 and 143, which had lost a large plasmid. When the large plasmid encoding virulence and the form I antigen from *Sh. sonnei*, which had been similarly marked, were transferred to the avirulent *E. coli* strains, invasiveness was not restored.

It may be concluded that invasive strains of shigellae and *E. coli* possess large plasmids which encode or regulate some of the functions required for the penetration of epithelial cells. Further study of the relationships between the large plasmids carried by the different *Shigella* species and EIEC is required. Virulence depends on a number of factors and the interaction of chromosomal and plasmid genes is important. In earlier experiments, attempts to confer virulence on K-12 by the transfer of chromosomal markers from *Sh. flexneri* had been unsuccessful. However, by incorporation of such markers with a large plasmid from *Sh. flexneri*, virulent K-12 derivatives were obtained (Sansonetti *et al.*, 1981a). Different combinations of chromosomal and plasmid genes resulted in transconjugants which could be distinguished in the various tests for virulence; some were only able to invade tissue culture cells, others were also Serény test positive. Gene cloning experiments should identify the essential DNA sequences which are necessary for the penetration of, and multiplication in, epithelial cells in these different models and provide information on the biological functions involved in virulence.

E. coli **Causing Extra-Intestinal Infections**

In addition to causing intestinal infections in man and animals, strains of *E. coli* are an important cause of extra-intestinal infections. Those responsible for generalized or septicaemic infections are certainly invasive but their properties are distinct from EIEC discussed in the preceding section. Although *E. coli* causing extra-intestinal infections are considered together in this section, the organisms possess different pathogenic mechanisms necessary for survival in such dissimilar habitats as the urinary tract and vascular system, as discussed in many chapters of this symposium. Usually, putative virulence factors have been recognized because they are possessed by a higher proportion of strains from disease than by faecal strains from healthy individuals. Genetic studies of a number of these factors are described.

Haemolysin Production

Several haemolysins may be produced by strains of *E. coli* (Smith, 1963; Walton and Smith, 1969). α-Haemolysin is released extra-cellularly, whereas β-haemolysin is cell-bound. Many studies, including those by Minshew *et al.* (1978a,b), Brooks *et al.* (1980) and Green and Thomas (1981), showed that α-haemolysin was produced by a larger proportion of strains isolated from extra-intestinal infections than from faeces of healthy individuals. The role of α-haemolysin in pathogenicity has not been established, although in several animal models, haemolytic strains have been reported to be more virulent than non-haemolytic ones, as will be discussed later. It has been considered that α-haemolysin may make iron available to the virulent organism from lysed erythrocytes in host tissue fluids where iron is usually in an unavailable form (Linggood and Ingram, 1982). Additional functions for haemolysins in pathogenicity are discussed in Chapter 7.

Smith and Halls (1967c) first demonstrated that in faecal strains from pigs the ability to produce α-haemolysin was carried on a plasmid (Hly). A conjugative plasmid coding for production of β-haemolysin has also been reported (Goebel and Schrempf, 1971). Plasmids coding for Hly have been shown to belong to a number of different incompatibility groups and both fi^+ and fi^- plasmids have been described (Smith and Halls, 1967c; Le Minor and Le Coueffic, 1975; Monti-Bragadin et al., 1975; de la Cruz et al., 1979). Studies of the molecular weights of haemolysin plasmids (Goebel et al., 1974; de la Cruz et al., 1979), of their degree of relatedness in DNA hybridization experiments (Royer-Pokora and Goebel, 1976; de la Cruz et al., 1980b) and of their restriction enzyme cleavage patterns (de la Cruz et al., 1979) have confirmed the diversity of the plasmids. It was, therefore, of interest to know whether the genes encoding Hly activity in these plasmids were related or whether different gene products were responsible for the common phenotype of ability to lyse erythrocytes. De la Cruz et al. (1980a) cloned a DNA fragment coding for Hly from a plasmid (pHly 152) carried by a murine strain of E. coli. The plasmid was conjugative and belonged to incompatibility group I_2. A fragment with a molecular weight of about 3.5×10^6 coded for full haemolytic activity. RNA probes transcribed from different parts of the cloned fragment showed homology with Hly plasmids belonging to four different incompatibility groups from porcine and human strains of E. coli, demonstrating that they shared a common sequence controlling α-haemolysin production. Hybridization studies were also performed with an RNA probe from the whole pHly 152 plasmid; there was extensive homology with one other Hly plasmid, but with the other Hly plasmids examined the fragments in common were few or restricted to the Hly determinant.

Noegel et al. (1979, 1981) and Goebel and Hedgpeth (1982) also studied plasmid pHly 152. The plasmid, of molecular weight 41×10^6, was mapped by restriction enzyme analysis and, as previously mentioned, a region, with a molecular weight of about 3.5×10^6, was shown to code for full haemolytic activity. At least three cistrons in this region were involved in the synthesis and secretion of α-haemolysin and their gene products were identified in minicells. The hlyA gene of 3.2 kb coded for a protein of 107 kd which appeared to be a precursor of α-haemolysin; hlyC encoded a protein of 16–18 kd which was necessary for the conversion of the precursor to active haemolysin; hlyB was involved in the secretion of α-haemolysin from the cell. One of the gene products of minicells harbouring the complete plasmid pHly 152 was a protein with a molecular weight of 58,000 which was assumed to be α-haemolysin by comparison with purified active extra-cellular α-haemolysin.

Stark and Shuster (1982) also used restriction enzyme analysis to study a non-conjugative Hly plasmid, pHly 185, carried by an E. coli strain of porcine origin. Restriction enzyme digests of plasmid derivatives following Tn5 insertion allowed two sites controlling structure and secretion to be identified in an arrangement similar to that found in pHly 152. However the *Eco*RI and *Hin*dIII fragments obtained with pHly 152 and pHly 185 were different.

Various groups have studied the virulence of strains which have been cured of the Hly plasmid to assess the role of the plasmid in virulence. Although haemolysin production is the phenotype that is identified, other additional unknown functions may be specified by the genes carried on the plasmid. Loss of the Hly plasmid had no effect on the ability of strains to cause diarrhoea in piglets or weaned pigs and it was concluded that haemolysin production was not a virulence factor in diarrhoeal disease (Smith and Linggood, 1971b). In tests related to the ability of strains to cause septicaemia, loss of the Hly plasmid had an effect. Strains possessing the Hly plasmid were more virulent for mice when tested by intraperitoneal injection than their Hly⁻ derivatives although the response varied depending on the Hly plasmid tested (Smith and Linggood, 1971b). Strains of *E. coli* from porcine oedema which were able to cause fatal haemorrhagic lung oedema in mice after nasal inhalation were avirulent when cured of the Hly plasmid (Emödy *et al.*, 1980). One strain tested, P673, produced α- and β-haemolysin; a derivative which produced only β-haemolysin was avirulent and virulence was recovered when the Hly plasmid was reintroduced. Transfer of the plasmid from strain P673 into a K-12 strain resulted in good haemolysin production; the Hly⁺ K-12 derivative was highly virulent in a chick embryo death test but only weakly virulent in the mouse lung test. Clearly other properties of the wild-type strain were lacking from K-12 and were important in some *in vivo* tests. Significant killing of rabbit peritoneal leucocytes by the α-haemolytic strain was also noted (Emödy *et al.*, 1980). Strain P673 was nephropathogenic in the mouse after intravenous injection; using this model, Waalwijk *et al.* (1982) associated loss of virulence with loss of the Hly plasmid; virulence was restored by reintroduction of the plasmid. Again K-12 was not virulent in this model when it acquired the Hly plasmid.

In a number of the genetic studies reported above, only a small proportion of the haemolytic strains transferred an Hly plasmid. Although in some cases this was probably due to the non-conjugative nature of the Hly plasmid, it was also possible that in some strains the Hly genes were chromosomal. Hull *et al.* (1982) failed to detect an Hly plasmid in a haemolytic strain, J96, isolated from a patient with pyelonephritis. However, in mating experiments with Hfr strains of *E. coli*, Hly was transferred and they concluded that the Hly genes were chromosomally determined in the region of the *ilv* gene cluster. Although Hly⁺ and Hly⁻ derivatives of the wild-type strain showed an association between haemolysis and virulence in a chick embryo test, a haemolytic K-12 transconjugant did not become virulent in this test. This again suggests that other factors are necessary.

To determine whether haemolysin or some other plasmid-coded product was associated with virulence, Welch *et al.* (1981) cloned an 11.7-kb DNA fragment coding for haemolysin production from strain J96, forming a hybrid plasmid pSF4000. K-12 carrying plasmid pSF4000 was not virulent in a rat intra-abdominal sepsis model; in contrast, after introduction of pSF4000 an avirulent non-haemolytic wild-type strain, J198, was lethal in this test. When haemolysin production was prevented by the insertion of the transposon *Tn*1 into the structural gene, pSF4000 no longer conferred virulence on J198, confirming that haemolysin was responsible for the increase in lethality. However, these workers noted that the time courses of infection with strain J198 which had been transformed with pSF4000 and J96 were not identical. Also, when other cloned haemolysin genes were introduced into J198 the behaviour of the derivatives in the same model was different, although the genes were apparently related by restriction enzyme analysis.

Chromosomal genes determining Hly were cloned from four *E. coli* strains belonging to O serogroups 4, 6, 18 and 75 (Berger *et al.*, 1982). Three cistrons were identified with functions similar to those described for Hly genes encoded by plasmid pHly152. Restriction enzyme analysis showed similarities between the four cloned chromosomal genes and pHly152 in the *hly*B and *hly*C regions, but all five clones differed in the *hly*A region.

Goebel *et al.* (1981) prepared gene probes carrying *hly*A, *hly*B or *hly*C, all derived from plasmid-determined Hly genes, and examined hybridization with the DNA of a strain carrying chromosomal Hly genes. There was considerable hybridization with *hly*B, less with *hly*C and little or none with *hly*A. These results and those of Berger *et al.* (1982) indicate conservation of *hly*B and *hly*C but diversity of the *hly*A genes which are considered to determine the precursor for active α-haemolysin.

ColV Plasmids

Smith (1974) reported that production of colicin V was a common property of strains causing bacteraemia in man and animals. Production of colicin V was plasmid-determined and acquisition of the plasmid by several strains of *E. coli* resulted in increased virulence when they were injected intravenously into chickens or intraperitoneally into mice. ColV plasmids of *E. coli* isolated from humans, pigs and chickens had similar properties (Smith, 1974). Wild-type strains of *E. coli* cured of the ColV plasmid were less virulent than the original strain carrying the ColV plasmid (Smith and Huggins, 1976; Smith, 1978). A crude colicin V preparation was not lethal when injected into chickens and the increase in virulence appeared to be due to increased survival in blood and body tissues of strains carrying the ColV plasmid. Other plasmids tested by Smith (1974), including ColI and ColE plasmids, did not increase virulence of their host strains.

Several surveys have since reported a higher incidence of colicin V production by strains isolated from extra-intestinal infections than from faecal isolates (Minshew *et al.*, 1978a,b; Davies *et al.*, 1981). Although colicin V was not toxic to animals, Ozanne *et al.* (1977) reported that it increased vascular permeability and also affected phagocytosis. However, recent genetic studies have shown that other products encoded on ColV plasmids have a role in increasing the pathogenicity of invasive strains. Quackenbush and Falkow (1979) inactivated the genes responsible for colicin synthesis on a non-conjugative ColV plasmid by insertion of the *Tn1* transposon. Strains carrying such altered plasmids were more lethal to mice when injected intraperitoneally than strains lacking the plasmid, showing that production of colicin V was not itself necessary for the increased virulence. Further studies have attempted to identify the plasmid-encoded properties which do enhance virulence.

Binns *et al.* (1979) associated the plasmid ColV,I-K94 with increased resistance to the bactericidal effects of serum. Fragments of the plasmid were cloned and the hybrid plasmids were tested for their effect on the virulence of a non-colicinogenic strain of O serogroup 78 (KH933) isolated from calf bacteraemia. Strains of this serogroup commonly cause generalized infection in domestic animals, and this *E. coli* strain was chosen as a suitable genetic background for the cloned ColV,I-K94 fragments rather than K-12. Cloned fragments coding for colicin production or immunity to colicin V did not increase virulence of strain KH933 when injected intramuscularly in day-old chicks. A DNA fragment which did not carry the genes for colicin production increased the virulence of strain KH933. Enhanced virulence was correlated with increased survival of strain KH933 in rabbit serum and a locus, *iss* of 5.3 kb, specifying increased survival was mapped at a site close to the genes for colicin V.

Another property conferred by ColV plasmids from *E. coli* strains isolated from human, pig and chicken infections was an iron uptake system (Williams 1979). This system enabled ColV$^+$ strains to grow in the presence of transferrin, which converted iron to a relatively unavailable form, whereas the growth of ColV$^-$ derivatives was markedly reduced. Mutations which affected the iron uptake mechanism in a ColV plasmid, ColV-K30, decreased the virulence of the strain when it was injected intraperitoneally in mice although colicin V synthesis was unaffected (Williams and Warner, 1980). Mutations affecting colicin synthesis had no effect either on virulence properties or iron uptake. The ColV,I-K94 plasmid did not possess the iron uptake system, indicating that ColV plasmids differ in the properties they encode (Williams, 1979). Interactions between the products of chromosomal and ColV plasmid genes in iron transport have been described by Braun *et al.* (1982) and Braun and Burkhardt (1982).

Another property conferred on *E. coli* strains by a ColV plasmid was described by Clancy and Savage (1981). They associated the presence of a ColV plasmid in strains of K-12 with the ability to adhere *in vitro* to discs of mouse intestinal tissue. Loss of the ColV plasmid was accompanied by a decrease in numbers of bacteria adhering and also loss of the sex fimbriae encoded by the plasmid. Whether such adhesive mechanisms play a role in pathogenicity *in vivo* remains to be established. The iron uptake mechanism and other functions mediated by ColV plasmids are discussed further in Chapter 7, as are chromosomally located systems involved in iron uptake.

Vir Plasmid

Smith (1974) reported the presence of a plasmid, which he termed Vir, in a strain of *E. coli* causing bacteraemia in a lamb; possession of the plasmid was associated with a specific surface antigen and synthesis of a heat-labile, non-dialysable toxin lethal to chickens, mice and rabbits when injected intravenously. The plasmid could be transferred at high frequency to strains of *E. coli, Salmonella typhimurium, S. typhi* and *Sh. sonnei*.

Lopez-Alvarez and Gyles (1980) found only 2 of 46 *E. coli* strains isolated from septicaemic disease of humans, chickens and calves possessed the Vir phenotype; both strains were from calves. They transferred a conjugative plasmid, pJL1, from one strain; it was an fi^- plasmid with a molecular weight of 92 $\times$ 10^6 and belonged to the F1V incompatibility group. Acquisition of pJL1 by K-12 resulted in production of the Vir antigen and of the chick lethal toxin. When this plasmid was transferred to a non-fimbriate *E. coli* strain, two types of fimbriae were produced; Lopez-Alvarez and Gyles (1980) concluded that those of one type were sex fimbriae and the second type were fimbriae responsible for the surface antigen associated with the Vir phenotype. The Vir antigen was not produced in bacteria grown at 20°C; similarly, synthesis of other fimbrial antigens such as K88, K99 and CFA/I does not occur at this temperature.

Another Vir plasmid pJL2, from a strain of O serogroup 78 isolated from bovine septicaemia, was transferred to K-12 only after mobilization (Lopez-Alvarez *et al.*, 1980); this plasmid with a molecular weight of 92 $\times$ 10^6 was fi^+. Restriction enzyme analysis showed that pJL1 and pJL2 were related but not identical. The Vir phenotype had not been transferable from another strain of O serogroup 78 characterized by Lopez-Alvarez *et al.* (1980), either directly or by mobilization. No plasmid DNA was detected in this strain, so it is probable that the Vir phenotype can be encoded by chromosomal or plasmid genes.

Adhesive properties associated with the Vir plasmid were reported by Morris *et al.* (1982). A strain of O serogroup 15 isolated from lamb septicaemia was able to adhere *in vitro* to calf epithelial tissue in the presence of mannose. A Vir plasmid was transferred from this strain to K-12 and to a strain of O serogroup 9; transconjugants which acquired Vir also expressed the adhesive ability. The role of the Vir phenotype in pathogenicity is discussed further in Chapter 4.

K1 Capsular Antigen

In 1974 Robbins *et al.* reported that 84% of *E. coli* strains causing neonatal meningitis in the United States possessed the capsular polysaccharide antigen K1. Similar surveys in other countries have supported the idea that the K1 antigen enhanced the ability of *E. coli* strains to invade the central nervous system of the neonate (review by Schiffer *et al.*, 1976). K1-positive strains have also been associated with neonatal septicaemia and urinary tract infections of children (Sarff *et al.*, 1975; Kaijser *et al.*, 1977). A number of functions have been ascribed to the K1 antigen in virulence (see Chapter 5) and genetic studies may elucidate these. Experiments by Ørskov *et al.* (1976) identified the genes controlling K1 biosynthesis, *kps*A, as chromosomal, mapping near *ser*A. A large chromosome fragment of 35 kb carrying these genes was cloned by Silver *et al.* (1981) from *E. coli* **O**18:**K**1:**H**7; the cloned genes functioned in K-12, resulting in the production of capsular material identical chemically and antigenically to that of the wild-type K1$^+$ strain. Although possession of the K1 genes increased the serum resistance of the K-12 derivative, it lacked the invasiveness of the wild-type strain in an infant rat test; it was concluded that other properties lacking from K-12 were necessary for invasiveness.

Timmis *et al.* (1981) also used cloning techniques to transfer the K1 biosynthesis genes to K-12 and noted significantly increased serum resistance to K1$^+$ derivatives of K-12. An association between K1 and serum resistance had been reported by Gemski *et al.* (1980) in a study of rough, K1$^+$ strains isolated from blood samples. Variants lacking the K1 antigen, selected by their resistance to K1-specific phages (Gross *et al.*, 1977), were found to be significantly more sensitive to human serum than wild-type strains. Mating experiments with Hfr donors showed that an increase in serum resistance was conferred by the presence of the K1 antigen but not by other K antigens tested, such as K27 (Opal *et al.*, 1982).

The virulence of K1$^+$ wild-type strains, which also carried a ColV plasmid, was compared to that of derivatives lacking one or both properties (Cabello, 1979; Smith and Huggins, 1980). Loss of either resulted in decreased virulence in animal models although loss of the K1 antigen had a greater effect. Loss of the ColV plasmid from the K1-negative derivative did not result in a further decrease in virulence. Loss of the K1 antigen but not of the ColV plasmid was associated with a decrease in serum resistance in both studies. Mutations affecting the somatic antigen O18ac also decreased virulence and serum resistance (Smith and Huggins, 1980). Another derivative of the wild-type strain was obtained after mutagen treatment which had a markedly reduced virulence, although remaining **O**18ac$^+$, K1$^+$ and ColV$^+$; this indicated the role of other as yet undefined virulence properties (Smith and Huggins, 1980). This was also concluded from studies in which one or more of these properties (O18ac, K1 or ColV) were transferred to K-12 and, although each affected the lethality of K-12 to varying degrees, acquisition of all three did not result in virulence equivalent to that of the wild-type strain (Smith and Huggins, 1980).

Other Factors Conferring Serum Resistance

Survival in host tissues and body fluids is clearly important for invasive bacteria, and a number of the virulence factors already described, such as ColV and K1, have been implicated in increasing the serum resistance of the bacterium. There have been many studies to determine which components on the bacterial surface participate in serum resistance and both qualitative and quantitative changes in the lipopolysaccharides, capsules and membrane proteins have been considered important. Often mutants, with decreased or increased serum resistance, have been compared with the wild-type strains. However, the interrelationships of these cell surface components are still uncertain, as discussed in Chapters 7 and 25.

Although plasmid-mediated drug resistance is important when considering the treatment of bacterial disease, R plasmids have not been considered in this symposium as virulence factors. Nevertheless, R plasmids may directly alter the pathogenic properties of *E. coli* by increasing serum resistance. Reynard and Beck (1976) reported that K-12 strains which harboured the F-like plasmids R1 or R100 were more resistant to rabbit serum than strains lacking the plasmids. Both R plasmids conferred resistance to chloramphenicol, streptomycin and sulphonamides; in addition, R100 conferred tetracycline resistance and R1 resistance to ampicillin and kanamycin. There has been some disagreement on the extent to which R plasmids alter the serum sensitivity of K-12 strains but it seems probable that the effect is strain- and serum-dependent (Reynard and Beck, 1976; Fietta *et al.*, 1977; Reynard *et al.*, 1978; Taylor and Hughes, 1978; Ogata and Levine, 1980).

The effect of R plasmids in increasing serum resistance is probably more clearly seen in strains which, unlike K-12, possess complete lipopolysaccharide O side chains. For example, Taylor and Hughes (1978) showed that R1 and R100 increased the survival in serum of a wild-type strain, which had been cured of the R plasmids it originally carried. To confirm the role of other cell surface components in serum resistance, Taylor and Robinson (1980) used a recipient strain with mutations in chromosomal loci affecting both O and K antigen biosynthesis, which had been derived from a non-motile strain of *E. coli* O8:K27. In Hfr mating, acquisition of the ability to synthesize K27 caused no increase in serum resistance; when the ability to synthesize O8 side chains was acquired (with or without K27) the strain was subject to delayed killing by serum. Plasmids R1 and

R100 were introduced into all three classes of recombinants and increased survival in serum only when the O8 antigen was synthesized.

Several plasmids belonging to different incompatibility groups and with various drug resistance patterns have been shown to increase serum resistance (Fietta *et al.*, 1977; Reynard *et al.*, 1978). The conjugative plasmid R6-5 which, like R1 and R100, belongs to the FII incompatibility group, increased the resistance to human serum of both K-12 and a wild-type strain (Moll *et al.*, 1980); it conferred resistance to chloramphenicol, kanamycin, streptomycin and sulphonamides. By cloning techniques, the gene conferring serum resistance was located on a segment which controlled conjugational transfer and was identified as *tra*T, one of the two surface exclusion genes of R6-5. The *tra*T gene product was shown to be an outer-membrane protein with a molecular weight of 25,000 (Moll *et al.*, 1980). Ogata *et al.* (1982) have isolated the serum resistance gene from R100 by gene cloning and shown by nucleotide sequence analysis that the composition of the predicted protein synthesized would be almost identical to the *tra*T protein from the F factor.

Plasmids R1, R100, R6-5 and ColV-K94 are all F-like plasmids and have been shown to share extensive DNA homology in heteroduplex studies (Sharp *et al.*, 1973). However the *iss* gene, conferring serum resistance in ColV,I-K94, mapped outside the transfer genes of that plasmid. Binns *et al.* (1982) showed that *iss* genes conferred a similar level of serum resistance to *tra*T of plasmid R100, and when both were present there was no additive effect. Although they may act by a similar mechanism, the *iss* and *tra*T genes did not hybridize with each other and their gene products were immunologically distinct (Binns *et al.*, 1982; Ogata *et al.*, 1982).

Fimbriae of E. coli from Urinary Tract Infections Causing Mannose-Resistant Haemagglutination

Production of fimbriae by ETEC was shown to be important in their adhesion to intestinal epithelial cells and therefore their ability to cause diarrhoea. Such fimbriae were often recognized by their ability to give mannose-resistant haemagglutination (MR-HA) of erythrocytes of certain animal species, but were antigenically diverse. Strains of *E. coli* isolated from urinary tract infections commonly produced fimbriae causing MR-HA of human erythrocytes, as shown in a number of studies (Minshew *et al.*, 1978a; Ljungh *et al.*, 1979; Hagberg *et al.*, 1981; Kallenius *et al.*, 1981). These fimbriate strains were able to adhere to uroepithelial cells in the presence of mannose. The role of such fimbriae in pathogenicity is discussed in Chapter 4. Most of the fimbrial antigens of ETEC have been shown to be plasmid-mediated; in contrast, in studies of strains from urinary tract infection the genes encoding fimbriae were chromosomal.

Strain J96 of *E. coli* **O**4, which had been isolated from a case of pyelonephritis, produced both fimbriae giving MR-HA of human erythrocytes and type 1 fimbriae (associated with mannose-sensitive haemagglutination of guinea-pig erythrocytes). By cloning techniques, fragments of whole cell DNA from strain J96 were introduced into a non-fimbriate strain of K-12 (Hull *et al.*, 1981). K-12 derivatives which gave MR-HA were fimbriate and adhered to primary monkey kidney cells in the presence of mannose. Hull *et al.* (1981) concluded that the genes for MR-HA were chromosomal as fragments from the recombinants were not represented among the restriction enzyme fragments obtained with plasmid DNA from the parent strain. Recombinants which produced type 1 fimbriae were also obtained, and it was shown that the two fimbrial types were genetically and functionally distinct.

Chromosomal genes encoding MR-HA of human erythrocytes have been cloned from strains of O serogroup 6 isolated from human urinary tract infections (Clegg, 1982; Berger *et al.*, 1982). By further restriction enzyme analysis, the genes for MR-HA and fimbriae were located by Clegg (1982) on a fragment with a maximum size of 6.9 kb on a plasmid pDC5 of 12.3 kb. The fimbriae produced by the originally non-fimbriate recipient when these genes were introduced were agglutinated by a specific serum prepared against the fimbriae of the wild-type strain. The fimbriae were not expressed at 18°C in either the recipient or the donor strain.

Immunological studies of the fimbriae and identification of the fimbrial receptors on the host cell have shown that a number of different types of fimbriae are responsible for MR-HA in *E. coli* strains from extra-intestinal sources. Extension of the genetic studies of these adhesive factors should assist in showing their role in pathogenicity and also relationships between them.

Conclusions

Genetic studies of *E. coli* virulence have provided a great deal of information on the genetic organization and control of the different virulence properties and some of this knowledge is now being applied to the development of new tests for pathogenicity, to epidemiological studies and to production of potential vaccine strains. Plasmid, phage and chromosomal genes code for functions associated with pathogenicity in *E. coli*. A single phenotypic property such as haemolysin production is plasmid-mediated in many strains but in others the genes are chromosomally located. With other properties such as production of ST and LT by ETEC the genes have been shown to be plasmid-specified apart from one strain in which the LT genes were phage-determined. This situation contrasts with that found in *Vibrio cholerae* strains in which the cholera toxin (CT) genes are on the chromosome even though CT and LT are very similar in structure and function.

Although the genes encoding a specific virulence factor may be plasmid-determined, the properties of the host strain are very important in the full expression of virulence. This has been demonstrated with several systems such as strains of K-12 carrying plasmids coding for serum resistance or invasiveness. When animal models are used for testing, K-12 strains often show much lower virulence than the wild-type strains. In other examples, some plasmid-determined functions are expressed only in certain bacterial hosts; for example, production of CFA/II antigens CS1 and CS2 was not detected in K-12 and many wild-type strains. Therefore, the choice of *E. coli* strain for genetic experiments is very important and K-12 is often not particularly suitable even though it is usually the first choice from a genetic viewpoint.

Studies of virulence in *E. coli* have shown that several factors are involved in the virulence of a particular strain. For example, enterotoxigenic *E. coli* must be able to adhere to and colonize the intestine as well as produce enterotoxins. Thus, the control of these properties in such organisms must involve several different genes located on plasmids and the chromosome. It is interesting in this respect that in ETEC isolated from humans, plasmids code for both adhesion systems, such as CFA/I and CFA/II, and enterotoxin production. This is an efficient adaptation of these organisms to their environment in the human host.

Genetic and molecular studies of plasmids coding for properties contributing to virulence in *E. coli* have been used to study the distribution of particular plasmid-determined characters. For example, the plasmids coding for production of CFA/I and ST identified in several serotypes isolated in different parts of the world were very closely related. In addition to examining the distribution of certain plasmids, genetic studies have been applied to investigations of the epidemiology of particular pathogenic strains or clones. Possible *in vivo* transfer of plasmids, coding for properties associated with pathogenicity, between *E. coli* strains should also be considered in epidemiological studies of plasmids. Transfer of an Ent plasmid which also conferred drug resistance was detected in pigs and the presence of antibiotic in the feed did not affect the transfer (Gyles *et al.*, 1978). Experiments by Williams (1977) demonstrated transfer of a ColV plasmid from ingested strains of *E. coli* to resident bacteria in the human intestine in the absence of selective pressure.

Mutants lacking a specific property have proved useful in studies of pathogenicity because they can be compared with the wild-type strains and mutations affecting pathogenicity can be identified. In recent years several different genetic methods for the isolation of mutants have been developed, for example the inactivation of specific genes by insertion of drug resistance transposons.

Gene cloning has been particularly useful in the detailed analysis of the organization and function of the genes determining properties associated with pathogenicity. Such studies have led to the development of highly specific gene probes for use in hybridization experiments for detection of these specific sequences in wild-type strains. It is hoped that this type of test, together with immunological assays, will result in replacing some of the present animal tests and other laborious assays. An obvious limitation in the use of probes is that they will detect strains in which the genes are present but not expressed. Strains carrying plasmids with cloned genes have been used in the purification of gene products such as enterotoxins. The vector plasmids are usually present in multiple copies so that there is often increased synthesis of the required product compared to that made by the original strain. Analysis of products determined by cloned genes in minicells has been applied in the identification of proteins involved in a property associated with pathogenicity.

Many genetic studies have the aim of developing vaccine strains for the prevention of disease caused by *E. coli*. Strains carrying plasmids with cloned K88 and K99 genes have been used for the purification of fimbriae as a vaccine preparation in young animals. The amounts of antigen produced were considerably greater than those made by the corresponding wild-type strains and there was also the advantage that the strains with the cloned genes were non-toxigenic.

A mutant of *S. typhi* lacking galactose epimerase has been isolated and used as a living attenuated oral typhoid vaccine; it was safe and highly effective in a field trial (Germanier and Fürer, 1975; Wahdan *et al.*, 1980). Formal *et al.* (1981) proposed a further development for the use of this strain. A plasmid coding for production of the *Sh. sonnei* form I antigen was transferred into the *S. typhi galE* strain and the derivative produced the form I antigen of *Sh. sonnei*. It was suggested that the *S. typhi galE* mutant could be a carrier strain for other potentially protective antigenic determinants such as *E. coli* enterotoxins and adhesins.

The application of genetic studies of virulence in *E. coli* and other organisms clearly has great potential for the prevention and control of disease.

Acknowledgments

We thank Dr. Moyra McConnell and Dr. Geraldine Willshaw for helpful comments and Mrs. Janet Vaughan for preparing the manuscript.

References

Baldini, M. M., Kaper, J. B., Levine, M. M., Candy, D. C. A. and Moon, H. W. (1983). Plasmid-mediated adhesion in enteropathogenic *Escherichia coli*. *Journal of Pediatric Gastroenterology and Nutrition* **2**, 534–538.

Berger, H., Hacker, J., Juarez, A., Hughes, C. and Goebel, W. (1982). Cloning of the chromosomal determinants encoding hemolysin production and mannose-resistant hemagglutination in *Escherichia coli*. *Journal of Bacteriology* **152**, 1241–1247.

Binns, M. M., Davies, D. L. and Hardy, K. G. (1979). Cloned fragments of the plasmid ColV,I-K94 specifying virulence and serum resistance. *Nature (London)* **279**, 778–781.

Binns, M. M., Mayden, J. and Levine, R. P. (1982). Further characterization of complement resistance conferred on *Escherichia coli* by the plasmid genes *tra*T of R100 and *iss* of ColV,I-K94. *Infection and Immunity* **35**, 654–659.

Boedeker, E. C. (1983). *In* "Summary of a Workshop on enteropathogenic *Escherichia coli* (EPEC)" (Eds. R. Edelman and M. M. Levine), *Journal of Infectious Diseases* **147**, 1108–1118.

Bradley, D. E. (1981). Conjugative pili of plasmids in *Escherichia coli* K12 and *Pseudomonas* species. *In* "Molecular Biology, Pathogenicity and Ecology of Bacterial Plasmids" (Eds. S. B. Levy, R. C. Clowes and E. L. Koenig), pp. 217–226. Plenum, New York.

Bramucci, M. G., Twiddy, E. M., Baine, W. B. and Holmes, R. K. (1981). Isolation and characterization of hypertoxinogenic (*htx*) mutants of *Escherichia coli* KL320 (pCG86). *Infection and Immunity* **32**, 1034–1044.

Braun, V. and Burkhardt, R. (1982). Regulation of the ColV plasmid-determined iron (III)–aerobactin transport system in *Escherichia coli*. *Journal of Bacteriology* **152**, 223–231.

Braun, V., Burkhardt, R., Schneider, R. and Zimmermann, L. (1982). Chromosomal genes for ColV plasmid-determined iron (III)–aerobactin transport in *Escherichia coli*. *Journal of Bacteriology* **151**, 553–559.

Brooks, H. J. L., O'Grady, F., McSherry, M. A. and Cattell, W. R. (1980). Uropathogenic properties of *Escherichia coli* in recurrent urinary-tract infection. *Journal of Medical Microbiology* **13**, 57–68.

Burgess, M. N., Bywater, R. J., Cowley, C. M., Mullan, N. A. and Newsome, P. M. (1978). Biological evaluation of a methanol soluble heat-stable *Escherichia coli* enterotoxin in infant mice, pigs, rabbits and calves. *Infection and Immunity* **21**, 526–531.

Cabello, F. C. (1979). Determinants of pathogenicity of *Escherichia coli* K1. *In* "Plasmids of Medical, Environmental and Commercial Importance" (Eds. K. N. Timmis and A. Pühler), pp. 155–166. Elsevier/North Holland Biomedical Press, Amsterdam.

Candy, D. C. A., Chadwick, J., Leung, T., Phillips, A., Harries, J. T. and Marshall, W. C. (1978). Adhesion of Enterobacteriaceae to buccal epithelial cells. *Lancet* **2**, 1157–1158.

Cantey, J. R. and Blake, R. K. (1977). Diarrhea due to *Escherichia coli* in the rabbit: A novel mechanism. *Journal of Infectious Diseases* **135**, 454–462.

Cantey, J. R., Lushbaugh, W. B. and Inman, L. R. (1981). Attachment of bacteria to intestinal epithelial cells in diarrhea caused by *Escherichia coli* strain RDEC-1 in the rabbit: Stages and role of capsule. *Journal of Infectious Diseases* **143**, 219–230.

Chiarini, A. and Giammanco, A. (1978). L'adesivitá quale momento patogenetico nei batteri gram-negativio. Interazioni *in vitro* fra stipiti di *Escherichia coli* e colture di cellule HeLa. *Annali Sclavo* **20**, 337–349.

Clancy, J. and Savage, D. C. (1981). Another colicin V phenotype: *In vitro* adhesion of *Escherichia coli* to mouse intestinal epithelium. *Infection and Immunity* **32**, 343–352.

Clegg, S. (1982). Cloning of genes determining the production of mannose-resistant fimbriae in a uropathogenic strain of *Escherichia coli* belonging to serogroup O6. *Infection and Immunity* **38**, 739–744.

Cravioto, A., Gross, R. J., Scotland, S. M. and Rowe, B. (1979a). Mannose-resistant haemagglutination of human erythrocytes by strains of *Escherichia coli* from extraintestinal sources: Lack of correlation with colonisation factor antigen (CFA/I). *FEMS Microbiology Letters* **6**, 41–44.

Cravioto, A., Gross, R. J., Scotland, S. M. and Rowe, B. (1979b). An adhesive factor found in strains of *Escherichia coli* belonging to the traditional infantile enteropathogenic serotypes. *Current Microbiology* **3**, 95–99.

Cravioto, A., Scotland, S. M. and Rowe, B. (1982). Hemagglutination activity and colonization factor antigens I and II in enterotoxigenic and non-enterotoxigenic strains of *Escherichia coli* isolated from humans. *Infection and Immunity* **36**, 189–197.

Dallas, W. S. and Falkow, S. (1979). The molecular nature of heat-labile enterotoxin (LT) of *Escherichia coli*. *Nature (London)* **277**, 406–407.

Dallas, W. S. and Falkow, S. (1980). Amino acid sequence homology between cholera toxin and the *Escherichia coli* heat-labile toxin. *Nature (London)* **288**, 499–501.

Dallas, W. S., Gill, D. M. and Falkow, S. (1979a). Cistrons encoding *Escherichia coli* heat-labile toxin. *Journal of Bacteriology* **139**, 850–858.

Dallas, W. S., Moseley, S. and Falkow, S. (1979b). The characterization of an *Escherichia coli* plasmid determinant that encodes for the production of a heat-labile enterotoxin. *In* "Plasmids of Medical, Environmental and Commercial Importance" (Eds. K. N. Timmis and A. Pühler), pp. 113–122. Elsevier/North Holland Biomedical Press, Amsterdam.

Davies, D. L., Falkiner, F. R. and Hardy, K. G. (1981). Colicin production by clinical isolates of *Escherichia coli*. *Infection and Immunity* **31**, 574–579.

De, S. N., Bhattacharya, K. and Sarkar, J. K. (1956). A study of the pathogenicity of strains of *Bacterium coli* from acute and chronic enteritis. *Journal of Pathology and Bacteriology* **71**, 201–209.

Dean, A. G., Ching, Y.-C., Williams, R. G. and Harden, L. B. (1972). Test for *Escherichia coli* enterotoxin using infant mice. Application in a study of diarrhoea in children in Honolulu. *Journal of Infectious Diseases* **125**, 407–411.

de Graaf, F. K. and Mooi, F. R. (1982). The K88 and K99 plasmids of enterotoxigenic *Escherichia coli* strains. *In* "Resistance and Pathogenic Plasmids" (Eds. P. Pohl and J. Leunen), pp. 191–203. C.E.C. Seminar 1981, National Institute for Veterinary Research, Brussels.

de Graaf, F. K., Wientjes, F. B. and Klaasen-Boor, P. (1980). Production of K99 antigen by enterotoxigenic *Escherichia coli* strains of antigen groups O8, O9, O20 and O101 grown at different conditions. *Infection and Immunity* **27**, 216–221.

De La Cruz, F., Zabala, J. C. and Ortiz, J. M. (1979). Incompatibility among α-hemolytic plasmids studied after inactivation of the α-hemolysin gene by transposition of Tn802. *Plasmid* **2**, 507–519.

De La Cruz, F., Müller, D., Ortiz, J. M. and Goebel, W. (1980a). Hemolysin determinant common to *Escherichia coli* hemolytic plasmids of different incompatibility groups. *Journal of Bacteriology* **143**, 825–833.

De La Cruz, F., Zabala, J. C. and Ortiz, J. M. (1980b). The molecular relatedness among α-hemolytic plasmids from various incompatibility groups. *Plasmid* **4**, 76–81.

De Wilde, M., Yaebaert, M. and Harford, N. (1981). DNA sequence of the ST_{A2} enterotoxin gene from an *Escherichia coli* strain of human origin. *In* "Molecular Biology, Pathogenicity and Ecology of Bacterial Plasmids" (Eds. S. B. Levy, R. C. Clowes and E. L. Koenig), p. 596. Plenum, New York.

Dougan, G., Dowd, G. and Kehoe, M. (1983a). Organization of K88ac-encoded polypeptides in the *Escherichia coli* cell envelope: Use of minicells and outer membrane protein mutants for studying assembly of pili. *Journal of Bacteriology* **153**, 364–370.

Dougan, G., Kehoe, M., Dowd, G., Sellwood, R. and Winther, M. (1983b). Studies on the expression and organisation of the K88ac adherence antigen. *Developments in Biological Standardization* **53**, 183–187.

Echeverria, P. and Murphy, J. R. (1980). Enterotoxigenic *Escherichia coli* carrying plasmids coding for antibiotic resistance and enterotoxin production. *Journal of Infectious Diseases* **142**, 273–278.

Echeverria, P., Verhaert, L., Ulyangco, C. V., Komalarini, S., Ho, M. T., Ørskov, F. and Ørskov, I. (1978). Antimicrobial resistance and enterotoxin production among isolates of *Escherichia coli* in the Far East. *Lancet* **2**, 589–592.

Echeverria, P., Seriwatana, J., Chityothin, O., Chaicumpa, W. and Tirapat, C. (1982). Detection of enterotoxigenic *Escherichia coli* in water by filter hybridisation with three enterotoxin gene probes. *Journal of Clinical Microbiology* **16**, 1086–1090.

Edwards, P. R. and Ewing, W. H. (1972). "Identification of Enterobacteriaceae." Burgess, Minneapolis, Minnesota.

Elwell, L. P. and Shipley, P. L. (1980). Plasmid-mediated factors associated with virulence of bacteria to animals. *Annual Review of Microbiology* **34**, 465–496.

Emödy, L., Pal, T., Safonova, N. V., Kuch, B. and Golutva, N. K. (1980). Alpha-haemolysin: An additive virulence factor in *Escherichia coli*. *Acta Microbiologica Academiae Scientiarum Hungaricae* **27**, 333–342.

Evans, D. G. and Evans, D. J. (1978). New surface-associated heat-labile colonization factor antigen (CFA/II) produced by enterotoxingenic *Escherichia coli* of serogroups O6 and O8. *Infection and Immunity* **21**, 638–647.

Evans, D. G., Silver, R. P., Evans, D. J., Chase, D. G. and Gorbach, S. L. (1975). Plasmid-controlled colonization factor associated with virulence in *Escherichia coli* enterotoxigenic for humans. *Infection and Immunity* **12**, 656–667.

Evans, D. J., Evans, D. G. and DuPont, H. L. (1979). Hemagglutination patterns of enterotoxigenic and enteropathogenic *Escherichia coli* determined with human, bovine, chicken and guinea-pig erythrocytes in the presence and absence of mannose. *Infection and Immunity* **23**, 336–346.

Farkas-Himsley, H., Jessop, J. and Corey, P. (1978). Development and standardization of a cytotoxic micro-assay for detection of enterotoxins: Survey of enterotoxins from *Escherichia coli* of infant origin. *Microbios* **18**, 195–212.

Fietta, A., Romero, E. and Siccardi, A. G. (1977). Effect of some R factors on the sensitivity of rough *Enterobacteriaceae* to human serum. *Infection and Immunity* **18**, 278–282.

Formal, S. B. and Hornick, R. B. (1978). Invasive *Escherichia coli*. *Journal of Infectious Diseases* **137**, 641–644.

Formal, S. B., Baron, L. S., Kopecko, D. J., Washington, O., Powell, C. and Life, C. A. (1981). Construction of a potential bivalent vaccine strain: Introduction of *Shigella sonnei* form I antigen genes into the *galE Salmonella typhi* Ty21a typhoid vaccine strain. *Infection and Immunity* **34**, 746–750.

Franklin, A. and Möllby, R. (1981). Different plasmids coding for heat-stable enterotoxins in porcine *Escherichia coli* strains of O-group 149. *FEMS Microbiology Letters* **11**, 299–302.

Franklin, A., Söderlind, O. and Möllby, R. (1981). Plasmids coding for enterotoxins, K88 antigen and colicins in porcine *Escherichia coli* strains of O-group 149. *Medical Microbiology and Immunology* **179**, 63–72.

Gaastra, W. and de Graaf, F. K. (1982). Host-specific fimbrial adhesins of non-invasive enterotoxigenic *Escherichia coli* strains. *Microbiological Reviews* **46**, 129–161.

Gaastra, W., Mooi, F. R., Stuitje, A. R. and de Graaf, F. K. (1981). The nucleotide sequence of the gene encoding the K88ab protein subunit of porcine enterotoxigenic *Escherichia coli*. *FEMS Microbiology Letters* **12**, 41–46.

Gaastra, W., Klemm, P. and de Graaf, F. K. (1983). The nucleotide sequence of the K88ad protein subunit of porcine enterotoxigenic *Escherichia coli*. *FEMS Microbiology Letters* **18**, 177–183.

Geary, S. J., Marchlewicz, B. A. and Finkelstein, R. A. (1982). Comparison of heat-labile enterotoxins from porcine and human strains of *Escherichia coli*. *Infection and Immunity* **36**, 215–220.

Gemski, P., Cross, A. S. and Sadoff, J. C. (1980). K1 antigen-associated resistance to the bactericidal activity of serum. *FEMS Microbiology Letters* **9**, 193–197.

Germanier, R. and Fürer, E. (1975). Isolation and characterization of *galE* mutant Ty21a of *Salmonella typhi:* A candidate strain for a live, oral, typhoid vaccine. *Journal of Infectious Diseases* **131**, 553–558.

Goebel, W. and Hedgpeth, J. (1982). Cloning and functional characterization of the plasmid-encoded hemolysin determinant of *Escherichia coli*. *Journal of Bacteriology* **151**, 1290–1298.

Goebel, W. and Schrempf, H. (1971). Isolation and characterization of supercoiled circular deoxyribonucleic acid from beta-hemolytic strains of *Escherichia coli*. *Journal of Bacteriology* **106**, 311–317.

Goebel, W., Royer-Pokora, B., Lindenmaier, N. and Bujard, H. (1974). Plasmids controlling synthesis of hemolysin in *Escherichia coli:* Molecular properties. *Journal of Bacteriology* **118**, 964–973.

Goebel, W., Noegel, A., Rdest, U., Müller, D. and Hughes, C. (1981). Epidemiology and genetics of hemolysin formation in *Escherichia coli. In* "Molecular Biology, Pathogenicity and Ecology of Bacterial Plasmids" (Eds. S. B. Levy, R. C. Clowes and E. L. Koenig), pp. 43–50. Plenum, New York.

Goldschmidt, M. C. and DuPont, H. L. (1976). Enteropathogenic *Escherichia coli:* Lack of correlation of serotype with pathogenicity. *Journal of Infectious Diseases* **133**, 153–156.

Green, C. P. and Thomas, V. L. (1981). Hemagglutination of human type O erythrocytes, hemolysin production, and serogrouping of *Escherichia coli* isolates from patients with acute pyelonephritis, cystitis, and asymptomatic bacteriuria. *Infection and Immunity* **31**, 309–315.

Gross, R. J., Scotland, S. M. and Rowe, B. (1976). Enterotoxin testing of *Escherichia coli* causing epidemic infantile enteritis in the U.K. *Lancet* **1**, 629–631.

Gross, R. J., Cheasty, T. and Rowe, B. (1977). Isolation of bacteriophages specific for the K1 polysaccharide antigen of *Escherichia coli. Journal of Clinical Microbiology* **6**, 548–550.

Gross, R. J., Cravioto, A., Scotland, S. M., Cheasty, T. and Rowe, B. (1978). The occurrence of colonisation factor (CF) in enterotoxigenic *Escherichia coli. FEMS Microbiology Letters* **3**, 231–233.

Gurwith, M. J., Wiseman, D. A. and Chow, P. (1977). Clinical and laboratory assessment of the pathogenicity of serotyped enteropathogenic *Escherichia coli. Journal of Infectious Diseases* **135**, 736–743.

Gyles, C. L. (1979). Limitations of the infant mouse test for *Escherichia coli* heat-stable enterotoxin. *Canadian Journal of Comparative Medicine* **43**, 371–379.

Gyles, C. L. and Barnum, D. A. (1969). A heat-labile enterotoxin from strains of *Escherichia coli* enteropathogenic for pigs. *Journal of Infectious Diseases* **120**, 419–426.

Gyles, C. L., So, M. and Falkow, S. (1974). The enterotoxin plasmids of *Escherichia coli. Journal of Infectious Diseases* **130**, 40–49.

Gyles, C. L., Palchaudhuri, S. and Maas, W. K. (1977). Naturally occurring plasmid carrying genes for enterotoxin production and drug resistance. *Science* **198**, 198–199.

Gyles, C. L., Falkow, S. and Rollins, L. (1978). *In vivo* transfer of an *Escherichia coli* enterotoxin plasmid possessing genes for drug resistance. *American Journal of Veterinary Research* **39**, 1438–1441.

Hagberg, L., Jodal, V., Korhonen, T. K., Lidin-Janson, G., Lindberg, V. and Edén, C. S. (1981). Adhesion, hemagglutination and virulence of *Escherichia coli* causing urinary tract infections. *Infection and Immunity* **31**, 564–570.

Harford, N., De Wilde, M. and Cabezon, T. (1981). Cloning of two distinct but related ST enterotoxin genes from porcine and human strains of *Escherichia coli. In* "Molecular Biology, Pathogenicity and Ecology of Bacterial Plasmids" (Eds. S. B. Levy, R. C. Clowes and E. L. Koenig), p. 611. Plenum, New York.

Harnett, N. M. and Gyles, C. L. (1982). Enterotoxin plasmids in bovine and porcine atypical enteropathogenic *Escherichia coli. In* "Resistance and Pathogenic Plasmids" (Eds. P. Pohl and J. Leunen), pp. 241–273. CEC Seminar 1981, National Institute for Veterinary Research, Brussels.

Harris, J. R., Wachsmuth, I. K., Davis, B. R. and Cohen, M. L. (1982). High-molecular-weight plasmid correlates with *Escherichia coli* enteroinvasiveness. *Infection and Immunity* **37**, 1295–1298.

Honda, T., Tsuji, T., Takeda, Y. and Miwatani, T. (1981). Immunological nonidentity of heat-labile enterotoxins from human and porcine enterotoxigenic *Escherichia coli. Infection and Immunity* **34**, 337–340.

Hull, R. A., Gill, R. E., Hsu, P., Minshew, B. H. and Falkow, S. (1981). Construction and expression of recombinant plasmids encoding type 1 or D-mannose resistant pili from a urinary tract infection *Escherichia coli* isolate. *Infection and Immunity* **33**, 933–938.

Hull, S. I., Hull, R. A., Minshew, B. H. and Falkow, S. (1982). Genetics of hemolysin of *Escherichia coli*. *Journal of Bacteriology* **151**, 1006–1012.

Jacob, A. E., Shapiro, J. A., Yamamoto, L., Smith, D. I., Cohen, S. N. and Berg, D. (1977). Plasmids studied in *Escherichia coli* and other enteric bacteria. In "DNA Insertion Elements, Plasmids and Episomes" (Eds. A. I. Bukhari, J. A. Shapiro and S. L. Adhya), pp. 607–704. Cold Spring Harbor Laboratory, Cold Spring Harbor, New York.

Johnson, W. M., Lior, H. and Bezanson, G. S. (1983). Cytotoxic *Escherichia coli* O157:H7 associated with haemorrhagic colitis in Canada. *Lancet* **1**, 76.

Kaijser, B., Hanson, L. A., Jodal, V., Lidin-Janson, G. and Robbins, J. B. (1977). Frequency of *E. coli* K antigens in urinary-tract infections in children. *Lancet* **1**, 663–664.

Kallenius, G., Möllby, R., Svenson, S. B., Helin, I., Hultberg, H., Cedergren, B. and Winberg, J. (1981). Occurrence of P-fimbriated *Escherichia coli* in urinary tract infections. *Lancet* **2**, 1369–1372.

Kehoe, M., Sellwood, R., Shipley, P. and Dougan, G. (1981). Genetic analysis of K88-mediated adhesion of enterotoxigenic *Escherichia coli*. *Nature (London)* **291**, 122–126.

Kehoe, M., Winther, M., Dowd, G., Morrissey, P. and Dougan, G. (1982). Nucleotide sequence homology between the K88 and the K99 adhesion system of enterotoxigenic *Escherichia coli*. *FEMS Microbiology Letters* **14**, 129–132.

Kehoe, M., Winther, M. and Dougan, G. (1983). Studies on the expression of a cloned K88ac adhesion antigen determinant: Identification of a new adhesion cistron and role of a vector encoded promoter. *Journal of Bacteriology* **155**, 1071–1077.

Keusch, G. T. and Donta, S. T. (1975). Classification of enterotoxins on the basis of activity in cell culture. *Journal of Infectious Diseases* **131**, 58–63.

Klemm, P. (1982). Primary structure of the CFA/I fimbrial protein from human enterotoxigenic *Escherichia coli* strains. *European Journal of Biochemistry* **124**, 339–348.

Klipstein, F. A., Rowe, B., Engert, R. F., Short, H. B. and Gross, R. J. (1978). Enterotoxigenicity of enteropathogenic serotypes of *Escherichia coli* isolated from infants with epidemic diarrhea. *Infection and Immunity* **21**, 171–178.

Konowalchuk, J., Speirs, J. I. and Stavric, S. (1977). Vero response to a cytotoxin of *Escherichia coli*. *Infection and Immunity* **18**, 775–779.

Konowalchuk, J., Dickie, N., Stavric, S. and Speirs, J. I. (1978). Properties of an *Escherichia coli* cytotoxin. *Infection and Immunity* **20**, 575–577.

Kopecko, D. J., Washington, O. and Formal, S. B. (1980). Genetic and physical evidence for plasmid control of *Shigella sonnei* form I cell surface antigen. *Infection and Immunity* **29**, 207–214.

LaBrec, E. H., Schneider, H., Magnani, T. J. and Formal, S. B. (1964). Epithelial cell penetration as an essential step in the pathogenesis of bacillary dysentery. *Journal of Bacteriology* **88**, 1503–1518.

Lafcuillc, B., Darfcuillc, A., Pctit, S., Joly, B. and Cluzel, R. (1981). Étude de l'attachement aux entérocytes humains *in vitro* de *Escherichia coli* isolés de selles diarrhéiques chez l'enfant. *Annales de Microbiologie (Paris)* **132B**, 57–67.

Lathe, R., Hirth, P., De Wilde, M., Harford, N. and Lecocq, J.-P. (1980). Cell-free synthesis of enterotoxin of *E. coli* from a cloned gene. *Nature (London)* **284**, 473–474.

Le Minor, S. and Le Coueffic, E. (1975). Étude sur les hémolysines des Enterobacteriaceae. *Annales de Microbiologie (Paris)* **126B**, 313–332.

Linggood, M. A. and Ingram, P. L. (1982). The role of alpha haemolysin in the virulence of *Escherichia coli* for mice. *Journal of Medical Microbiology* **15**, 23–30.

Ljungh, A., Faris, A. and Wadström, T. (1979). Hemagglutination by *Escherichia coli* in septicemia and urinary tract infections. *Journal of Clinical Microbiology* **10**, 477–481.

Lopez-Alvarez, J. and Gyles, C. L. (1980). Occurrence of the Vir plasmid among animal and human strains of invasive *Escherichia coli*. *American Journal of Veterinary Research* **41**, 769–774.

Lopez-Alvarez, J., Gyles, C. L., Shipley, P. L. and Falkow, S. (1980). Genetic and molecular characteristics of Vir plasmids of bovine septicemic *Escherichia coli*. *Journal of Bacteriology* **141**, 758–769.

Maas, W. K. (1981). Genetics of bacterial virulence. *Symposium of the Society for General Microbiology* **31**, 341–360.

McConnell, M. M., Willshaw, G. A., Smith, H. R., Scotland, S. M. and Rowe, B. (1979). Tranposition of ampicillin resistance to an enterotoxin plasmid in an *Escherichia coli* strain of human origin. *Journal of Bacteriology* **139**, 346–355.

McConnell, M. M., Smith, H. R., Willshaw, G. A., Scotland, S. M. and Rowe, B. (1980). Plasmids coding for heat-labile enterotoxin production isolated from *Escherichia coli* O78: Comparison of properties. *Journal of Bacteriology* **143**, 158–167.

McConnell, M. M., Smith, H. R., Willshaw, G. A., Field, A. M. and Rowe, B. (1981). Plasmids coding for colonization factor antigen I and heat-stable enterotoxin production isolated from enterotoxigenic *Escherichia coli:* Comparison of their properties. *Infection and Immunity* **32**, 927–936.

McNeish, A. S., Turner, P., Fleming, J. and Evans, N. (1975). Mucosal adherence of human enteropathogenic *Escherichia coli*. *Lancet* **2**, 946–948.

Mazaitis, A. J., Maas, R. and Maas, W. K. (1981). Structure of a naturally occurring plasmid with genes for enterotoxin production and drug resistance. *Journal of Bacteriology* **145**, 97–105.

Merson, M. H., Ørskov, F., Ørskov, I., Jack, R. B., Huq, I. and Koster, F. T. (1979). Relationship between enterotoxin production and serotype in enterotoxigenic *Escherichia coli*. *Infection and Immunity* **23**, 325–329.

Meynell, E., Meynell, G. G. and Datta, N. (1968). Phylogenetic relationships of drug resistance factors and other transmissible bacterial plasmids. *Bacteriological Reviews* **32**, 55–83.

Minshew, B. H., Jorgensen, J., Counts, G. W. and Falkow, S. (1978a). Association of hemolysin production, hemagglutination of human erythrocytes, and virulence for chicken embryos of extra-intestinal *Escherichia coli* isolates. *Infection and Immunity* **20**, 50–54.

Minshew, B. H., Jorgensen, J., Swanstrum, M., Grootes-Reuvecamp, G. A. and Falkow, S. (1978b). Some characteristics of *Escherichia coli* strains isolated from extraintestinal infections of humans. *Journal of Infectious Diseases* **137**, 648–654.

Moll, A., Manning, P. A. and Timmis, K. N. (1980). Plasmid-determined resistance to serum bactericidal activity; a major outer membrane protein, the *tra*T gene product, is responsible for plasmid-specified serum resistance in *Escherichia coli*. *Infection and Immunity* **28**, 359–367.

Monti-Bragadin, C., Samer, L., Rottini, G. D. and Pani, B. (1975). The compatibility of *Hly* factor, a transmissible element which controls α-haemolysin production in *Escherichia coli*. *Journal of General Microbiology* **86**, 367–369.

Mooi, F. R., de Graaf, F. K. and van Embden, J. D. A. (1979). Cloning, mapping and expression of the genetic determinant that encodes for the K88ab antigen. *Nucleic Acids Research* **6**, 849–865.

Mooi, F. R., Harms, N., Bakker, D. and de Graaf, F. K. (1981). Organization and expression of genes involved in the production of the K88ab antigen. *Infection and Immunity* **32**, 1155–1163.

Mooi, F. R., Wouters, C., Wijfjes, A. and de Graaf, F. K. (1982). Construction and characterization of mutants impaired in the biosynthesis of the K88ab antigen. *Journal of Bacteriology* **150**, 512–521.

Moon, H. W. and Whipp, S. C. (1970). Development of resistance with age by swine intestine to effects of enteropathogenic *Escherichia coli*. *Journal of Infectious Diseases* **122**, 220–223.

Morris, J. A., Thorns, C. J., Scott, A. C. and Sojka, W. J. (1982). Adhesive properties associated with the Vir plasmid: A transmissible pathogenic characteristic associated with strains of invasive *Escherichia coli*. *Journal of General Microbiology* **128**, 2097–2103.

Moseley, S. L., Huq, I., Alim, A. R. M. A., So, M., Samadpour-Motalebi, M. and Falkow, S. (1980). Detection of enterotoxigenic *Escherichia coli* by DNA colony hybridisation. *Journal of Infectious Diseases* **142**, 892–898.

Moseley, S. L., Echeverria, P., Seriwatawa, J., Tirapat, C., Chaicumpa, W., Sakuldaipeara, T. and Falkow, S. (1982). Identification of enterotoxigenic *Escherichia coli* by colony hybridisation using three enterotoxin gene probes. *Journal of Infectious Diseases* **145**, 863–869.

Mullany, P., Field, A. M., McConnell, M. M., Scotland, S. M., Smith, H. R. and Rowe, B. (1983). Expression of plasmids coding for colonization factor antigen II (CFA/II) and enterotoxin production in *Escherichia coli*. *Journal of General Microbiology* **129**, 3591–3601.

Murray, B. E., Evans, D. J., Penaranda, M. E. and Evans, D. G. (1983). CFA/I-ST plasmids: Comparison of enterotoxigenic *Escherichia coli* (ETEC) of serogroups O25, O63, O78 and O128 and mobilisation from an R factor-containing epidemic ETEC isolate. *Journal of Bacteriology* **153**, 566–570.

Noegel, A., Rdest, U., Springer, W. and Goebel, W. (1979). Plasmid cistrons controlling synthesis and excretion of the exotoxin α-haemolysin of *Escherichia coli*. *Molecular and General Genetics* **175**, 343–350.

Noegel, A., Rdest, U. and Goebel, W. (1981). Determination of the functions of hemolytic plasmid pHly152 of *Escherichia coli*. *Journal of Bacteriology* **145**, 233–247.

O'Brien, A. D., Laveck, G. D., Thompson, M. R. and Formal, S. B. (1982). Production of *Shigella dysenteriae type 1*-like cytotoxin by *Escherichia coli*. *Journal of Infectious Diseases* **146**, 763–769.

Ogata, R. T. and Levine, R. P. (1980). Characterization of complement resistance in *Escherichia coli* conferred by the antibiotic resistance plasmid R100. *Journal of Immunology* **125**, 1494–1498.

Ogata, R. T., Winters, C. and Levine, R. P. (1982). Nucleotide sequence analysis of the complement resistance gene from plasmid R100. *Journal of Bacteriology* **151**, 819–827.

Opal, S., Cross, A. and Gemski, P. (1982). K antigen and serum sensitivity of rough *Escherichia coli*. *Infection and Immunity* **37**, 956–960.

Ørskov, I. and Ørskov, F. (1966). Episome-carried surface antigen K88 of *Escherichia coli*. I. Transmission of the determinant of the K88 antigen and influence on the transfer of chromosomal markers. *Journal of Bacteriology* **91**, 69–75.

Ørskov, I. and Ørskov, F. (1977). Special O:K:H serotypes among enterotoxigenic *E. coli* strains from diarrhea in adults and children. *Medical Microbiology and Immunology* **163**, 99–110.

Ørskov, I., Sharma, V. and Ørskov, F. (1976). Genetic mapping of the K1 and K4 antigens (L) of *Escherichia coli*. *Acta Pathologica Microbiologica Scandinavica Section B* **84B**, 125–131.

Ozanne, G., Mathieu, L. G. and Baril, J. P. (1977). Production of colicin V *in vitro* and *in vivo* and observations on its effects in experimental animals. *Infection and Immunity* **17**, 497–503.

Penaranda, M. E., Mann, M. B., Evans, D. G. and Evans, D. J. (1980). Transfer of an ST:LT:CFA/II plasmid into *Escherichia coli* K-12 strain RR1 by cotransformation with pSC301 plasmid DNA. *FEMS Microbiology Letters* **8**, 251–254.

Petrovskaya, V. G. and Licheva, T. A. (1982). A provisional chromosome map of *Shigella* and the regions related to pathogenicity. *Acta Microbiologica Academiae Scientiarum Hungaricae* **29**, 41–53.

Picken, R., Mazaitis, A. J., and Maas, W. K. (1982). Mapping of Tn3 insertions into an Ent plasmid resulting in increased production of heat labile enterotoxin. *Plasmid* **8**, 98.

Quackenbush, R. L. and Falkow, S. (1979). Relationship between colicin V activity and virulence in *Escherichia coli*. *Infection and Immunity* **24**, 562–564.

Reis, M. H. L., Affonso, M. H. T., Trabulsi, L. R., Mazaitis, A. J., Maas, R. and Maas, W. K. (1980). Transfer of a CFA/I-ST plasmid promoted by a conjugative plasmid in a strain of *Escherichia coli* of serotype O128ac:H12. *Infection and Immunity* **29**, 140–143.

Reynard, A. M. and Beck, M. E. (1976). Plasmid-mediated resistance to the bactericidal effects of normal rabbit serum. *Infection and Immunity* **14**, 848–850.

Reynard, A. M., Beck, M. E. and Cunningham, R. K. (1978). Effects of antibiotic resistance plasmids on the bactericidal activity of normal rabbit serum. *Infection and Immunity* **19**, 861–866.

Robbins, J. B., McCracken, G. H., Gotschlich, E. C., Ørskov, F., Ørskov, I. and Hanson, L. A.

(1974). *Escherichia coli* K1 capsular polysaccharide associated with neonatal meningitis. *New England Journal of Medicine* **290**, 1216–1220.

Robins-Browne, R. M., Levine, M. M., Rowe, B. and Gabriel, E. M. (1982). Failure to detect conventional enterotoxins in classical enteropathogenic (serotyped) *Escherichia coli* strains of proven pathogenicity. *Infection and Immunity* **38**, 798–801.

Rowe, B. (1979). The role of *Escherichia coli* in gastroenteritis. *Clinics in Gastroenterology* **8**, 625–644.

Royer-Pokora, B. and Goebel, W. (1976). Plasmids controlling synthesis of hemolysin in *Escherichia coli*. II. Polynucleotide sequence relationship among hemolytic plasmids. *Molecular and General Genetics* **144**, 177–183.

Sanchez, J., Bennett, P. M. and Richmond, M. H. (1982). Expression of *elt*-B, the gene encoding the B subunit of the heat-labile enterotoxin of *Escherichia coli,* when cloned in pACYC184. *FEMS Microbiology Letters* **14**, 1–5.

Sansonetti, P. J., David, M. and Toucas, M. (1980). Corrélation entre la perte d'ADN plasmidique et le passage de la phase I virulente à la phase II avirulente chez *Shigella sonnei. Comptes Rendus Hebdomadaires des Séances de l'Académie des Sciences, Série D* **290**, 879–882.

Sansonetti, P. J., Formal, S. B., Hale, T. L. and Kopecko, D. J. (1981a). Bases génétiques de la pénétration de *Shigella flexneri* dans les cellules épithéliales. *Annales d'Immunologie (Paris)* **132D**, 183–189.

Sansonetti, P. J., Kopecko, D. J. and Formal, S. B. (1981b). *Shigella sonnei* plasmids: Evidence that a large plasmid is necessary for virulence. *Infection and Immunity* **34**, 75–83.

Sansonetti, P. J., d'Hauteville, H., Formal, S. B. and Toucas, M. (1982a). Plasmid-mediated invasiveness of 'Shigella-like' *Escherichia coli. Annales de Microbiologie (Paris)* **132A**, 351–355.

Sansonetti, P. J., Kopecko, D. J. and Formal, S. B. (1982b). Involvement of a plasmid in the invasive ability of *Shigella flexneri. Infection and Immunity* **35**, 852–860.

Santos, D. S., Palchaudhuri, S. and Maas, W. K. (1975). Genetic and physical characteristics of an enterotoxin plasmid. *Journal of Bacteriology* **124**, 1240–1247.

Sarff, L. D., McCracken, G. H., Schiffer, M. S., Glode, M. P., Robbins, J. B., Ørskov, I. and Ørskov, F. (1975). Epidemiology of *Escherichia coli* K1 in healthy and diseased newborns. *Lancet* **1**, 1099–1104.

Satterwhite, T. K., Evans, D. G., DuPont, H. L. and Evans, D. J. (1978). Role of *Escherichia coli* colonisation factor antigen in acute diarrhoea. *Lancet* **2**, 181–184.

Schiffer, M. S., Oliveira, E., Glode, M. P., McCracken, G. H., Sarff, L. M. and Robbins, J. B. (1976). A review: Relation between invasiveness and the K1 capsular polysaccharide of *Escherichia coli. Pediatric Research* **10**, 82–87.

Schmitt, R., Mattes, R., Schmid, K. and Altenbuchner, J. (1979). Raf plasmids in strains of *Escherichia coli* and their possible role in enteropathogeny. *In* "Plasmids of Medical, Environmental and Commercial Importance" (Eds. K. N. Timmis and A. Pühler), Developments in Genetics, Vol. 1, pp. 199–210. Elsevier/North Holland Biomedical Press, Amsterdam.

Scotland, S. M., Gross, R. J. and Rowe, B. (1977). Serotype-related enterotoxigenicity in *Escherichia coli* O6.H16 and O148.H28. *Journal of Hygiene* **79**, 395–403.

Scotland, S. M., Gross, R. J., Cheasty, T. and Rowe, B. (1979). The occurrence of plasmids carrying genes for both enterotoxin production and drug resistance in *Escherichia coli* of human origin. *Journal of Hygiene* **83**, 531–538.

Scotland, S. M., Day, N. P. and Rowe, B. (1980a). Production of a cytotoxin affecting Vero cells by strains of *Escherichia coli* belonging to traditional enteropathogenic serogroups. *FEMS Microbiology Letters* **7**, 15–17.

Scotland, S. M., Day, N. P., Willshaw, G. A. and Rowe, B. (1980b). Cytotoxic enteropathogenic *Escherichia coli. Lancet* **1**, 90.

Scotland, S. M., Day, N. P., Cravioto, A., Thomas, L. V. and Rowe, B. (1981). Production of heat-labile or heat-stable enterotoxins by strains of *Escherichia coli* belonging to serogroups O44, O114 and O128. *Infection and Immunity* **31**, 500–503.

Scotland, S. M., Day, N. P. and Rowe, B. (1983a). Acquisition and maintenance of enterotoxin plasmids in wild-type strains of *Escherichia coli*. *Journal of General Microbiology* **129**, 3111–3120.

Scotland, S. M., Richmond, J. E. and Rowe, B. (1983b). Adhesion of enteropathogenic strains of *Escherichia coli* (EPEC) to HEp-2 cells is not dependent on the presence of fimbriae. *FEMS Microbiology Letters* **20**, 191–195.

Scotland, S. M., Smith, H. R., Willshaw, G. A. and Rowe, B. (1983c). Vero cytotoxin production in strain of *Escherichia coli* is determined by genes carried on bacteriophage. *Lancet* **ii**, 216.

Serény, B. (1957). Experimental keratoconjunctivitis shigellosa. *Acta Microbiologica Academiae Scientiarum Hungaricae* **4**, 367–376.

Sharp, P. A., Cohen, S. N. and Davidson, N. (1973). Electron microscope heteroduplex studies of sequence relations among plasmids of *Escherichia coli*. II. Structure of drug resistance (R) factors and F factors. *Journal of Molecular Biology* **75**, 235–255.

Shipley, P. L., Gyles, C. L. and Falkow, S. (1978). Characterization of plasmids that encode for the K88 colonization antigen. *Infection and Immunity* **20**, 559–566.

Shipley, P. L., Dallas, W. S., Dougan, G. and Falkow, S. (1979). Expression of plasmid genes in pathogenic bacteria. *In* "Microbiology—1979" (Ed. D. Schlessinger), pp. 176–180. American Society for Microbiology, Washington, D.C.

Shipley, P. L., Dougan, G. and Falkow, S. (1981). Identification and cloning of the genetic determinant that encodes for the K88ac adherence antigen. *Journal of Bacteriology* **145**, 920–925.

Silva, M. L. M., Maas, W. K. and Gyles, C. L. (1978). Isolation and characterisation of enterotoxin-deficient mutants of *Escherichia coli*. *Proceedings of the National Academy of Sciences of the U.S.A.* **75**, 1384–1388.

Silva, R. M., Toledo, M. R. F. and Trabulsi, L. R. (1982a). Plasmid-mediated virulence in *Shigella* species. *Journal of Infectious Diseases* **146**, 99.

Silva, R. M., Toledo, M. R. F. and Trabulsi, L. R. (1982b). Correlation of invasiveness with plasmid in enteroinvasive strains of *Escherichia coli*. *Journal of Infectious Diseases* **146**, 706.

Silver, R. P., Finn, C. W., Vann, W. F., Aaronson, W., Schneerson, R., Kretschmer, P. J. and Garon, C. F. (1981). Molecular cloning of the K1 capsular polysaccharide genes of *E. coli*. *Nature (London)* **289**, 696–698.

Sivaswamy, G. and Gyles, C. L. (1976). The prevalence of enterotoxigenic *Escherichia coli* in the feces of calves with diarrhea. *Canadian Journal of Comparative Medicine* **40**, 241–246.

Smith, H. R., Cravioto, A., Willshaw, G. A., McConnell, M. M., Scotland, S. M., Gross, R. J. and Rowe, B. (1979). A plasmid coding for the production of colonisation factor antigen I and heat-stable enterotoxin in strains of *Escherichia coli* of serogroup O78. *FEMS Microbiology Letters* **6**, 255–260.

Smith, H. R., Willshaw, G. A. and Rowe, B. (1982). Mapping of a plasmid, coding for colonization factor antigen I and heat-stable enterotoxin production, isolated from an enterotoxigenic strain of *Escherichia coli*. *Journal of Bacteriology* **149**, 264–275.

Smith, H. R., Scotland, S. M. and Rowe, B. (1983). Plasmids coding for production of colonization factor antigen II and enterotoxin production in strains of *Escherichia coli*. *Infection and Immunity* **40**, 1236–1239.

Smith, H. W. (1963). The haemolysins of *Escherichia coli*. *Journal of Pathology and Bacteriology* **85**, 197–211.

Smith, H. W. (1974). A search for transmissible pathogenic characters in invasive strains of *Esche-*

richia coli: The discovery of a plasmid-controlled toxin and a plasmid-controlled lethal character closely associated, or identical, with colicine V. *Journal of General Microbiology* **83**, 95–111.

Smith, H. W. (1976). The exploitation of transmissible plasmids to study the pathogenesis of *Escherichia coli* diarrhoea. *Society for Applied Bacteriology Symposium Series* **4**, 227–242.

Smith, H. W. (1978). Transmissible pathogenic characteristics of invasive strains of *Escherichia coli*. *Journal of the American Veterinary Medical Association* **173**, 601–607.

Smith, H. W. and Gyles, C. L. (1970a). The relationship between two apparently different enterotoxins produced by enteropathogenic strains of *Escherichia coli* of porcine origin. *Journal of Medical Microbiology* **3**, 387–401.

Smith, H. W. and Gyles, C. L. (1970b). The effect of cell-free fluids prepared from cultures of human and animal enteropathogenic strains of *Escherichia coli* on ligated intestinal segments of rabbits and pigs. *Journal of Medical Microbiology* **3**, 403–409.

Smith, H. W. and Halls, S. (1967a). Observations by the ligated intestinal segment and oral inoculation methods on *Escherichia coli* infections in pigs, calves, lambs and rabbits. *Journal of Pathology and Bacteriology* **93**, 499–529.

Smith, H. W. and Halls, S. (1967b). Studies of *Escherichia coli* enterotoxin. *Journal of Pathology and Bacteriology* **93**, 531–543.

Smith, H. W. and Halls, S. (1967c). The transmissible nature of the genetic factor in *E. coli* that controls haemolysin production. *Journal of General Microbiology* **47**, 153–161.

Smith, H. W. and Halls, S. (1968). The transmissible nature of the genetic factor in *Escherichia coli* that controls enterotoxin production. *Journal of General Microbiology* **4**, 301–312.

Smith, H. W. and Huggins, M. B. (1976). Further observations on the association of the colicine V plasmid of *Escherichia coli* with pathogenicity and with survival in the alimentary tract. *Journal of General Microbiology* **92**, 335–350.

Smith, H. W. and Huggins, M. B. (1978). The influence of plasmid-determined and other characteristics of enteropathogenic *Escherichia coli* on their ability to proliferate in the alimentary tracts of piglets, calves and lambs. *Journal of Medical Microbiology* **11**, 471–492.

Smith, H. W. and Huggins, M. B. (1980). The association of the O18, K1 and H7 antigens and the ColV plasmid on a strain of *Escherichia coli* with its virulence and immunogenicity. *Journal of General Microbiology* **121**, 387–400.

Smith, H. W. and Linggood, M. A. (1971a). The transmissible nature of enterotoxin production in a human enteropathogenic strain of *Escherichia coli*. *Journal of Medical Microbiology* **4**, 301–305.

Smith, H. W. and Linggood, M. A. (1971b). Observations on the pathogenic properties of the K88, Hly and Ent plasmids of *Escherichia coli* with particular reference to porcine diarrhoea. *Journal of Medical Microbiology* **4**, 467–485.

Smith, H. W. and Linggood, M. A. (1972). Further observations on *Escherichia coli* enterotoxins with particular regard to those produced by atypical piglet strains and by calf and lamb strains: The transmissible nature of these enterotoxins and of a K antigen possessed by calf and lamb strains. *Journal of Medical Microbiology* **5**, 243–250.

Smith, H. W. and Parsell, Z. (1975). Transmissible substrate-utilising ability in enterobacteria. *Journal of General Microbiology* **87**, 129–140.

Smyth, C. J. (1982). Two mannose-resistant haemagglutinins of enterotoxigenic *Escherichia coli* of serotype O6:K15:H16 or H⁻ isolated from travellers' and infantile diarrhoea. *Journal of General Microbiology* **128**, 2081–2096.

Smyth, C. J. (1984). Serologically distinct fimbriae on enterotoxigenic *Escherichia coli* of serotype O6: K15: H16 or H-. *FEMS Microbiology Letters* **21**, 51–57.

Smyth, C. J., Kaijser, B., Black, E., Faris, A., Möllby, R., Söderlind, U., Stintzing, G., Wadström, T. and Habte, D. (1979). Occurrence of adhesins causing mannose-resistant haemagglutination of bovine erythrocytes in enterotoxigenic *E. coli*. *FEMS Microbiology Letters* **5**, 85–90.

So, M. and McCarthy, B. J. (1980). Nucleotide sequence of the bacterial transposon Tn1681 encoding a heat-stable (ST) toxin and its identification in enterotoxigenic *Escherichia coli* strains. *Proceedings of the National Academy of Sciences of the U.S.A.* **77**, 4011–4015.

So, M., Crandall, J. F., Crosa, J. H. and Falkow, S. (1974). Extrachromosomal determinants which contribute to bacterial pathogenicity. *In* "Microbiology—1974" (Ed. D. Schlessinger), pp. 16–26. American Society for Microbiology, Washington, D.C.

So, M., Crosa, J. H. and Falkow, S. (1975). Polynucleotide sequence relationships among Ent plasmids and the relationship between Ent and other plasmids. *Journal of Bacteriology* **121**, 234–238.

So, M., Boyer, H. W., Betlach, M. and Falkow, S. (1976). Molecular cloning of an *Escherichia coli* plasmid determinant that encodes for the production of heat-stable enterotoxin. *Journal of Bacteriology* **128**, 463–472.

So, M., Dallas, W. S. and Falkow, S. (1978a). Characterization of an *Escherichia coli* plasmid encoding for the synthesis of heat-labile toxin: Molecular cloning of the toxin determinant. *Infection and Immunity* **21**, 405–411.

So, M., Heffron, F. and Falkow, S. (1978b). Method for the genetic labelling of cryptic plasmids. *Journal of Bacteriology* **133**, 1520–1523.

So, M., Heffron, F. and McCarthy, B. J. (1979). The *E. coli* gene encoding heat-stable toxin is a bacterial transposon flanked by inverted repeats of IS1. *Nature (London)* **277**, 453–455.

So, M., Atchison, R., Falkow, S., Moseley, S. and McCarthy, B. J. (1981). A study of the dissemination of Tn1681: A bacterial transposon encoding a heat-stable toxin among enterotoxigenic *Escherichia coli* isolates. *Cold Spring Harbor Symposia on Quantitative Biology* **45**, 53–58.

Spicer, E. K. and Noble, J. A. (1982). *Escherichia coli* heat-labile enterotoxin nucleotide sequence of the A subunit gene. *Journal of Biological Chemistry* **257**, 5716–5721.

Spicer, E. K., Kavanaugh, W. M., Dallas, W. S., Falkow, S., Konisberg, W. H. and Schafer, D. E. (1981). Sequence homologies between A subunits of *E. coli* and *V. cholerae* enterotoxins. *Proceedings of the National Academy of Sciences of the U.S.A.* **78**, 50–54.

Stark, J. M. and Shuster, C. W. (1982). Analysis of hemolytic determinants of plasmid pHly185 by Tn5 mutagenesis. *Journal of Bacteriology* **152**, 963–967.

Stieglitz, H., Fonseca, R., Olarte, J. and Kupersztoch-Portnoy, Y. M. (1980). Linkage of heat-stable enterotoxin activity and ampicillin resistance in a plasmid isolated from an *Escherichia coli* strain of human origin. *Infection and Immunity* **30**, 617–620.

Takeda, Y. and Murphy, J. R. (1978). Bacteriophage conversion of heat-labile enterotoxin in *Escherichia coli*. *Journal of Bacteriology* **133**, 172–177.

Taylor, J., Wilkins, M. P. and Payne, J. M. (1961). Relation of rabbit gut reaction to enteropathogenic *Escherichia coli*. *British Journal of Experimental Pathology* **42**, 43–52.

Taylor, P. W. and Hughes, C. (1978). Plasmid carriage and the serum sensitivity of enterobacteria. *Infection and Immunity* **22**, 10–17.

Taylor, P. W. and Robinson, M. K. (1980). Determinants that increase the serum resistance of *Escherichia coli*. *Infection and Immunity* **29**, 278–280.

Thomas, L. V. and Rowe, B. (1982). The occurrence of colonisation factors (CFA/I, CFA/II and E8775) in enterotoxigenic *Escherichia coli* from various countries in South East Asia. *Medical Microbiology and Immunology* **171**, 85–90.

Thomas, L. V., Cravioto, A., Scotland, S. M. and Rowe, B. (1982). A new fimbrial antigenic type (E8775) which may represent a colonization factor in enterotoxigenic *Escherichia coli* in humans. *Infection and Immunity* **35**, 1119–1124.

Timmis, K. N., Manning, P. A., Echarti, C., Timmis, J. K. and Moll, A. (1981). Serum resistance in *E. coli*. *In* "Molecular Biology, Pathogenicity, and Ecology of Bacterial Plasmids" (Eds. S. B. Levy, R. C. Clowes and E. L. Koenig), pp. 133–144. Plenum, New York.

Van Embden, J. D. A., de Graaf, F. K., Schouls, L. M. and Teppema, J. S. (1980). Cloning and expression of a deoxyribonucleic acid fragment that encodes for the adhesive antigen K99. *Infection and Immunity* **29**, 1125–1133.

Waalwijk, C., van den Bosch, J. F., MacLaren, D. M. and de Graaf, J. (1982). Hemolysin plasmid coding for the virulence of a nephropathogenic *Escherichia coli* strain. *Infection and Immunity* **35**, 32–37.

Wachsmuth, I. K., Falkow, S. and Ryder, R. W. (1976). Plasmid-mediated properties of heat-stable enterotoxin-producing *Escherichia coli* associated with infantile diarrhea. *Infection and Immunity* **14**, 403–407.

Wade, W. G., Thom, B. T. and Evans, N. (1979). Cytotoxic enteropathogenic *Escherichia coli*. *Lancet* **2**, 1235–1236.

Wahdan, M. H., Serie, C., Germanier, R., Lackany, A., Cerisier, Y., Guerin, N., Sallam, S., Geoffroy, P., Sadek El Tantawi, A. and Guesry, P. (1980). A controlled field trial of live oral typhoid vaccine Ty21a. *Bulletin of the World Health Organization* **58**, 469–474.

Walton, J. R. and Smith, D. H. (1969). New hemolysin (γ) produced by *Escherichia coli*. *Journal of Bacteriology* **98**, 304–305.

Welch, R. A., Dellinger, E. P., Minshew, B. and Falkow, S. (1981). Haemolysin contributes to virulence of extra-intestinal *E. coli* infections. *Nature (London)* **294**, 665–667.

Williams, P. H. (1977). Plasmid transfer in the human alimentary tract. *FEMS Microbiology Letters* **2**, 91–95.

Williams, P. H. (1979). Novel iron uptake system specified by ColV plasmids: An important component in the virulence of invasive strains of *Escherichia coli*. *Infection and Immunity* **26**, 925–932.

Williams, P. H. and Warner, P. J. (1980). ColV plasmid-mediated, colicin V-independent iron uptake system of invasive strains of *Escherichia coli*. *Infection and Immunity* **29**, 411–416.

Williams, P. H., Evans, N., Turner, P., George, R. H. and McNeish, A. S. (1977). Plasmid mediating mucosal adherence in human enteropathogenic *Escherichia coli*. *Lancet* **1**, 1151.

Williams, P. H., Sedgwick, M. I., Evans, N., Turner, P. J., George, R. H. and McNeish, A. S. (1978). Adherence of an enteropathogenic strain of *Escherichia coli* to human intestinal mucosa is mediated by a colicinogenic conjugative plasmid. *Infection and Immunity* **22**, 393–402.

Willshaw, G. A., Barclay, E. A., Smith, H. R., McConnell, M. M., and Rowe, B. (1980). Molecular comparison of plasmids encoding heat-labile enterotoxin isolated from *Escherichia coli* strains of human origin. *Journal of Bacteriology* **143**, 168–175.

Willshaw, G. A., Smith, H. R., McConnell, M. M., Barclay, E. A., Krnjulac, J. and Rowe, B. (1982a). Genetic and molecular studies of plasmids coding for colonization factor antigen I and heat-stable enterotoxin in several *Escherichia coli* serotypes. *Infection and Immunity* **37**, 858–868.

Willshaw, G. A., Smith, H. R., McConnell, M. M. and Rowe, B. (1982b). Cloning and expression of plasmid regions encoding colonisation factor antigen I (CFA/I) in *Escherichia coli*. *Society for General Microbiology, Quarterly* **9**, M8.

Willshaw, G. A., Smith, H. R. and Rowe, B. (1983). Cloning of regions encoding colonisation factor antigen I and heat-stable enterotoxin in *Escherichia coli*. *FEMS Microbiology Letters* **16**, 101–106.

Wilson, M. W. and Bettelheim, K. A. (1980). Cytotoxic *Escherichia coli* serotypes. *Lancet* **1**, 201.

Yamamoto, T. and Yokota, T. (1980). Cloning of deoxyribonucleic acid regions encoding a heat-labile and heat-stable enterotoxin originating from an enterotoxigenic *Escherichia coli* strain of human origin. *Journal of Bacteriology* **143**, 652–660.

Yamamoto, T. and Yokota, T. (1981). *Escherichia coli* heat-labile enterotoxin genes are flanked by repeated deoxyribonucleic acid sequences. *Journal of Bacteriology* **145**, 850–860.

Yamamoto, T., Yokota, T. and Kaji, A. (1981). Molecular organization of heat-labile enterotoxin genes originating in *Escherichia coli* of human origin and construction of heat-labile toxoid-producing strain. *Journal of Bacteriology* **148,** 983–987.

Yamamoto, T., Tamura, T. A., Ryoji, M., Kaji, A., Yokota, T. and Takano, T. (1982a). Sequence analysis of the heat-labile enterotoxin subunit B gene originating in human enterotoxigenic *Escherichia coli*. *Journal of Bacteriology* **152,** 506–509.

Yamamoto, T., Tamura, T., Yokota, T. and Takano, T. (1982b). Overlapping genes in the heat-labile enterotoxin operon originating from *Escherichia coli* human strain. *Molecular and General Genetics* **188,** 356–359.

9

The Immune Response to *Escherichia coli*

P. PORTER, J. R. POWELL, W. D. ALLEN AND MARGARET A.
LINGGOOD

*Department of Immunology, Unilever Research, Colworth Laboratory, Sharnbrook,
Bedford, UK*

Introduction

The mucosal immune system is part of the overall consideration of microbial
pathogenicity. We are, therefore, concerned primarily with the role of IgA,
rather than IgG or IgM, in mucosal defence. It is the objective of this paper to
consider neither the elegant studies concerning cellular (lymphocyte) differentiation and traffic nor the transport and secretion of antibodies at mucosal surfaces.
Such studies have provided a detailed understanding of the host response and the
means to monitor it effectively. More pertinent, perhaps, to the subject of this
symposium on *E. coli* infection is an evaluation of the host–pathogen relationship taking into account plasmid-mediated virulence factors and effector
mechanisms of immune protection.

It is clearly necessary to consider the implications of surface immune mechanisms in all tissues of the mucosal immune system, since *E. coli* infection may
occur in the respiratory tract, reproductive tract, urinary tract, gut, mammary
gland and ocular glands, particularly in the farm animal species. Our own
investigations have been concerned with pigs, cattle and poultry because *E. coli* is
the cause of major economic losses in agriculture. This is largely due to intensive
farming practices and the indiscriminate use of drugs, which favours the accumulation of transmissible drug-resistance and associated plasmid-mediated
virulence factors. This is obviously detrimental to any host species stressed by
over-population and over-nutrition. Clearly, in this situation the gut is the focus of
attention but integrated with it is consideration of the role of the mammary gland of
the mammalian species, since this is the only source of passive immunity for their
progeny, which are reared in a highly contaminated environment.

The gut–mammary link in mucosal immunity has been the subject of attention
with regard to defining the origin of secretory antibodies in milk and colostrum.

THE VIRULENCE OF ESCHERICHIA COLI

It was first demonstrated in the pig that the presence of anti-*E. coli* antibodies in colostrum and milk were not the result of transudation or selective transport from the blood (Porter, 1969), since there is no correlation between blood and co-lostral isotypes. Furthermore, from studies of enterovirus infection in sows, it was concluded that mammary IgA antibodies arose as a direct consequence of intestinal infection without the intermediate presence of serum antibodies (Saif *et al.*, 1972). Similarly, in lactating women, intestinal *E. coli* infection gives rise to IgA antibodies in the milk (Goldblum *et al.*, 1975).

Apart from the interesting perspectives that these observations provide, to-gether with the opportunity for developing unifying hypotheses of mucosal im-munity (McDermott and Bienenstock, 1979), it is also important to consider the role of maternal immunoglobulins in the gut of the suckling offspring. This, in turn, brings us full circle to a consideration of passive and active immunity in the developing organism and thus to a consideration of the most desirable features of maternal immunization.

Role of Maternal Passive Immunity in the Neonatal Gut

In human colostrum and milk, IgA is the dominant antibody, accounting for approximately 80% of the total immunoglobulin (Table 1). There is little, if any, augmentation of the blood-borne passive immunity in the neonate by absorption of maternal colostral immunoglobulins. The pig differs from the human in that there is no *pre-partum* transfer of maternal antibodies to the offspring by the transplacental route. In order to support the physiological need for a normal high serum IgG profile, sow's colostrum and early milk is rich in IgG (Table 1) and the piglets are able to absorb this over the first day or so of life until a peculiar feature of intestinal closure sets in (Porter and Hill, 1970). Thereafter, as in the human infant, there is no further increase of blood-borne passive immunity. At this stage the piglet and the human infant are broadly similar, requiring maternal immunoglobulin only for the role it plays in protection of the lumen of the

Table 1. *Levels of immunoglobulins in the colostrum of various species*

Species	Immunoglobulin (mg/ml)		
	IgG	IgA	IgM
Sheep	60	2.0	4.1
Cow	50–90	4.5	6.0
Pig	57	10	2.7
Rabbit	2.4	4.5	0.1
Man	0.2	18	0.8

alimentary tract. The lactation profile of immunoglobulin concentration in the sow becomes rather like that of the human female, being dominated by IgA (Porter *et al.*, 1970a).

Since maternal sIgA is absorbed by neither pigs nor humans the relevant question becomes that of the function of these antibodies in the lumen of the bowel. From a teleological standpoint it would seem irrelevant for the mammary gland to participate in a specialized link with the maternal gut if its product had little bearing on protection of the suckling offspring. However, from an anti-bacterial point of view, sIgA might be considered to be suspect. It does not fix complement by the conventional pathway but only operates via the alternative pathway of complement activation (Burdon, 1973). In addition, bacteriolysis by sIgA requires lysozyme as well as complement (Adinolphi *et al.*, 1966; Hill and Porter, 1974).

Since the digestive system is noted for anti-complementary activity, lytic functions are highly unlikely to participate in passive protection. However, sIgA is excellent in agglutination and can participate in bacteriostasis (Porter *et al.*, 1977), both of which roles will assist in preventing the colonization of the upper small intestine. Indeed, the resistance of sIgA to enzyme degradation and its mode of passage along the alimentary tract of the piglet (Porter *et al.*, 1970b) provides for a continuous passive bathing of the mucosal surface with maternal IgA. Together with bacteriostasis and agglutination, natural functions in the gut such as mucus secretion and peristalsis will facilitate the release of attaching bacterial colonies and enhance their removal. Nevertheless, it is pertinent to consider specific anti-adhesion functions and it is particularly in this respect that the neonatal piglet provides an interesting model.

The K88 antigen of *E. coli* was identified as a plasmid-associated adhesion factor by Ørskov and Ørskov (1966). Thereafter, Rutter and Jones (1973) provided early evidence that immunization of pregnant sows induced colostral antibodies which protected the suckling offspring from oral challenge with virulent strains of *E. coli*. However, the parenteral administration of adjuvanted vaccines gives rise to IgG-class antibodies, not mucosal antibodies such as IgA or IgM (Porter *et al.*, 1978). There is no doubt that IgG antibodies, like any other class of antibody, will inhibit attachment of K88 *E. coli* to enterocyte membrane receptors; however, it is fundamental that there must be the assurance that no other untoward phenomena follow from 'non-natural' routes of immunization. Anti-*E. coli* antibodies arising in the milk of a sow responding to intestinal colonization with *E. coli* differ markedly in immunoglobulin profile from those of a sow parenterally immunized with *E. coli* antigens. In the former, the activity is dominated by IgA and IgM with IgG assuming a very minor role. In the latter, IgG dominates with IgA participating at the most insignificant level.

As stated earlier, sIgA can participate only in passive defence of the suckling offspring through a specific antibody role in the intestinal lumen. In the pig, on the other hand, IgG is absorbed in large amounts into the blood stream and, with a half-life of approximately 20 days, will, thereafter, have the opportunity to exert a long-term physiological effect on the host (Porter and Hill, 1970). One of the consequences can be a disturbing immunosuppression. Muscoplatt *et al.* (1977) demonstrated that maternal antibody can regulate the immune response in neonatal piglets. Furthermore, since IgG does not act at mucosal sites its systemic function will leave the piglet with greater susceptibility to *E. coli* infection, particularly at the time of weaning. It is, therefore, important to examine the peculiar needs of the host in its environment before resorting to immunization schedules with emphasis only on the antigen and not the preferable route of administration.

In European agriculture, K88 vaccines have been used extensively for vaccination of pregnant sows, the objective being hyperimmunization by the parental route. If IgG antibody were the desired activity and K88 antigen of overriding significance, then *E. coli* disease in pigs should by now be a thing of the past. However, it is becoming recognized that significant changes in the virulence determinants of porcine *E. coli* are emerging, which suggests that this conventional anti-adhesion approach to vaccination may be too simplistic. The changes relate to the detection of new variants of K88 (Guinee and Jansen, 1979) and the implication of the previously calf-specific adhesin K99 in strains pathogenic for pigs (Moon *et al.*, 1977). Furthermore, whilst K88 was previously associated with neonatal enteritis, it now predominates in post-weaning diarrhoea which arises with greater virulence (Soderlind *et al.*, 1982).

Against this background of host immunity relative to microbial virulence, it becomes pertinent to enquire into the natural mechanisms operative in the host–pathogen relationship. It is, perhaps most relevant to begin with an examination of the changing mucosal immune function of the gut beginning from the time of birth.

Ontogeny of Intestinal Antibody Synthesis in the Pig

Studies of the response to *E. coli* infection in germfree and conventional piglets have indicated a pattern of cellular activity in the lamina propria dominated by IgM production in the early phase (Allen and Porter, 1977). IgM, like IgA, is transported across the epithelium in vesicles, presumably based on secretory component as a receptor in the enterocyte membrane (Allen *et al.*, 1976). The development of the immunoblast population in the lamina is peculiarly orientated towards the duodenum, whereas the Peyer's patches (PP), the source of antigenically primed cells (Cebra *et al.*, 1979), are located predominantly in the ileum. In humans specialized cells, M cells (Owen and Jones, 1974), can be identified within the PP as sampling cells for antigen. After antigen priming, lymphocytes leave the PP via the mesenteric lymphatics, eventually entering the blood stream in the thoracic duct and ultimately entering the lamina propria, a major site for their activity. It is appropriate that the population dynamics are such that disproportionate immunoblast activity is orientated to the upper small intestine, the region most sensitive to enterotoxin (Allen and Porter, 1977). The mechanism might be attributable to greater vascularity of this area. In this respect, Parrott and Ottaway (1980) have suggested that infiltration of mucosal tissues may be associated with an early inflammatory response, enabling an increased number of precursor immunoblasts to infiltrate the challenge tissue where they could be immobilized and then proliferate after contact with antigen.

There is now abundant evidence in the literature showing that there is an immunological exchange between the gut and mammary gland. This results in colostral and milk antibody activity reflecting the immunological experience of the maternal intestinal tract (Bohl *et al.*, 1973; Goldblum *et al.*, 1975; Montgomery *et al.*, 1980). Recently, it has been shown that plasma cell recruitment, at least in the mammary gland, is additionally under hormonal control (Lamm *et al.*, 1979).

Owing to the focal role of the gut in mucosal immune responses, we have been interested in the potential of oral immunization to generate colostral and milk antibodies which are protective to the neonate.

Live versus Inactivated *E. coli* Antigens as Vaccines for Stimulation of Colostral Immunity

Intestinal colonization is the natural route of the stimulus that influences the level of antibody secretion in the colostrum of the sow (Saif *et al.*, 1972). Paradoxically, it is also the major source of infection of the neonate. Starting from this accepted standpoint, a systematic investigation of the immunologlobulin-class antibody distribution was carried out as an essential step in defining the best characteristics of maternal response essential for passive protection of the neonate.

Four sows were given live *E. coli* **O**149:**K**91, **K**88ac at a dose rate of 5×10^{10} viable organisms per day for 3 days by the oral route. Two sows received the doses on days 88, 89 and 90 of gestation and the other two animals on days 100, 101 and 102 of gestation, procedures similar to those advocated by Kohler (1974) for autogenous vaccination against neonatal colibacillosis. The crucial test to set against the data in this study is the efficacy of the maternal antibody in providing a solid state of passive immunity against neonatal infection. Neonatal enteric colibacillosis can consistently be reproduced in model infection experiments with a high mortality rate (Saunders *et al.*, 1963). In the studies referred to here the mortality rate in the control group of animals was 76% (Chidlow *et al.*, 1979).

Table 2. *Neonatal protection after sow vaccination by oral and parenteral schedules*

Sow immunization schedule	Litters	Number of piglets	Mean antibody titre		Antibody distribution anti-O	Percentage mortality in infection test
			Anti-K88	Anti-O		
Untreated control	4	46	2	64	Not measurable	76
Oral live						
(1) 88–90 days	2	14	1024	2048	IgM 79% IgA 16% IgG 5%	7
(2) 100–102 days	2	12	128	4096	IgA 46% IgM 35% IgG 19%	25
Oral priming i/m booster	4	42	16	8192	IgM 89% IgA 6% IgG 5%	2
Double injection	2	23	1024	2048	IgG 86% IgM 9% IgA 5%	36

Progeny from the sows receiving the two oral live *E. coli* immunization schedules demonstrated marked passive protection with mortalities down to 7 and 25%, respectively. There was a marked change in immunoglobulin profile of the antibody function associated with the different timing of the infection schedules (Table 2). This is possibly associated with the changing hormonal status of the animals during this critical stage of gestation approaching the time of colostrum formation. It appeared that the preferred basis for passive immunity against neonatal enteric colibacillosis was an antibody profile dominated by IgM.

In a study of sow immunization by *E. coli* antigenic products given in the feed there appeared to be the correct characteristics of antibody distribution within the immunoglobulin classes. However, the level of antibody activity was insufficient to provide a consistent protective function. In a previous study (Chidlow *et al.*, 1979), this response was boosted by one appropriately timed intramuscular injection of *E. coli* antigens, with adequate retention of the required antibody distribution characteristics.

In comparison, the response of sows to heat-inactivated *E. coli* antigens of the same pathogenic strain was studied. The antigens were incorporated in the diet by means of a carrier premix and fed to the animals in order to establish a daily intake of free antigen, derived from approximately 2×10^9 *E. coli* bacteria, from day 50 of gestation until parturition. Prior investigation, in fistulated animals, had established that this exceeded by approximately 40-fold the minimum daily dose required to stimulate intestinal antibody secretion. This was calculated to provide the optimum primary antigenic stimulus to facilitate the production of humoral IgM antibody after a single parenteral antigen dose. The peak antibody response of pre-parturient sows was regulated to coincide with colostrum formation, so that high levels of IgM antibody were made available via the colostrum for neonatal defence. With the dominating position of IgM in the distribution of antibody (Table 2), this response was comparable with that in natural circumstances indicated by the best of the intestinal colonization schedules.

Again, the new schedule was subject to the crucial test of functional success in the neonatal infection model and this time the mortality level was reduced to only 2%. Thus, exploitation of the gut as a target organ for primary antigenic stimulation to elicit a predominantly circulating IgM antibody response on subsequent parenteral administration of non-viable antigens provides protective function superior to that mediated by intestinal colonization of the dam. It was significant that this protective function was not identifiable with the presence of anti-K88 antibodies, a feature we stressed previously (Porter *et al.*, 1978).

The oral/parenteral protocol described above has the important advantage of being without the inherent dangers of environmental contamination that arise from bacterial proliferation in the maternal gut. Indeed, because it is a major source of infection for the neonate one should seek to control it. One particular feature emerged from these studies, which has probably not been contemplated by advocates of live oral vaccine regimes. The progeny of infected sows showed significantly reduced birth weight compared with the normal and they grew at a slower rate when compared with piglets of sows receiving the oral/parenteral regime and which were reared under the same conditions.

In normal veterinary practice, attempts to confer passive protection on the neonate by means of colostral antibody are made by hyperimmunizing the sow with multiple intramuscular injections of *E. coli* vaccines. This approach was also compared with the natural response, and four sows were each injected by the intramuscular route at days 70 and 95 of gestation. In terms of anti-O and anti-K88 antibody titre, activity compared favourably with that induced with the live oral schedules but the greater part of the activity was associated with IgG, as would be expected. Passive protection attributable to this antibody class, was substantially inferior, with mortality in the neonatal infection model reaching 36% (Table 2).

In vivo 'Curing' of K88 Plasmids in the Immune Sow and Suckling Offspring

Enteropathogenicity of *E. coli* is intimately linked with host-specific adhesion factors, some of which are coded for by transmissible plasmids. Acquisition of such plasmids by strains of *E. coli* confers upon an organism the ability to adhere to the mucosa of the anterior regions of the small intestine, where the tissues are extremely sensitive to the effects of enterotoxin.

Under normal conditions of management, the sow is the major reservoir of infection for her offspring. Piglets born without the benefits of transplacental transfer of immunity meet the major infectious challenge of their new environment during the very act of seeking to suckle the colostrum which should sustain them with protective antibody. Using the combined oral/parenteral vaccination schedule for the maternal sow, we have observed plasmid 'curing' in an *E. coli* infection induced in the sow some 3 days before parturition and passive transfer of 'curing' antibodies to her offspring (Linggood and Porter, 1980). As part of the trial of the maternal vaccination schedule, an immunized sow was dosed with large numbers of the *E. coli* O149:K91,K88ac pathogen during the 3 days preceding parturition so that the piglets would receive the infection naturally from the sow soon after birth. Loss of the K88 plasmid from the pathogen occurred rapidly in the gut of the immunized sow and, shortly after parturition, 70% of the O149 strain present in her faeces was $K88^-$. The O149 strain was also isolated from the faeces of the piglets. In 10 of the 11 piglets the $K88^+/K88^-$ ratio fluctuated but followed a general trend with initially higher numbers of the $K88^+$ form but with the proportion of the $K88^-$ variant increasing, until they formed practically 100% of the O149 population (Table 3). However, in the 11th piglet the reverse happened—initially, the $K88^+$ variant

Table 3. *Elimination of K88ac from an O149:K91, K88ac strain of E. coli in vivo and in vitro*

Sow	Piglet mortality	Loss of K88 *in vivo*		Loss of K88 *in vitro* (72 h)	
		In sow	In piglets (72 h)	+10% colostrum	+10% milk
39 (immunized)	1 of 11	70%	65%[a] 0%[b]	100%	63%
42 (control)	9 of 13	0%	0%	6%	0%

[a]Average of 10 protected piglets.
[b]Piglet that developed diarrhoea.

formed only about 30% of the O149 strain present in the faeces but 2 days later the K88$^+$ form constituted 100% of the O149 population and, unlike its healthy, thriving littermates, the 11th piglet developed severe fatal diarrhoea.

An unimmunized sow was maintained in the same unit and her litter was naturally infected with the K88$^+$/K88$^-$ mixture excreted by the immunized sow. In this unprotected litter, the K88$^+$ type rapidly came to form 100% of the O149 population in all the piglets, and half of these died from the infection during the first 48 h.

The colostrum and milk from immunized sows contain a wide range of protecting antibodies, bactericidal and bacteriostatic antibodies, anti-adhesive antibodies, antitoxins, etc., but K88-'curing' antibodies are clearly also present since the process of plasmid elimination begun in the sow continued in her offspring. In 10 of 11 piglets from the immunized sow, the proportion of the O149:K91K88$^-$ strain continued to increase in spite of the advantage that the K88$^+$ form would be expected to have, due to its ability to adhere to intestinal epithelial cells. There was rapid overgrowth of the K88$^+$ form in the 11th piglet of the litter, which may not have received sufficient colostral antibody from the sow, and in the unprotected litter in the adjacent pen, which had received the same initial *E. coli* challenge.

Further evidence that a 'curing' factor was present in the colostrum and milk of the immunized sow was obtained from the *in vitro* culture of the O149 strain in broth containing 10% of the immune fractions. After growth for 72 h in the presence of immune whey from the colostrum and 1st-day milk, the K88 plasmid had been eliminated from 100 and 63%, respectively, of the O149 colonies tested, while negligible plasmid loss had occurred from the strain when growth in the presence of wheys from the control sow.

Oral Immunization and Drug Resistance Transfer in *E. coli*

A notable feature of the extensive field trials undertaken with the oral/parenteral vaccination schedule for maternal sows in pig herds (Chidlow *et al.*, 1979) was a

Table 4. *Effect of oral/parenteral immunization of sows with an E. coli vaccine on antibiotic therapy in piglets*

		Pigs treated per 100 pigs born		
Schedule	Number of pigs	1 day	2 days	3 days
Vaccinated	11566	21.1[a]	5.9[a]	1.3[a]
Control	9591	55.4	39.5	26.6

[a]$P = 0.01$.

dramatic reduction in the need for therapeutic intervention with antibiotics. In trials involving more than 23,000 pigs, neonatal mortality was reduced by more than 50% over the controls. This reduction was set against a background of medication in which the control litters had a four times greater need for treatment with antibiotics than did the litters from vaccinated sows. A further, more important, point emerged from the data which was a signal for the discovery of other antibody-mediated plasmid effects. The results were examined to determine the extent to which animals were subjected to medication—more particularly, whether there was a beneficial response to 1 day of treatment or whether there was a requirement for 2, 3 or more days of medication to overcome *E. coli* infection. The new data (Table 4) demonstrate the highly significant reduction in need for treatment of the progeny from the vaccinated sows and, moreover, demonstrates the significantly better responses to treatment in the vaccinated animals compared with the controls.

The total reduction in need for therapeutic assistance in the vaccinated group was compounded from fewer animals needing treatment, together with a superior responsiveness to therapy in those requiring treatment. Since the investigations were conducted simultaneously with paired litters in the same environment, the synergy of drug responsiveness with vaccination is a real phenomenon, not an artefact resulting from different conditions, age, time or management.

In the light of the data derived from the studies of adhesion plasmids, we pursued the obvious enquiry as to whether the colostral antibodies of sows could also induce the loss of R factors, but without success. However, perhaps more pertinent to the situation in young animals is whether maternal antibodies might prevent the spread of drug resistance, that is block the transfer of R factors. With this objective, we set up an *in vitro* model in which a high rate of transfer of the neomycin plasmid was achieved between a porcine enterpathogenic *E. coli* O149:K91,K88ac neo[+] and a non-pathogenic, streptomycin-resistant mutant strain of *E. coli* P4. After 24 h of mixed culture the numbers of transconjugants were counted on an agar medium containing streptomycin and neomycin.

In control media, the number of P4 str[+] neo[+] colony-forming units arising from a mixed culture of the pathogen and non-pathogen was more than 10^7 per millilitre. This exceedingly high rate of transfer provided a sensitive background against which to identify the effect of any antibody capable of interfering with transfer. Studies were undertaken with wheys and sera from sows immunized late in gestation by natural infection with a live pathogen **O**149:**K**91,**K**88ac (Porter *et al.*, 1978). Significant reductions in the development of P4 str$^+$ neo$^+$ were recorded when the serum, colostrum or milk samples of immunized sows or serum samples from their suckling piglets were added to the mating medium. Immune colostral wheys fractionated by gel filtration to give IgM- IgA- and IgG-rich fractions were used to evaluate the specific activity of the antibody classes derived from the immunization schedule. All three classes of immunoglobulin participated in the neomycin plasmid-transfer blocking effect but IgG was much less effective than the mucosal immunoglobulins IgM and IgA.

Conclusions

Secretory antibodies can interfere with microbial pathogenicity in a variety of ways and solid immunity to the pathogen will probably only be obtained as a result of several of these mechanisms acting in conjunction. The possible defence mechanisms will include agglutination and lysis of the bacteria, bacteriostasis, inhibition of adhesion and toxin neutralization as well as elimination of adhesion determinants. The latter mechanism differs from the others listed in that it results in the excretion of a defective form of the pathogen, which will be less successful at establishing infection in young animals. The other antibody-mediated defence mechanisms may confer solid protection on the initially infected animal but the pathogens excreted in the faeces will be fully virulent and capable of causing disease in non-immune animals, thus completing the cycle of infection, excretion and re-infection.

The concept of an immunological mechanism capable of genetically modifying a pathogen and so rendering it less virulent is intriguing, since bacterial modification is normally to the advantage of the pathogen, so enabling it to breach the immune defences of the host animal. The *in vivo* elimination of adhesion factors and the blocking of R factor transfer from pathogenic strains moves the host–pathogen relationship strongly in favour of the host. This antibody-mediated activity could be of great significance as it will lead to a reduction in the concentration of virulence plasmids in the microbial population, not only in the intestinal tracts of individual animals but in the environment as a whole. This has obvious relevance to intensive agriculture with high animal population densities and may also be relevant to man in his urban environment.

References

Adinolfi, M., Glynn, A. H., Lindsay, M. and Milne, G. M. (1966). Serological properties of IgA antibodies to *E. coli* present in human colostrum. *Immunology* **10**, 517–526.
Allen, W. D. and Porter, P. (1977). The relative frequency and distribution of immunoglobulin

bearing cells in intestinal mucosae of neonatal and weaned pigs and their significance in the development of secretory immunity. *Immunology* **32**, 819–824.

Allen, W. D., Smith, C. G. and Porter P. (1976). Evidence for secretory transport mechanism of intestinal immunoglobulin: The ultrastructural distribution of IgM. *Immunology* **34**, 449–457.

Bohl, E. H., Saif, C. J., Gupta, R. K. P. and Frederick, G. T. (1973). Secretory antibodies in milk of swine against transmissible gastroenteritis virus. *Advances in Experimental Medicine and Biology* **45**, 337–342.

Burdon, D. W. (1973). The bactericidal action of immunoglobulin A. *Journal of Medical Microbiology* **6**, 131–139.

Cebra, J. J., Crandall, D. A., Gearhart, P. J., Robertson, S. M., Tseng, J. and Watson, P. M. (1979). Cellular events concerned with the initiation, expression and control of the mucosal immune response. In "Immunology of Breast Milk" (Eds. P. L. Ogra and D. Dayton), pp. 1–18. Raven Press, New York.

Chidlow, J. W., Blades, J. H. and Porter, P. (1979). Sow vaccination by combined oral and intramuscular antigen: A field study on maternal protection against neonatal *Escherichia coli* enteritis. *Veterinary Record* **105**, 437–440.

Goldblum, R. M., Ahlsledt, S., Carlson, B., Hanson, L. A., Jodal, V., Lindin-Janson, G. and Sohl-Akerland, H. (1975). Antibody-forming cells in human colostrum after oral immunisation. *Nature (London)* **257**, 797–799.

Guinee, P. A. M. and Jansen, W. H. (1979). Behaviour of *Escherichia coli* K antigens K88ab, K88ac and K88ad in immuno-electrophoresis, double diffusion and haemagglutination. *Infection and Immunity* **23**, 700–705.

Hill, I. R. and Porter, P. (1974). Studies of bactericidal activity to *E. coli* of porcine serum and colostral immunoglobulins and the role of lysozyme with secretory IgA. *Immunology* **26**, 1239–1250.

Kohler, E. M. (1974). Protection of pigs against neonatal enteric colibacillosis with colostrum and milk from orally vaccinated cows. *American Journal of Veterinary Research* **35**, 331–338.

Lamm, M. E., Weisz-Carrington, P., Roux, M. E., McWilliams, M. and Phillips-Quagliata, J. M. (1979). Mode of induction of an IgA response in the breast and other secretory sites by oral antigen. *Immunology of Breast Milk* (Eds. P. L. Ogra and D. Dayton), pp. 105–114. Raven Press, New York.

Linggood, M. A. and Porter, P. (1980). The antibody mediated elimination of adhesion determinants from enteropathogenic strains of *E. coli*. *In* "Microbial Adhesion to Surfaces" (Eds. R. C. W. Berkeley, J. M. Lynch, J. Melling, P. R. Rutter and B. Vincent), pp. 441–453. Ellis Horwood, Chichester.

McDermott, M. R. and Bienenstock, J. (1979). Evidence for a common mucosal immunologic system. *Journal of Immunology* **122**, 1892–1898.

Montgomery, P. C., Lemaitre-Coelho, I. M. and Vaerman, J. P. (1980). Molecular diversity of secretory IgA anti-DNP antibodies elicited in rat bile. *Journal of Immunology* **125**, 518–522.

Moon, H. W., Nagy, B., Isaacson, R. E. and Orskov, I. (1977). Occurrence of K99 antigen on *Escherichia coli* isolated from pigs and colonisation of pig ileum by K99+ enterotoxigenic *E. coli* from calves and pigs. *Infection and Immunity* **15**, 614–620.

Muscoplatt, C. C., Setcavage, I. M. and Kim, N. B. (1977). Regulation of immune responses in neonatal piglets by maternal antibody. *International Archives of Allergy and Applied Immunology* **54**, 165–171.

Ørskov, I. and Ørskov, F. (1966). Episome carried surface antigen K88 of *E. coli* 1. Transmission of the determinant of the K88 antigen and its influence on the transfer of chromosomal markers. *Journal of Bacteriology* **91**, 69–75.

Owen, R. C. and Jones, A. L. (1974). Specialised lymphoid follicle epithelial cells in the human and non-human primate: A possible antigen uptake site.*In* "Scanning electron microscopy. III. Proceedings of a Workshop on Advances in Biomedical Applications of a Scanning Electron Microscope Publication," p. 697. I.T.T. Research Institute, Chicago, Illinois.

Parrott, D. M. V. and Ottaway, C. A. (1980). The control of lymphoblast migration to the small intestine. *In* "Mucosal Immune System in Health and Disease," p. 81. Proceedings of the Eighty-First Ross Conference on Paediatric Research, Columbus, Ohio.

Porter, P. (1969). Porcine colostral IgA and IgM antibodies to *E. coli* and their intestinal absorption by the neonatal piglet. *Immunology* **17**, 617–626.

Porter, P. and Hill, I. R. (1970). Serological changes in immunoglobulins IgG, IgA and IgM and *E. coli* antibodies in the young pig. *Immunology* **18**, 565–573.

Porter, P., Noakes, D. E. and Allen, W. D. (1970a). Secretory IgA and antibodies to *E. coli* in porcine colostra and milks and their significance in the alimentary tract of young pigs. *Immunology* **18**, 245–257.

Porter, P., Noakes, D. E. and Allen, W. D. (1970b). Intestinal secretion of immunoglobulins and antibodies to *E. coli* in the pig. *Immunology* **18**, 909–920.

Porter, P., Kenworthy, R. and Allen, W. D. (1974). Intestinal antibody secretion in the young pig in response to oral immunisation with *E. coli*. *Immunology,* **27**, 841–853.

Porter, P., Parry, S. H. and Allen, W. D. (1977). Significance of immune mechanisms in relation to enteric infection of the gastrointestinal trace in animals. *Ciba Foundation Symposium* **46** (new series), 55–67.

Porter, P., Linggood, M. A. and Chidlow, J. W. (1978). Elimination of *E. coli* K88 adhesion determinant by antibody in porcine gut and mammary secretions following oral immunisation. *In* "Secretory Immunology and Infection" (Eds. J. R. McGhee and J. Mestecky), pp. 133–142. Elsevier, Amsterdam.

Rutter, J. M. and Jones, G. W. (1973). Protection against enteric disease caused by *E. coli*—a model for vaccination with a virulence determinant. *Nature (London)* **242**, 531–533.

Saif, L. J., Bohl, E. H. and Gupta, R. K. P. (1972). Isolation of porcine immunoglobulins and determination of the immunoglobulin classes of transmissible gastroenteritis viral antibodies. *Infection and Immunity* **6**, 289–301.

Saunders, C. N., Stevens, A. D., Spence, J. B. and Sojka, W. (1963). *Escherichia coli* infection: Reproduction of the disease in naturally reared piglets. *Research in Veterinary Science* **4**, 333–346.

Soderlind, O., Olsson, E., Smyth, C. J. and Mollby, R. (1982). The effect of parenteral vaccination of dams on the intestinal *Escherichia coli* flora in piglets with diarrhoea. *Infection and Immunity* **36**, 900–906.

Part II

Methods

10

Haemagglutination Methods in the Study of *Escherichia coli*

D. C. OLD

Department of Medical Microbiology, University of Dundee Medical School, Ninewells Hospital, Dundee, UK

Introduction

The ability of bacteria to adhere to different kinds of substrates is likely to be advantageous at particular times and in particular circumstances, especially if that initial act of adherence should lead to successful colonization of the organism's particular ecological niche. It is not surprising, therefore, that many members of most species and genera of Enterobacteriaceae, including pathogens, commensals and saprophytes, form a wide range of surface materials, some of them appendages, that enable them to adhere to diverse substrates both *in vitro* and *in vivo*. For some adhesive factors, a specific role in the colonization of epithelial surfaces has been established. Thus, the demonstration that K88 fimbriae were important virulence determinants in *Escherichia coli* strains which induced diarrhoea in piglets (Smith and Linggood, 1971) was quickly followed by equally convincing demonstrations of specific adhesive roles for other fimbriae such as K99, 987P, CFA/I, CFA/II, P and X in different strains of *E. coli* pathogenic for man or animals (see Gaastra and de Graaf, 1982). However, the function and role of many other adhesive factors, including those widely distributed in commensal and saprophytic species, remain to be demonstrated (Duguid and Old, 1980).

Adhesins of E. coli

Many of the adhesive factors (adhesins) described in *E. coli* bestow on bacteria the ability to adhere to a wide range of diverse cellular substrates including erythrocytes. For the many adhesins that exhibit the property of haemagglutination, their specific adhesion to different species of erythrocytes, that is, their behaviour as haemagglutinins (HAs), has afforded a useful basis for their classi-

287

fication in *E. coli*. Not all adhesins, however, have haemagglutinating activity. For example, some *E. coli* strains recovered from piglets with diarrhoea possess the antigen 987P, but neither 987P-fimbriated bacteria nor purified 987P fimbriae agglutinate any of a wide range of erythrocyte species tested though they determine intestinal colonization *in vivo* (Isaacson *et al.*, 1977; Nagy *et al.*, 1977; Isaacson and Richter, 1981). Furthermore, antigenic analyses have described the presence of both non-haemagglutinating and haemagglutinating classes of fimbriae on some *E. coli*, though it is not always clear whether the former are *in vivo*-produced variants of the latter (Klemm *et al.*, 1982). It is obvious that adhesins, silent with respect to haemagglutinating properties, must be classified by other methods (Ørskov and Ørskov, 1983).

Though early documented observations showed both the specificity and diversity of haemagglutinating strains of *E. coli* (Guyot, 1908; Kauffmann, 1948), the first comprehensive and adequate study of the HAs of *E. coli*, and one that still forms the essential basis of current classification, was that of Duguid *et al.* (1955), who described three major groups (I–III) among haemagglutinating strains of *E. coli*.

Mannose-sensitive haemagglutinin. Group I strains grown under appropriate conditions produced fimbriae of external diameter 7–8 nm and channelled on negative staining, the type 1 fimbriae of Duguid *et al.* (1966). Type 1 fimbriate strains of enterobacteria, including *E. coli*, show the same patterns of haemagglutinating specificity for the erythrocytes of different animal species: those of guinea-pigs, hens, horses and monkeys are strongly agglutinated; those of men, including all common blood group types, are moderately agglutinated; those of sheep are weakly agglutinated; and those of oxen are not agglutinated (detailed reactions for these and other species in Duguid *et al.*, 1955, 1979; Duguid and Old, 1980). Type 1 fimbriae-associated haemagglutination is inhibited in the presence of D-mannose and related analogues (Duguid and Gillies, 1957; Old, 1972) and the HA is described as being mannose-sensitive, that is, MS-HA (Duguid, 1959).

Though all type 1 fimbriated strains of enterobacteria show a qualitatively similar pattern of reactions with the erythrocytes of different animal species, different bacterial strains and even different cultures of the same strain may show quantitative differences in the strength of their reactions with different species of erythrocytes. Thus cultures not showing a reaction with human erythrocytes because their content of MS-HA is small may wrongly be thought to possess an MS-HA of different specific affinities from that in MS-HA-rich cultures that do agglutinate human erythrocytes. Accordingly, it is not clear that studies purporting to have described an atypical form of MS-HA unreactive with human erythrocytes and associated particularly with strains of classical enteropathogenic (EPEC) serotypes (HA type III of D. J. Evans *et al.*, 1979) have taken account of

Table 1. *Distribution of MS-HA and MRE-HAs in Escherichia coli strains in different studies*

No. of strains examined	% of strains producing:		% of strains of HA type:				Reference
	MS-HA	MRE-HA	MS+MRE+	MS+MRE−	MS−MRE+	MS−MRE−	
306	84	53	47	37	6	10	Crichton (1980)
200	74	35	28	46	8	19	Duguid *et al.* (1979)
2117	43	36	14	30	22	35	D. J. Evans *et al.* (1979, 1980c, 1981)

that quantitative factor, which is especially relevant since their tests for MS-HA were made with agar-grown cultures in which MS-HA is generally poorly produced. MS-HA is the commonest adhesive factor in *E. coli,* being present in 75–85% of strains (Table 1).

Mannose-resistant and eluting haemagglutinins. Whereas the pattern of reaction with different erythrocytes given by different group I (MS-HA$^+$) strains was essentially constant, the patterns of haemagglutination given by strains of groups II and III were more complex. With reference to the 14 erythrocyte species tested (Duguid *et al.,* 1955), many different patterns of haemagglutination were observed for the different strains of groups II and III, though the pattern of haemagglutination given by any strain was characteristic of that strain. The patterns of haemagglutination ranged from narrow-spectrum types reacting with only one species of erythrocyte, to broad-spectrum types reacting with 6–11 species of erythrocytes. Red cells of different animal species may vary in their reactivity to any one of the HAs produced by strains of groups II or III. However, the reactions with the least susceptible erythrocytes will not be obtained except with cultures grown, and tests made, under optimal conditions. Regardless, however, of whether they were of narrow or broad spectrum, the HAs of strains of groups II and III were active in the presence of D-mannose, that is were mannose-resistant and showed the property of 'elution'—their interactions with some species of erythrocytes were more stable at lower (4°C) than at higher (20–35°C) temperatures of HA testing. These HAs were described by Duguid (1964) as mannose-resistant and eluting; MRE-HAs. Because the interactions between MRE-HA$^+$ bacteria and some erythrocyte species are demonstrably stable at ambient temperatures of testing, some workers have not attempted or considered defining MR-HAs in *E. coli* as 'eluting', yet it is a useful additional property not shared by the other classes of MR-HA, such as MR/K-HA, MR/P-HA and MR/Y-HA described in other members of Enterobacteriaceae (Duguid, 1959; Duguid and Old, 1980; MacLagan and Old, 1980; Adegbola and Old, 1982; Old and Adegbola, 1982). Again, as the temperature of the haemagglutination test is lowered, there may be a shift in the number of erythrocyte species agglutinated so that the pattern of haemagglutination is broadened. The MRE-HAs in *E. coli* strains of groups II and III (Duguid *et al.,* 1955) were associated with, respectively, fimbrial and non-fimbrial materials (Duguid *et al.,* 1955, 1979; D. J. Evans *et al.,* 1980a; Ip *et al.,* 1981). That finding together with the observed differences in haemagglutination patterns and in antigenic properties of the MRE fimbriae (Clegg, 1978; Duguid *et al.,* 1979) made it clear that, although the MRE-HAs shared the same general properties, their classification in a group as 'mannose-resistant' was one of convenience only, serving primarily to distinguish them from MS-HA. The recent elegant studies of the MRE-HAs associated with P fimbriae on pyelonephritogenic strains of *E. coli,* which identified the disaccharide α-D-Gal*p*-(1–4)-β-D-Gal*p* (Gal-Gal) as the minimal receptor structure for these P fimbriae (Källenius *et al.*, 1980a,b, 1981, 1982; Korhonen *et al.*, 1982b), represent the first important subgrouping by specific receptor recognition within the heterogeneous MRE-HAs.

Haemagglutination Tests

The MS-HA and many of the MRE-HAs of *E. coli* can be detected rapidly by simple tests with inexpensive, readily available reagents. This is but one of the methods which allows the MRE-HAs and their associated antigens to be classified. Many HA typing schemes are in operation but I will describe only a few that have stood the test of time. Many currently practised modifications of the original schemes have been so shortened or streamlined that, as will become apparent, the information they yield is of limited value.

Methodology

Erythrocytes. For the demonstration of MS-HA the use of erythrocytes of guinea-pigs (or fowls or horses) alone is usually sufficient. However, to establish the different types of MRE-HA by their patterns of haemagglutinating activity, it is wise to use as many readily available species of erythrocytes as possible. In the comprehensive HA typing system of Duguid *et al.* (1955, 1979) as many as 14 species of erythrocytes were used, including those of dogs, hens (that is, domesticated fowls), frogs, guinea-pigs, horses, men (all common blood groups), monkeys (rhesus), mice, oxen, pigs, rabbits, rats, sheep and toads. The sources of these bloods included laboratory breeds of animals, unselected breeds of animals after slaughter at the abattoir or during veterinary care, commercial suppliers and, for human erythrocytes, the regional blood transfusion service. Because it was difficult to obtain some of the 14 species either on a regular basis or in sufficient quantity, we have for routine screening purposes generally used 7 commonly available erythrocyte species—fowl, guinea-pig, horse, man, ox, pig and sheep—supplemented on occasion with those of mouse, rabbit and rat. The red cells were separated by low-speed centrifugation from citrated, defibrinated or heparinized blood (dependent on source), washed thrice with 0.15 M NaCl (saline) and suspended to 3% (v/v) in saline (Duguid *et al.*, 1955) or in phosphate-buffered saline (0.15 M NaCl and 0.02 M phosphate buffer, pH 7.2) (Källenius and Möllby, 1979). Erythrocyte suspensions stored at 4°C remain stable and give reproducible results for up to 1 wk after preparation.

Occasional variations are noted in the reactions given by some MRE-HA[+] strains with the blood from different individuals or breeds of animals, but for most MRE-HAs and MS-HA the reactions are generally constant (Duguid *et al.*, 1955, 1979; Källenius *et al.*, 1980a).

Types of test. Two basic kinds of HA test are commonly used: (1) the *rocked tile test,* performed with mechanical rocking of a porcelain tile (Duguid *et al.,* 1955, 1979) or manual rocking of a glass slide (D. G. Evans *et al.,* 1977), and (2) the *static settling test,* in which erythrocytes and bacteria in the presence or absence of D-mannose are allowed to settle at 4°C in plastic 'microtitre' trays (Jones and Rutter, 1974) or in glass tubes (Old and Adegbola, 1982). The rocked tests are suitable for detection of haemagglutinating activity due to MS-HA and most of the MRE-HAs, and the static tests for the very weak MRE-HAs such as K88 or K99.

For the demonstration of MS-HA activity, aliquots ($\sim$0.02 ml) of saline, erythrocyte suspension (guinea-pig, fowl or horse) and bacterial suspension ($\sim$5–10 $\times$ 10^{10} bacteria ml^{-1}) in saline are mixed in depressions (25 mm) of a 12-cavity, glazed-procelain tile (115 $\times$ 85 mm) (Gallenkamp, East Kilbride) and rocked for 10 min (30 times per minute) at ambient temperature ($\sim$20°C) on a mechanical rocking device (Rotatest and suspension mixer, Luckham Ltd., Burgess Hill, Sussex). Haemagglutination was read as strongly (+ + +) or moderately (+ +) positive when coarse clumping occurred in, respectively, 20 s or 2 min, and as weakly (+) positive when fine agglutination occurred in 10 min (Duguid *et al.,* 1979). The absence of haemagglutination in parallel tests in which 2% (w/v) D-mannose or α-methyl-D-mannoside in saline was substituted for saline confirmed the MS nature of the HA activity.

For the demonstration of MRE-HA activity, aliquots ($\sim$0.02 ml) of D-mannose–saline, erythrocyte suspensions (as many as available) and bacterial suspension ($\sim$1 $\times$ 10^{12} bacteria ml^{-1} in saline) are mixed in the depressions of a tile rocked mechanically for 20 min in a cold room or domestic refrigerator at 4°C. Reactions (Fig. 1) with each of the (7–10) erythrocyte species are read *immediately* on removal from the cold because the characteristic 'elution' is so rapid from some erythrocyte species that it can be missed with delays in reading. For example, the MRE-HA reactions with sheep erythrocytes given by *E. coli* strains possessing F antigens 7–12 were readily demonstrable when performed by me as described above (unpublished), yet are described as 'immediately eluting' in tests performed at room temperature (Ørskov and Ørskov, 1983). The reactions of some MRE-HA$^+$ strains with some species of erythrocytes, for example, *E. coli* strains with fimbrial antigens F7–F12, and erythrocytes of man, though stable at room temperature, show the characteristic 'elution' from human erythrocytes as the tile is gently heated to $\sim$50°C by passing it through a bunsen flame, and are reformed on subsequent rocking at 4°C.

Microscope slides, preferably cavity slides (Smyth, 1982), may be used instead of the porcelain tile. Bacterial and erythrocyte suspensions are mixed at room temperature for 1 min, placed thereafter on ice and, with intermittent manual rotation, observed after a further 2 min. In this system (D. G. Evans *et al.,* 1977), in which the degrees of MS-HA activity are assessed in the absence of

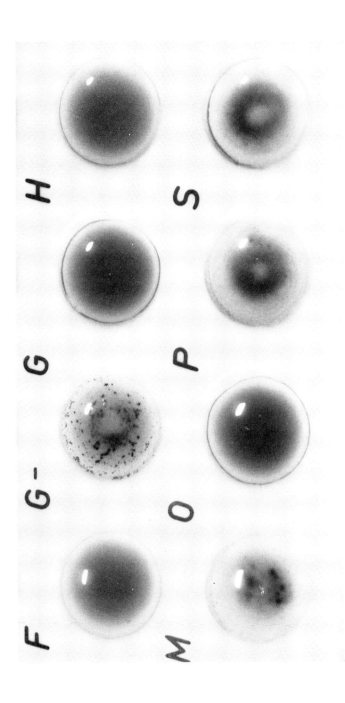

Fig. 1. Rocked tile test of MS⁺ MRE⁺ strain of *E. coli* showing agglutination of red cells of guinea-pig in the absence of D-mannose (G⁻); in the presence of D-mannose, red cells of man (M), pig (P) and sheep (S) were also agglutinated but not those of fowl (F), guinea-pig (G), horse (H) or ox (O).

D-mannose and those of MRE-HA activity in the presence of D-mannose with the erythrocytes of fowl, guinea-pig, man, cattle and monkey (African green), the reactions which appear within 2 min in minimally rocked tests probably detect only quantitatively strong HA activities and fail to detect the weaker activities detected in the rocked tile test.

The agglutination in the tile (or slide) test requires a strong enough ad-hesiveness of the MRE-HA$^+$ bacteria for the sensitive erythrocytes to withstand the shearing movements generated on rocking. However, the MRE-HA reactions of some *E. coli* strains, for example, those given by antigens K88 or K99, are so weak that their HA activities are not demonstrable in the rocked tests; for such weakly haemagglutinating antigens, a 'microhaemagglutination' test such as that of Jones and Rutter (1974) is necessary. With all reagents and trays chilled prior to use, doubling dilutions of bacterial suspension (5×10^{10} bacteria ml^{-1} in saline) are prepared in saline or D-mannose–saline in round-bottom wells of microtitre trays (e.g., from Flow Laboratories, Irvine, Ayrshire). To each well, aliquots (0.025 ml) of 3% (v/v) erythrocyte suspensions (as available) are added and the trays sealed with acetate tape as in the modification of Källenius and Möllby (1979). Trays are incubated at 4°C for 1 h, and haemagglutination observed as a complete and even sheet of agglutinated red cells.

Regardless of whether the tile or tray method is used, the system can be quantitated so that the haemagglutination power or titre is determined after the appropriate procedure as the reciprocal of the lowest concentration of bacteria giving definite, visible agglutination of the erythrocytes (Duguid and Gillies, 1957; Jones and Rutter, 1974). Titres based on comparison with a 'standard' haemagglutinating culture may, however, be misleading unless the standard culture is maximally haemagglutinating, that is, as strongly haemagglutinating as any other culture might be.

Cultural conditions. Type 1 fimbriate bacteria vary spontaneously between fimbrial and non-fimbrial phases (Brinton, 1959), which are subject to selection by different cultural conditions (Duguid and Old, 1980), and the amount of fimbrial synthesis is also influenced directly by particular cultural conditions (Duguid and Old, 1980). These variations must be considered in choosing the cultural methods designed to detect MS-HA activity.

For *E. coli* and most other type 1 fimbriate enterobacteria, production of type 1 fimbriae and MS-HA occurs best in nutrient broth cultures incubated statically in air or microaerophilically [as in an anaerobic jar from which ~90% (v/v) of the air is removed by vacuum but not replaced] with subcultures made every 48–72 h (Duguid *et al.*, 1955, 1966; Duguid and Gillies, 1957) up to six times (Duguid *et al.*, 1979; Fein, 1981). Under such conditions, type 1 fimbriate bacteria form fimbrial pellicles (see Clegg and Old, 1979) at the broth–air interface in the period after 6 hr of incubation (Duguid *et al.*, 1966; Old *et al.*, 1968). The growth advantage of fimbriae-dependent pellicle formation demonstrated with type 1 fimbriate enterobacteria (Old and Duguid, 1970, 1971; Duguid and Old, 1980) is important in the outgrowth of the fimbriate-phase bacteria. Because bacteria reach the air–broth interface to form the pellicle by aerotactic locomotion (Duguid *et al.*, 1966; Old and Duguid, 1970), non-motile strains of *E. coli* and other enterobacteria form their fimbrial pellicles at a later time in any culture than motile strains, and require more serial subcultures for the selection of their fimbriate-phase bacteria from the population.

Production of type 1 fimbriae and MS-HA is generally poor or absent in young, logarithmic-phase broth cultures in air, in broth cultures aerated with shaking, in anaerobically grown cultures and in cultures incubated on solid media (Duguid and Gillies, 1957; Duguid and Wilkinson, 1961). Synthesis of MS-HA is also subject to catabolite repression by many carbohydrates (Saier *et al.*, 1978), the presence of which may influence the production of type 1 fimbriae, the formation of fimbriae-dependent pellicles and hence the selection of the fimbrial phase (Old *et al.*, 1968; Old and Duguid, 1970; Duguid and Old, 1980).

Failure to take note of these factors known to influence the expression of MS-HA and subsequent selection of type 1 fimbriate bacteria may result in the characterization of MS-HA$^+$ strains as MS-HA$^-$. We have generally tested six serial broth cultures before considering a strain as MS-HA$^-$, and although the formation of type 1 fimbriae is generally independent of the growth temperature in the range 20–37°C occasional strains are found that are MS-HA$^-$ at 37°C but MS-HA$^+$ at 30°C or lower, for example, exceptional strains of *Salmonella typhi* (Duguid *et al.*, 1966). Two different series of strains in which MS-HA was assessed after the above recommended procedure of serial broth cultures revealed 74–84% of them to be MS-HA$^+$ (Table 1), results in keeping with the high percentage (86%) found in a third series of 1050 *E. coli* strains of faecal and urinary origins (P. B. Crichton and D. C. Old, unpublished). The latter study revealed that failure to use serial broth culture would have led to misidentification as MS-HA$^-$ of the strains (~24%) which formed MS-HA in a second, or later, culture. The very low estimate of 43% for MS-HA production in the third series of Table 1 is probably associated with the fact that agar-grown cultures only were screened (D. J. Evans *et al.*, 1979, 1980c). Inappropriate cultural conditions probably accounted for similar low estimates of MS-HA (Varian and Cooke, 1980) and for irregularity of adherence to uroepithelial cells (Svanborg Edén and Hansson, 1978) of *E. coli* from other collections.

It is generally agreed that MR adhesins or MRE-HAs are formed better at 37°C than at lower temperatures, and absence of production of MR adhesins at 18°C is a useful property characterizing this class of adhesins. Growth on a solid medium at 37°C for 24 h is perhaps the most commonly recommended procedure. Among the media used and proved satisfactory for this purpose are: phosphate-buffered (PB) agar, that is, Oxoid nutrient agar with 0.36% (w/v) KH_2PO_4 and 0.64% (w/v) Na_2HPO_4 (pH 7.0) as described by Duguid et al. (1979); colonization factor antigen (CFA) agar, that is, 1% (w/v) Difco casamino acids, 0.15% (w/v) Difco yeast extract with 0.005% (w/v) $MgSO_4$, 0.005% (w/v) $MnCl_2$ and 2% (w/v) agar (pH 7.4) as described by D. G. Evans et al. (1977); Minca medium, that is, 0.136% (w/v) KH_2PO_4, 1.01% (w/v) $Na_2HPO_4 \cdot 2H_2O$, 0.1% (w/v) glucose, 0.1% (w/v) Difco casamino acids, 1.2% (w/v) agar and trace elements (see Guinée et al., 1976); improved Minca, supplemented with 1% (w/v) Iso-VitaleX (BBL, Cockeysville, Md., USA) instead of glucose, as described by Guinée et al. (1977). Others have successfully used synthetic or semi-synthetic media with or without vitamin supplements for adhesins such as K88 (Middleldorp and Witholt, 1981) or K99 (de Graaf et al., 1980; Francis et al., 1982).

In a comparative study of the different media for the production of the fimbrial MR adhesins K88, K99 and 987P by E. coli strains of animal origin, Francis et al. (1982) showed that no one medium supported optimal production of all antigens by all strains. Similar difficulties with the MRE-HAs of E. coli strains of human origin have been observed; thus, the MRE-HA activity associated with fimbrial antigen E8775 was produced well on CFA agar but poorly on PB agar; those associated with P fimbriae were better formed on PB agar than on CFA agar; and that associated with an X-fimbriated strain (SS10), though not produced on CFA agar or PB agar, was produced well in nutrient broth (D. C. Old, unpublished). Although nutrient broth generally leads to poor production of MR adhesin (de Graaf et al., 1980), we have found that phosphate-buffered Oxoid nutrient broth (pH 7.0), as above for PB agar, allowed development of MRE-HAs by many human E. coli strains of urinary origin (Crichton, 1980). Again, if, as seems likely (Gaastra and de Graaf, 1982), the MRE-HAs are subject to phase variation in a fashion similar to that of type 1 fimbriate strains, the use of serial culture on the chosen broth or agar media should be attempted (Guinée et al., 1977; Väisänen et al., 1981). Much work is needed before a set of cultural conditions suitable for a majority of MR adhesins or MRE-HAs can be recommended. In the meantime, therefore, it would seem wise to follow the recommendation of Francis et al. (1982), so that a strain is not designated as deficient in the ability to produce MR adhesin until a combination of at least two media has been assessed for any strain. Our own preferred combination for the MRE-HAs of E. coli of human origin would be PB and CFA agars. After incubation at 37°C for 18–24 h, the growth from confluently inoculated PB or CFA agars is harvested with a loop or swab into saline (0.6 ml) to a density of $\sim 1 \times 10^{12}$ bacteria ml^{-1}. This very high concentration of bacteria is essential for MRE-HA testing if weak reactions with less susceptible bloods are not to be missed.

Interpretation of HA Tests

Strains Forming One Haemagglutinin

With the 30–40% of *E. coli* strains that formed MS-HA only (Table 1), its identification should not prove difficult provided the conditions for its production have not been overlooked. The interaction between MS-HA$^+$ bacteria and erythrocytes of the most sensitive species of erythrocytes (guinea-pig > fowl > horse) is strong, occurs equally well in HA tests performed at different temperatures from 4 to 37°C and, provided high concentrations of bacteria are tested, may be detected even in cultures poor in MS-HA content; the presence of MS-HA is confirmed by its inhibition by D-mannose.

There are problems, on the other hand, in testing for MRE-HA activity even in tests with strains that produce MRE-HA only. The interaction between MRE-HA$^+$ bacteria and susceptible erythrocytes is generally less stable than that of MS-HA. Detection of MRE-HA is best attempted, therefore, at temperatures (0–4°C) low enough to prevent the elution of bacteria from agglutinated erythrocytes that occurs on elevation of tile temperature. Tests should also be made with dense suspensions of bacteria ($\sim 10^{12}$ bacteria ml^{-1}) to facilitate recognition of the weaker reactions given with less sensitive erythrocyte species, otherwise the patterns of HA activity defined, reflecting the strongest reactions only, will be incomplete.

Strains Forming MS- and MRE-Haemagglutinins

In the case of those strains (14–47%, Table 1) on which both MS- and MRE-HAs are present together, care must be taken to identify separately the properties of each HA. The presence and properties of the MRE-HAs in these multiply haemagglutinating strains are readily demonstrated in tests made at 4°C with a range of erythrocytes in the presence of D-mannose to eliminate MS-HA activity. The separate presence of MS-HA in multiply haemagglutinating strains is demonstrated most readily in the case of strains forming MRE-HAs unreactive with one of the species of erythrocytes (e.g., guinea-pig or fowl or horse or mouse) with which MS-HA is highly reactive and which may, therefore, be used to demonstrate the MS reaction. If strains produce MRE-HAs reacting with all erythrocyte species highly susceptible to MS-HA, MS-HA can still be demonstrated (1) by showing that the agglutinated bacteria 'elute' in tests with D-mannose but not in its absence, (2) in tests with MS$^+$MRE$^+$ suspensions after treatment with heat (65°C for 30 min) or 0.5% (w/v) formaldehyde (37°C for 4 h) to destroy their MRE-HA activity or (3) in tests with bacteria after serial nutrient broth culture at 18°C, which should yield MS$^+$MRE$^-$ suspensions (Duguid and Old, 1980).

Strains Forming Multiple MRE-Haemagglutinins

It is probable that many strains of *E. coli* form two or more kinds of MRE-HA, the combined HA activities of which are those at present attributed to that of one HA. The demonstration of the separate presence of different MRE-HAs on a strain is difficult and has generally been established from the use of variants forming one or other of the adhesins (Morris *et al.*, 1980, 1982; Chanter, 1983). Thus, animal strains of *E. coli* carrying both antigens K99 and F41 led to confusion about the HA properties of each adhesin (Morris *et al.*, 1977, 1978). Purification of the fimbriae associated with these MR adhesins, however, indicated that K99 fimbriae agglutinated erythrocytes of horse strongly and those of sheep weakly in settling tests (Isaacson, 1977; de Graaf and Roorda, 1982); F41 fimbriae, on the other hand, agglutinated erythrocytes of man and guinea-pig strongly and those of horse and sheep moderately (de Graaf and Roorda, 1982). However, two *E. coli* strains, carrying F41 alone or together with K99, examined by me by the rocked tile method at 4°C (which does not detect the HA properties of K99) gave strong MRE-HA activity with erythrocytes of man, pig, sheep and rat; those of guinea-pig, horse and rabbit were not agglutinated (D. C. Old, unpublished). It is difficult, therefore, to interpret these results in view of the haemagglutination reactions of purified F41 fimbriae (de Graaf and Roorda, 1982), and the question must be asked whether the F41$^+$ strains examined by me possess a previously undetected MRE-HA, that is in addition to K99 and F41. Of course, the different test systems used may reveal HA properties for purified F41 fimbriae quite different from those of F41$^+$ bacteria. Similar discrepancies between the haemagglutinating activities of purified F9 fimbriae and F9-fimbriated cultures have been noted (Korhonen *et al.*, 1982a).

A most convincing demonstration that *E. coli* strains may possess several types of MR adhesin with different HA activities came from the Källenius group (Svenson *et al.*, 1982). Their earlier studies had shown that P fimbriation occurred with high frequency in *E. coli* strains from children with non-obstructive pyelonephritis (Källenius *et al.*, 1980b, 1981). The P fimbriae recognize and bind to human cell receptors containing the α-D-Gal*p*-(1–4)-β-D-Gal*p* saccharide entity (Gal-Gal), a part of the glycosphingolipids related to P blood group antigens of human erythrocytes. In haemagglutination tests, therefore, P-fimbriated bacteria agglutinated human erythrocytes of the blood groups P_1 and P_2^k but not those of the rare p̄ phenotype lacking the specific glycosphingolipids of P blood group antigens. The (Gal-Gal) disaccharide inhibits agglutination of human erythrocytes by P-fimbriated bacteria (Källenius *et al.*, 1982). The recent development of a P-receptor-specific particle agglutination (PPA) test, in which 'particles' (*sic*) are coated with P-receptor glycoside containing Gal-Gal, provided a highly specific diagnostic test for the presence of P fimbriae (Svenson *et al.*, 1982). The studies showed that urinary strains of *E. coli* may possess other types of fimbriae that give MR agglutination of human erythrocytes, including those of

Table 2. *Haemagglutination and PPA[a] reactions in urinary Escherichia coli strains producing MR adhesin[b]*

No. of strains	Mannose-resistant agglutination of erythrocytes of:					Agglutination in PPA test	Type of MR fimbriae present
	Man; P-type			Ox	Guinea-pig		
	P_1	P^k_2	$\bar{p}$				
58	+	+	−	−	−	+	P
7	+	+	+	−	−	−	X
6	+	+	+	−	−	+	P and X
16	+	+	+	+	+	+	P and other[c]
22	+	+	+	+	+	−	Other
88	−	−	−	—	—	−	None

[a]P-receptor particle agglutination test (Svenson *et al.*, 1982).
[b]Data modified from Svenson *et al.* (1982).
[c]'Other' indicates MR fimbriae other than P or X.

the $\bar{p}$ type, sometimes along with the MR agglutination of erythrocytes of guinea-pig or ox (Svenson *et al.*, 1982). The data from HA tests, therefore, must be interpreted with great caution because these other types of fimbriae may be present alone or together with P fimbriae. The results from an extended study by Svenson *et al.* (1982), using combined haemagglutination and PPA testing, are shown for 197 urinary strains of *E. coli* (Table 2). None of the 58 *E. coli* strains judged to be P-fimbriated by their inability to give MR agglutination of human $\bar{p}$ erythrocytes while agglutinating P_1 or P^k_2 erythrocytes was negative in the PPA test, the specificity of which was confirmed by the ability of a synthetic (Gal-Gal)–glycoside to inhibit the PPA reaction. Fifty-one of the other strains agglutinated $\bar{p}$ as well as P_1 and P^k_2 erythrocytes, and on the basis of HA assay alone might have been thought not to possess P fimbriae. However, the ancillary use of the PPA test revealed that 22 of these 51 strains did indeed possess P fimbriae along with some other fimbriae responsible for the MR agglutination of human $\bar{p}$ erythrocytes.

A second class of fimbriae, the X fimbriae, has been described as being of probable significance in colonization of the urinary tract by *E. coli* strains (Väisänen *et al.*, 1981). Tested in HA tests at densities of $\sim 2 \times 10^9$ bacteria ml^{-1}, X-fimbriated bacteria agglutinated human erythrocytes of P_1, P^k_2 and $\bar{p}$ types but not those of guinea-pig or cattle (Svenson *et al.*, 1982). As judged by HA reactions, therefore, 13 strains possessed X fimbriae, either alone (PPA-negative) or together with P fimbriae (PPA-positive) (Table 2). Thirty-eight strains which agglutinated in MR fashion the erythrocytes of man, guinea-pig and ox (Table 2) formed some MRE-HA other than those associated with P or X fimbriae, though 16, giving positive reactions in the PPA test, must have carried

P fimbriae also. Thus, these observations established that at least 20% of the 109 MR-HA$^+$ strains carried two kinds of fimbriae with MR-HA activities. Another 88 strains showed no MR-HA activity with these erythrocytes at the density tested (Table 2).

Thus, in summary, with HA typing alone only 54 (50%) of the MR-HA$^+$ strains would have been correctly identified as P-fimbriated, whereas the use of the PPA test in conjunction with HA typing revealed a further 22 strains with P fimbriae and some other MR haemagglutinin (Table 2). Thus it becomes clear that detection of MR agglutination of human erythrocytes by urinary strains of *E. coli* does not imply that they are P-fimbriated. Indeed, even with the inclusion of the rare human p̄ erythrocytes in the HA assay, only those strains producing P fimbriae alone, or with another type of fimbriae unreactive with human erythrocytes, would have been correctly identified as P-fimbriated. The diagnostic potential of these specific particle tests is immense and the development of additional PPA tests made specific for other colonizing fimbriae as their receptors become identified is an exciting prospect.

These same studies, however, reveal just how important it is in testing for MRE-HA activity to use dense suspensions of bacteria grown under the cultural conditions appropriate for MRE-HA expression. Thus, although X-fimbriated strains are said not to agglutinate erythrocytes of guinea-pig or ox (Svenson *et al.*, 1982), strain SS10 described as possessing only X fimbriae (G. Källenius and S. Svenson, personal communication) in my hands gave MRE agglutination of erythrocytes of guinea-pig, man, ox, pig and rat when tested at $\sim 5 \times 10^{11}$ bacteria ml^{-1} (unpublished results); that MRE-HA pattern is common in fimbriated, urinary isolates of *E. coli* (Crichton, 1980). Thus, the distinction of X- and non-X-fimbriated bacteria by the pattern of their reactions with these erythrocytes does not appear to be valid. Nevertheless, the general observations that many strains of *E. coli* were multiply fimbriated are no doubt correct (Svenson *et al.*, 1982).

Tests in microtitre trays with suspensions at ~ 1–2×10^9 bacteria ml^{-1} indicated that P-fimbriated bacteria gave MR agglutination of human erythrocytes only of eight species screened (Källenius and Möllby, 1979). However, two P-fimbriated strains of *E. coli* (ER2 and JR1) tested at $\sim 10^{11}$ bacteria ml^{-1} in the rocked tile test at 4°C agglutinated erythrocytes of man, pig, sheep and rabbit (D. C. Old, unpublished results); this MRE-HA type is pattern 1 of Duguid *et al.* (1979), the most common MRE-HA type and one present in $\sim 33\%$ of MRE-HA$^+$ strains of *E. coli* from urinary tract infection (Crichton, 1980). A further interesting observation was the finding that this MRE-HA pattern was also the predominant one present in most (65%) of the biochemically atypical strains of the 'Alkalescens–Dispar' types (Crichton *et al.*, 1981) which have been implicated as opportunistic pathogens in urinary infection (Kauffmann, 1969). Tests with diluted cultures of our MRE-HA$^+$ strains of this man, pig, sheep (MPS) type showed loss of the sheep activity at $< 10^{10}$ bacteria ml^{-1}.

These observed differences in the haemagglutination patterns given by P- or X-fimbriated *E. coli* depend no doubt on the density of bacterial suspension tested, but show how difficult it is to compare results from different studies even for the same strains.

A wide variety of MRE-HAs has been described in *E. coli* (Duguid *et al.*, 1955, 1979; Crichton, 1980), though a specific role in infection has as yet been described for very few of them. In order to correct some of the confusion in the literature, it would seem good practice to provide for any MRE-HA as complete as possible a description of its haemagglutinating properties obtained with as wide a range of erythrocytes as is available. The practice of screening for the presence of MRE-HAs and describing them with reference to only one or two species of erythrocytes is unacceptable. First, strains of *E. coli* with MRE-HAs reacting weakly or not at all with the chosen erythrocytes will be misidentified as MRE-HA$^-$; thus, for example, the use of, say, the erythrocytes of guinea-pig and man (Van den Bosch *et al.*, 1980) or those of fowl and man (Minshew *et al.*, 1978) would have resulted in ~20% of our MRE-HA$^-$ strains being mistyped (P. B. Crichton and D. C. Old, unpublished results). Second, strains of diverse MRE-HA types are grouped as similar with the use of a limited range of erythrocytes; thus, for example, the use of only human erythrocytes would class together as MR-human (1) urinary strains of *E. coli* with P, X or M fimbriae (Väisänen *et al.*, 1982), and those with fimbrial antigens F7–F12; (2) ETEC strains with CFA/I or E8775 fimbriae, and strains with non-fimbrial narrow-spectrum 'human-only' MRE-HA activity (Duguid *et al.*, 1979; Ip *et al.*, 1981; Sheladia *et al.*, 1982).

Haemagglutination Patterns and Fimbrial Antigens

In Enterobacteriaceae type 1 and type 3 fimbriae are widely distributed (Duguid and Old, 1980; Adegbola and Old, 1982; Old and Adegbola, 1982). Each of these fimbrial types shows its own characteristically constant pattern of haemagglutinating activity, type 1 associated with MS-HA (as described above) and type 3 with MR/K-HA, the haemagglutinin active on tannic acid-treated ox erythrocytes (Duguid, 1959; Duguid and Old, 1980; Old and Adegbola, 1982). Nevertheless, type 1 fimbriae of different species show little or no sharing of fimbrial antigens (Gillies and Duguid, 1958; Duguid and Campbell, 1969; Nowotarska and Mulcyzk, 1977), and those in a single species like *E. coli* may share only one of several fimbrial antigens (Gillies and Duguid, 1958). Similar antigenic heterogeneity has been demonstrated both between and among type 3 fimbriated strains of *Serratia, Klebsiella* and *Enterobacter* (Adegbola and Old, 1982; Old and Adegbola, 1983) and among type MR/P fimbriated strains of *Serratia* of different MR/P-HA patterns (Adegbola and Old, 1982).

Similarly, it must be realized that different strains of *E. coli* with identical patterns of MRE-HA activity may possess fimbriae with quite distinct fimbrial antigens. Identity of MRE-HA pattern is not synonymous with homogeneity of fimbrial antigens and problems result when HA results are incorrectly used as predictors of particular kinds of fimbriae, as shown in the following examples.

Two important fimbrial antigens have been described in ETEC strains and shown to be important in intestinal colonization in man: CFA/I which gives MRE-HA of erythrocytes of man and ox (D. G. Evans *et al.* 1977) and CFA/II which gives MRE-HA of erythrocytes of ox only (Evans and Evans, 1978), HA types I and II in the extended HA typing scheme of D. J. Evans *et al.* (1979, 1980c). That fimbrial antigens such as CFA/I and CFA/II must be defined serologically, however, was adequately demonstrated in the comprehensive HA study of Cravioto *et al.* (1982) in which 916 strains of *E. coli* of diverse origin were examined (Table 3). Among 205 ETEC strains, 42 gave MR-HA of erythrocytes of man and ox and 49 those of ox only; among these 91 strains, putatively CFA/I$^+$ or CFA/II$^+$ on the basis of their HA reactions, 8 strains were not confirmed serologically as possessing either of these fimbrial antigens. Furthermore, two strains putatively CFA/I$^+$ on the basis of their haemagglutination reactions were confirmed serologically as CFA/II$^+$ (Table 3). Of the other 711 strains of faecal and extra-intestinal origin, MR haemagglutination of human or bovine erythrocytes was commonly present, yet none of 274 MR-HA$^+$ strains was confirmed serologically as CFA/I$^+$ or CFA/II$^+$. Accordingly, MR-HA of erythrocytes of either or both of man and ox was not correlated with the presence of CFA/I or CFA/II, except strains already classified as ETEC from evidence of their toxigenic status or their antigen type. Again in ETEC strains other fimbrial antigens such as E8775 have been described (Thomas and Rowe, 1982; Thomas *et al.*, 1982) which also give MR-HA of erythrocytes of man and ox and yet are serologically distinct from CFA/I and CFA/II. Furthermore, among strains labelled CFA/II$^+$ on the basis of their MR-HA of bovine erythrocytes, different fimbrial and non-fimbrial antigens have been defined (Smyth, 1982; Cravioto *et al.*, 1982).

Table 3. *Lack of correlation between MR haemagglutination of bovine and human erythrocytes and presence of colonization factors CFA/I and CFA/II[a]*

		No. of strains showing MR agglutination of erythrocytes of:			No. confirmed serologically as:	
No.	Group	Man	Ox	Man; ox	CFA/I$^+$	CFA/II$^+$
205	ETEC	0	49	42	36	47
186	EPEC, faecal	12	6	7	0	0
149	Non-EPEC, faecal	20	2	11	0	0
122	CSF, extra-intestinal	62	12	7	0	0
124	Urine, extra-intestinal	45	4	8	0	0
130	Blood, extra-intestinal	60	4	14	0	0

[a]Data modified from Cravioto *et al.* (1982).

New fimbrial antigens F7–F12 in *E. coli* strains have been defined by crossed immunoelectrophoresis (CIE) (Ørskov and Ørskov, 1983). Strains C1212, C1254, 3669 and C1979 possessing, respectively, fimbrial antigens F7, F8, F9 and F12, when examined by me in rocked tile tests at 4°C, gave MRE-HA of erythrocytes of man, pig, sheep and rabbit, that is, the MRE-HA pattern of authentic P-fimbriated strains; they were also type 1 fimbriate and produced MS-HA (D. C. Old, unpublished). It is possible that each of these urinary isolates showing this MRE-HA pattern possesses P fimbriae which are antigenically distinct in strains of different O:K:H serotypes (Ørskov and Ørskov, 1983). On the other hand, it is equally likely that several types of MR fimbriae give this, the commonest, MRE-HA pattern (Duguid *et al.*, 1979). The use of the highly specific PPA test in such cases would be invaluable, especially since some of these strains probably carry, as well as type 1 fimbriae, two kinds of MR fimbriae (Korhonen *et al.*, 1982a; Klemm *et al.*, 1982; Ørskov and Ørskov, 1983).

It should be clear, therefore, from these few examples that the reported observations in some studies purporting to discuss the role of adhesins in *E. coli* may be insufficiently precise to define the types of HA present and, indeed, to state whether several, or no, HAs are present. Even when HA tests have been adequately performed, HA patterns themselves provide no more than an indication of some of the adhesive activities present and, hence, should be used only in preliminary screening procedures. Fimbrial antigens should be analyzed by some serological technique.

Adhesins with Weak Haemagglutinating Activity

Thus far, the difficulties involved in separately identifying the presence and properties of the different MR adhesins in any culture have been discussed for adhesins with readily demonstrable haemagglutination properties. The description of MR adhesins requiring special methods for demonstration of their HA activities is even more problematic; and for the study of non-haemagglutinating adhesins, other techniques such as *in vivo* and *in vitro* adhesion to epithelial cells are needed.

A settling test, such as the microhaemagglutination method in plastic trays, will demonstrate the weak MRE-HA activities of some MR adhesins of importance in animal strains in *E. coli* (Jones and Rutter, 1974; Burrows *et al.*, 1976) and has been successfully exploited to unravel the several HA activities of strains with multiple weak MR adhesins (Morris *et al.*, 1982; Chanter, 1983). It is, however, perhaps too sensitive a technique for use in inexperienced hands as a primary, routine screening procedure, though it has a clear role as an applied system with which to prove the HA properties of fimbriate strains that are known to be non-haemagglutinating in rocked tests.

Two particular sources of error should suffice to indicate the discretion with which HA results obtained in settling tests should be interpreted. First, some strains of *E. coli* adhere strongly to the plastic of microtitre trays and the pattern of sedimentation of red cells onto bacteria already adherent to plastic in the wells of the tray may give a dispersed effect that mimics that of truly agglutinated erythrocytes. This kind of problem has also been experienced in interpreting the results of adhesion studies of *E. coli* strains to cellular monolayers in microtitre trays (Vosbeck and Huber, 1982). Because some of these adhesions to plastic probably occur randomly as a result of chance charge effects on some areas of the plastic, they are difficult to control. Second, agglutination of erythrocytes demonstrable in settling but not in shaken tests may not be associated with adherence of bacteria to the erythrocytes. Thus, in the course of screening non-fimbriate MS$^-$MRE$^-$ strains of *E. coli* for the presence of weak haemagglutinating activities, we found that most of the ~30 strains tested gave MR agglutination of erythrocytes of pig and sheep in settling tests in glass or plastic, but not in the conventional rocked tile test at 4°C (P. B. Crichton and D. C. Old, unpublished).

Similar MR agglutination of pig and sheep erythrocytes by *Salmonella typhimurium* strains of diverse biotypes was obtained after their growth on PB agar and occasionally in broth (Tavendale *et al.*, 1983). However, MR-HA sediments examined by phase-contrast microscopy did not show bacteria attached to red cells or red cells to each other; again, cell-free filtrates and supernates had MR-HA titres as high as the positive bacterial suspensions and washed bacteria were MR-HA$^-$. These findings suggested that a diffusible product produced particularly well on agar-grown cultures acted as an MR 'haemagglutinin' though it did not necessarily indicate the presence of a bacterial 'adhesin'. A similar phenomenon also demonstrated in settling tests but due, apparently, to a different kind of diffusible 'haemagglutinin' has been reported for other enterobacteria (Jones *et al.*, 1980). If these diffusible HAs, and others with different HA specificities, are widely distributed, as seems likely, the need for caution in the interpretation of positive HA results by settling tests is apparent. These observations show that a 'positive' dispersed pattern of sedimentation in a settling test may not indicate either adhesion of bacteria to erythrocytes or haemagglutination, that is, clumping of erythrocytes.

MR activity with pig and sheep erythrocytes is common in MRE-HA$^+$ strains of *E. coli* (Table 4): in these strains, however, it is cell-bound and associated with activity against human erythrocytes.

Electron Microscopy

When the adhesive properties of any group of enterobacteria are defined by means of haemagglutination tests, it is essential that parallel electron micro-

Table 4. *Haemagglutination properties[a] of Escherichia coli possessing important fimbrial antigens*

Strain no. or designation	MR fimbrial antigen	Production of:	
		MS-HA	MRE-HA, pattern
JR1	P	−	Man, pig, sheep, rabbit
ER2	P	+	Man, pig, sheep, rabbit
SS10	X	−	Guinea-pig, man, ox, pig, rat, mouse
H10407	CFA/I	+	Man, ox, pig
PB-176	CFA/II	−	Ox
C1212-77	F17	+	Man, pig, sheep, rabbit
C1254-77	F8	+	Man, pig, sheep, rabbit
3669	F9	+	Man, pig, sheep, rabbit
C1976-79	F11	+	Man, pig
C1979-79	F12	+	Man, pig, sheep, rabbit
E8775-A	E8775	−	Fowl, man
O101:K99; F41	F41	+	Man, pig, sheep, rat

[a]In rocked tile tests performed at 4°C and bacterial suspensions at ~10^{12} bacteria ml^{-1}.

scopic studies are made on bacterial suspensions the HA properties of which are known. Especially when tests are made with bacteria grown under a variety of conditions to express or suppress formation of one or the other of the HA activities, it is possible to establish any correlation existing between the presence of different HAs and that of fimbriae. By such a combination of techniques, the following points have been established in *E. coli:* (1) MS-HA is correlated with the presence of type 1 fimbriae (Duguid *et al.*, 1955, 1979); (2) some broad-spectrum MRE-HAs are associated with diverse types of fimbriae (Duguid *et al.*, 1979; Duguid and Old, 1980; Gaastra and de Graaf, 1982); (3) the narrow-spectrum MRE-HAs, that is the monospecific HAs of D. J. Evans *et al.* (1980c), are non-fimbrial (Duguid *et al.*, 1955, 1979; D. J. Evans *et al.*, 1980a; Ip *et al.*, 1981); (4) weakly haemagglutinating adhesins such as K88 or K99 are fibrillar or fimbrial (Stirm *et al.*, 1967; Burrows *et al.*, 1976); (5) some non-haemagglutinating MR adhesins, such as 987P, are fimbrial (Nagy *et al.*, 1977; Isaacson and Richter, 1981). Furthermore, there is increasing evidence to suggest that weakly or non-haemagglutinating MR adhesins may be more common in *E. coli* strains of human origin than originally realised (Klemm *et al.*, 1982; Korhonen *et al.*, 1982a; Smyth, 1982; Ørskov and Ørskov, 1983). The use of the electron microscope is particularly helpful in establishing the presence of these adhesins of weak or non-haemagglutinating status, though not in multiply adhesive strains.

Direct electron microscopy combined with negative staining techniques makes the task of distinguishing fimbriae of distinct morphologies relatively straightforward. However, when two or more types of morphologically similar fimbriae are produced by a strain, it is essential to use immuno-electron microscopy (IEM) for their satisfactory characterization. Thus, on strains of *E. coli* of serotypes O6:K15:H16 or H⁻ of a common MR-HA type, Smyth (1983) distinguished morphologically indistinguishable types of fimbriae by their differential staining reactions with appropriately absorbed fimbrial anti-sera made specific for each fimbrial type. We have used IEM with similar success to demonstrate in multiply haemagglutinating strains of *Enterobacter, Klebsiella* and *Serratia* the presence of serologically different forms of type 3 fimbriae associated with the MR/K-HA, and different serological types of MRP fimbriae associated with the MR/P-HAs of different patterns (Adegbola and Old, 1982, 1983; Old and Adegbola, 1983). A major benefit of IEM is its direct demonstration of what is coated with antibody and, though it may be less sensitive in detecting minor differences in fimbrial antigens than other serological techniques such as immunodiffusion (Cravioto *et al.*, 1982) or CIE (Ørskov and Ørskov, 1983), the results from these indirect techniques are less easily related to structural materials.

The value of HA typing, therefore, in characterizing the adhesins of *E. coli* is greatly enhanced when it is used in conjunction with morphological, serological or chemical techniques. Used alone, it may yield erroneous information and will certainly do so if it is assumed that MR-HA patterns do anything more than identify the presence of haemagglutinating adhesins.

HA Typing in Epidemiological Studies

Many different studies have contributed to our understanding of the important associations existing in animal and human strains of *E. coli* between O:K:H serotypes, enterotoxigenicity and adhesins, the associations influenced to some extent by the stability of carriage of some plasmids in strains of particular serotypes (reviewed by Gaastra and de Graaf, 1982). The HA typing scheme of Evans *et al.* (1979, 1980c), based on both MS- and MRE-HA patterns obtained in tests with erythrocytes of cattle, chickens, guinea-pigs, men and monkeys (African green), established the existence of 7 major HA types and ~24 minor, variant HA types. Their studies revealed certain general trends in the associations between HA types, enterotoxigenicity, serotypes and clinical sources of strains.

Thus, ETEC strains with CFA/I or CFA/II antigens belonged, respectively, to HA types I and II; it has already been indicated, however, that most strains of HA types I and II will not be confirmed serologically as CFA/I⁺ or CFA/II⁺ unless they happen to belong also to ETEC serotypes. HA types III and IV generally formed MS-HA but few formed MRE-HA, and they included strains of the classical EPEC serotypes and faecal isolates of non-EPEC serotypes, results in agreement with those of other workers indicating that MRE-HA production is not common in these strains (Duguid *et al.*, 1979; Cravioto *et al.*, 1982). HA type V strains, without particular clinical association, possessed monospecific

MR-HAs, equivalent to the narrow-spectrum types 6–10 of Duguid *et al.* (1979) which are non-fimbrial (Duguid *et al.*, 1955; Ip *et al.*, 1981). Strains of HA types VI and VII possessed broad-spectrum MRE-HAs, the former group being found in *E. coli* strains from extra-intestinal sites such as blood, cerebrospinal fluid and, in particular, urine (D. J. Evans *et al.*, 1980b,c, 1981).

Used in that way, HA typing provided valuable supplementary information which was correctly used in conjunction with other epidemiological data about serotype and enterotoxigenicity. It also drew attention to the need to include a description of fimbrial antigens in serotyping *E. coli* and acted as a major stimulus of the current progress in that field (Ørskov and Ørskov, 1983). It is, however, unfortunate that others have applied that HA typing scheme independently of other data, for, used in that way, the information it yields is limited.

Our application of HA typing to the epidemiological study of *E. coli* has been more simple. Type 1 fimbriation was included as a property in our biotyping scheme for *E. coli* because it was a remarkably stable character, discriminating among 1242 cultures a minority of only 17% that failed to form type 1 fimbriae (Crichton and Old, 1979, 1982). Our routine screening of strains of *E. coli* for their MRE-HA patterns with seven erythrocyte species (fowl, guinea-pig, horse, man, ox, pig, sheep) revealed that ~50–60% of strains were MRE-HA$^+$. Although many MRE-HA patterns were noted, most MRE-HA$^+$ strains produced one of a limited number (10–15) of common MRE-HA patterns (Duguid *et al.*, 1979; Crichton, 1980). Accordingly, though HA typing provides additional information that is useful in describing strains, especially those from patients with long-term urinary tract infection, it is insufficiently discriminating to be used as a primary typing technique. For primary discrimination, therefore, we have relied on data from biotyping and resistotyping techniques, which, used in conjunction, are sufficient to answer most questions regarding strain identity (Crichton and Old, 1980; Old *et al.*, 1980; Wilson *et al.*, 1981). The data from HA typing have generally supported those from these two systems and most MRE-HA characters have proved to be stable *in vivo* and *in vitro* over many years (Crichton and Old, 1980). However, with some series of replicate isolates of a strain of *E. coli* recovered from patients over years, we have noticed that a minority of the cultures, though otherwise identical in biotype, resistotype, serotype and antibiogram type to the majority, had lost or gained an MRE-HA character (Old *et al.*, 1980; Crichton and Old, 1980). Thus, because any typing character is subject to variation *in vivo*, the results of HA typing, even used as an ancillary tool, must be used with discretion.

Concluding Remarks

Haemagglutination typing of *E. coli* is useful in two ways: (1) as a means to identify haemagglutinating strains and indicate some of the adhesins that should be investigated for a role in colonization of epithelial surfaces and (2) as a tool in epidemiological studies to characterize particular types of strains involved in different kinds of infections.

For the former, HA typing is a simple, inexpensive technique by which to classify and identify the major, that is, the MS-HA and MRE-HA, adhesins of *E. coli,* and to demonstrate their separate properties when produced together. However, because many of the more common MRE-HA patterns are associated with adhesins of diverse antigenicity, the antigenic identity of fimbriae cannot be deduced from HA patterns. HA typing is not by itself sufficient in studies of adhesive strains producing two or more MRE-HAs, and quite inappropriate for studying strains with non-haemagglutinating adhesins. Used in conjunction with electron microscopic, serological and chemical techniques, HA typing has aided the classification of both fimbrial and nonfimbrial HAs and provided a basis for the study of their other *in vitro* and *in vivo* adhesive properties.

Because many surface-associated adhesins important in colonization of epithelial surfaces in man and animals show readily demonstrable haemagglutinating properties, it might be thought that HA typing would be a useful technique for epidemiological studies. However, HA typing discriminates too few types for its successful use as a primary typing system for subspecies identification in *E. coli*. It may, however, when used as an ancillary typing technique in conjunction with a primary system such as serotyping or biotyping, provide some additional strain discrimination. HA types prove remarkably stable over extended periods both *in vitro* and *in vivo* and so their description aids strain discrimination in, for example, long-term urinary tract infection.

Acknowledgments

I thank Professor J. P. Duguid for helpful comments and Drs. G. Källenius, J. Morris, F. Ørskov, I. Ørskov, B. Rowe and S. Svenson for gifts of strains of known fimbrial types.

References

Adegbola, R. A. and Old, D. C. (1982). New fimbrial hemagglutinin in *Serratia* species. *Infection and Immunity* **38**, 306–315.

Adegbola, R. A. and Old, D. C. (1983). Fimbrial haemagglutinins of *Enterobacter* species. *Journal of General Microbiology* **129**, 2175–2180.

Brinton, C. C. (1959). Non-flagellar appendages of bacteria. *Nature (London)* **183**, 782–786.

Burrows, M. R., Sellwood, R. and Gibbons, R. A. (1976). Haemagglutinating and adhesive properties associated with the K99 antigen of bovine strains of *Escherichia coli*. *Journal of General Microbiology* **96**, 269–275.

Chanter, N. (1983). Partial purification and characterization of two non K99 mannose-resistant haemagglutinins of *Escherichia coli* B41. *Journal of General Microbiology* **129**, 235–243.

Clegg, S. (1978). The adhesive and antigenic properties of enterobacterial haemagglutinins. Ph.D. Thesis, Dundee University.

Clegg, S. and Old, D. C. (1979). Fimbriae of *Escherichia coli* K-12 strain AW405 and related bacteria. *Journal of Bacteriology* **137**, 1008–1012.

Cravioto, A., Scotland, S. M. and Rowe, B. (1982). Hemagglutination activity and colonization factor antigens I and II in enterotoxigenic and non-enterotoxigenic strains of *Escherichia coli* isolated from humans. *Infection and Immunity* **36**, 189–197.

Crichton, P. B. (1980). Differential typing of *Escherichia coli* strains. Ph.D. Thesis, Dundee University.

Crichton, P. B. and Old, D. C. (1979). Biotyping of *Escherichia coli*. *Journal of Medical Microbiology* **12**, 473–486.

Crichton, P. B. and Old, D. C. (1980). Differentiation of strains of *Escherichia coli:* Multiple typing approach. *Journal of Clinical Microbiology* **11**, 635–640.

Crichton, P. B. and Old, D. C. (1982). A biotyping scheme for the subspecific discrimination of *Escherichia coli*. *Journal of Medical Microbiology* **15**, 233–242.

Crichton, P. B., Ip, S. M. and Old, D. C. (1981). Hemagglutinin typing as an aid in identification of biochemically atypical *Escherichia coli* strains. *Journal of Clinical Microbiology* **14**, 599–603.

de Graaf, F. K. and Roorda, I. (1982). Production, purification and characterization of the fimbrial antigen F41 isolated from calf enteropathogenic *Escherichia coli* strain B41M. *Infection and Immunity* **36**, 751–758.

de Graaf, F. K., Wientjes, F. B. and Klaasen-Boor, P. (1980). Production of K99 antigen by enterotoxigenic *Escherichia coli* strains of antigen groups 08, 09, 020, and 0101 grown at different conditions. *Infection and Immunity* **27**, 216–221.

Duguid, J. P. (1959). Fimbriae and adhesive properties in Klebsiella strains. *Journal of General Microbiology* **21**, 271–286.

Duguid, J. P. (1964). Functional anatomy of *Escherichia coli* with special reference to entero-pathogenic *E. coli*. *Revista Latino-americana de Microbiologia* **7**, Supplementos 13–14, 1–16.

Duguid, J. P. and Campbell, I. (1969). Antigens of the type-1 fimbriae of salmonellae and other enterobacteria. *Journal of Medical Microbiology* **2**, 535–553.

Duguid, J. P. and Gillies, R. R. (1957). Fimbriae and adhesive properties in dysentery bacilli. *Journal of Pathology and Bacteriology* **74**, 397–411.

Duguid, J. P. and Old, D. C. (1980). Adhesive properties of Enterobacteriaceae. *In* "Bacterial Adherence" (Ed. E. H. Beachey), Receptors and Recognition Series B, Vol. 6, pp. 185–217. Chapman and Hall, London.

Duguid, J. P. and Wilkinson, J. F. (1961). Environmentally induced changes in bacterial morphology. *Symposium of the Society for General Microbiology* **11**, 69–99.

Duguid, J. P., Smith, I. W., Dempster, G. and Edmunds, P. N. (1955). Non-flagellar filamentous appendages ("fimbriae") and haemagglutinating activity in *Bacterium coli*. *Journal of Pathology and Bacteriology* **70**, 335–348.

Duguid, J. P., Anderson, E. S. and Campbell, I. (1966). Fimbriae and adhesive properties in salmonellae. *Journal of Pathology and Bacteriology* **92**, 107–138.

Duguid, J. P., Clegg, S. and Wilson, M. I. (1979). The fimbrial and non-fimbrial haemagglutinins of *Escherichia coli*. *Journal of Medical Microbiology* **12**, 213–227.

Evans, D. G. and Evans, D. J. (1978). New surface-associated heat-labile colonization factor antigen (CFA/II) produced by enterotoxigenic *Escherichia coli* of serogroup O6 and O8. *Infection and Immunity* **21**, 638–647.

Evans, D. G., Evans, D. J. and Tjoa, W. (1977). Hemagglutination of human group A erythrocytes by enterotoxigenic *Escherichia coli* isolated from adults with diarrhoea: Correlation with colonization factor. *Infection and Immunity* **18**, 330–337.

Evans, D. J., Evans, D. G. and du Pont, H. L. (1979). Hemagglutination patterns of enterotoxigenic and enteropathogenic *Escherichia coli* determined with human, bovine, chicken and guinea-pig erythrocytes in the presence and absence of mannose. *Infection and Immunity* **23**, 336–346.

Evans, D. J., Clegg, S. and Evans, D. G. (1980a). Fimbrial antigens and pathogenic *Escherichia coli*. *Lancet* **1**, 201.

Evans, D. J., Evans, D. G. and Clegg, S. (1980b). Lethality of bacteremia-associated *Escherichia coli* for mice in relation to serotype and hemagglutination (HA) type. *FEMS Microbiology Letters* **9**, 171–174.

Evans, D. J., Evans, D. G., Young, L. S. and Pitt, J. (1980c). Hemagglutination typing of *Escherichia coli:* Definition of seven hemagglutination types. *Journal of Clinical Microbiology* **12**, 235–242.

Evans, D. J., Evans, D. G., Höhne, C., Noble, M. A., Haldane, E. V., Lior, H. and Young, L. S. (1981). Hemolysin and K antigens in relation to serotype and hemagglutination type of *Escherichia coli* isolated from extraintestinal infections. *Journal of Clinical Microbiology* **13**, 171–178.

Fein, J. E. (1981). Screening of uropathogenic *Escherichia coli* for expression of mannose-sensitive adhesins: Importance of culture conditions. *Journal of Clinical Microbiology* **13**, 1088–1095.

Francis, D. H., Remmers, G. A. and De Zeeuw, P. S. (1982). Production of K88, K99, and 987P antigens by *Escherichia coli* cultured on synthetic and complex media. *Journal of Clinical Microbiology* **15**, 181–183.

Gaastra, W. and de Graaf, F. K. (1982). Host-specific fimbrial adhesins of noninvasive enterotoxigenic *Escherichia coli* strains. *Microbiological Reviews* **46**, 129–161.

Gillies, R. R. and Duguid, J. P. (1958). The fimbrial antigens of *Shigella flexneri*. *Journal of Hygiene* **56**, 303–318.

Guinée, P. A. M., Jansen, W. H. and Agterberg, C. M. (1976). Detection of the K99 antigen by means of agglutination and immunoelectrophoresis in *Escherichia coli* isolates from calves and its correlation with enterotoxigenicity. *Infection and Immunity* **13**, 1369–1377.

Guinée, P. A. M., Veldkamp, J. and Jansen, W. H. (1977). Improved Minca medium for the detection of K99 antigen in calf enterotoxigenic strains of *Escherichia coli*. *Infection and Immunity* **15**, 676–678.

Guyot, G. (1908). Ueber die bakterielle Hämagglutination (Bakterio-Haemagglutination). *Zentralblatt für Bakteriologie, Parasitenkunde, Infektionskrankenheiten und Hygiene, Abteilung 1: Originale* **47**, 640–653.

Ip, S. M., Crichton, P. B., Old, D. C. and Duguid, J. P. (1981). MRE haemagglutinins and fimbriae in *Escherichia coli*. *Journal of Medical Microbiology* **14**, 223–226.

Isaacson, R. E. (1977). K99 surface antigen of *Escherichia coli:* Purification and partial characterization. *Infection and Immunity* **15**, 272–279.

Isaacson, R. E. and Richter, P. (1981). *Escherichia coli* 987P pilus: Purification and partial characterization. *Journal of Bacteriology* **146**, 784–789.

Isaacson, R. E., Nagy, B. and Moon, H. W. (1977). Colonization of porcine small intestine by *Escherichia coli:* Colonization and adhesion factors of pig enteropathogens that lack K88. *Journal of Infectious Diseases* **135**, 531–539.

Jones, G. W. and Rutter, J. M. (1974). The association of K88 antigen with haemagglutinating activity in porcine strains of *Escherichia coli*. *Journal of General Microbiology* **84**, 135–144.

Jones, G. W., Richardson, L. A. and Vanden-Bosch, J. L. (1980). Phases in the interaction between bacteria and animal cells. In "Microbial Adhesion to Surfaces" (Eds. R. C. W. Berkeley, J. M. Lynch, J. Melling, P. R. Rutter and B. Vincent), pp. 211–219. Ellis Horwood, Chichester.

Källenius, G. and Möllby, R. (1979). Adhesion of *Escherichia coli* to human periurethral cells correlated to mannose-resistant agglutination of human erythrocytes. *FEMS Microbiology Letters* **5**, 295–299.

Källenius, G., Möllby, R., Svenson, S. B., Winberg, J. and Hultberg, H. (1980a). Identification of a carbohydrate receptor recognised by uropathogenic *Escherichia coli*. *Infection, Supplement* **3**, 288–293.

Källenius, G., Möllby, R., Svenson, S. B., Winberg, J., Lundblad, A., Svensson, S. and

Cedergren, B. (1980b). The P^k antigen as receptor for the haemagglutinin of pyelonephritic *Escherichia coli*. *FEMS Microbiology Letters* **7**, 297–302.

Källenius, G., Möllby, R., Svenson, S. B., Helin, I., Hultberg, H., Cedergren, B. and Winberg, J. (1981). Occurrence of P-fimbriated *Escherichia coli* in urinary tract infections. *Lancet* **2**, 1369–1372.

Källenius, G., Svenson, S. B., Möllby, R., Korhonen, T., Winberg, J., Cedergren, B., Helin, I. and Hultberg, H. (1982). Carbohydrate receptor structures recognised by uropathogenic *E. coli*. *Scandinavian Journal of Infectious Diseases, Supplementum* **33**, 52–60.

Kauffmann, F. (1948). On haemagglutination by *Escherichia coli*. *Acta Pathologica et Microbiologica Scandinavica* **25**, 502–506.

Kauffmann, F. (1969). "The Bacteriology of *Enterobacteriaceae*," 2nd ed. Munksgaard, Copenhagen.

Klemm, P., Ørskov, I. and Ørskov, F. (1982). F7 and type 1-like fimbriae from three *Escherichia coli* strains isolated from urinary tract infections: Protein, chemical and immunological aspects. *Infection and Immunity* **36**, 462–468.

Korhonen, T. K., Väisänen, V., Kallio, P., Nurmiaho-Lassila, E.-L., Ranta, H., Siitonen, A., Svenson, S. B. and Svanborg Edén, C. (1982a). The role of pili in the adhesion of *Escherichia coli* to human urinary tract epithelial cells. *Scandinavian Journal of Infectious Diseases, Supplementum* **33**, 26–31.

Korhonen, T. K., Väisänen, V., Saxén, H., Hultberg, H. and Svenson, S. B. (1982b). P-antigen-recognizing fimbriae from human uropathogenic *Escherichia coli* strains. *Infection and Immunity* **37**, 286–291.

MacLagan, R. M. and Old, D. C. (1980). Haemagglutinins and fimbriae in different serotypes and biotypes of *Yersinia enterocolitica*. *Journal of Applied Bacteriology* **49**, 353–360.

Middeldorp, J. M. and Witholt, B. (1981). K88-mediated binding of *Escherichia coli* outer membrane fragments to porcine intestinal epithelial cell brush borders. *Infection and Immunity* **31**, 42–51.

Minshew, B. H., Jorgensen, J., Counts, G. W. and Falkow, S. (1978). Association of hemolysin production, hemagglutination of human erythrocytes, and virulence for chicken embryos of extra-intestinal *Escherichia coli* isolates. *Infection and Immunity* **20**, 50–54.

Morris, J. A., Stevens, A. E. and Sojka, W. J. (1977). Preliminary characterization of cell-free K99 antigen isolated from *Escherichia coli* B41. *Journal of General Microbiology* **99**, 353–357.

Morris, J. A., Stevens, A. E. and Sojka, W. J. (1978). Anionic and cationic components of the K99 surface antigen from *Escherichia coli* B41. *Journal of General Microbiology* **107**, 173–175.

Morris, J. A., Thorns, C. J. and Sojka, W. J. (1980). Evidence for two adhesive antigens on the K99 reference strain *Escherichia coli* B41. *Journal of General Microbiology* **118**, 107–113.

Morris, J. A., Thorns, C. J., Scott, A. C., Sojka, W. J. and Wells, G. A. (1982). Adhesion in vitro and in vivo associated with an adhesive antigen (F41) produced by a K99 mutant of the reference strain *Escherichia coli* B41. *Infection and Immunity* **36**, 1146–1153.

Nagy, B., Moon, H. W. and Isaacson, R. E. (1977). Colonization of porcine intestine by enterotoxigenic *Escherichia coli*: Selection of piliated forms in vivo, adhesion of piliated forms to epithelial cells in vitro, and incidence of pilus antigen among porcine enteropathogenic *E. coli*. *Infection and Immunity* **16**, 344–352.

Nowotarska, M. and Mulczyk, M. (1977). Serologic relationship of fimbriae among *Enterobacteriaceae*. *Archivum Immunologiae et Therapiae Experimentalis* **25**, 7–16.

Old, D. C. (1972). Inhibition of the interaction between fimbrial haemagglutinins and erythrocytes by D-mannose and other carbohydrates. *Journal of General Microbiology* **71**, 149–157.

Old, D. C. and Adegbola, R. A. (1982). Haemagglutinins and fimbriae of *Morganella*, *Proteus* and *Providencia*. *Journal of Medical Microbiology* **15**, 551–564.

312   D. C. OLD

Old, D. C. and Adegbola, R. A. (1983). A new mannose-resistant haemagglutinin in *Klebsiella*. *Journal of Applied Bacteriology* **55**, 165–172.

Old, D. C. and Duguid, J. P. (1970). Selective outgrowth of fimbriate bacteria in static liquid medium. *Journal of Bacteriology* **103**, 447–456.

Old, D. C. and Duguid, J. P. (1971). Selection of fimbriate transductants of *Salmonella typhimurium* dependent on motility. *Journal of Bacteriology* **107**, 655–658.

Old, D. C., Corneil, I., Gibson, L. F., Thomson, A. D. and Duguid, J. P. (1968). Fimbriation, pellicle formation and the amount of growth of salmonellas in broth. *Journal of General Microbiology* **51**, 1–16.

Old, D. C., Crichton, P. B., Maunder, A. J. and Wilson, M. I. (1980). Discrimination of urinary strains of *Escherichia coli* by five typing methods. *Journal of Medical Microbiology* **13**, 437–444.

Ørskov, I. and Ørskov, F. (1983). Serology of *Escherichia coli* fimbriae. *Progress in Allergy* **33**, 80–105.

Saier, M. H., Schmidt, M. R. and Leibowitz, M. (1978). Cyclic-AMP-dependent synthesis of fimbriae in *Salmonella typhimurium*: Effects of *cya* and *pts* mutations. *Journal of Bacteriology* **134**, 356–358.

Sheladia, V. L., Chambers, J. P., Guevara, J. and Evans, D. J. (1982). Isolation, purification and partial characterization of type V-A hemagglutinin from *Escherichia coli* GV-12, 01:H⁻. *Journal of Bacteriology* **152**, 757–761.

Smith, H. W. and Linggood, M. A. (1971). Observations on the pathogenic properties of the K88, Hly and Ent plasmids of *Escherichia coli* with particular reference to porcine diarrhoea. *Journal of Medical Microbiology* **4**, 467–485.

Smyth, C. J. (1982). Two mannose-resistant haemagglutinins on enterotoxigenic *Escherichia coli* of serotype O6:K15:H16 or H⁻ isolated from travellers' and infantile diarrhoea. *Journal of General Microbiology* **128**, 2081–2096.

Smyth, C. J. (1983). Serologically distinct haemagglutinating fimbriae of enterotoxigenic *Escherichia coli* of serotype O6:K15:H16 or H⁻. *Society for General Microbiology, 96th Ordinary Meeting, 1983*.

Stirm, S., Ørskov, F., Ørskov, I. and Birch-Andersen, A. (1967). Episome-carried surface antigen K88 of *Escherichia coli*. III. Morphology. *Journal of Bacteriology* **93**, 740–748.

Svanborg Edén, C. and Hansson, H. A. (1978). *Escherichia coli* pili as possible mediators of attachment to human urinary tract epithelial cells. *Infection and Immunity* **21**, 229–237.

Svenson, S. B., Källenius, G., Möllby, R., Hultberg, H. and Winberg, J. (1982). Rapid identification of P-fimbriated *Escherichia coli* by a receptor specific particle agglutination test. *Infection* **4**, 209–214.

Tavendale, A., Jardine, C. K. H., Old, D. C. and Duguid, J. P. (1983). Haemagglutinins and adhesion of *Salmonella typhimurium* to HEp2 and HeLa cells. *Journal of Medical Microbiology* **16**, 371–380.

Thomas, L. V. and Rowe, B. (1982). The occurrence of colonisation factors (CFA/I, CFA/II and E8775) in enterotoxigenic *Escherichia coli* from various countries in south east Asia. *Medical Microbiology and Immunology* **171**, 85–90.

Thomas, L. V., Cravioto, A., Scotland, S. M. and Rowe, B. (1982). New fimbrial antigenic type (E8775) that may represent a colonization factor in enterotoxigenic *Escherichia coli* in humans. *Infection and Immunity* **35**, 1119–1124.

Väisänen, V., Elo, J., Tallgren, L. G., Siitonen, A., Makela, P. H., Svanborg Edén, C., Källenius, G., Svenson, S. B., Hultberg, H. and Korhonen, T. (1981). Mannose-resistant haemagglutination and P antigen recognition are characteristic of *Escherichia coli* causing primary pyelonephritis. *Lancet* **2**, 1366–1369.

Väisänen, V., Korhonen, T. K., Jokinen, M., Gahmberg, C. G. and Ehnholm, C. (1982). Blood group M specific haemagglutinin in pyelonephritogenic *Escherichia coli*. *Lancet* **1**, 1192.

Van den Bosch, J. F., Verboom-Sohmer, U., Postma, P., de Graaf, J. and MacLaren, D. M. (1980). Mannose-sensitive and mannose-resistant adherence to human uroepithelial cells and urinary virulence of *Escherichia coli*. *Infection and Immunity* **29**, 226–233.

Varian, S. A. and Cooke, E. M. (1980). Adhesive properties of *Escherichia coli* from urinary tract infections. *Journal of Medical Microbiology* **13**, 111–119.

Vosbeck, K. and Huber, U. (1982). An assay for measuring specific adhesion of an *Escherichia coli* strain to tissue culture cells. *European Journal of Clinical Microbiology* **1**, 22–28.

Wilson, M. I., Crichton, P. B. and Old, D. C. (1981). Characterisation of urinary isolates of *Escherichia coli* by multiple typing: A retrospective analysis. *Journal of Clinical Pathology* **34**, 424–428.

11

Biotyping of *Escherichia coli:* Methods and Applications

PAMELA B. CRICHTON AND D. C. OLD

Department of Medical Microbiology, University of Dundee Medical School, Ninewells Hospital, Dundee, UK

Introduction

In the laboratory, isolates of Enterobacteriaceae may be identified to the level of genus and species by a recognizable overall pattern of reactions in conventional tubed biochemical media (Edwards and Ewing, 1972). For example, the majority of *Escherichia coli* are lactose-fermenting, indole-positive and citrate- and urease-negative (Table 1). An alternative approach is to use the positive and negative results obtained in commercially available biochemical systems such as API (API Laboratory Products, Rayleigh, Essex, UK) to provide a numerical profile, from which the species is identified by reference to a computer-based profile index.

Subspecific Identification

For a greater understanding of normal host–bacterium interactions involving *E. coli,* or of situations in which that species is implicated as a cause of disease in the community, determining its presence in animate hosts or on inanimate vehicles of infection demands discrimination of subtypes within the species. Such discrimination requires the identification of genetic markers by which isolates of the species from the same or different sites can be recognized as identical, related or non-identical in type. The success of typing methods used for that purpose depends primarily on the stability of the typing characters used.

Serotyping. Until recently, antigenic analysis was probably the most widely accepted method for identification of individual strains of *E. coli,* as, for example, in the investigation of urinary tract infections, in which *E. coli* is the most

Table 1. *Biochemical reactions of Escherichia coli*

| Test or substrate | Percentage positive from referred source (and number of cultures or strains examined) | |
	Edwards and Ewing, 1972 (1021 cultures)	Crichton and Old, 1979 (917 strains)
Indole	99	NT[a]
Citrate	0	NT
Urease	0	NT
Hydrogen sulphide	0	NT
Adonitol	5	6
L-Arabinose	100	100
D-Cellobiose	3	1
Inositol	1	0
Lactose	90	96
Maltose	86	96
D-Sorbitol	93	96
D-Trehalose	98	99

[a]NT, not tested for strain discrimination because results are almost uniform within the species.

frequently incriminated pathogen. Many workers, however, have attempted to serotype with only O antisera raised against the commonest types, an approach that is unsound because not only are many strains autoagglutinable or non-typable (Grüneberg *et al.*, 1968) but also a particular O serogroup may include many different strains (Gargan *et al.*, 1982). These problems lead to errors in interpretation.

Compared with the genus *Salmonella,* less is known about structural relationships between O antigens of *E. coli,* and it is possible that a minor *in vivo* variation in the cell wall lipopolysaccharide may alter O specificity, and smooth-to-rough variation may occur, rendering the strain O-untypable. Furthermore, owing to the interest in *E. coli* serotyping and its application to many problems in medical microbiology (Bettelheim and Taylor, 1969; Bettelheim *et al.*, 1974; Sakazaki *et al.*, 1974; Ørskov and Ørskov, 1975; Ørskov *et al.*, 1976), the number of new O antigens recognized has, of necessity, been limited and now only the O antigens of strains that are of particular medical, epidemiological or scientific interest are added to the antigenic scheme (Ørskov *et al.*, 1977).

Hence, even in a laboratory with a full range of O antisera, many isolates may be untypable (Old *et al.*, 1980) and precise strain characterization should include identification of the H (flagellar) and K (capsular or envelope) antigens. More recently included in the antigenic scheme for *E. coli* are the protein MRE fimbrial F antigens, of which 11 have so far been described (Ørskov and Ørskov, 1983) though others are known. Several MS fimbrial antigens have been described in *E. coli* but have not been designated by specific numbers (Gillies and Duguid, 1958). Complete, that is O:K:H:F, serotyping obviously provides excellent strain discrimination but the expense and time required for its performance effectively restrict its use to a limited number of places, including national reference laboratories with the resources necessary for production and testing of the vast range of antisera involved. Accordingly, in small laboratories alternative techniques are needed which provide accurate, inexpensive and speedy identification of individual strains for studies of the ecology or epidemiology of *E. coli*.

Criteria for a successful typing system. If typing procedures used to differentiate members of a bacterial species are to be helpful, they must satisfy several criteria: (1) tests should show high discriminating ability and, ideally, epidemiologically unrelated strains should be grouped into almost equal numbers of positive and negative types with respect to any particular marker; (2) the results obtained should be stable and reproducible; (3) a high proportion of strains should be typable; (4) if tests are to be used routinely, ease of performance should also receive consideration.

Tests for Biotyping of *E. coli*

Importance of Discriminating Power

For some years we have been involved in the development of a scheme of biochemical and physiological tests that provides accurate discrimination of strains of *E. coli* from patients with infections of the urinary tract and other sites (Crichton and Old, 1979, 1980, 1982, 1983; Old *et al.*, 1980; Wilson *et al.*, 1981). For incorporation in such a biotyping scheme, it is important to distinguish tests that are more appropriate for species identification, and hence not generally worthy of inclusion in the scheme, and those for differentiation of types within the species, and hence worthy of inclusion. Table 1 indicates that, for *E. coli*, tests in the former category include those for indole production (99% positive), adonitol fermentation (95% negative) and hydrogen sulphide production (100% negative) because the majority of strains are clearly positive or negative for these characters. Additionally, tests for the fermentation of arabinose, cellobiose, inositol, lactose, maltose, sorbitol and trehalose offer very poor type discrimination. On those rare occasions when strains expressing a particular minority character are encountered, for example, adonitol-fermenting strains (Hinton *et al.*, 1982) or hydrogen sulphide-producing strains (Magalhães and Vance, 1978), tests that identify the less commonly observed property may themselves provide additional information helpful in subspecific discrimination of types. However, although helpful in these rare cases, the inclusion of such tests on a regular basis cannot be justified and should not be considered in a biotyping scheme that aims for maximal discrimination of *all* isolates of *E. coli*.

Reproducibility of Results

It has been recognized that, to ensure reproducibility of results on different occasions of testing the same strain, it is necessary to define the optimal inoculum (Murray, 1978). In addition, there should be established for each test an optimal, that is 'definitive', time of reading after an incubation period chosen to separate clearly strains that are genotypically positive for the character tested from those that are genotypically negative but give positive results because of spontaneous mutations arising after prolonged incubation (Duguid *et al.*, 1975). Thus, for example, in their early biotyping work with *E. coli*, Bettelheim and Taylor (1969) failed to define times of reading and incubated tests for up to 1 wk. This no doubt explained many of the anomalies in the interpretation of their results and their conclusion that biotyping was of little value for primary strain identification. In a revised scheme (Bettelheim *et al.*, 1974), periods of incubation for individual tests were defined and, correspondingly, the reproducibility of biotypes was greatly improved. Furthermore, biotyping methods must be carefully chosen. Crichton and Old (1982) found it necessary to exclude, for example, the testing, in liquid media, of salicin and aesculin, substrates which are hydrolysed in *E. coli* by an inducible β-glucosidase (Miskin and Edberg, 1978) and for which it was impossible to define a satisfactory end-point in liquid media.

For the biotyping scheme presented here, the methods and substrates have been carefully selected with the above difficulties in mind.

Methods

Dehydrated media from commercial sources were prepared according to the instructions of the manufacturer. Agar media were used in 15 to 20-ml amounts in plastic petri dishes 8.5 cm in diameter.

Preparation of inoculum. Fresh or stored isolates were plated on MacConkey agar (Oxoid Ltd., Basingstoke, Hants.) and examined for purity after overnight incubation in air at 37°C. These conditions were used for all cultures. Half of a well-isolated colony was inoculated to 10 ml nutrient broth (Oxoid nutrient broth no. 2) in a cotton-wool-stoppered test tube (15 × 1.5 cm) and incubated statically for 48 h; the other half was spread on Oxoid nutrient agar and, after 24-h incubation, the resulting growth was suspended in physiological saline (NaCl 8.5 g litre^{-1}) to a density of 10^9 bacteria ml^{-1}. A standard volume (~0.02 ml) of that suspension was used as inoculum for both liquid and solid media delivered from, respectively, a capillary pipette or inoculating wire.

Media. Unless otherwise stated, culture media prepared *de novo* were sterilized by autoclaving at 121°C for 15 min.

For fermentation tests, the basal medium contained Oxoid peptone water 15 g and, as indicator, Bromcresol Purple [British Drug Houses (BDH) Ltd., Poole, Dorset] 0.02 g per litre of deionized water. The following carbohydrates were prepared as 10% (w/v) solutions in deionized water: dulcitol, L-rhamnose and L-sorbose (Sigma London Chemical Co. Ltd., Poole, Dorset) and D-raffinose (BDH). All were steamed for 1 h at 100°C and were added to the sterile basal medium at a final concentration of 0.5% (w/v). The completed media were dispensed in 4-ml volumes in sterile, screw-capped 6-ml bottles and steamed at 100°C for 30 min.

The basal medium for amino acid decarboxylation tests contained Oxoid bacteriological peptone 5 g, Oxoid yeast extract 3 g, D-glucose of Analar grade (BDH) 1 g, and Bromcresol Purple 0.02 g per litre of deionized water; L-lysine monohydrochloride or L-ornithine monohydrochloride (Sigma) was added to the basal medium at a final concentration of 0.5% (w/v). Controls were basal medium without any addition. The media were adjusted to pH 6.7 and dispensed in 5-ml amounts in screw-capped 6-ml bottles and sterilized by autoclaving. It was not found necessary to overlay decarboxylase media with mineral oil (Hinton *et al.*, 1982) but caps were screwed down firmly before incubation of the tests.

Aesculin agar (pH 7.4), containing aesculin of laboratory reagent grade (BDH) 1 g, ferric citrate 0.5 g and Oxoid peptone water 15 g per litre of deionized water, was gelled by the addition of 2% (w/v) Bacto-Agar (Difco Laboratories, West Molesey, Surrey) and sterilised by autoclaving. The basal mineral salts agar (pH 7.0), for the detection of auxotrophic strains, contained K_2HPO_4 7 g, KH_2PO_4 3 g, $(NH_4)_2SO_4$ 1 g, $MgSO_4·7H_2O$ 0.1 g and Difco Bacto-Agar 15 g per litre of deionized water. A sterile solution of D-glucose was added to give a final concentration of 0.3% (w/v). Growth requirements of auxotrophs were determined by inoculating a series of glucose–mineral salts plates each containing a different amino acid or vitamin at an appropriate concentration (see Meynell & Meynell, 1965 and Table 6). Motility medium contained Oxoid tryptone water 15 g and Difco Bacto-Agar 3 g per liter of deionized water. The melted agar was thoroughly mixed, dispensed in 10-ml volumes in cotton-wool-stoppered test tubes (15 × 1.5 cm) and autoclaved.

Interpretation of results. Positive results in fermentation tests were indicated by a colour change, from purple to yellow, in the medium after incubation periods that differed according to the test carbohydrate: rhamnose and sorbose tests at 24 h, dulcitol at 48 h, raffinose at 72 h. Tests for amino acid decarboxylation were read as positive when the medium became purple (alkaline) at 48 h and the control medium remained acidic (yellow) due to glucose fermentation. Aesculin hydrolysis on solid medium was indicated by growth with numerous brownish-black papillae and a surrounding brownish-black precipitate at 72 h.

Glucose–mineral salts agar was examined for the presence or absence of growth at 24 h.

In motility agar, growth that spread uniformly from the stab inoculum at 24 h indicated a positive result; any other growth pattern was recorded as negative. After serial 48-h culture in aerobic, static broth, strains were examined for their production of type 1 fimbriae (Duguid *et al.*, 1966), indicated macroscopically by mannose-sensitive haemagglutination. Full details of the test have been published elsewhere (Duguid *et al.*, 1955, 1979; see also Chapter 10).

Primary Biotyping Tests

When the biotype profiles of 599 strains (1242 cultures) of *E. coli,* collected from different sources (for details, see Crichton and Old, 1982) and species-identified by standard procedures (Edwards and Ewing, 1972), were assessed, results indicated that tests with raffinose, sorbose, ornithine and dulcitol should be used to assign strains to primary biotypes because of their high discriminating power (dividing strains almost equally into positive and negative types) and their reproducibility when read at the definitive times (Table 2). These four tests, which are easy to perform and interpret, allow the identification of 16 primary biotypes (1–16) according to the different combinations of positive and negative reactions (Table 3).

Secondary Biotyping Tests

In our biotyping scheme, tests for motility and aesculin hydrolysis were included at the secondary level because, although their results afforded almost optimal

Table 2. *Biotyping results for 599 strainsᵃ of Escherichia coli*

Test (and biotype character)		Percentage of 599 strains positive in test
Fermentation of raffinose	(raf)	52
Fermentation of sorbose	(sor)	54
Decarboxylation of ornithine	(orn)	59
Fermentation of dulcitol	(dul)	67
Fermentation of rhamnose	(rha)	87
Decarboxylation of lysine	(lys)	90
Hydrolysis of aesculin	(aes)	45
Motilty	(mot)	56
Production of type 1 fimbriae	(fim)	83
Prototrophy	(pro)	88

ᵃWhen more than one isolate of a strain was available, the biotype of the first isolate was used in calculations.

Table 3.   *Results of the 16 primary biotypes of Escherichia coli*

Primary biotype no.	Result[a] in biotype test[b] with			
	raf	sor	orn	dul
1	+	+	+	+
2	+	+	+	−
3	+	+	−	+
4	+	+	−	−
5	+	−	+	+
6	+	−	+	−
7	+	−	−	+
8	+	−	−	−
9	−	+	+	+
10	−	+	+	−
11	−	+	−	+
12	−	+	−	−
13	−	−	+	+
14	−	−	+	−
15	−	−	−	+
16	−	−	−	−

[a]Plus signs indicate a positive and minus signs a negative result.
[b]For explanation of biotype test characters, see Table 2.

type differentiation among the 599 strains (Table 2), a degree of technical exper-
tise was required for their interpretation on a minority of occasions. Although
aiding strain differentiation, tests that discriminated less than 15% of the minority negative or positive type were also used as secondary tests: those for rhamnose fermentation (13%), lysine decarboxylation (10%) and growth requirements
(12%) (Table 2). Most strains (88%) were prototrophic and grew luxuriously on
glucose–mineral salts agar in 24 h; only 74 of the 599 strains required supple-
mentation of the medium with growth factors (for details, see Crichton and Old,
1982), the commonest of which were nicotinamide (required for growth by 42%
of the auxotrophs) and thiamine (20%). Because these two requirements were
relatively common, they were designated in the biotype by specific secondary
characters (x and y, respectively); other less common growth factors were collec-
tively indicated by subtype character z (Table 4).

The test for the presence of type 1 fimbriae, which differentiated 17% of the
minority negative type was also included at the secondary level because the
availability of reagents to some might make it relatively difficult to perform. We
had previously included that stable, chromosomally determined property (Maccacaro and Hayes, 1961) as a character in haemagglutinin (HA) typing, a method
that usefully extends the information gained from biotyping (Crichton and Old,
1980, 1983; Old *et al.*, 1980; Wilson *et al.*, 1981).

The subtype designations for each of the secondary biotyping tests are ex-

Table 4. *Subtype designations and results in secondary biotyping tests*

Subtype designation	Result
a	Positive results in all secondary tests
b	L-Rhamnose not fermented in 24 h
c	L-Lysine not decarboxylated in 48 h
d	Aesculin not hydrolysed in 72 h
e	Non-motile in 24 h, or motile variants produced after 24 h
f	Type 1 fimbriae not produced
x	Nicotinamide required for growth
y	Thiamine required for growth
z	A growth factor other than nicotinamide or thiamine required

plained in Table 4. Full biotypes are indicated by a primary biotype number followed by the appropriate subtype letters. For example, a strain which degrades raffinose, sorbose, ornithine and dulcitol and gives positive results in all secondary tests except that for motility has the full biotype designation 1e.

Strain Discrimination Provided by Biotyping

It was possible to obtain a biotype designation for each of the 599 strains of *E. coli* in our collection, when biotyped by means of the two-tier system incorporating tests for raffinose, sorbose, dulcitol and rhamnose fermentation, lysine and ornithine decarboxylation, aesculin hydrolysis, motility, prototrophy and type 1 fimbriation. Representatives of each of the theroretically attainable primary types were detected (Table 5); strains of these 16 primary biotypes were assigned to 213 full biotypes according to their reactions in secondary tests (Table 5). Primary biotype 1 accounted for the greatest number of strains (18%) and these 110 strains were subdivided into 23 full biotypes, none of which contained more than 29 strains (see Crichton and Old, 1982).

When the API 20E Enterobacteriaceae system was used by Davies (1977) in a study of 574 urinary strains of *E. coli,* 55 biotypes were detected. However, that API system is produced primarily for identification of species, and 16 of its 21 tests provide little or no discrimination of types of *E. coli;* thus it was not surprising that almost 70% of the strains examined by him were assigned to only 6 biotypes. Buckwold *et al.* (1979) subsequently formulated a scheme of nine tests, including five (rhamnose, lysine, raffinose, sucrose and motility) independently selected by Crichton and Old (1979) and four others (melibiose, adonitol, lactose and β-haemolysis) not recommended by us because of their poor discrimination, duplication of information or possible plasmid control. That scheme separated 959 strains into 78 biotypes, but again only a few (11) types accounted

Table 5. *Number of full biotypes detected in 599 strains^a of Escherichia coli*

Primary biotype no.	Number of strains	No. of full biotypes^b
1	110	23
2	42	12
3	49	14
4	12	8
5	49	10
6	21	12
7	19	13
8	11	10
9	81	20
10	9	7
11	10	6
12	13	9
13	21	13
14	18	11
15	65	20
16	69	25
Any	599	213

^{a}See footnote, Table 2.
^{b}For details of full biotypes observed, see Crichton and Old (1982).

for a large proportion (67%) of the strains. The system presented by us provides improved discrimination because the 13 commonest full biotypes detected, each of which contained 10 or more strains, accounted for only 36% of the 599 strains examined.

Reliability

The reliability of our two-tier biotyping scheme was demonstrated by observations from our collection of 599 strains that identical biotype profiles were obtained in the following circumstances: (1) when many colonies from a pure culture of a strain were tested simultaneously; (2) when a single culture was tested at its time of isolation and on different occcasions after prolonged storage on non-selective medium; (3) when sequential isolates from the urinary tract or other anatomical sites of an individual patient were tested. *In vivo* variation of a biotype character was shown in second or subsequent isolates of only five strains (<1%), three showing variation to auxotrophy and the other two changes in the lysine or ornithine decarboxylase characters. Thus, contrary to the biotype findings of other workers (Bettelheim and Taylor, 1969; Hinton *et al.*, 1982), the reliability of the ornithine test (>99%), in our hands, justified its inclusion at the primary level.

Other Observations

Among lactose-fermenting strains of *E. coli* there exists a strong correlation between the results of sucrose and raffinose fermentation (Crichton and Old, 1979), although the rare sucrose-negative and raffinose-positive phenotype has been encountered (see Table 7 and text). The additional differentiation among sucrose-fermenting strains provided by the sucrose-weak and sucrose-strong phenotypes, demonstrable on sucrose–eosin–methylene blue agar (Wilson *et al.*, 1981), could not readily be included, even as a secondary test, in our biotyping system. Accordingly, because the sucrose test proved less reliable in its performance (Crichton and Old, 1979) and generally gave the same result as the raffinose test, the former was excluded from our scheme. Others have included both of these substrates in biotyping schemes without having greatly improved type discrimination (Buckwold *et al.*, 1979; Hinton *et al.*, 1982). The technical difficulties experienced by us in the testing of certain substrates such as salicin and aesculin in liquid medium have been confirmed by others (Buckwold *et al.*, 1979; Hinton *et al.*, 1982), but the use of salicin has been recommended in some schemes. For the assessment of motility we have favoured the use of semi-solid agar cultures, rather than observations by direct microscopy, the latter being unreliable (Edwards and Ewing, 1972) and unnecessarily time-consuming for routine handling of large numbers of cultures.

Other workers, not all of whom were apparently aware of the need for definitive times of reading, have recorded the usefulness of some, but not all, of our biotyping tests for strain discrimination (Mushin and Ashburner, 1964; Bettelheim *et al.*, 1974; Sakazaki *et al.*, 1974; Van der Waaij *et al.*, 1975; Myerowitz *et al.*, 1977; Scotland *et al.*, 1977; Magalhães and Vance, 1978; Buckwold *et al.*, 1979; Gargan *et al.*, 1982; Hinton *et al.*, 1982; Achtman *et al.*, 1983). Furthermore, in some of these biotyping schemes the test substrates routinely included some which were highly discriminating for all strains of *E. coli* and others which were probably appropriate for specific, rather than subspecific, identification because strains of the minority type were so rarely encountered. Again, we have preferred to exclude from our biotyping scheme plasmid-determined characters, such as β-haemolysin production, not only because they are subject to loss or gain during epidemic spread of the strain, but also because the cultural conditions for expression of the character and tests for its demonstration *in vitro* may be ill-defined.

Our two-tier biotyping scheme offers excellent strain discrimination, reliability in practice and great flexibility. Thus, as further substrates are described which appear to offer additional discrimination of types, they can be incorporated into the scheme at the secondary level. For example, we have confirmed the observations of Gargan *et al.* (1982) that the substrate 5-ketogluconate is highly discriminatory for typing of *E. coli* strains and propose to designate non-utilizing strains in our scheme by the symbol 'g'.

Usefulness of Biotyping

A few examples will serve to demonstrate that when our biotyping methods were used routinely to test *E. coli* isolated from the urinary tract or from other sites, the results obtained provided considerable discrimination of types and were helpful in indicating relationships among sequential isolates of cultures from individual patients. To increase the precision of strain identification isolates were tested after biotyping with a limited range (24) of commerical O antisera (for details, see Crichton and Old, 1979) or, in certain cases, against a complete range of O, K and H antisera at a national reference laboratory (Old *et al.*, 1980).

For example, each of five specimens of urine from patient F1 (Table 6) yielded non-fimbriate isolates of *E. coli* of the same biotype (15f), indicating the presence over a period of 40 days of a single *E. coli* strain of stable biotype, and this conclusion was supported by the subsequent demonstration that all were isolates of serogroup **O**5. Studies of many urinary isolates have demonstrated the stability *in vivo* of biotype characters over periods of up to 4.5 years (Crichton and Old, 1983).

E. coli isolated from two urine specimens (days 1 and 53) from patient F2 (Table 6) were of the same serogroup (**O**75) and in this, and many other examples, the use of limited serotyping alone for strain characterization would have given the wrong answer; the use of biotyping distinguished two distinct strains different in two primary and two secondary characters. Two isolates from specimens of urine from F3 on days 1 and 16 were of the same biotype, which was different in two primary and four secondary characters (ornithine, dulcitol, lysine, aesculin, motility and type 1 fimbriation) from that of the isolate recovered on day 57 (Table 6); furthermore, each of these two biotypes differed from that of the strain recovered on days 73 and 74. Hence, biotyping revealed the presence over a 10-wk period of three distinct strains, each of which was untypable with a limited range of O antisera. Finally, *E. coli* from urine specimens from patient F4 during and after pregnancy were those of three different strains (Table 6). The strain isolated on days 1–18, however, was of special interest for it was identical, in all biotype characteristics, to cultures isolated from her child (N1) born on day 47 of the mother's clinical history. In that case, therefore, biotyping methods revealed a probable source of the child's infecting strain.

Table 6. *Biotyping results for 25 isolates[a] of Escherichia coli from four patients*

Patient	No. of isolates	Time of isolation (days)	Biotype character										Biotype designation	O group[b]
			raf	sor	orn	dul	rha	lys	aes	mot	fim	pro		
F1	5	1, 3, 8, 21, 40	−	−	−	+	+	+	+	+	−	+	15f	5
F2	2	1, 53	−	+	+	+	+	+	+	+	+	−	9x[c]	75
			+	−	+	+	+	+	−	+	+	+	5d	75
F3	5	1, 16, 57, 73, 74	−	−	+	−	+	+	+	−	−	+	14ef	NT
			−	−	−	+	+	−	−	+	+	+	15cd	NT
			−	−	−	−	+	−	+	+	+	−	16cz[c]	NT
F4	8	1, 16, 17, 18, 10, 15, 60, 65	−	+	+	+	+	+	+	−	+	+	9e	4
			+	+	+	−	+	+	−	+	+	+	2d	7
			−	+	+	+	+	+	−	+	−	−	10dy[c]	22
N1	5	1, 2	−	+	+	+	+	+	+	−	+	+	9e	4

[a] Isolates from F1–F4 were from urine; those from N1 were from blood, ear, eyes and nose.

[b] Isolates from F3 were non-typable (NT) with a limited range of 24 O antisera; details are published elsewhere (Crichton and Old, 1979).

[c] Concentrations of growth factors were: x, nicotinamide at 1 μg ml^{-1}; y, thiamine at 1 μg ml^{-1}; z, methionine at 10 μg ml^{-1}.

Many other examples of the value of this biotyping system have been detailed elsewhere (Crichton and Old, 1979, 1980, 1982, 1983; Old *et al.*, 1980; Wilson *et al.*, 1981).

Multiple Typing

As with any other typing system used alone, even the present biotyping scheme, which provides such accurate and reliable strain characterization, has limitations in that variation of typing characters, though rare, may occur *in vivo*. Consequently, isolates of clonal origin that differ in a single typing marker may be wrongly classed as different, just as, conversely, isolates from widely dispersed sources may be wrongly classed as identical if they belong to the same common biotype. These difficulties may be resolved and the true relationships between isolates more accurately assessed by the combined use of several typing systems. In addition to biotype and serotype, properties including antibiotic resistance markers and production of haemagglutinins, or haemolysin or colicin, may be exploited as supplementary typing systems.

Resistotyping

However, to supplement biotyping we have favoured resistotyping, a method based upon the selective toxicity of a carefully selected set of chemicals used at critical concentrations (see Elek and Higney, 1970). The chemicals affording discrimination of types were: (A) sodium arsenate; (B) phenylmercuric nitrate; (C) 4:4'-diamidinodiphenylamine dihydrochloride; (D) boric acid; (E) acriflavine; (F) 4-chlororesorcinol; (G) copper sulphate; (H) Malachite Green. The methods used were those of Wilson *et al.* (1981) modified from the original system (Elek and Higney, 1970). The selective toxicity has such fine margins that we prepared a series of plates for each chemical encompassing a narrow range around the optimal concentration that discriminates. Five reference strains of known behaviour with the eight chemicals were inoculated along with 20 test isolates to each plate in each series; after 18 h incubation, tests were read as resistant, partially resistant or sensitive at that concentration of chemicals allowing correct classification of the reference strains. The biological titration of chemicals against reference strains ensured the optimal discrimination of test strains and, theoretically, the reproducibility of results. In practice, however, it was essential that isolates for direct comparison were tested together. A strain of resistotype ACdEf was resistant to chemicals A, C, E, partially resistant to D, F and sensitive to B, G, H. Isolates different in one partial resistance, for example, CDEF and CDEf, were considered identical (Elek and Higney, 1970).

When we used biotyping and resistotyping in conjunction, agreement between

Table 7. *Typing results for 26 urinary isolates of Escherichia coli from five patients*

Patient	No. of isolates	Time of isolation (days)	Biotype character										Biotype	Resistotype	MRE haemagglutinin	O group[a]
			raf	sor	orn	dul	rha	lys	aes	mot	fim	pro				
F5	2	1, 143	+	+	+	−	+	+	+	−	+	+	2e	CDEFh	+ (MPS)	75
			+	+	+	+	+	+	+	+	+	+	1a	BCDEF	+ (M)	75
F6	4	1, 7, 65, 442	−	−	−	+	+	−	−	+	+	+	15cd	ABCDEfGH	−	NT
			+	+	+	−	+	+	−	+	+	+	2d	acdefG	+ (0)	NT
			−	−	−	−	−	+	+	+	+	+	16b	ABCDEF	−	NT
F7	4	1, 120, 160, 178,	−	−	−	+	+	+	−	−	−	+	15def	acDEh	−	NT
			−	−	−	+	+	+	−	−	+	+	15de	aBCDEFG	−	86
F8	2	1, 54	+	+	+	+	+	+	−	−	+	−	1dey	CDEFh	+ (FGMOP)	NT
			+	+	+	+	+	−	−	−	+	−	1cdey			
F9	14	1, 20, 86, 98, 216, 217, 218, 219, 267, 281, 313, 435, 446, 467	+	+	+	+	+	+	−	−	+	+	1de	ABCDEH ABCDEh AbCDEh	−	NT

[a]Isolates from patients F6, F7 and F8 were non-typable (NT) with a limited range of O antisera (see Table 6); those from F9 were non-typable at a reference laboratory (Dr. B. Rowe, personal communication).

results of the two systems was good, so that a change in one was generally paralleled by a change in the other. In a retrospective study of 110 paired urinary isolates from 110 individual patients (Wilson *et al.*, 1981) each culture was typable by both methods, and their joint use allowed identification of 96% of the paired strains as similar or dissimilar and was especially valuable in distinguishing among strains of common, urinary O serogroups. For the few cases in which the relationships between strains were in doubt, the biotype–resistotype profiles were extended mainly by use of mannose-resistant and eluting HA patterns (Duguid *et al.*, 1979; see also Chapter 10) or, on occasion, by determination of colicin production (Ozeki *et al.*, 1962) or sensitivity (Barker and Old, 1979).

Some of the benefits derived from combined typing are exemplified by results shown in Table 7. For example, two cultures (days 1, 143) from the urine of F5 belonged to a common serogroup (**O**75) and yet were clearly representative of two distinct strains different in biotype, resistotype and MRE-HA type (Table 7). These three methods used together also discriminated, independently of serotyping, three strains present in the urine of F6 over a period of 442 days. When, rarely, a patient's isolates belonged to very similar but non-identical biotypes, the question of strain identity was ususally resolved without difficulty from the available, supplementary data. Thus, though the first two urinary isolates (days 1, 120) from F7 showed only one biotype character difference (type 1 fimbriation) from that of the subsequent isolates, the presence of two distinct strains was clarified by their differences in three major (B, F, G) and two minor (C, H) resistotype characters. The two cultures from F8, on the other hand, though different in only one biotype character (lysine), were judged to be clonal, their resistotypes, HA types and colicin types being identical.

Finally, with reference to the 14 isolates from F9's urine, the criteria of Elek and Higney (1970) would ordinarily allow that cultures of resistotype ABCDEh showed acceptable strain identity with those of resistotype ABCDEH on the one hand, and those of resistotype AbCDEh on the other; a difference of two partial resistances in the latter two resistotypes, however, would ordinarily indicate strain non-identity. However, all 14 isolates were of the same biotype (1de) and showed the same, rare, sucrose-negative and raffinose-positive phenotype and were, hence, probably identical when all data were assessed. Serotyping offered no help in this example, for all 14 strains were untypable with available O, K and H antisera (Dr. B. Rowe, personal communication).

Other examples of our success with combined typing techniques, used independently of serotyping, have been presented in detail elsewhere (Crichton and Old, 1980, 1982, 1983, Old *et al.*, 1980; Wilson *et al.*, 1981).

Concluding Remarks

The biotyping system presented here is highly discriminatory and reliable and defines types of *E. coli* that are extremely stable both *in vivo* and *in vitro*.

Our interest in biotyping arose because we did not have access to the complete serotyping necessary for accurate strain identification; the development of the present scheme means that small laboratories, in which complete serotyping is impracticable, can still attempt studies on the ecology or epidemiology of *E. coli*. We do not recommend that biotyping should replace serotyping, but it has proved to be valuable for distinguishing strains of *E. coli* when used independently of serotyping and even more valuable when used in conjunction with other good techniques such as resistotyping.

For biotyping, the methods used are simple and relatively inexpensive and the stability of the types achieved renders it useful for both prospective and retrospective studies. We suggest that biotyping should now be given greater consideration as a system for differentiating strains of *E. coli,* though for that purpose our proposed scheme need not necessarily be used in its entirety. Its flexibility allows for future expansion to include additional substrates or for its contraction so that only some of our selected tests may be used.

Despite the excellent strain discrimination and reliability provided, biotyping, like any other typing method used in isolation, is limited in its accuracy by the rare occurrence of *in vivo* variations, and should be supplemented by other systems. Although we have favoured the use of resistotyping, others may prefer different ancillary techniques such as limited O serogrouping, colicin production or plasmid analysis.

References

Achtman, M., Mercer, A., Kusecek, B., Pohl, A., Heuzen Roeder, M., Aaronson, W., Sutton, A. and Silver, R. P. (1983). Six widespread bacterial clones among *Escherichia coli* K1 isolates. *Infection and Immunity* **39**, 315–335.

Barker, R. and Old, D. C. (1979). Biotyping and colicine typing of *Salmonella typhimurium* strains of phage type 141 isolated in Scotland. *Journal of Medical Microbiology* **12**, 265–276.

Bettelheim, K. A. and Taylor, J. (1969). A study of *Escherichia coli* isolated from chronic urinary infection. *Journal of Medical Microbiology* **2**, 225–236.

Bettelheim, K. A., Teoh-Chan, C. H., Chandler, M. E., O'Farrell, S. M., Rahamin, L., Shaw, E. J. and Shooter, R. A. (1974). Further studies of *Escherichia coli* in babies after normal delivery. *Journal of Hygiene* **73**, 277–285.

Buckwold, F. J., Ronald, A. R., Harding, G. K. M., Marrie, T. J., Fox, L. and Cates, C. (1979). Biotyping of *Escherichia coli* by a simple multiple-inoculation agar plate technique. *Journal of Clinical Microbiology* **10**, 275–278.

Crichton, P. B. and Old, D. C. (1979). Biotyping of *Escherichia coli*. *Journal of Medical Microbiology* **12**, 473–486.

Crichton, P. B. and Old, D. C. (1980). Differentiation of strains of *Escherichia coli:* Multiple typing approach. *Journal of Clinical Microbiology* **11**, 635–640.

Crichton, P. B. and Old, D. C. (1982). A biotyping scheme for the subspecific discrimination of *Escherichia coli*. *Journal of Medical Microbiology* **15**, 233–242.

Crichton, P. B. and Old, D. C. (1983). Characterization of *Escherichia coli* strains from long-term urinary tract infections by combined typing techniques. *Journal of Urology* **129**, 160–162.

Davies, B. I. (1977). Biochemical typing of urinary *Escherichia coli* strains by means of the API 20E *Enterobacteriaceae* system. *Journal of Medical Microbiology* **10**, 293–298.

Duguid, J. P., Smith, I. W., Dempster, G. and Edmunds, P. N. (1955). Non-flagellar filamentous appendages (''fimbriae'') and haemagglutinating activity in *Bacterium coli*. *Journal of Pathology and Bacteriology* **70**, 335–348.

Duguid, J. P., Anderson, E. S. and Campbell, I. (1966). Fimbriae and adhesive properties in salmonellae. *Journal of Pathology and Bacteriology* **92**, 107–138.

Duguid, J. P., Anderson, E. S., Alfredsson, G. A., Barker, R. and Old, D. C. (1975). A new biotyping scheme for *Salmonella typhimurium* and its phylogenetic significance. *Journal of Medical Microbiology* **8**, 149–166.

Duguid, J. P., Clegg, S. and Wilson, M. I. (1979). The fimbrial and non-fimbrial haemagglutinins of *Escherichia coli*. *Journal of Medical Microbiology* **12**, 213–227.

Edwards, P. R. and Ewing, W. H. (1972). ''Identification of *Enterobacteriaceae*,'' 3rd ed. Burgess, Minneapolis, Minnesota.

Elek, S. D. and Higney, L. (1970). Resistogram typing - a new epidemiological tool: Application to *Escherichia coli*. *Journal of Medical Microbiology* **3**, 103–110.

Gargan, R., Brumfitt, W. and Hamilton-Miller, J. M. T. (1982). A concise biotyping system for differentiating strains of *Escherichia coli*. *Journal of Clinical Pathology* **35**, 1366–1369.

Gillies, R. R. and Duguid, J. P. (1958). The fimbrial antigens of *Shigella flexneri*. *Journal of Hygiene* **56**, 303–318.

Grüneberg, R. N., Leigh, D. A. and Brumfitt, W. (1968). *Escherichia coli* serotypes in urinary tract infection: Studies in domiciliary, antenatal and hospital practice. *In* ''Urinary Tract Infection'' (Eds. F. O'Grady and W. Brumfitt), pp. 68–79. Oxford University Press, London and New York.

Hinton, M., Allen, V. and Linton, A. H. (1982). The biotyping of *Escherichia coli* isolated from healthy farm animals. *Journal of Hygiene* **88**, 543–555.

Maccacaro, G. A. and Hayes, W. (1961). The genetics of fimbriation in *Escherichia coli*. *Genetical Research* **2**, 394–405.

Magalhães, M. and Vance, M. (1978). Hydrogen sulphide-positive strains of *Escherichia coli* from swine. *Journal of Medical Microbiology* **11**, 211–214.

Meynell, G. G. and Meynell, E. (1965). ''Theory and Practice in Experimental Bacteriology,'' pp. 35–36. Cambridge University Press, London and New York.

Miskin, A., and Edberg, S. C. (1978). Esculin hydrolysis reaction by *Escherichia coli*. *Journal of Clinical Microbiology* **7**, 251–254.

Murray, P. R. (1978). Standardization of the Analytab Enteric (API 20E) system to increase accuracy and reproducibility of the test for biotype characterization of bacteria. *Journal of Clinical Microbiology* **8**, 46–49.

Mushin, R. and Ashburner, F. M. (1964). Ecology and epidemiology of coliform infections. II. The biochemical reactions and drug sensitivity of coliform organisms. *Medical Journal of Australia* **1**, 303–308.

Myerowitz, R. L., Albers, A. C., Yee, R. B. and Ørskov, F. (1977). Relationship of K1 antigen to biotype in clinical isolates of *Escherichia coli*. *Journal of Clinical Microbiology* **6**, 124–127.

Old, D. C., Crichton, P. B., Maunder, A. J. and Wilson, M. I. (1980). Discrimination of urinary strains of *Escherichia coli* by five typing methods. *Journal of Medical Microbiology* **13**, 437–444.

Ørskov, F. and Ørskov, I. (1975). *Escherichia coli* O: H serotypes isolated from human blood. *Acta Pathologica et Microbiologica Scandinavica, Section B: Microbiology* **83B**, 595–600.

Ørskov, F., Ørskov, I., Evans, D. J., Sack, R. B., Sack, D. A. and Wadström, T. (1976). Special *Escherichia coli* serotypes among enterotoxigenic strains from diarrhoea in adults and children. *Medical Microbiology and Immunology* **162**, 73–80.

Ørskov, I. and Ørskov, F. (1983). Serology of *Escherichia coli* fimbriae. *Progress in Allergy* **33**, 80–105.

Ørskov, I., Ørskov, F., Jann, B. and Jann K. (1977). Serology, chemistry and genetics of O and K antigens of *Escherichia coli*. *Bacteriological Reviews* **41**, 667–710.

Ozeki, H., Stocker, B. A. D. and Smith, S. M. (1962). Transmission of colicinogeny between strains of *Salmonella typhimurium* grown together. *Journal of General Microbiology* **28**, 671–687.

Sakazaki, R., Tamura, K. and Nakamura, K. (1974). Further studies on enteropathogenic *Escherichia coli* associated with diarrhoeal diseases in children and adults. *Japanese Journal of Medical Science and Biology* **27**, 7–18.

Scotland, S. M., Gross, R. J. and Rowe, B. (1977). Serotype-related enterotoxigenicity in *Escherichia coli* 06.H16 and 0148.H28. *Journal of Hygiene* **79**, 395–403.

Van der Waaij, D., Speltie, T. M., Guinée, P. A. M. and Agterberg, C. (1975). Serotyping and biotyping of 160 *Escherichia coli* strains: Comparative study. *Journal of Clinical Microbiology* **1**, 237–238.

Wilson, M. I., Crichton, P. B. and Old, D. C. (1981). Characterisation of urinary isolates of *Escherichia coli* by multiple typing: A retrospective analysis. *Journal of Clinical Pathology* **34**, 424–428.

12

In Vitro Techniques for the Attachment of *Escherichia coli* to Epithelial Cells

C. J. THORNS AND J. A. MORRIS

Central Veterinary Laboratory, Weybridge, Surrey, UK

Introduction

In the course of the pathogenesis of *E. coli* neonatal diarrhoea, enteropathogenic *E. coli* adhere to the mucosal epithelium of the small intestine. This attachment would be expected to promote the bacterial colonization of the mucosa because the constant flow of mucus and digesta over the mucosal surface would tend to remove non-adherent bacteria. Many of these adhesive antigens are fimbriae, and non-fimbriate variants are less effective at colonizing the small intestine (Jones and Rutter, 1972; Nagy *et al.*, 1977). Thus, certain fimbriae can be regarded as virulence determinants because bacterial colonization of the mucosa appears to be essential in the pathogenesis of *E. coli* neonatal diarrhoea (Smith and Linggood, 1971).

E. coli enteropathogenic for piglets produce the fimbrial antigens K88 (Ørskov *et al.*, 1961), K99 (Moon *et al.*, 1977) or 987P (Nagy *et al.*, 1977), while the majority of calf and lamb enteropathogens produce K99 (Ørskov *et al.*, 1975) and/or F41 (Morris *et al.*, 1982a). *E. coli* enteropathogenic for man produce CFA/I or CFA/II (Evans and Evans, 1978), while F7 has been found on some uropathogenic strains (Ørskov *et al.*, 1980). In addition, many enteropathogenic and non-enteropathogenic *E. coli* produce fimbriae referred to as type I or common fimbriae. These have certain adhesive properties when examined *in vitro*, but, in contrast to the other fimbriae, this adhesive activity can be blocked by the presence of D-mannose (Duguid and Gillies, 1957).

While studying immunity of piglets to *E. coli,* Wilson and Hohmann (1974) noticed that enteropathogenic strains adhered to isolated intestinal epithelial cells. Subsequently, *in vitro* adherence of *E. coli* to epithelial tissues has been used to study enteropathogenic strains from piglets (Sellwood *et al.*, 1975),

THE VIRULENCE OF ESCHERICHIA COLI

ISBN 0-12-677520-6

cattle (Burrows *et al.*, 1976) and man (McNeish *et al.*, 1975), uropathogenic strains from man (Svanborg Edén and Hanson, 1978) and certain invasive strains from calves and lambs (Morris *et al.*, 1982b). *In vitro* techniques involve (1) the isolation of epithelial cells that still retain their structural integrity, (2) the culture of *E. coli* in media that will optimise the expression of relevant fimbriae and (3) the estimation of adherent and non-adherent *E. coli* after incubation of bacteria with epithelial cells. However, attachment in these model systems must not be taken in isolation but should be carefully evaluated in relation to *in vivo* observations (Arbuthnott and Smyth, 1979).

Methods

In Vitro Attachment of E. coli to Pig, Calf or Lamb Brush-Border Epithelial Cells (Modified from Sellwood et al., 1975)

PREPARATION OF BRUSH BORDERS
1. Take about 1 m of posterior small intestine from a freshly killed animal 2–4 days old (adult pigs may also be used).
2. Flush out the intestinal contents with 0.15 M NaCl at 20°C.
3. Tie the length of intestine at one end and half fill with ethylenediaminetetraacetate (EDTA) buffer pH 6.8. Tie the other end, and immerse the length of intestine in sucrose buffer pH 6.8. Incubate at 4°C for 30 min.
4. Discard intestinal contents and refill the lumen with sucrose buffer. Detach the brush-border epithelial cells into the sucrose buffer by gentle massage for 2–3 min. Keep the contents at 4°C.
5. Centrifuge the suspension at 1000 g for 10 min at 4°C and resuspend pellet in 10–20 volumes of hypotonic EDTA solution.
6. Homogenize the cell suspension with a Teflon-tipped tissue grinder (clearance 0.015–0.023 cm), moving the pestle up and down six times while it is rotated at approximately 600 rpm.
7. Centrifuge the homogenate at 1000 g for 10 min at 4°C.
8. Resuspend the pellet in cold hypotonic EDTA and repeat the homogenization and centrifugation procedure until supernatant is clear of all debris and the pellet is smooth and gelatinous in appearance.
9. Wash the pellet once by centrifugation at 300 g for 5 min in Krebs–Henseleit buffer and resuspend in the same buffer to a concentration of about 10^6 cells ml^{-1}.
10. Store brush-border epithelial cells for up to a year at −20°C in an equal volume of Krebs–Henseleit buffer and glycerol.

EPITHELIAL CELL ADHESION TEST

1. Wash the brush-border epithelial cells with phosphate-buffered saline (PBS) and centrifuge at 300 g for 10 min at 20°C. Resuspend in 2 ml of PBS to a concentration of 10^6 ml^{-1}.
2. Centrifuge broth cultures or harvest agar cultures into PBS and then centrifuge at 1000 g for 10 min at room temperature.
3. Resuspend the bacteria in PBS to about 10^9 ml^{-1}.
4. Add 100 μl of bacterial suspension (10^8 ml^{-1}) to 100 μl of brush-border suspension (10^7 ml^{-1}) in a half-dram vial, mix well and incubate at 37°C for 1 h.
5. Place a drop of the reaction mixture on a glass slide under a coverslip and examine by differential interference or phase-contrast microscopy at ×400 magnification.
6. Count the number of bacteria adhering to the brush-border edge of 20 well-defined epithelial cells and determine the mean number of bacteria adhering per epithelial cell.

In Vitro Attachment of E. coli to Calf, Lamb, Rabbit or Mouse Intestinal Villi (Girardeau, 1980)

PREPARATION OF VILLI

1. Remove the entire small intestine from a freshly killed animal (2–7 days old) and place in Krebs buffer pH 7.2, at 4°C.
2. Open the intestine longitudinally and gently wash it with Krebs buffer at 4°C.
3. Gently scrape the mucosa with a glass slide to detach the villi and suspend them in cold Krebs buffer.
4. After allowing them to settle, wash the villi in cold Krebs buffer until clear supernatant fluid is obtained.
5. Resuspend the villi in Hanks–DMSO (dimethyl sulfoxide) medium and place 1⁻ ml aliquots at −20°C for 24 h followed by storage at −70°C.

VILLUS ADHESION TEST

1. Thaw the villi at 20°C and place into Krebs buffer with 1% (v/v) formaldehyde for 1 h.
2. Wash twice with Krebs buffer and resuspend in the same buffer.
3. Transfer about 20 villi into 1.5 ml of Krebs buffer. Add the bacterial suspension to the villi so as to obtain an absorbency of about 1.2 at 620 nm.
4. Gently mix the reactants for 20 min at 20°C.
5. Gently remove the villi with an automatic pipette and place on a glass slide under a coverslip.

6. Examine by differential interference or phase-contrast microscopy at magnification ×400. Any bacteria adhering to the brush-border edge of the villi will be clearly visible. Adherent bacteria can be expressed as number of bacteria per unit length.

In Vitro Attachment of E. coli to Calf Tracheal and Oesophageal Epithelial Cells (Morris et al., 1982b)

PREPARATION OF EPITHELIAL CELLS
1. Remove the trachea or oesophagus from 1–3-wk-old calves and open longitudinally under cold 100 mM phosphate buffer pH 7.2.
2. Wash the mucosa with a jet of buffer from a pipette and place in fresh buffer.
3. Gently scrape off epithelium fragments with the edge of a glass slide and collect into fresh buffer.
4. Store the suspensions at 4°C for up to 1 wk.

ADHESION TEST
1. Incubate 100 μl of epithelial cell suspension (about 2×10^6 ml^{-1}) with 100 μl of bacterial suspension (about 2×10^9 ml^{-1}) at 37°C for 1 h.
2. Place a drop of the reaction mixture on a glass slide under a coverslip. Examine in the same way as for isolated brush-border epithelial cells. Usually only qualitative observations are possible with adherence to squamous epithelium from the oesophagus since the number of adherent bacteria is too high to count accurately.

Note: To inhibit attachment due to type 1 fimbriae, D-mannose should be included in the tests at a final concentration of 0.5% (w/v).

Materials

BUFFERS
1. *EDTA buffer pH 6.8*
   0.096 M NaCl, 0.008 M KH$_2$PO$_4$, 0.0056 M Na$_2$HPO$_4$, 0.0015 M KCl, 0.01 M EDTA.
2. *Sucrose buffer pH 6.8*
   Same as (1) except that EDTA is replaced by 0.3 M sucrose.
3. *Hypotonic EDTA solution pH 7.4*
   0.005 M EDTA (pH adjusted with 0.5 M Na$_2$CO$_3$).
4. *Krebs–Henseleit buffer 7.4*
   0.12 M NaCl, 0.014 M KCl, 0.025 M NaHCO$_3$, 0.001 M KH$_2$PO$_4$.

5. *Hanks–DMSO Medium*
50 ml Hanks medium with lactalbumin buffer (Gibco); 10 ml foetal calf serum (Gibco), 30 ml DMSO; 20 ml glycerol.

MEDIA
1. *Minca agar (Guinee et al., 1976)*
This medium is used for the expression of K88, K99 and F41 surface-adhesive antigens. It consists of KH_2PO_4, 1.36 g; $Na_2HPO_4 \cdot 2H_2O$, 10.1 g; glucose (BDH), 1 g; trace salts solution, 1 ml; casamino acids (Difco), 1 g; agar (Difco), 12 g and distilled water, 1000 ml.
Trace salts solution
$MgSO_4 \cdot 7H_2O$, 10 g; $MnCl_2 \cdot 4H_2O$, 1 g; $FeCl_3 \cdot 6H_2O$, 0.135 g; $CaCl_2 \cdot 2H_2O$, 0.4 g; distilled water, 1000 ml.
2. *Improved Minca medium (Guinee et al., 1977) for K99 expression*
Same as Minca medium but replace glucose with 1% (v/v) IsoVitaleX (BBL).
3. *Tryptone soya agar (Oxoid, CM131) and tryptone soya broth (Oxoid, CM129)*
These media are used for the expression of 987P surface-adhesive antigen. On agar at least two types of colony morphology are often observed. The smaller colonies, which are more transparent in direct transmitted light, should be tested as these contain a higher proportion of fimbriated bacteria. In broth cultures, organisms from the pellicle should be used.
4. *Sheep blood agar*
This medium is used for the expression of 'Vir' surface-adhesive antigen. Blood agar base (Oxoid) plus 5% (v/v) whole defibrinated sheep blood.

References

Arbuthnott, J. P. and Smyth, C. J. (1979). Bacterial adhesion in host/pathogen interactions in animals. *In* "Adhesion of Microorganisms to Surfaces" (Eds. D. C. Ellwood, J. Melling and P. Rutter), pp. 165–198. Academic Press, London.

Burrows, M. R., Sellwood, R. and Gibbons, R. A. (1976). Haemagglutinating and adhesive properties associated with the K99 antigen of bovine strains of *Escherichia coli*. *Journal of General Microbiology* **96**, 269–275.

Duguid, J. P. and Gillies, R. R. (1957). Fimbriae and adhesive properties in dysentery bacilli. *Journal of Pathology and Bacteriology* **74**, 397–411.

Evans, D. G. and Evans, D. J. (1978). New surface associated heat labile colonization factor antigen (CFA/II) produced by enterotoxigenic *Escherichia coli* of serogroups O6 and O8. *Infection and Immunity* **21**, 638–647.

Girardeau, J. P. (1980). A new *in vitro* technique for attachment to intestinal villi using enteropathogenic *Escherichia coli*. *Annales de Microbiologie (Paris)* **131B**, 31–37.

Guinee, P. A. M., Jansen, W. H. and Agterberg, C. M. (1976). Detection of the K99 antigen by means of agglutination and immunoelectrophoresis in *Escherichia coli* isolates from calves and its correlation with enterotoxigenicity. *Infection and Immunity* **13**, 1369–1377.

Guinee, P. A. M., Veldkamp, J. and Jansen, W. H. (1977). Improved Minca medium for the detection of K99 antigen in calf enterotoxigenic strains of *Escherichia coli*. *Infection and Immunity* **15**, 676–678.

Jones, G. W. and Rutter, J. M. (1972). Role of the K88 antigen in the pathogenesis of neonatal diarrhoea caused by *Escherichia coli* in pigs. *Infection and Immunity* **6**, 918–927.

McNeish, A. S., Turner, P., Fleming, J. and Evans, N. (1975). Mucosal adherence of human enteropathogenic *Escherichia coli*. *Lancet* **2**, 946–948.

Moon, H. W., Nagy, B., Isaacson, R. E. and Ørskov, I. (1977). Occurrence of K99 antigen on *Escherichia coli* isolated from pigs and colonization of pigs ileum by K99+ enterotoxigenic *E. coli* from calves and pigs. *Infection and Immunity* **15**, 614–620.

Morris, J. A., Thorns, C. J., Scott, A. C., Sojka, W. J. and Wells, G. A. (1982a). Adhesion *in vitro* and *in vivo* associated with an adhesive antigen (F41) produced by a K99 mutant of the reference strain *Escherichia coli* B41. *Infection and Immunity* **36**, 1146–1153.

Morris, J. A., Thorns, C. J., Scott, A. C. and Sojka, W. J. (1982b). Adhesive properties associated with the Vir plasmid: A transmissible pathogenic characteristic associated with strains of invasive *Escherichia coli*. *Journal of General Microbiology* **128**, 2097–2103.

Nagy, B., Moon, H. W. and Isaacson, R. E. (1977). Colonization of porcine intestine by enterotoxigenic *Escherichia coli*. Selection of piliated forms *in vivo*, adhesion of piliated forms to epithelial cells *in vitro* and incidence of a pilus antigen among porcine enteropathogenic *E. coli*. *Infection and Immunity* **6**, 344–352.

Ørskov, I., Ørskov, F., Sojka, W. J. and Leach, J. N. (1961). Simultaneous occurrence of *E. coli* B and L antigens in strains from diseased swine. Influence of cultivation temperature on two new *E. coli* K antigens K87 and K88. *Acta Pathologica et Microbiologica Scandinavica* **53**, 404–422.

Ørskov, I., Ørskov, F., Smith, H. W. and Sojka, W. J. (1975). The establishment of K99, a thermolabile, transmissible *Escherichia coli* K antigen, previously called "Kco," possessed by calf and lamb enteropathogenic strains. *Acta Pathologica et Microbiologica Scandinavica, Section B: Microbiology* **83B**, 31–36.

Ørskov, I., Ørskov, F. and Birch-Andersen, A. (1980). Comparison of *Escherichia coli* fimbrial antigen F7 with type 1 fimbriae. *Infection and Immunity* **27**, 657–666.

Sellwood, R., Gibbons, R. A., Jones, G. W. and Rutter, J. M. (1975). Adhesion of enteropathogenic *Escherichia coli* to pig intestinal brush borders: the existence of two pig phenotypes. *Journal of Medical Microbiology* **8**, 405–411.

Smith, H. W. and Linggood, M. A. (1971). Observations on the pathogenic properties of K88, Hly and Ent plasmids of *Escherichia coli* with particular reference to porcine diarrhoea. *Journal of Medical Microbiology* **11**, 471–492.

Svanborg Edén, C. and Hansson, H. A. (1978). *Escherichia coli* pili as possible mediators of attachment to human urinary tract epithelial cells. *Infection and Immunity* **21**, 229–237.

Wilson, M. R. and Hohmann, A.W. (1974). Immunity to *Escherichia coli* in pigs: Adhesion of enteropathogenic *Escherichia coli* to isolated intestinal epithelial cells. *Infection and Immunity* **10**, 776–782.

13

Method for Investigating Bacterial Adherence to Isolated Uroepithelial Cells and Uromucoid

M. J. HARBER, SUSAN CHICK, RUTH MACKENZIE AND A. W. ASSCHER

Department of Renal Medicine, Welsh National School of Medicine, KRUF Institute, Royal Infirmary, Cardiff, Wales

Introduction

The interaction between bacteria and epithelial cells or mucus is an important area for study since adhesion to mucosal surfaces appears to be a pre-requisite for the initiation of infection at these sites. Bladder epithelium is covered with a layer of mucopolysaccharide (Parsons *et al.*, 1975; Fukushi *et al.*, 1979) and urine contains free mucus in the form of Tamm-Horsfall protein. This has been shown to bind bacteria with mannose-sensitive adhesins (Ørskov *et al.*, 1980), while organisms with mannose-resistant adhesins attach more readily to uncoated uroepithelial cells (Chick *et al.*, 1981). Unfortunately, it is not possible to make this important distinction with standard adherence assays (Svanborg-Edén *et al.*, 1977; Schaeffer *et al.*, 1979). Therefore, a modified method which fulfills this requirement is described.

The method involves a combination of techniques used previously by Gibbons and van Houte (1971), Fowler and Stamey (1977) and Källenius and Winberg (1978) to study bacterial adherence to different types of epithelial cell. Efficient separation of adherent from non-adherent bacteria is achieved by vacuum filtration and the uroepithelial cell preparations are then stained in order to render uromucoid clearly visible under the light microscope.

Methods

A mixture of 10^5 urinary epithelial cells and 10^8 bacteria, each suspended in 1 ml of phosphate-buffered saline (PBS) pH 6.8, is incubated for 1 h at 37°C with continuous rotation. The suspension is then filtered under vacuum through a

339

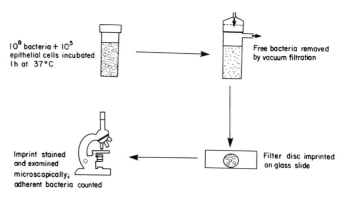

Fig. 1. Flow diagram of the uroepithelial cell adherence assay.

5-μm Nuclepore polycarbonate membrane filter and washed with 15 ml of PBS. The filter membrane is pressed onto a glass microscope slide to give an imprint containing uroepithelial cells and the membrane itself is discarded. The imprint on the slide is treated with 0.25% Toluidine Blue in 1% borax to stain the uromucoid and the preparation is then ready for examination by light micros-copy. A scheme of the method is shown in Fig. 1.

Adherence to uroepithelial cells is quantified by determining the mean number of adherent bacteria per epithelial cell for 30 cells which are not overlaid with mucus. Adherence to mucus cannot be quantified but, since it is essentially an all-or-none phenomenon, a score of 1 may be given to bacterial strains which show definite attachment to mucus, and a score of 0 to strains which do not. This scoring method allows data to be analysed statistically by the Mann-Whitney U test.

Micrographs illustrating the adherence of different strains of *Escherichia coli* to uroepithelial cells and uromucoid are presented in Fig. 2.

Materials

Buffer

Phosphate-buffered saline (PBS) pH 6.8 contains: KH_2PO_4, 3.45 g; Na_2HPO_4, 4.45 g; NaCl, 5.0 g; KCl, 0.2 g and distilled water, 1000 ml. When required, D-mannose should be incorporated into this buffer at a concentration of 2.5% (w/v).

Epithelial Cells

Uroepithelial cells may be obtained from the urine deposit of healthy females and should be washed three times in PBS pH 6.8 before use. Ideally, only cell

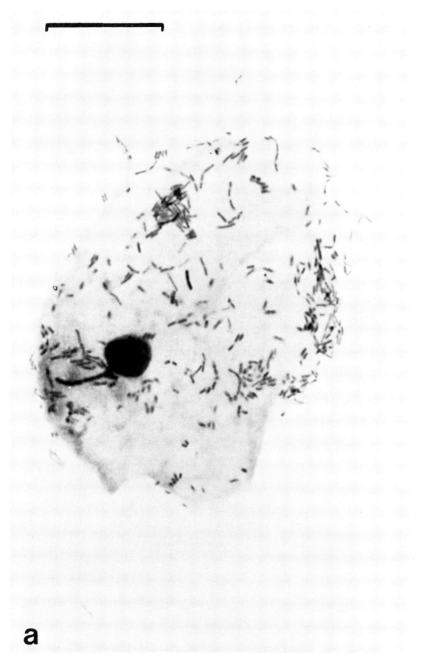

a

Fig. 2. Micrographs showing (a) an *E. coli* strain with mannose-resistant fimbriae adhering to a uroepithelial cell and (b) an *E. coli* strain with mannose-sensitive fimbriae adhering to uromucoid. The bar markers represent 20 μm. (*continued*)

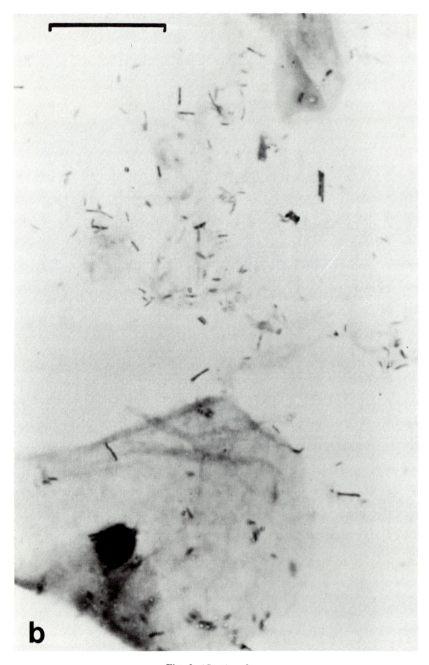

Fig. 2. (*Continued*)

preparations with a very low number of endogenous adherent bacteria (≤4 per epithelial cell) should be used for adherence studies and this number should be taken into consideration when the results of adherence assays are calculated.

Bacteria

Any appropriate bacterial strains may be used as long as the following points are borne in mind. First, if adherence data obtained *in vitro* are to be related to bacterial pathogenicity *in vivo* it is important to use bacteria freshly isolated from a clinical source since the expression of bacterial adhesins changes markedly under different growth conditions. Urine is a poor medium for the expression of fimbriae (Ofek *et al.*, 1981; Harber *et al.*, 1982) but the expression of mannose-sensitive fimbriae may be promoted by sequential subculture in static nutrient broth while the expression of mannose-resistant fimbriae is favoured by sub-culture on solid media. Hence adherence data obtained with subcultured bacteria are unlikely to reflect the adherence properties of the same organisms *in vivo*. Second, it should be noted that fimbriae are not necessarily required for the attachment of bacteria to mucosal or epithelial surfaces and non-fimbrial adherence may be a more important factor in the colonization of periurethral and bladder epithelia (Harber *et al.*, 1984).

References

Chick, S., Harber, M. J., Mackenzie, R. and Asscher, A. W. (1981). Modified method for studying bacterial adhesion to isolated uroepithelial cells and uromucoid. *Infection and Immunity* **34**, 256–261.

Fowler, J. E., Jr. and Stamey, T. A. (1977). Studies of introital colonization in women with recurrent urinary infections. V11. The role of bacterial adherence. *Journal of Urology* **117**, 472–476.

Fukushi, Y., Orikasa, S. and Kagayama, M. (1979). An electron microscopic study of the interaction between vesical epithelium and *E. coli*. *Investigative Urology* **17**, 61–68.

Gibbons, R. J. and van Houte, J. (1971). Selective bacterial adherence to oral epithelial surfaces and its role as an ecological determinant. *Infection and Immunity* **3**, 567–573.

Harber, M. J., Chick, S., Mackenzie, R. and Asscher, A. W. (1982). Lack of adherence to epithelial cells by freshly isolated urinary pathogens. *Lancet* **1**, 586–588.

Harber, M. J. Mackenzie, R. and Asscher, A. W. (1984). "Virulence factors of *Escherichia coli*." *In* "Pyelonephritis, Vol. V, Urinary Tract Infections" (Eds. H. Losse, A. W. Asscher, A. E. Lison and V. T. Andriole), pp. 43–50. Thieme, Stuttgart and New York.

Källenius, G. and Winberg, J. (1978). Bacterial adherence to periurethral epithelial cells in girls prone to urinary tract infections. *Lancet* **2**, 540 543.

Ofek, I., Mosek, A. and Sharon, N. (1981). Mannose-specific adherence of *Escherichia coli* freshly excreted in the urine of patients with urinary tract infections, and of isolates subcultured from the infected urine. *Infection and Immunity* **34**, 708–711.

Ørskov, I., Ferencz, A. and Ørskov, F. (1980). Tamm-Horsfall protein or uromucoid is the normal urinary slime that traps type 1 fimbriated *Escherichia coli*. *Lancet* **1**, 887.

Parsons, C. L., Greenspan, C. and Mulholland, S. G. (1975). The primary antibacterial defence mechanism of the bladder. *Investigative Urology* **13,** 72–76.
Schaeffer, A. J., Amundsen, S. K. and Schmidt, L. N. (1979). Adherence of *Escherichia coli* to human urinary tract epithelial cells. *Infection and Immunity* **24,** 753–759.
Svanborg Edén, C., Eriksson, B. and Hanson, L. Å. (1977). Adhesion of *Escherichia coli* to human uroepithelial cells *in vitro*. *Infection and Immunity* **18,** 767–774.

14

Serotyping of *Escherichia coli*

R. J. GROSS AND B. ROWE

Division of Enteric Pathogens, Central Public Health Laboratory, London, UK

Introduction

The serotyping scheme for *Escherichia coli* is based on that described by Kauff-mann (1947) and depends on the identification of the heat-stable lipopolysac-charide somatic or O antigens, the flagellar H and the surface or capsular K antigens. So far over 160 O antigens and more than 50 H antigens have been described and numerous different O:H combinations have been found.

The term K antigen was originally used for surface antigens that cause 'O inagglutinability'—the phenomenon in which the agglutination of living orga-nisms in an antiserum prepared with a heated vaccine is inhibited or blocked by K antigen. Kauffmann divided K antigens into three classes according to the effect of heat on the agglutinability, antigenicity and antibody-binding power of the bacterial strains that possess them. The L-type K antigens are heat-labile; after heating at 100°C for 1 h a bacterial suspension may be agglutinated by O antiserum but not by K antiserum, and the suspension no longer has K anti-genicity or the ability to bind K antibody. Pure K antiserum can be prepared for L-type K antigens by using living or unheated organisms as a vaccine and absorbing the resulting antiserum with a heated suspension of the homologous strain. The A-type K antigens are unaffected by heating at 100°C. After heating at 121°C for 2 h, a suspension of organisms may be agglutinated by O antiserum but not by K antiserum, and the suspension no longer has K antigenicity. Nev-ertheless, the ability to bind K antibody remains. To prepare a pure A-type K antiserum it is, therefore, necessary to absorb with a K⁻ variant of the homolo-gous strain or with an organism with the same O antigen but different K antigen.

The existence of many L- and A-type K antigens is not in doubt but the evidence for the existence of the third group, the B-type K antigens, is less conclusive. Their presence was originally inferred from the rise in titre in OK antiserum observed when a suspension was heated at 100°C. More recently, Ørskov and Ørskov (1970, 1972) and Ørskov *et al.* (1971) carried out immu-

345

noelectrophoresis of bacterial extracts and showed that all the serotypes of *E. coli* fell into a few well-defined groups according to the electrophoretic characters of their O and K antigens. However, many of the strains believed to possess B-type K antigens did not possess detectable K antigen.

As a result of these studies the existence of many of the 100 or more K antigens described is now in doubt and even reference laboratories rarely attempt full K antigen identification. Fortunately, O:H serotyping is usually adequate for epidemiological purposes and we shall describe in detail methods for the identification of O and H antigens. Identification of K antigens may be of value in particular studies and we shall describe methods for their identification only briefly.

For further details of the serology, chemistry and genetics of O and K antigens the review of Ørskov *et al.* (1977) is recommended.

O Antigen Identification

The basic technique used in O antigen identification is that of tube agglutination. The exact method may be varied depending on the tubes or agglutination trays available and on the availability of suitable pipetting devices or machinery. The procedure described here was used in our own laboratory for many years but has now been replaced by a method using microtitre trays and semi-automatic multiple-pipetting equipment.

1. Selection of Vaccine Strains

A list of representative type strains for *E. coli* O antigens has been agreed internationally (see Materials). Vaccine suspensions should be prepared from these strains.

2. Preparation of Vaccine Suspensions

(a). Strains of *E. coli* often become rough or degraded during laboratory storage. Before preparation of vaccine suspensions type strains should, therefore, be streaked out on nutrient or blood agar and smooth colonies selected.

(b). The selected colonies are inoculated onto two nutrient agar slopes (6 × 5/8 inch tubes) or onto a 10-cm nutrient agar plate and incubated overnight at 37°C.

(c). The growth from the slopes or the plate is washed off in about 10 ml saline and heated at 100°C for $2\frac{1}{2}$ h. The suspension is then centrifuged and the supernatant discarded. The deposit is washed in 10 ml saline and centrifuged again.

(d). The heated, washed deposit is resuspended in saline containing 0.3%

(v/v) commercial formalin and the volume adjusted to give a concentration of about 10⁹/ml.

3. Immunization of Rabbits

(a). Injections are given via the marginal ear vein, as follows:

Day 1	Inject 0.5 ml
5	Inject 1.0 ml
10	Inject 2.0 ml
15	Inject 2.0 ml
20	Inject 2.0 ml
25	Bleed 40 ml
30	Bleed 40 ml
35	Bleed out

(b). After the separation of the serum the products of all three bleedings are pooled and merthiolate added to a concentration of 1/10,000 (w/v) as a preservative.

4. Testing of Antisera

(a). Antigen suspensions are prepared from all the standard type strains by growing overnight broth cultures at 37°C and heating for 1 h at 100°C. Formalin is added to a concentration of 0.3% (v/v) as a preservative.

(b). Samples of antisera are prepared for testing by diluting 1/50 in saline containing 1/10,000 (w/v) merthiolate or 0.5% (w/v) phenol as a preservative.

(c). The diluted antisera are first screened for cross-reactions by setting up single-tube tests against all the test strains. A 0.3-ml portion of antiserum (1/50 dilution) is added to 0.3 ml antigen suspension in Dreyer's tubes or WHO plastic agglutination trays. For each antigen suspension a control tube is required containing 0.3 ml saline and 0.3 ml antigen suspension. The tests are incubated at 50°C overnight and the tubes or wells then examined for agglutination. Control tubes should show no agglutination.

(d). The antisera are then tested to determine their agglutination titre with any antigen suspension which showed agglutination in the screening step.

Doubling dilutions are prepared starting with the 1/50 antiserum dilution used above; saline is used as the diluent and 0.3-ml volumes are used. Antigen suspension (0.3 ml) is then added to each tube. A control tube is required.

The tests are again incubated at 50°C overnight and then examined for agglutination. The agglutination titre is defined as the highest dilution of antiserum in which agglutination is seen. Control tubes should show no agglutination.

5. Preparation of Monospecific Antisera

(a). There are numerous cross-reactions between *E. coli* O antigens and many antisera required absorption to make them specific. Type strains are used for absorptions.

(b). Absorbing suspensions are prepared as follows: smooth colonies are selected from cultures on nutrient or blood agar and inoculated into nutrient broth. After 5–6 h at 37°C a few drops of broth culture are placed onto 15-cm nutrient agar plates and spread evenly with a sterile glass spreader. The plates are then incubated overnight at 37°C.

It is difficult to estimate the number of plates required but a rough guide is one plate per 1000 titre per 1 ml unabsorbed serum.

(c). The plates are harvested into 10 ml saline with a suitable scraper and heated at 100°C for 1 h. The suspension is then centrifuged and the supernatant discarded.

(d). The heated organisms are mixed with the antiserum and the mixture incubated for 2 h at 50°C.

(e). After centrifugation the clear supernatant is removed and retested against the cross-reacting antigen suspensions and the homologous antigen suspension. (Note: antiserum stores best in the unabsorbed state and it is, therefore, advisable to prepare only sufficient absorbed antiserum for a few weeks work.)

6. Preparation of Pooled O Antisera

(a). The antisera in each pool are grouped according to their major antigenic relationships (see Materials). It is not necessary to use absorbed antisera for the preparation of pooled antisera.

(b). The dilution of each constituent antiserum is adjusted in the pool so that its titre is about 1:16 to 1:32. Saline containing 1/10,000 (w/v) merthiolate is used as the diluent.

7. Identification of O Antigens

(a). The O antigen suspensions for identification are prepared by heating overnight broth cultures for 1 h at 100°C and adding commercial formalin to a final concentration of 0.3% (v/v) as a preservative.

(b). A 0.3-ml portion of O antigen suspension is added to 0.3 ml of each O antiserum pool in Dreyer's tubes or WHO agglutination trays; a saline control is required for each suspension. The tests are incubated at 50°C overnight and then examined for agglutination. There should be no agglutination in the control tubes. Because of cross-reactions there may be agglutination in more than one O antiserum pool.

(c). The O antigen suspension is then tested against the individual O antisera which make up the pool(s) in which agglutination occurred. For this purpose 1/50 dilutions of the individual antisera are prepared in saline containing 1/10,000 (w/v) merthiolate or 0.5% (w/v) phenol. A 0.3-ml portion of antigen suspension is added to 0.3 ml of each diluted antiserum in Dreyer's tubes or WHO plastic agglutination trays; a saline control is required for each suspension. It is not necessary to use absorbed antisera at this stage. The tests are incubated overnight at 50°C and examined for agglutination.

(d). Serial doubling dilutions are prepared for each antiserum in which agglutination occurred in order to determine the agglutination titre. The 1/50 dilutions of antiserum are used as the initial antiserum dilution and saline is used as the diluent. Volumes of 0.3 ml are again used and a saline control is required for each antigen suspension. The tests are incubated overnight at 50°C and then examined for agglutination. The titre is defined as the highest dilution in which agglutination occurred.

(e). At this stage the titres obtained with the antigen suspension of the strain under investigation are compared with the homologous titres of the antisera obtained previously.

    (i). If the strain under investigation was agglutinated to within one doubling dilution of the homologous titre by one antiserum but gave no reaction, or only minor reactions, with other antisera then the strain is considered to be identified.

    (ii). If the strain under investigation gave no reaction, or only minor reactions, with all the antisera then the strain is considered to be unidentified. However, in case the strain possesses a heat-resistant (A-type) K antigen which is preventing O agglutination, a new O antigen suspension is prepared. A new overnight broth culture is grown and autoclaved at 121°C for $2\frac{1}{2}$ h; formalin is added to a concentration of 0.3% (v/v) as a preservative. The identification process is then repeated from 7(b).

    (iii). If the strain under investigation was agglutinated to within one doubling dilution of the homologous titre by more than one O antiserum then serial dilutions are set up using the absorbed antisera prepared previously and the agglutination titres determined. Agglutination should now be obtained with only one antiserum.

H Antigen Identification

1. Vaccine Strains

Reference laboratories in various countries have agreed on a standard set of type strains for *E. coli* H antigens. At present these antigens are numbered from H1 to

H56 with the absence of H50, which was deleted from the scheme. There are therefore 55 type strains and these are listed in Materials.

2. Preparation of Vaccine Suspensions

(a). Rough and degraded variants frequently arise in strains of *E. coli* stored in the laboratory. Strains used for the preparation of vaccines should therefore first be streaked onto nutrient or blood agar for the selection of smooth colonies.

(b). For the preparation of H antisera it is necessary to use cultures in which the majority of cells are motile. To obtain such cultures it is necessary to passage the selected colonies from the type strains through Craigie tubes containing semi-solid agar (see Materials). If it is difficult to obtain motile cultures at 37°C, improved motility may be obtained at 30°C.

(c). Nutrient broth cultures are inoculated from Craigie tube cultures and grown at 37°C for 4–5 h. Commercial formalin is then added to a concentration of 0.3% (v/v).

3. Immunization Schedule

Rabbits are immunized via the marginal ear vein using the formalised nutrient broth cultures as vaccines, as follows:

Day 1	Inject 0.5 ml
5	Inject 1.0 ml
10	Inject 2.0 ml
15	Inject 2.0 ml
20	Inject 2.0 ml
25	Bleed 40 ml
30	Bleed 40 ml
35	Bleed out

4. Testing of Antisera

(a). After separation of serum the products of all bleedings are pooled and merthiolate added as a preservative [final concentration 1/10,000 (w/v].

(b). Dilutions for testing are prepared at 1/100 in saline containing 1/10,000 (w/v) merthiolate or 0.5% (w/v) phenol as a preservative.

(c). The H antigen suspensions are prepared from all the type strains by the method described above for the preparation of vaccines.

(d). Antisera are tested against all 55 antigen suspensions. The tests are performed in Dreyer's tubes and for each suspension a control tube is required. For each test 0.3 ml antiserum (1/100 dilution) and 0.3 ml antigen suspension are added to a Dreyer's tube. Control tubes contain 0.3 ml saline and 0.3 ml antigen suspension.

(e). The tubes are incubated for 2 h at 50°C in a water-bath. The presence of agglutination in the tube containing antiserum indicates a positive reaction. All control tubes should show uniform turbidity.

(f). Whenever a positive reaction occurs the agglutination titre should be determined. Starting with the 1/100 antiserum dilution, doubling dilutions are prepared in Dreyer's tubes using saline as the diluent; a control tube containing saline alone should be included. An equal volume of antigen suspension is then added to each tube and the tubes are incubated for 2 h at 50°C. The agglutination titre is the highest dilution in which agglutination occurs. The control tubes should show uniform turbidity.

5. *Preparation of Absorbed Antisera*

Antisera with unwanted cross-reactions should be absorbed to remove these cross-reactions.

(a). For the preparation of an absorbing suspension a motile broth culture is first required (see preparation of vaccine suspensions above). This broth culture is used to inoculate 15-cm nutrient agar plates by placing 4–6 drops nutrient broth on each plate and spreading evenly with a sterile glass spreader. The plates are then incubated overnight at 37°C.

It is difficult to estimate the number of plates required for each absorption but a rough guide is 1 plate per 1000 titre per 1 ml unabsorbed serum.

(b). The plates are harvested into Bridge's solution (see Materials) with a suitable scraper. The volume of Bridge's solution used should be equal to the volume of antiserum to be absorbed. (Note: since antisera keep best in the unabsorbed state it is good policy to absorb only enough for a few weeks work.) The absorbing suspension is added to the unabsorbed antiserum and incubated for 2 h at 50°C. The suspension is then centrifuged and the clear supernatant taken off.

(c). The absorbed antiserum is retested against the cross-reacting antigen suspensions and the homologous antigen suspension.

6. *Preparation of Pooled H Antisera*

(a). A suitable composition for pooled H antisera is shown in Materials.

(b). Since cross-reacting antisera are grouped together it is not necessary to use absorbed antisera for the preparation of pooled H antisera.

(c). The dilution of each component is adjusted so that the pooled serum has a titre of about 1:32 against each appropriate antigen suspension. Saline containing 1/10,000 (w/v) merthiolate is used as the diluent.

7. *H Identification*

(a). The H antigen suspensions for identification are prepared in the same way as vaccine suspensions (see above).

(b). A 0.3-ml portion of H antigen suspension is added to 0.3 ml of each H antiserum pool in Dreyer's tubes; a saline control is required. The tubes are incubated at 50°C for 2 h in a water-bath and examined for the presence of agglutination. The saline control should show uniform turbidity.

(c). The H antigen suspension is then tested against the individual H antisera which make up the pool in which agglutination occurred. For this purpose dilutions of the individual antisera are prepared in saline containing 1/10,000 (w/v) merthiolate such that the homologous titre is about 1:32. Absorbed antisera are used where necessary.

A 0.3-ml portion of antigen suspension is added to 0.3 ml of each diluted antiserum in Dreyer's tubes; a saline control is again required. The tubes are incubated at 50°C for 2 h and examined for agglutination.

(d). The H antigen is considered to be identified when agglutination occurs in one of the individual antisera. Titrations are not considered necessary at this stage.

K Antigen Identification

1. Vaccine Strains

Full K antigen typing is not recommended although it may be desirable to test for the presence of particular K antigens in research studies. A list of K antigen type strains is shown in Materials. It must be emphasized that the K antigens indicated are those originally described in these strains and, as stated in the introduction, the existence of many of these is doubtful.

2. Preparation of Vaccine

(a). Strains of E. coli often lose their K antigens on storage and the selection of suitable colonies for vaccine production is difficult. K$^+$ colonies are usually raised, opaque white, glossy and slightly mucoid on blood or nutrient agar, while K$^-$ colonies tend to be flatter, greyer and non-mucoid. If O antisera are available these may be used to test colonies for O inagglutinability by slide agglutination; O inagglutinability indicates the presence of K antigen.

(b). Either a nutrient broth suspension or a wash-off from nutrient agar may be used as a vaccine. In either case the suspension should *not* be heated but formalin should be added to a final concentration of 0.3%. Some workers use live suspensions (without formalin) for the final two injections.

3. Immunization of Rabbits

(a). Injections are given via the marginal ear vein and blood is taken according to the schedule for O antiserum preparation.

(b). It should be noted that unheated suspensions are sometimes toxic for rabbits and it may be necessary to use diluted suspensions.

4. Testing and Absorption of Antisera

(a). The O and H antibodies may be removed from O:K:H antisera by absorption. Ideally, the absorbing suspensions should be prepared from K^- colonies isolated from the vaccine strains; in this case the suspensions are used unheated. In the case of L-type K antigens it may be possible to use a heated suspension prepared from K^+ colonies but further absorption may then be required to remove H antibody.

(b). The general methods for absorption described for O antisera may be followed except that absorbing suspensions are unheated where appropriate.

(c). Antigen suspensions for testing purposes are prepared with wash-offs from nutrient agar suspended in saline containing 0.3% commercial formalin or in Bridge's solution (see Materials).

(d). The K titres may be determined by titration in the same way as O titres (see above) but it is recommended that K agglutination tests should be incubated at 37°C for 2 h, then at room temperature or 4°C overnight.

5. Counter-Current Immunoelectrophoresis

Semjen *et al.* (1977) have described a simple electrophoretic method (CIE) which is suitable for the identification of acidic polysaccharide K antigens and is more reliable than simple agglutination. Suitable K antigens include K1 to K57, K62, K74, K82, K83, K84, K87, K92 to K98, K100 to K103.

(a). *Preparation of antigen.* The growth from four 14-cm nutrient agar plates is suspended in 10 ml saline and heated at 100°C for 1 h. The heated suspension is used as a crude extract and it is unnecessary to centrifuge.

(b). *Preparation of antisera.* Unabsorbed antisera prepared as above are used.

(c). *Electrophoresis.* Glass slides 10 × 10 cm are covered with 1% (w/v) agarose in veronal buffer, pH 8.6, 0.06 *M* (see Materials). Holes 4 mm in diameter are cut in the agar in pairs 4 mm apart. Antigen (20 μl) diluted 1/100 in veronal buffer is placed in the wells destined to be nearest the cathode. The same volume of antiserum diluted 1/4 is placed in the wells nearest the anode. The electrophoresis is then run at 2.5 V/cm for 1 h.

(d). *Results.* The slides are allowed to cool for 15 min and then examined for a precipitin line between the wells.

(e). *Interpretation.* In most cases only a single line corresponding to K antigen is seen. However, a few strains with anodic O antigens (e.g., O83:K24, O139:K82, O150:K93 and O81:K97) may also give a line corresponding to O antigen. If there is any doubt it may be necessary to perform control tests using O antiserum in place of O:K antiserum.

Materials

1. E. coli O Antigen Type Strains

Original no.	O	K[a]	H	DEP[b] ref. no.
U5/41	1	1	7	Sc 13
U9/41	2	1	4	Sc 393
U14/41	3	2ab	2	Sc 4
U4/41	4	3	5	Sc 18
U1/41	5	4	4	Sc 394
Bi7458/41	6	2ac	1	Sc 9
Bi7509/41	7	1	—	Sc 23
G3404/41	8	8	4	Sc 398
Bi316/42	9	9	12	Sc 399
Bi8337/41	10	5	4	Sc 389
Bi623/42	11	10	10	Sc 66
Bi626/42	12	5	—	Sc 46
Su4321/41	13	11	11	Sc 63
Su4411/41	14	7	—	Sc 458
F7902/41	15	14	4	Sc 48
F11119/41	16	1	—	Sc 49
K12a	17	16	18	Sc 395
F10018/41	18ab	76	14	Sc 59
Ewing 3219/54	18ac	77	—	Sc 385A
F8188/41	19ab	?	7	Sc 396
P7a	20	17	—	Sc 304
E19a	21	20	—	Sc 52
E14a	22	13	1	Sc 53
E39a	23	18	15	Sc 57
E41a	24	?	—	Sc 397
E47a	25	19	12	Sc 54
H311b	26	60	—	Sc 239
F9884/41	27	?	—	Sc 457
K1a	28ab	?	—	Sc 418
Katwijk	28ac	?	—	Sc 483
Su4338/41	29	?	10	Sc 456
P2a	30	?	—	Sc 316
P6a	32	?	19	Sc 318
E40	33	?	—	Sc 319
H304	34	?	10	Sc 320
E77a	35	?	10	Sc 302
H502a	36	?	9	Sc 321
H510c	37	?	10	Sc 322
F11621/41	38	?	26	Sc 462
H7	39	?	—	Sc 323
H316	40	?	4	Sc 463
H710c	41	?	40	Sc 324
P11a	42	?	37	Sc 83

Original no.	O	K[a]	H	DEP[b] ref. no.
Bi7455/41	43	?	2	Sc 56
H702c	44	?	18	Sc 153
H61	45	?	10	Sc 339
P1c	46	?	16	Sc 340
U8/41	48	?	—	Sc 343
U12/41	49	?	12	Sc 344
U18/41	50	?	4	Sc 310
U19/41	51	?	24	Sc 407
U20/41	52	?	10	Sc 325
Bi7327/41	53	?	3	Sc 70
Su3972/41	54	?	2	Sc 345
Su3912/41	55	?	—	Sc 245
Su3684/41	56	?	—	Sc 346
F8198/41	57	?	—	Sc 347
F8962/41	58	?	27	Sc 348
F9095/41	59	?	19	Sc 349
F10167a/41	60	?	33	Sc 350
F10167b/41	61	?	19	Sc 351
F10524/41	62	?	30	Sc 236
F10598/41	63	?	—	Sc 327
K6b	64	?	—	Sc 384
K11a	65	?	—	Sc 329
P1a	66	?	25	Sc 330
P7d	68	?	4	Sc 332
P9b	69	?	38	Sc 84
P9c	70	?	42	Sc 226
P10a	71	?	12	Sc 334
P12a	73	?	31	Sc 464
E3a	74	?	39	Sc 85
E3b	75	?	5	Sc 465
E5d	76	?	8	Sc 336
E10	77	?	—	Sc 337
E38	78	80	—	Sc 237
E49	79	?	40	Sc 402
E71	80	?	26	Sc 338
H5	81	?	—	Sc 362
H14	82	?	—	Sc 353
H17a	83	?	31	Sc 354
H19	84	?	21	Sc 355
H23	85	?	1	Sc 356
H35	86	?	25	Sc 275
H40	87	?	12	Sc 357
H53	88	?	25	Sc 358
H68	89	?	16	Sc 311
H77	90	?	—	Sc 233
H307b	91	?	—	Sc 359
H308a	92	?	33	Sc 360
H308b	93	?	—	Sc 312

(*continued*)

Original no.	O	K[a]	H	DEP[b] ref. no.
H311a	95	?	33	Sc 363
H319	96	?	19	Sc 313
H320a	97	?	—	Sc 364
H501d	98	?	8	Sc 365
H504c	99	?	33	Sc 366
H509a	100	?	2	Sc 367
H510a	101	?	33	Sc 368
H511	102	?	40	Sc 235
H515b	103	?	8	Sc 466
H519	104	?	12	Sc 369
H520b	105ab	?	8	Sc 413
NCTC9115	105ac	?	18	Sc 376
H521a	106	?	33	Sc 371
H705	107	?	27	Sc 372
H708b	108	?	10	Sc 373
H709c	109	?	19	Sc 374
H711c	110	?	39	Sc 258
Stoke W	111	58	—	Sc 459
1411/50	112ab	68	18	Sc 293
Guanabara (M194)	112ac	66	—	Sc 506
6182/50 = 32w	113	75	21	Sc 375
26w	114	90	32	Sc 461
27w	115	?	18	Sc 414
28w	116	?	10	Sc 377
30w	117	?	4	Sc 467
31w	118	?	—	Sc 468
34w	119	?	27	Sc 460
35w	120	?	6	Sc 378
39w	121	?	10	Sc 379
43w	123	?	16	Sc 381
Ewing 227	124	72	30	Sc 496
Canioni	125	70	19	Sc 100
E611 Taylor	126	71	2	Sc 102
Ewing 4932/53	127	63	—	Sc 503
Cigleris	128	67	2	Sc 121
Seeliger 178/54	129	?	11	Sc 118
Ewing 4866/53	130	?	9	Sc 117
S239	131	?	26	Sc 116
N87	132	?	28	Sc 115a
N282	133	?	29	Sc 114
4370-53	134	?	35	Sc 101
Coli Pecs	135	?	—	Sc 112
1111/55 (Sakazaki)	136	78	—	Sc 430
RVC1787 (Rees)	137	79	41	Sc 119
CDC62-57	138	81	—	Sc 218
CDC63-57	139	82	1	Sc 219
CDC149-51	140	?	43	Sc 220
RVC2907	141	85	4	Sc 287

Original no.	O	K[a]	H	DEP[b] ref. no.
C771	142	86	6	Sc 288
Ewing 4608-58	143	?	—	Sc 223
Ewing 1624-56	144	?	—	Sc 224
1385(3) (Taylor)	145	?	—	Sc 451
CDC2950-54 = OX4	146	?	21	Sc 408
G1253	147	89,88ac	19	Sc 409
E519/66	148	?	28	Sc 416
CS1483	149	91	10	Sc 446
1935	150	93	6	Sc 511
880-67	151	?	50	Sc 508
1184-68	152	?	—	Sc 512
14097	153	?	7	Sc 513
E1541/68	154	94	4	Sc 449
E1529/68	155	?	9	Sc 448
E1585/68	156	?	47	Sc 450
A2 (Abottstown)	157	88ac	19	Sc 644
E1020/72	158	?	20	Sc 659
E2476/72	159	?	23	Sc 658
E110/69	160	?	34	Sc 476
E223/69	161	?	54	Sc 477
10b1/1	162	?	10	Sc 703
SN3B/1	163	?	19	Sc 704
145/46	164	?	—	Sc 647
E78634	165	?	—	Sc 793
Ewing 3866-54	166	?	4	Sc 794
E10702	167	?	5	Sc 771
E10710	168	?	16	Sc 773
Ewing 179$_2$-54	169	?	8	Sc 796
Ewing 745-54	170	?	1	Sc 807

[a]The K antigens listed here are those originally described in these strains and are subject to the reservations expressed in the introduction.

[b]DEP is the Division of Enteric Pathogens, Central Public Health Laboratory, Colindale, London, U.K.

2. *E. coli Pooled O Antisera*

A	B	C	D	E	F	G	H	J	K	L	M	N	P	Q	R
1	2	3	5	6	8	10	11	13	14	21	22	27	79	80	136
24	20	4	7	28ab	9	49	39	17	45	32	55	28ac	107	84	140
31	36	18ab	65	29	12	57	43	19	62	33	86	42	117	87	142
35	41	18ac	70	30	15	100	58	44	85	37	90	63	132	88	143
69	50	23	71	46	16	105ab	61	73	112ab	38	111	64	153	89	144
97	51	25	114	48	40	105ac	76	77	112ac	52	119	66	155	95	160
99	53	26	121	54	60	110	78	106	120	81	126	82	156	96	161
138	74	34	125	59	93	151	92	115	130	83	127	98	157	103	162
148	118	56	146	75	101	152	104	129	137	91	128	113	158	109	163
149		68		124	123		139	133	145	131		168	159	116	165
150		102		141				135		134		169			166
154		108		164				147				170			167

3. E. coli H Antigen Type Strains

H	O:K	Original ref. no.	DEP[b] ref. no.	H	O:K	Original ref. no.	DEP[b] ref. no.
1	O2K2ab	Sul242	Sc 16	29	O19K?	N282	Sc 75
2	O43K?	Bi7455/41	Sc 56	30	O38K?	N157	Sc 76
3	O53K?	Bi7327/41	Sc 70	31	O3K?	K15	Sc 78
4	O2K1	U9/41	Sc 298	32	O114K?	K10	Sc 88
5	O4K3	U4/41	Sc 18	33	O11K?	K181	Sc 79
6	O2K1	A20a	Sc 15	34	O86K61	BP12665	Sc 80
7	O1K1	U5/41	Sc 13	35	O134K?	4370/53	Sc 101A
8	O2K?	Ap.320c	Sc 17	36	O86abK64	5017/53	Sc 82
9	O8K25	Bi7575/41	Sc 65	37	O42K?	P11a	Sc 83
10	O11K10	Bi623/42	Sc 66	38	O69K?	P9b	Sc 84
11	O13K11	Su4321/41	Sc 469	39	O74K?	E3a	Sc 401
12	O9K9	Bi316/42	Sc 399	40	O79K?	E49	Sc 86
13[a]	O17K?	P6c	Sc 61	41	O137K79	RVC1787	Sc 277
14	O18K?	F10018/41	Sc 59	42	O70K?	P9c	Sc 234
15	O23K18	E39a	Sc 57	43	O140K?	149/51	Sc 227
16	O6K15	F8316/41	Sc 19	44	O3K?	781/55	Sc 228
17	O15K?	P12b	Sc 261	45	O52K?	4106/54	Sc 229
18	O17K16	K12a	Sc 60	46	O26K60	5306/56	Sc 230
19	O9K?	A18d	Sc 67	47	O86K?	1755/58	Sc 231
20	O8K?	H330b	Sc 295	48	O16K?	P48	Sc 286
21	O8K?	U11a/44	Sc 64	49	O6K13	2147-59	Sc 415
22[a]	O8K?	A231a	Sc 91	51	O8,060K?	C218-70	Sc 682
23	O45K?	K42	Sc 68	52	O11K?	C2189-69	Sc 509
24	O51K?	K72	Sc 69	53	O148K?	E480/68	Sc 445
25	O15K?	N234	Sc 71	54	O161K?	E223/69	Sc 477
26	O131K?	S239	Sc 72	55	O75K?	E2987/73	Sc 725
27	O15K?	K50	Sc 73	56	O139K82	SN3N/1	Sc 714
28	O132K?	N87	Sc 74				

[a]These strains are *Citrobacter* sp.
[b]DEP is the Division of Enteric Pathogens.

4. H. Antiserum Pools

A	B	C	D	E	F
1	4	9	20	34	43
2	5	13	22	35	44
3	6	14	24	36	45
11	7	15	25	37	46
12	8	16	26	38	47
21	10	17	27	39	48
30	23	18	28	40	49
31	55	19	29	41	51
32		54	33	42	52
		56			53

5. E. coli K Antigen Type Strains

K no.	Strain	O	K	H	DEP[a] ref. no.
1	U9/41	2	1	4	Sc 735
2	U14/41	3	2	2	Sc 4
3	U4/41	4	3	5	Sc 18
4	U1/41	5	4	4	Sc 8
5	Bi8337/41	10	5	4	Sc 389
6	Bi7457/41	4	6	5	Sc 5
7	Pus3432/41	7	7	4	Sc 422
8	G3404/41	8	8	4	Sc 398
9	Bi316/42	9	9	12	Sc 399
10	Bi623/42	11	10	10	Sc 66
11	Su4321/41	13	11	11	Sc 63
12	Su65/42	4	12	—	Sc 481
13	Su4344/41	6	13	1	Sc 22
14	F7902/41	15	14	4	Sc 48
15	F8316/41	6	15	16	Sc 19
16	K12a	17	16	18	Sc 60
17	P7a	20	17	—	Sc 304
18	E39a	23	18	15	Sc 57
19	E47a	25	19	12	Sc 54
20	E19a	21	20	—	Sc 52
21	H38	23	21		Not available
22	H67	23	22	—	Sc 264
23	H54	25	23	1	Sc 268
24	H45	22	24	31	Sc 470
25	Bi7575/41	8	25 B[b]	9	Sc 65
26	Bi449/42	9a	26 A	—	Sc 487
27	E56b	8	27 A	—	Sc 502
28	K14a	9a,b	28 A	—	Sc 423
29	Bi161/42	9	29 A	—	Sc 36
30	E69	9	30 A	12	Sc 499
31	Su3973/41	9	31 A	—	Sc 500
32	H36	9	32 A	19	Sc 498
33	Ap.289	9	33 A	—	Sc 505
34	E75	9	34 A	—	Sc 44
35	A140a	9	35 A	—	Sc 424
36	A198a	9	36	19	Sc 33
37	A84a	9	37 A	—	Sc 38
38	A262a	9	38 A	—	Sc 425
39	A121a	9	39 A	9	Sc 492
40	A51d	8	40	9	Sc 31
41	A433a	8	41	11	Sc 29
42	A295b	8	42 A	—	Sc 501
43	A195a	8	43	11	Sc 439
44	A168a	8	44	—	Sc 484
45	A169a	8	45	9	Sc 25
46	A236a	8	46	30	Sc 494

K no.	Strain	O	K	H	DEP[a] ref. no.
47	A282a	8	47	2	Sc 24
48	A290a	8	48	9	Sc 28
49	A180a	8	49	21	Sc 26
50	PA80c	8	50	—	Sc 495
51	A183a	1	51	—	Sc 490
52	A103	4	52	—	Sc 488
53	PA236	6	53	—	Sc 21
54	A12b	6	54	10	Sc 10
55	N24c	9	55 A	—	Sc 45
56	H17b	2	56 B	7	Sc 426
57	H509d	9	57 B	32	Sc 427
58	Stoke W	111	58 B	—	Sc 459
59	Aberdeen 1064	55	59 B	6	Sc 491
60	F41(Hall)	26	60 B	—	Sc 273
61	E990 Taylor	86	61 B	—	Sc 106
62	F1961 Ørskov	86	62	2	Sc 428
63	4932-53 Ewing	127a	63 B	—	Sc 503
64	5017-53 Ewing	86ab	64 B	36	Sc 82
65	2160-53 Ewing	127ab	65 B	4	Sc 429
66	Guanabara	112ac	66 B	—	Sc 506
67	Cigleris Taylor	128	67 B	2	Sc 121
68	1411-50	112ab	68 B	18	Sc 293
69	34w	119	69 B	27	Sc 460
70	Canioni	125	70 B	19	Sc 100
71	E611 Taylor	126	71 B	2	Sc 102
72	227 Ewing	124	72 B	30	Sc 496
73	Katwijk	28	73 B	—	Sc 483
74	H702c	44	74	18	Sc 153
75	6182/50	113	75 B	21	Sc 375
76	10018/41	18ab	76 B	14	Sc 59
77	D-M3219-54	18ac	77 B	7	Sc 385
78	1111-55	136	78 B	—	Sc 430
79	RVC1787	137	79	41	Sc 277
80	E38	78	80 B	—	Sc 237
81	62-57	138	81 B	—	Sc 218
82	63-57	139	82 B	1	Sc 219
83	134-51	20ab	83 B	26	Sc 431
84	2292-55	20ab	84 B	26	Sc 432
85	RVC2907	141	85 B	4	Sc 287
86	C771	142	86 B	6	Sc 288
87	145	(32)	87 B	45	Sc 433
88	E68	141	85 B, 88[c]ab	4	Sc 482
89	D357	147	89 B	19	Sc 493
90	26w	114	90ab B	32	Sc 461
91	D616	149	91 B	10	Sc 485
92	6181-66	73	92	34	Sc 646
93	1935	150	93	6	Sc 511
94	E1541/68	154	94	4	Sc 449

(continued)

K no.	Strain	O	K	H	DEP[a] ref. no.
95	E3b	75	95	5	Sc 465
96	E10	77	96	—	Sc 337
97	H5	81	97	—	Sc 362
98	H705	107	98	7	Sc 372
99	B41	101	99[b]	—	Sc 744
100	F147	75	100	5	Sc 743
101	1473	20	101	—	Sc 746
102	6CB10/1	8	102 A	—	Sc 747
103	8CE275/6	101	103 A	—	Sc 748

[a]DEP is the Division of Enteric Pathogens.
[b]The existence of many B antigens is doubtful (see Introduction).
[c]The K88 and K99 antigens are protein, fimbrial structures and are more appropriately considered with other fimbrial antigens.

6. Thiotone Motility Medium

Oxoid peptone P, 50 g; Difco beef extract, 15 g; gelatin, 400 g; NaCl, 25 g; shred agar, 20 g; distilled water, 5000 ml.

Heat water and dissolve all constituents. Steam and agitate frequently until gelatin dissolves. Boil for 2 min. Adjust pH to 6.9. Distribute as Craigie tubes.

Autoclave at 10 lb for 20 min.

Peptone P is the Oxoid equivalent of Thiotone (BBL)

7. Preparation of Bridge's Solution

Stock solution A: mercuric iodide, 1 g; potassium iodide, 4 g; distilled water, 100 ml.

Stock solution B: 2% commercial formalin in physiological saline brought to pH 7.6 with disodium hydrogen phosphate.

Solution for use: stock solution A, 10 ml; stock solution B, 2.5 ml; physiological saline, 90 ml.

8. Veronal Buffer

Sodium barbitone, 74.23 g; calcium lactate, 4.61 g; sodium azide, 0.60 g; distilled water to 3000 ml. Adjust to pH 8.6 with concentrated HCl; dilute to 0.60 M for use by adding equal volume of distilled water. For the preparation of 1% agarose use undiluted buffer and add equal volume of 2% agarose (w/v) in distilled water.

References

Kauffmann, F. (1947). The serology of the coli group. *Journal of Immunology* **57**, 71–100.
Ørskov, F. and Ørskov, I. (1970). The K antigens of *Escherichia coli*. *Acta Pathologica et Microbiologica Scandinavica, Section B: Microbiology and Immunology* **78B**, 593–604.

Ørskov, F. and Ørskov, I. (1972). Immunoelectrophoretic patterns of extracts from *Escherichia coli* O antigen test strains O1 to O157. Examinations in homologous OK sera. *Acta Pathologica et Microbiologica Scandinavica, Section B: Microbiology and Immunology* **80B**, 905–910.

Ørskov, F., Ørskov, I., Jann, B. and Jann, K. (1971). Immunoelectrophoretic patterns of extracts from all *Escherichia coli* O and K antigen test strains. Correlation with pathogenicity. *Acta Pathologica et Microbiologica Scandinavica, Section B: Microbiology and Immunology* **79B**, 142–152.

Ørskov, I., Ørskov, F., Jann, B. and Jann, K. (1977). Serology, chemistry and genetics of O and K antigens of *Escherichia coli*. *Bacteriological Reviews* **41**, 667–710.

Semjen, G., Ørskov, I. and Ørskov, F. (1977). K antigen determination of *Escherichia coli* by counter-current immunoelectrophoresis (CIE). *Acta Pathologica et Microbiologica Scandinavica*, **85B**, 103–107.

15

Isolation and Characterization of Lipopolysaccharides (O and R Antigens) from *Escherichia coli*

K. JANN

Max-Planck-Institut für Immunbiologie, Freiburg-Zähringen, Federal Republic of Germany

Introduction

Lipopolysaccharides (LPS) are integral components of the outer membrane of Gram-negative bacteria. They are composed of a hydrophobic moiety (lipid A), covalently bound to a hydrophilic carbohydrate moiety. The latter consists of an oligosaccharide part (core oligosaccharide) and a polysaccharide chain. Since the latter determines the serological O specificity of the bacteria, the polysaccharide moiety is called the O-specific polysaccharide. The LPS present in bacterial wild types (S forms) contains all three structural regions (lipid A, core and O-specific polysaccharide), whereas those of R mutants lack the O-specific polysaccharide and contain only lipid A and core. The core of R mutants may have the same structures as in the O antigens of wild-type S forms, or may express various degrees of incompleteness (complete R antigens or subclasses thereof). These characteristics, as well as their genetic determination, biochemical implications and structural expressions, have been described (Jann and Westphal, 1975; Lüderitz *et al.*, 1971; Westphal *et al.*, 1983). The lipid A moiety as well as the core oligosaccharide are charged by substitution with phosphate and/or ethanolamine phosphate. The O-specific polysaccharide of certain LPS contain hexuronic acids, phosphate or other negatively charged components.

From this brief description it is evident that LPS are a complex group of molecules. The most obvious subdivision is that into S- and R-LPS. These differ in charge and hydrophobicity, which is reflected in their solubility in various media and in their extractability from the bacterial surface. Such differences may even be more expressed when the O-specific polysaccharide of LPS is negatively charged. They are of great importance with respect to their extraction, less so with respect to their chemical characterization and play practically no role in the technical details of their serological characterization. In the following, the methods for extraction of S- and R-LPS will be presented separately and the influence of charge in the O-specific polysaccharides will be considered. The techniques of chemical and serological characterization of S- and R-LPS will be treated together.

Extraction of Lipopolysaccharides with Phenol/Water

This procedure has previously been applied to the extraction of S- and R-LPS (i.e. for O and R antigens). It is now mainly used to extract S-LPS from bacteria or isolated bacterial cell walls, whereas R-LPS are extracted by the method described below.

The extraction method described here is based on the fact that proteins and DNA are soluble in phenol and that polysaccharides as well as LPS have only a limited solubility in phenol but are readily soluble in water. In its original applications, the bacteria were treated with a cold mixture of phenol and water. It was shown later by Westphal and his colleagues (Westphal and Jann, 1965) that the miscibility of phenol and water at higher temperatures and their limited miscibility at lower temperatures can be used greatly to improve the extraction procedure. At temperatures above 65°C phenol and water are miscible in any proportion. On cooling, the homogeneous mixture separates into an upper aqueous phase, saturated with phenol, and a lower phenol phase, saturated with water. Aqueous phenol is a weak acid and has a high dielectric constant. This has proved advantageous for the dissociation of protein–nucleic acid and protein–(lipo)polysaccharide complexes and has wide applicability (Westphal and Jann, 1965). The separation of the mixture into distinct phases at lower temperature brings about a separation of the extracted substances according to their solubility. In this separation LPS prefer the aqueous phase, from which they can easily be recovered as sediment by subsequent ultracentrifugation. During ultracentrifugation, LPS with a negatively charged O-specific polysaccharide often remain in the supernatant. They can be obtained by fractional precipitation with cetyltrimethylammonium bromide (CTAB; Cetavlon) in the same fashion as capsular polysaccharides of *E. coli* (see Chapter 16).

Method

After cultivation in suitable media, the bacteria are killed by the addition of phenol (1% final concentration), centrifuged and washed with saline. They are then dried with acetone (or freeze-dried) or are used as bacterial paste, i.e. the sediment of the final centrifugation. The respective amounts are indicated below as dry or wet weight of bacteria, respectively.

Fifty grams of dry bacteria, or 500 g wet bacteria, are suspended in 1 litre water at 68°C (water-bath). One litre 90% phenol, prewarmed to 68°C, is added and the mixture is kept at 68°C with vigorous stirring for 10 min. After cooling to about 10°C in an ice-bath, the suspension is centrifuged at 4000–5000 g for 30 min. This results in the formation of two phases, with a precipitate between the layers and a bacterial pellet. The upper aqueous phase is collected by suction and the lower phenol phase, together with the pellet is treated with 1 litre water at 68°C as described above. The combined aqueous phases are dialysed to remove phenol and low-molecular-weight material. The dialysed, slightly opalescent solution is concentrated *in vacuo* to a volume of about 100 ml and centrifuged to remove traces of insoluble material. The solution is freeze-dried to give a white powder. This procedure yields 4–5 g crude extract (8–10%, based on dry bacteria).

To remove RNA from the preparation, the crude extract is dissolved in water (3%), care being taken first to triturate the material with a minimal amount of water, followed by slow addition of water with constant stirring. Solution may be facilitated by warming in a water-bath. The opalescent solution is centrifuged at 100,000 g for 3 hs. The sediment is redissolved in water (3%) and the centrifugation is repeated twice.

The first supernatant may contain acidic polysaccharides or acidic LPS and should by lyophilized separately. For their isolation, use is made of their negative charge. The freeze-dried material is dissolved (1%) in 0.25 M sodium chloride. From this solution the acidic LPS is obtained by fractional precipitation with CTAB as described in detail for the isolation of high-molecular-weight capsular polysaccharides (see below).

Extraction of Lipopolysaccharides with Phenol/Chloroform/Petroleum Ether

This method is ideal for the extraction of R-LPS and very hydrophobic S-LPS and SR-LPS (for definition, see Jann and Westphal, 1975). From bacterial preparations which contain S-LPS and R-LPS only the latter are extracted.

It was found that certain S-LPS and also some R-LPS have a more or less pronounced solubility in water-saturated phenol and this complicates the extraction of these antigens with phenol/water. Therefore, an extraction procedure was devised (Galanos *et al.*, 1969) which proved to be well-suited for the extraction of R-LPS. This method uses a mixture of phenol, chloroform and low-boiling petroleum ether (PCP) which extracts glycolipids. From the extract the R-LPS are then precipitated by the removal of the volatile solvents and careful addition of water. The PCP extraction procedure has the advantage of being very mild and yielding pure R-LPS which is more soluble in water than preparations obtained with the phenol/water extraction procedure.

Extraction of Lipopolysaccharide

For a successful extraction, the bacteria must be washed with water instead of saline. They should be well dried and powdered. The mixture used in the extraction consists of 90% aqueous phenol, chloroform and petroleum ether (boiling point 40–60°C) in a volume ratio of 2:5:8. It should be monophasic and clear. If water is present in the solid phenol, the mixture is turbid and should be clarified by careful addition of some solid phenol.

The pre-cooled extraction mixture (300 ml) is added to 50 g dry bacteria and the suspension is agitated with an Ultra-Turrax homogeniser (Janke and Kunkel, Staufen, FRG) for 5 min with cooling in an ice-bath. The bacteria are recovered by centrifugation (8000 g) and the supernatant is collected. The bacterial pellet is extracted again twice as above, each time with 300 ml extraction mixture. The combined supernatant solutions (light brown) are concentrated in a rotary evaporator at 30–35°C to remove chloroform and petroleum ether. If phenol begins to crystallize after the evaporation, sufficient water at about 30°C is added to dissolve the crystals and the mixture is kept at about 30°C. Water is then added dropwise until the R-LPS begins to precipitate and the warm mixture is allowed to stand for about 2 min. The R-LPS is collected by centrifugation at about 5000 g without cooling.

For further purification, the centrifuge tubes containing the pellet of crude R-LPS are allowed to drain upside down for 2–3 min. The pellet is then washed three times with small amounts (5–10 ml) of 80% phenol and the walls of the centrifugation tubes are carefully wiped with filter paper after each wash. The phenol is removed by three washes with ether and the R-LPS is dried *in vacuo*. The dry powder is dissolved in 50 ml water. The viscous solution is warmed to 45°C, carefully degassed *in vacuo* and centrifuged twice at 100,000 g for 4 h. The final pellet is dissolved in water and freeze-dried. The extraction yields about 1–1.5 g pure R-LPS (2–3%, based on the weight of dry bacteria).

Further Purification of S- and R-Lipopolysaccharides

Both S- and R-LPS contain, even after the purification described above, ionic contaminants such as ethanolamine, cadaverin, or ethanolamine phosphate. These, together with divalent ions (Ca^{2+}, Mg^{2+}), which affect the solubility of the LPS preparations, can be removed from the LPS preparations by electrodialysis (Galanos and Lüderitz, 1975). The resulting anionic LPS are converted into uniform salt forms by neutralisation with suitable bases, such as KOH, NaOH or trimethylammonium hydroxide. After freeze-drying, the LPS are obtained as white powders which dissolve readily in water to give clear solutions. This procedure is used for the preparation of standard endotoxin (LPS) for biological and medical purposes.

Since the procedure has been described in detail by Galanos and Lüderitz (1975), it is presented here only briefly. The apparatus for electrodialysis consists of three chambers made of plexiglass and separated by dialysis tubing. The outer compartments contain the electrodes and can be cooled with the aid of a silicon tubing cooling spiral. Water is filled into the outer compartment and a solution of LPS (8 mg ml^{-1}) in water into the inner one (250 ml). The electrodialysis is run, while cooling with ice-water, at a maximum current of 100 mA. The end of the experiment is indicated by a sharp fall in the electric current. Since LPS tend to stick to the anionic dialysis membrane, it is advisable to reverse the direction of current several times during the experiment. If this is done, the outer compartments should be refilled with fresh deionized water each time. After electrodialysis, the LPS is removed from the center compartment (if necessary, it is scraped from the inner face of the anionic dialysis membrane), dissolved in water and neutralized carefully (preferably with trimethylammonium hydroxide), dialysed against water, concentrated *in vacuo* and freeze-dried.

Physical and Chemical Characterization of S- and R-Lipopolysaccharides

Sodium dodecyl sulphate polyacrylamide gel electrophoresis (SDS–PAGE). Differences in the size of the O-specific polysaccharides of LPS molecules alter their hydrophobicity/hydrophilicity ratio. This can be demonstrated in SDS–PAGE. With this method (Jann *et al.*, 1975; Palva and Mäkelä, 1980) it can be determined whether preparations contain R-, SR- or S-LPS. The method can be performed in tubes or in slab gels and visualization can be achieved with the periodate–Schiff reagent or by pre-staining of the LPS with Procion Red (ICI) (Jann *et al.*, 1975) or with a silver stain (Tsai and Frasch, 1982). In the former case R-, SR- and S-LPS are seen in different and characteristic regions of the gel, R-LPS exhibiting the fastest and S-LPS the slowest mobility. The silver staining method is much more sensitive and reveals a ladder-like series of bands along the whole gel, with greatest intensities in the same regions as the bands visualised with the periodate–Schiff reagent or with Procion Red pre-stained preparations. This was also detected with biochemically radio-labelled LPS and interpreted as due to polysaccharide chains differing by molecular weight increments corresponding to individual repeating units (Palva and Mäkelä, 1980).

If dyed LPS are to be used, they can be prepared by adding 15 mg Procion Red (ICI) in 1.5 ml water to 15 mg LPS in 15 ml water. To this solution are added 30 mg sodium chloride (after 5 min) and 30 mg sodium bicarbonate (after 30 min). The dyed LPS are isolated by chromatography on Sephadex G25. It is also possible to use an aliquot of the mixture directly.

Before the electrophoretic run the LPS or dyed LPS (500 μg) are disaggregated by heating for 5 min at 100°C in a mixture of deionized water (70 μg), a cocktail of 10 mM Tris buffer (pH 8.5), 1 mM EDTA, 4% SDS, 30% sucrose and 2.5% dithioerythritol (50 μg) and a solution of a suitable tracking dye (30 μl).

The gel electrophoretic system used for the separations is that of Laemmli (1970). The gel can be prepared in 0.6 × 10 cm tubes or in 1-mm slabs of 14% gels (acrylamide:N,N'-methylenbisacrylamide = 30:0.8). The runs are performed at 0.5 mA per tube or 25 mA per slab.

After the electrophoretic run, the gel is fixed with a 40% ethanol–5% acetic acid solution overnight. For the application of the silverstain (Tsai and Frasch, 1982) the fixing solution is replaced with 0.7% periodic acid in 40% ethanol–5% acetic acid and the gel is kept in this mixture for 5 min. After three washes each with 1 litre of water, the gel is treated with a staining solution prepared with 2 ml concentrated ammonium hydroxide, 28 ml 0.1 N sodium hydroxide, 5 ml 20% (w/w) silver nitrate and 115 ml water (final volume of 150 ml). The gel is kept in this solution for 10 min with agitation on a rotator (about 70 rpm) and then washed three times with 1 litre of water each. Fixation is then carried out in 200 ml of a mixture of 50 mg citric acid and 0.5 ml 37% formaldehyde in 1 litre. The LPS bands in the gel stain dark brown within 2–5 min. The development is terminated when the stain reaches the desired intensity or when the clear background shows the first signs of discoloration. The gel is then washed in water and is ready for inspection or drying and photography.

Detection and Determination of Characteristic Components

All LPS contain hydroxy fatty acids (mostly β-hydroxymyristic acid, β-HMA) in the lipid A moiety and 2-keto-3-deoxymannosoctonic acid (KDO) in the linkage region between lipid A and carbohydrate moieties. These constituents can be used for the detection and determination of LPS. The fatty acids are converted into their methyl esters by transesterification with methanol–HCl and are then determined with gas–liquid chromatography. KDO is released from the LPS by mild acid hydrolysis, oxidised with sodium metaperiodate and the chromogen thus formed is determined in a reaction with thiobarbituric acid.

The determination of fatty acids (with reference to β-HMA) has been described (Wollenweber et al., 1983; Wollenweber and Rietschel, 1983). It is described here briefly. The LPS (2–5 mg) is heated in 1 ml of 10 mM hydrochloric acid in methanol at 85°C for 18 h in a sealed ampoule. After cooling, the mixture is concentrated to about half its volume under a stream of nitrogen. An equal volume of half-saturated sodium chloride solution is added and the methyl esters are extracted twice with two volumes each of a 1:1 mixture (v/v) of distilled petroleum ether (b.p. 40–60°C) and ethyl acetate. The combined extracts are concentrated under a stream of nitrogen to a minimal volume (not to dryness) and to this concentrate 50 μl of trifluoroacetic anhydride (Sigma) in acetonitrile (1:1, v/v) is added. The mixture is placed into a small glass tube (5 ml) with a Teflon-lined screw cap, which is then closed, and placed into a boiling-water bath for 2 min and kept at room temperature for 10 min. Acetonitrile (100 μl) is added and the mixture is directly subjected to gas chromatographic analysis. It is best to use glass capillary columns, but SE-30 (10% on GasChrom Q, 100–120 mesh), Castorwax (2.5% on Chromosorb W, 80–100 mesh) or EGSS-X (15% on GasChrom P, 100–120 mesh) may also be used. In all cases a nitrogen flow of 30 ml min^{-1} is applied in the gas chromatography. In our laboratory a Varian 3700 gas chromatograph, equipped with a flame ionization detector and combined with an automatic integrator (Hewlett-Packard 3380A) is used. The methyl esters are detected by their retention times in the gas chromatogram under standardized conditions. The runs should be calibrated with authentic fatty acid esters.

The determination of KDO is based on its conversion to 2,4-dioxobutyrate by oxidation with sodium metaperiodate and subsequent reaction with 2-thiobarbituric acid. A coloured product is formed which has an absorption maximum at 548 nm. The procedure was described by Karkhanis *et al.* (1978).

Before its determination, KDO has to be liberated from the LPS by mild acid hydrolysis. Since KDO is acid-labile, the optimal conditions for its hydrolytic release have first to be carefully established. Conditions which have proved to be satisfactory are heating of the LPS in 0.2 N sulphuric acid for 8–20 min or in acetate buffer at pH 4.5 at 100°C for 30 min.

To 50 μl of the hydrolysate (1 mg LPS per ml) are added 50 μl 0.1 N sodium metaperiodate in water, and after 10 min at room temperature, 200 μl 4% sodium metaarsenite in 0.5 N hydrochloric acid. After thorough mixing, 800 μl 0.6% of freshly prepared thiobarbituric acid (Sigma) in water is added and the mixture is heated at 100°C for 10 min. To the still hot (70–80°C) mixture 1 ml dimethyl sulphoxide is added and, after cooling to room temperature, the optical density of the reaction mixture is measured at 548 nm. A calibration curve is set up with the commercially available ammonium salt of KDO (Sigma), or with an LPS of known KDO content, in 0.2 N sulphuric acid. The colorimetric response is linear between 2 and 20 μg KDO.

It is possible to estimate the approximate amount of LPS in bacteria by treating the cells directly with dilute acid (e.g. 0.25 N sulphuric acid at 100°C for 10 min). After centrifugation, the KDO content of the supernatant is determined with the thiobarbituric acid reagent, as described above for LPS. In principle, this is a convenient method not only for the estimation of the LPS content of bacteria but, if properly standardized, also for an estimation of the number of bacteria in a sample. Since, as is now known, many pathogenic *E. coli* strains also have KDO in their capsules, the method can lead to false results. It is, therefore, advisable to monitor the bacteria in question for the presence of a capsular polysaccharide, by immunoelectrophoresis of saline extracts.

Serological Characterization of Lipopolysaccharides

For the serological characterization of R- and S-LPS, techniques are applied such as passive haemagglutination and haemolysis, gel precipitation methods, immunoelectrophoresis and immune reactions on solid supports, such as enzyme-linked immunosorption assay (ELISA) or radio-immune assay (RIA) with radio-labelled anti-antibody or protein A. The reader is referred to the relevant standard literature. Only the passive haemagglutination and immunoelectrophoresis are described here.

Passive Haemagglutination

In this technique, erythrocytes are used as carrier cells for the reaction of LPS with anti-LPS antibodies. Usually sheep erythrocytes are coated (modified) with LPS; this modification is often achieved with alkali-treated LPS. Since alkali may remove O-acetyl groups wbich may be parts of antigenic determinants, this treatment can result in a loss of serological activity of the LPS. Therefore, the LPS may be heated in aqueous solution rather than treated with alkali.

In preparation for erythrocyte modification, the LPS is heated in 0.25 M sodium hydroxide (1%) at 56°C for 60 min. The solution is neutralized, dialysed against water and the LPS is recovered by freeze-drying. Alternatively, the LPS may be heated in dionized water (1%) at 100°C for 30 min. After cooling, an aliquot of this solution is used directly.

For the modification of erythrocytes, sheep red blood cells are washed twice with saline and centrifuged. One millilitre of the sediment is suspended in 9.5 ml saline, and 250 µg LPS (alkali-treated or heated) in 0.5 ml saline is addcd. The suspension is incubated at 37°C for 30 min and centrifuged. The sedimented modified erythrocytes are washed three times with saline.

The haemagglutination is performed in microtitre plates (20 µl serum in two fold serial dilutions and 20 µl of a 0.5% suspension of modified erythrocytes per well). The mixtures are incubated at 37°C for 30 min and at room temperature for at least 3 h before the results are read.

Immunoelectrophoresis

This method is well suited for the differentiation of S-LPS which have acidic O-specific polysaccharides from those with neutral ones. It has been applied to a serological analysis of *E. coli* strains in relation to their pathogenicity (Ørskov *et al.*, 1971) and the results agree well with the chemical analysis of isolated LPS. The immunoelectrophoresis analysis can be performed directly on saline extracts of bacteria and an isolation of pure LPS is not necessary for an initial study.

The elctrophoresis runs are performed on microscopic slides with 3 ml of

1.3% agarose per slide (in barbiturate buffer of pH 8.6; 1.84 g diethylbarbituric acid and 10.3 g sodium diethylbarbiturate in 1 litre water). Using a Gelman immunoelectrophoretic unit, with double rows of three slides each, 3–4 mA per slide (about 210 V) is applied. After about 3 h, with a suitable dye as front marker, the electrophoresis is stopped. The center trough is cut out and filled with antiserum (about 100 μl). The slides are then kept in a moist chamber and inspected after various times. Precipitation arcs are usually visible after 1–2 days. In most cases staining with Coomassie Blue is not necessary.

References

Galanos, C. and Lüderitz, O. (1975). Electrodialysis of lipopolysaccharides and their conversion to uniform salt forms. *European Journal of Biochemistry* **54**, 603–610.

Galanos, C., Lüderitz, O. and Westphal, O. (1969). A new method for the extraction of R lipopolysacharides. *European Journal of Biochemistry* **9**, 245–249.

Jann, B., Reske, K. and Jann, K. (1975). Heterogeneity of lipopolysaccharides. Analysis of polysaccharide chain lengths by sodium dodecyl-sulfate-polyacryl-amide gel electrophoresis. *European Journal of Biochemistry* **60**, 239–246.

Jann, K. and Westphal, O. (1975). Microbial polysaccharides. *In* "The Antigens," Vol. 3 (Ed. M. Sela), pp. 1–125. Academic Press, New York.

Karkhanis, Y. D., Zeltner, J. K., Jackson, J. J. and Carlo, D. J. (1978). A new and improved microassay to determine 2-keto-3-deoxyoctonate in lipopolysaccharide of gram-negative bacteria. *Analytical Biochemistry* **85**, 595–601.

Laemmli, U. K. (1970). Cleavage of structural proteins during the assembly of the head of bacteriophage T4. *Nature (London)* **227**, 680–685.

Lüderitz, O., Westphal, O., Staub, A. M. and Nikaido, H. (1971). Isolation and chemical and immunological characterization of bacterial lipopolysaccharides. *In* "Microbial Toxins," Vol. 4 (Eds. G. Weinbaum, S. Kadis and S. J. Ajl), pp. 145–233. Academic Press, New York.

Ørskov, F., Ørskov, I., Jann, B. and Jann, K. (1971). Immunoelectrophoretic patterns of extracts from all Escherichia O and K test strains. Correlation with pathogenicity. *Acta Pathologica et Microbiologica Scandinavica, Section B: Microbiology and Immunology* **79B**, 142–152.

Palva, E. T. and Mäkelä, P. H. (1980). Lipopolysaccharide heterogeneity in *Salmonella typhimurium* analyzed by sodium dodecylsulfate/polyacrylamide gel electrophoresis. *European Journal of Biochemistry* **107**, 137–143.

Tsai, C.-M. and Frasch, C. E. (1982). A sensitive silver stain for detecting lipopolysaccharides in polyacrylamide gels. *Analytical Biochemistry* **119**, 115–119.

Westphal, O. and Jann, K. (1965). Bacterial lipopolysaccharides. Extraction with phenol-water and further application of the procedure. *Methods in Carbohydrate Chemistry* **5**, 80–91.

Westphal, O., Jann, K. and Himmelspach, K. (1983). Chemistry and immunochemistry of bacterial lipopolysaccharides as cell wall antigens and endotoxins. *Progress in Allergy* **33**, 9–39.

Wollenweber, H. W., Schlecht, S., Lüderitz, O. and Rietschel, E. T. (1983). Fatty acids in lipopolysaccharides of *Salmonella* species grown at low temperature. *European Journal of Biochemistry* **130**, 167–171.

16

Isolation and Characterization of Capsular Polysaccharides (K Antigens) from *Escherichia coli*

K. JANN

Max-Planck-Institut für Immunbiologie, Freiburg-Zähringen, Federal Republic of Germany

Introduction

Many strains of *Escherichia coli* are surrounded by an extracellular layer, the bacterial capsule, which consists of acidic polysaccharides. The capsules express the serological K specificity of *E. coli* and the capsular polysaccharides are the K antigens. Structural, immunochemical and genetic studies (Jann and Jann, 1983; Ørskov *et al.*, 1977) have revealed that the *E. coli* capsular polysaccharides are of basically two different types. Some K antigens have a high molecular weight and a low electrophoretic mobility. In their chemistry they resemble the K antigens of *Klebsiella*. Other K antigens have low molecular weight and high electrophoretic mobility. Their chemical properties resemble those of the capsular antigens of *Neisseria*. The differences between these different polysaccharides, especially with respect to molecular weight, have to be taken into account in the isolation procedures. Immunoelectrophoresis and its Cetavlon modification can be used to characterize each type of polysaccharide. The procedures for the isolation of high-molecular-weight and low-molecular-weight polysaccharides of *E. coli* capsules are presented separately.

Isolation of Capsular Polysaccharides

High-Molecular-Weight Polysaccharides

High-molecular-weight acidic capsular polysaccharides are usually isolated from bacteria grown on agar. The first steps of the procedure are identical with those for the isolation of S-LPS (see Chapter 15). From the supernatant of the first ultracentrifugation (100,000 *g* for 4 h), which in essence is a mixture of acidic

375

polysaccharides and RNA, the acidic capsular polysaccharides are obtained by fractional precipitation with cetyltrimethylammonium bromide (CTAB; Cetavlon). The CTAB salts of the polysaccharides are subsequently converted into other salts (e.g., sodium salts) by precipitation with ethanol from suitable salt solutions (e.g., sodium chloride solution). The general principles of the method and its application to the isolation of acidic polysaccharides have been described (Scott, 1960; Tarcsay et al., 1971; Westphal and Jann, 1965). The following description of the procedure starts with the material obtained by freeze-drying the supernatant of the ultracentrifugation of the aqueous phase from the phenol–water extraction of E. coli bacteria (see isolation of S-LPS; Chapter 15). From encapsulated E. coli about 200–250 mg material per gram dry weight of bacteria is obtained from the supernatant.

To a solution of 500 mg of the freeze-dried material in 50 ml 0.25 M sodium chloride, 1 g CTAB in 25 ml 0.25M sodium chloride is added with stirring, which results in the precipitation of CTAB–RNA salts. The mixture is stirred at room temperature for 15 min and the precipitate is removed by centrifugation at 10,000 g for 1 h. To the clear supernatant, containing the acidic capsular polysaccharide, water is added slowly until a second precipitate forms; usually about three volumes of water are needed. The mixture is allowed to stand in an ice-bath for at least 2 h and is then centrifuged at 9000 g. The viscous and glutinous sediment is dissolved in a minimum amount (about 5–8 ml) of 1 M sodium chloride (or any other salt solution of equal ionic strength, if desired) and from this solution the acidic polysaccharide in its respective salt form is precipitated by the addition of 8–10 volumes of ethanol. After standing in an ice-bath for at least 2 h, the precipitate is collected by centrifugation (9000 g) and the pellet is reprecipitated from salt solution with ethanol as before. The final pellet is dissolved in a minimum amount of water and dialysed against deionized water. If the dialysed solution is slightly opalescent it is clarified by centrifugation at 100,000 g for 2 h. The now clear supernatant is concentrated by rotary evaporation and the viscous solution is freeze-dried to give a white and fibrous powder. About 150–200 mg of sodium salt of the acidic capsular polysaccharides is usually obtained from 500 mg of the starting material (35–40%, corresponding to a yield of 6–7% of the acidic capsular polysaccharide based on dry bacteria).

Low-Molecular-Weight Polysaccharides

Attempts to isolate the low-molecular-weight acidic capsular polysaccharides from agar-grown bacteria have met with only limited success. It is preferable to isolate these polysaccharides from liquid bacterial cultures by precipitation with CTAB (Gotschlich et al., 1972; Jann et al., 1980). The bacteria are grown to the late exponential phase (4–6 h at 37°C) in DO medium containing casamino acids (2%), dialysable fraction of yeast extract (100 ml per litre from 100 g yeast extract) and glucose (0.2%). Usually 7–10 litres of bacterial culture are grown in 15 to 20-litre fermenters.

To the bacterial culture an equal volume of 0.2% aqueous solution of CTAB is added. The mixture is kept overnight and is then centrifuged. The bacterial paste (about 100 g) is suspended in 450 ml deionized water with the aid of an Ultra Turax homogenizer (Janke and Kunkel, Staufen, FRG), while the mixture is cooled in an ice-bath. To this suspension 110 g calcium chloride (hexahydrate form) is added with agitation to give a final concentration of 1 M calcium chloride. After the addition of 132 ml 96% ethanol (20% final concentration), the mixture is kept in an ice-bath and the precipitate is removed by centrifugation at 16,000 g for 20 min. Ethanol is added to the clear supernatant to a final concentration of 80% and the mixture is kept at 4°C for 16 h. The precipitate formed is collected by centrifugation at 16,000 g and the pellet is dissolved in 500 ml 10% saturated sodium acetate. The solution is agitated for 3–5 h at 4°C with 585 ml 80% phenol, buffered with sodium acetate to pH 6.5 (500 ml liquid 90% phenol and 85 ml aqueous 16% sodium acetate). The mixture is then centrifuged for phase separation (16,000 g, 20 min). The upper aqueous layer is carefully collected by suction, and the lower phenol phase and the interface precipitate are discarded. Ethanol is added to the aqueous phase to a final concentration of 80% and the mixture is kept overnight at 4°C. The precipitate formed is collected by centrifugation and reextracted at least once with acetate-buffered phenol as described above. The final pellet is dissolved in a small amount of 10% saturated ammonium acetate and the solution is centrifuged at 100,000 g for 4 h. The clear supernatant is dialysed against several changes of deionized water, concentrated *in vacuo* at a temperature not exceeding 40°C and freeze-dried. This procedure yields 50–150 mg capsular polysaccharide (sodium salt) per litre of bacterial culture, depending on the extent of capsular expression.

Characterization of Capsular Polysaccharides

For the general characterization of high- and low-molecular-weight capsular (K) polysaccharides of *E. coli* the reader is referred to a recent review (Jann and Jann, 1983). The low-molecular-weight K antigens contain a rather easily removed lipid moiety (phosphatic acid) at their reducing end (Gotschlich *et al.*, 1981; Schmidt and Jann, 1982). Therefore, these K antigens may be present as relatively small (10–40 kilodalton) polysaccharides or as micelle-forming amphiphatic molecules. The two forms have distinct physical properties, which can be observed in immunoelectrophoresis, ultracentrifugation or gel permeation chromatography (Schmidt and Jann, 1982). The phosphatic acid can be removed by milk alkali treatment (pH 10, room temperature) or by heating the polysaccharide for 1–2 h at pH 5.5–6 (the usual pH of deionized water). With the former treatment O-acetyl groups may be removed (with concomitant changes in serological specificity), and with the latter method labile glycosidic (KDO) linkages in the polysaccharide may also be split. Heating of the polysaccharide preparations in deionized water (pH 5.5–6) should therefore be carefully controlled with respect to temperature and duration.

Serological Characterization

For a serological characterization of the capsular (K) antigens the same methods may be applied as described for the LPS (see Chapter 15). The modification of the erythrocytes for passive haemagglutination and haemolysis can be carried out with the polysaccharides directly (200 μg ml^{-1}); heating or alkali treatment is unnecessary.

Immunoprecipitation is much better with K than with O antigens. This is probably due to a marked non-specific precipitation with O but not with K antigens. Serial dilutions (1:2) of a solution of the K antigen in saline, starting with about 500 μg polysaccharide per millilitre, are set up. To each dilution (0.2 ml), 0.1 ml of anti-K antiserum is added. The mixtures are kept at 37°C for 1 h and then at 4°C for 3 days, after which they are centrifuged in a bench-top centrifuge. The pellets are washed three times with 0.9% saline and the final pellets are dissolved in 50 μl 1 *M* sodium hydroxide, followed by 500 μl water each. The antibody contents of the solutions are determined either by measuring the optical density at 280 nm, or with the Folin reagent. In either case suitable protein references must be set up.

Immunoelectrophoresis is an important method for the characterization of the acidic capsular (K) polysaccharides. Since the electrophoretic mobility of a polymer depends on its charge and size, this method is well suited for the quick and easy differentiation between high- and low-molecular-weight acidic capsular (K) polysaccharides. In view of the different growth conditions suggested for *E. coli* with high- or low-molecular-weight K antigens, this test should be performed before any attempt to isolate polysaccharide is made, unless the nature of the capsular material is already known (Ørskov *et al.*, 1977; Jann and Jann, 1983). As mentioned with the LPS (see Chapter 15), it is not necessary to use pure K antigens for the immunoelectrophoretic analysis, but saline extracts, usually the growth from one plate suspended in 3 ml saline and heated for 30 min at 60°C, can be used. The immunoelectrophoretic runs are performed as described for the serological analysis of LPS.

A modification of the procedure was introduced by Ørskov (1976). Instead of antiserum, a 0.1% aqueous solution of CTAB is placed into the center trough. The acidic polysaccharides are visible as precipitation arcs of the insoluble polysaccharide–CTAB complexes.

Serological Semi-Quantitation of Acidic Capsular (K) Antigens

It is known that the Vi antigens of *Salmonella* inhibit the agglutination of sheep erythrocytes by anti-sheep erythrocyte antibodies. This phenomenon was used by Glynn *et al.* (1971) to determine the K antigen content of uropathogenic *E. coli*, which was related to the relative virulence of the respective strains.

Bacteria are grown overnight on agar, harvested and dried with acetone. The dried bacteria are then extracted with 0.9% saline. Known amounts of the extracts (0.1 ml) are added to 0.1-ml suspensions of sheep erythrocytes, followed by the addition of dilutions of anti-sheep erythrocyte antiserum in a chessboard fashion. The mixtures are incubated at 37°C for 30 min and then at room temperature. The agglutination titre is determined after several hours and the agglutination inhibition titre of the bacterial extract is taken as the reciprocal of the agglutination titre of the antiserum.

References

Glynn, A. A., Brumfitt, W. and Howard, C. J. (1971). K antigens of *Escherichia coli* and renal involvement in urinary tract infections. *Lancet* **1,** 514–516.

Gotschlich, E. C., Rey, M., Etienne, C., Sanborn, W. R., Train, R. and Cvetanovic, B. (1972). Immunological response observed in field studies in Africa with meningococcal vaccines. *Progress in Immunobiological Standardization* **5,** 458–491.

Gotschlich, E. C., Frazer, B. A., Nishimura, O., Robbins, J. B. and Lui, T. Y. (1981) Lipid on capsular polysaccharide antigens of gram-negative bacteria. *Journal of Biological Chemistry* **256,** 8915–8921.

Jann, K. and Jann, B. (1983). The K antigens of *Escherichia coli*. *Progress in Allergy* **33,** 53–79.

Jann, K., Jann, B., Schmidt, M. A. and Vann, W. F. (1980). Structure of the *Escherichia coli* K2 capsular antigen, a teichoic acid-like polymer. *Journal of Bacteriology* **143,** 1108–1115.

Ørskov, F. (1976). Agarose electrophoresis combined with second dimensional Cetavlon precipitation. A new method for demonstration of acidic polysaccharide K antigens. *Acta Pathologica et Microbiologica Scandinavica, Section B: Microbiology* **84B,** 319–320.

Ørskov, I., Ørskov, F., Jann, B. and Jann, K. (1977). Serology, chemistry and genetics of O and K antigens of *Escherichia coli*. *Bacteriological Reviews* **41,** 667–710.

Schmidt, A. M. and Jann, K. (1982). Phospholipid substitution of capsular (K) polysaccharide antigens from *Escherichia coli* causing extraintestinal infections. *FEMS Microbiology Letters* **14,** 69–74.

Scott, J. E. (1960). Aliphatic ammonium salts in the assay of acidic polysaccharides from tissue. *Methods in Biochemical Analysis* **8,** 145–197.

Tarcsay, L., Jann, B. and Jann, K. (1971). Immunochemistry of K antigens of *Escherichia coli*. The K87 antigen from *E. coli* O8:K87(B?):H19. *European Journal of Biochemistry* **23,** 505–514.

Westphal, O. and Jann, K. (1965). Bacterial lipopolysaccharides. Extraction with phenol-water and further application of the procedure. *Methods in Carbohydrate Chemistry* **5,** 80–91.

17

Isolation and Characterization of Fimbriae from *Escherichia coli*

K. JANN

Max-Planck-Institut für Immunbiologie, Freiburg-Zähringen, Federal Republic of Germany

Introduction

Many strains of *Escherichia coli* possess hair-like extracellular appendages called fimbriae (Duguid *et al.*, 1955) or pili (Brinton, 1959). They mediate bacterial adhesion to epithelial cells and mucosal linings of the host and are thus parameters of bacterial pathogenicity (Ofek and Beachey, 1980). Adhesion may or may not be inhibitable by D-mannose and, accordingly, is termed mannose-sensitive (MS) or mannose-resistant (MR). These terms are also applied to the respective fimbriae, so that one may speak of 1 MS and MR fimbriae, respectively. MS fimbriae are also called common or type fimbriae (Brinton, 1959; Ofek and Beachey, 1980) and constitute a group of closely related bacterial recognition structures which have mannose-rich glycoproteins as receptors. MR fimbriae are adapted to various glycolipids on the eucaryotic cells.

Some MR fimbriae from intestinally pathogenic *E. coli* have been characterized, such as the colonization factor antigens CFA/I and CFA/II from strains pathogenic in humans (Evans and Evans, 1978; Evans *et al.*, 1979), the K88 antigen which occurs in several serological variants from strains pathogenic in swine (Stirm *et al.*, 1966, 1967) and the K99 antigen from strains pathogenic in calves (Isaacson, 1977). More recently the 987P fimbriae have been isolated from pig pathogenic *E. coli* (Isaacson and Richter, 1981). The receptors of all these fimbriae seem to be neuraminic acid-containing gangliosides. For further information the reader is referred to a recent review by Gaastra and de Graaf (1982).

MR fimbriae from extra-intestinally pathogenic *E. coli* have been studied serologically and chemically (Jann *et al.*, 1981; Korhonen *et al.*, 1980; Ørskov and Ørskov, 1983; Wevers *et al.*, 1980). The receptor for many of these pili,

THE VIRULENCE OF ESCHERICHIA COLI

especially those from urinary tract infective strains, are globosides, e.g. of the blood group P system (Leffler and Svanborg Edén, 1981).

In recent studies, a common serological system for *E. coli* fimbriae has been established (Ørskov and Ørskov, 1983). In this, fimbriae are termed F antigens. F1–F6 are the common type 1 fimbriae, CFA/I, CFA/II, K88, K99 and 987P. The MR fimbriae isolated from invasive (notably uropathogenic) *E. coli* are termed F7, F8, etc.

With respect to isolation and purification, MS and MR fimbriae can generally be considered together. However, the growth conditions for their optimal expression differ, as will be described below.

The fertility-associated sex pili, or F pili, have been studied extensively (Achtmann *et al.*, 1978; Brinton *et al.*, 1964). In contrast to the fimbriae mentioned above, they do not seem to play a role in host–parasite interactions, but function in bacterial conjugation (Achtmann *et al.*, 1978). The F pili will not be discussed here.

Bacterial Growth Conditions for Expression of Fimbriae

The expression of common type 1 (MS) fimbriae is generally favoured by static growth of bacteria for longer periods in liquid culture (about 48–72 h). Growth on agar often suppresses or greatly diminishes fimbria formation. MR fimbriae are formed in liquid cultures as well as on agar. The formation of K99 fimbriae is inhibited by the presence of alanine in the medium (Giradeau *et al.*, 1982) and that of common type 1 fimbriae by the presence of glucose but this is not due to catabolite repression (Eisenstein and Dodd, 1982). These are the only reports on the effect of constituents of the growth medium on fimbrial expression. It has repeatedly been observed that fimbriae are not expressed at growth temperatures of 18–20°C. This is, therefore, a method for obtaining phenotypically unfimbriated bacteria. It should be mentioned, however, that the mucoid (M) antigen expressed by many *E. coli* K12 (rough) strains is preferentially formed at lower growth temperatures and may interfere with experiments designed with these unfimbriated bacteria.

For the preparation of MS fimbriae the bacteria are grown in liquid media containing yeast extract, peptone or casamino acids and sodium chloride (Merck standard I or Luria broth) or in brain heart infusion broth for at least 48 h at 37°C without aeration. The bacteria often form a pellicle at the surface of the culture. They are harvested by centrifugation at about 8000 *g*.

For *the preparation of MR fimbriae* the above media can be used with a cultivation time of about 16 h at 37°C. Alternatively, the bacteria may be grown overnight on agar which is prepared on the basis of the above media. They are harvested from the agar plates into 50 m*M* Tris–HCl buffer, pH 7.5, or into phosphate-buffered saline. Usually about 1–2 ml is used for one large agar plate (14 cm in diameter). The bacterial suspension is centrifuged at about 800 g. For the preparation of CFA antigens, an agar (CF–agar) is suggested which contains magnesium and manganese ions (Evans *et al.*, 1979).

Isolation and Purification of Fimbriae

The preparation of fimbriae can be achieved by treating bacterial suspensions with an Omnimixer, Waring blender, Vortex mixer or equivalent device. After removal of the bacteria, the fimbriae are isolated from the supernatant by ultra-centrifugation, or precipitation with ammonium sulphate. Precipitation with magnesium chloride (0.1 M final concentration) has been shown to cause less co-precipitation of contaminating material (Eshdat *et al.*, 1981). This procedure is, however, effective only with MS fimbriae. Although the isolation can be achieved by centrifugation at 150,000 g (Eshdat *et al.*, 1981), fimbriae are usually precipitated with ammonium sulphate (Salit and Gotschlich, 1977; Korhonen *et al.*, 1980; Wevers *et al.*, 1980). Fimbriae are frequently precipi-tated with final ammonium sulphate concentrations of 50–60% (Korhonen *et al.*, 1980), which is unnecessarily high and causes co-precipitation of outer mem-brane constituents which are difficult to remove later. In most cases a final ammonium sulphate concentration of 10–15% suffices. This should, however, be determined in small-scale pilot experiments. The efficacy of precipitation can be checked by sodium dodecyl sulphate polyacrylamide gel electrophoresis (SDS–PAGE; see below). It is advisable to inspect the bacteria with the electron microscope before the isolation procedure to ensure sufficient fimbriation.

Method

Bacteria from 100 agar plates (14 cm in diameter) or from 15 litres of a 48-h broth culture are suspended in 15 mM Tris–HCl buffer (pH 7.8) and agitated in an Omnimixer at position 2 for three 5-min periods, with intermittent cooling periods of 5 min, in an ice-bath. The bacteria are removed by centrifugation (8000 g, 20 min), resuspended in 15 ml Tris–HCl and treated again as described above. Ammonium sulphate is added to the combined supernatants to a final concentration of 12% saturation. After standing at 4°C for at least 4 h, the precipitated fimbriae are collected by centrifugation, resuspended in about 5 ml 50 mM Tris–HCl (pH 7.8) and precipitated with ammonium sulphate as above. In most cases this preparation exhibits typical fimbrial subunit bands on SDS–PAGE (see below) without noticeable bands characteristic of outer membrane proteins (with about 38,000 to 41,000 daltons). If desired, the fimbriae can be purified further as follows.

The method has been described in great detail by Korhonen *et al.* (1980) and is, therefore, only outlined here. To a suspension of fimbriae in 10 m*M* Tris–HCl buffer (pH 7.5) deoxycholate is added to a final concentration of 0.5% (by weight) and the mixture is dialysed against the same buffer for 48 h. The dialysed suspension is centrifuged for 20 min at 10,000 *g* and the supernatant is layered onto a 10–60% sucrose gradient in the above buffer. After centrifugation at 22,000 rpm (Beckman L3-50 centrifuge) for 20 h, the contents of the tubes are fractionated and the fractions tested for protein. The fractions containing fimbriae, often one of several visible bands, are collected, dialysed overnight against the Tris buffer without deoxycholate and concentrated to a small volume by ultrifiltration. If the preparation contains outer membrane proteins or lipopolysaccharide, the above procedure is repeated. It should be noted that during gradient centrifugation deoxycholate forms a micelle band at a density of about 1.2 g cm^{-3}.

Preparations of fimbriae containing flagella may be treated with 6 *M* urea (in 10 m*M* Tris–HCl, pH 7.5) and chromatographed on Sepharose 4B with Tris–urea. This results in the de-aggregation of flagella into subunits but not of fimbriae. De-aggregated flagella and intact flagella can be separated in this way but only if the flagella are a contamination and not a major constituent of the mixture.

Characterization of Fimbriae

The characterization of fimbriae with respect to functional and serological properties may be performed with fimbriated bacteria or with the isolated fimbriae. For the molecular characterization isolated and purified fimbriae must be used. Antisera to fimbriae can be obtained with whole fimbriated bacteria or isolated fimbriae (in Freund's adjuvant). The sera should be absorbed with the corresponding unfimbriated bacteria, grown at fimbria-restrictive temperatures (18–20°C). Since the methods have been described in detail, and may have to be adapted to the problem and the nature of the fimbriae studied, the procedures are here only briefly described and are augmented with references to papers in which details and modifications are given.

Electron Microscopy

Bacteria grown overnight on agar (for MR fimbriae) or for 48 h in static liquid culture (for MS fimbriae) are used. A drop of a bacterial suspension (about 10^{10} cells ml^{-1}) in phosphate-buffered saline is put on a Formvar-coated copper grid and, after sedimentation of the bacteria for about 1 min, is stained with phosphotungstic acid (1% in water of pH 6.4). This is followed by electron microscopic inspection. As an alternative, shadowing with platinum–carbon may be used.

Whenever possible, studies with fimbriated bacteria should be compared with the fimbriation as observed with electron microscopy.

Analysis of Agglutinating Properties

This is done with whole fimbriated bacteria and only rarely with purified fimbriae. For a preliminary test, bacteria from agar plates or from the pellet of a centrifuged liquid culture can be mixed directly with a suspension of erythrocytes (3% in saline). This can be done on white tiles or on microscope slides, with human erythrocytes in the presence of 50 m*M* α-methylmannoside (for MR adherence) and with guinea-pig erythrocytes or baker's yeast (*Saccharomyces cerevisiae*) with and without 50 m*M* α-methylmannoside (for MS agglutination). Erythrocytes from other species can also be used for further characterization (Evans *et al.*, 1980).

To quantitate and compare the agglutinating power of different fimbriated strains the bacteria can be titrated against erythrocytes on a micro scale. Twenty microlitres of a bacterial suspension of known density (counted in a micro-chamber) are diluted serially in a microtitre plate. To each dilution 20 μl of a suspension of erythrocytes (1% in saline) is added. After mixing the plate by tapping, the mixtures are covered to prevent evaporation and allowed to stand at room temperature. The agglutination titres are read microscopically after various times. This analysis is meaningful only if the state of fimbriation is first ascertained electron microscopically (see above).

SDS–Polyacrylamide Gel Electrophoresis

Fimbrial preparations are diluted 1:1 with a disintegration mixture (60 m*M* Tris buffer, pH 6.8, containing 2% SDS, 5% β-mercaptoethanol, 10% glycerol and 0.002% Bromphenol Blue). After heating at 100°C for 5 min, 25 μl of the mixture (containing 10–20 μg protein) is put into a slot of the slab gel. The gel is about 1 mm thick and consists of 5 volumes of separation gel (lower part) and 1 volume of stacking gel (upper part).

The separation gel has 13% acrylamide, containing 0.8% bisacrylamide and 0.1% SDS, prepared in 350 m*M* Tris–glycine buffer, pH 8.8. The stacking gel has 5% acrylamide containing 0.8% bisacrylamide and 0.1% SDS, prepared in 60 m*M* Tris–glycine buffer, pH 6.8. The buffer in the electrode chambers is 25 m*M* Tris, 190 m*M* glycine, pH 8.3, containing 0.1% SDS.

The electrophoresis is run at 18 mA per gel for 30 min at 25 mA per gel for about 3 h. It is stopped when the tracking dye (Bromphenol Blue) has almost reached the bottom of the separation gel.

After the run the gels are stained with a solution of 25% isopropanol, 10% acetic acid in water, containing 0.003% Coomassie Blue. After several hours in the staining solution, the gel is washed in 10% aqueous acetic acid until the background is colourless. It is then ready for drying and/or photography.

In all SDS–PAGE runs, suitable reference proteins, covering a molecular mass region of 12,000–40,000 daltons, should be included.

Preparation of Anti-Fimbriae Antisera

If purified fimbriae are used as immunogens, rabbits are immunised by four injections with the fimbriae incorporated in incomplete Freund's adjuvant. The first injection is with 500 μg protein, followed by three injections with 50 μg protein each. The first injection (2 ml) is given intramuscularly and sub-cutaneously. The second (0.2 ml), after 4 wk, and each of the others, at weekly intervals (0.2 ml), are administered intravenously. Test bleedings are made 1 wk after the last injection. The titre is checked by ELISA or RIA. Animals are bled by cardiac puncture.

It is also possible to use whole fimbriated bacteria as immunogens. They should be killed by ultraviolet irradiation and given at weekly intervals intravenously at doses of about 10^8 cells per rabbit.

The sera, which may also be obtained with different immunization protocols, should be absorbed with bacteria grown at about 18°C. For absorption of 1 ml of antiserum, bacteria from five agar plates (14 cm in diameter) are used.

Serological Methods

Solid-support immune assays such as ELISA, in which peroxidase-coupled antibody and diaminobenzidine–hydrogen peroxide is used, and RIA, with radio-iodinated protein A, have been described (Butler *et al.*, 1978; Johnson *et al.*, 1980; Kato *et al.*, 1976; Yolken and Leister, 1981). In both bases the microtitre wells in which the tests are performed are coated with suspensions of fimbriae (10–20 μg ml^{-1} in 50 mM carbonate buffer, pH 9.5; 50 μl per well) overnight at 4°C. After washing with phosphate-buffered saline (0.05% Tween 20), remaining active sites are saturated with bovine serum albumin (1% in phosphate-buffered saline–0.05% Tween 20) for 1 h at 37°C. The tests are then performed as described.

Crossed immunoelectrophoresis (CIE) and crossed-line immunoelectrophoresis (CLIE) have been described in detail (Weeke, 1973a,b; Larsen *et al.*, 1980). The gel consists of 1% agarose in barbital- or barbital–Tris–glycine buffer, pH 8.6. Either suspensions of purified fimbriae (20–30 μg ml^{-1}) or crude bacterial saline extracts can be used. During electrophoresis in the second dimension the intermediate gels may contain either identical or related fimbrial extract, anti-fimbrial antibody or no addition (see Ørskov and Ørskov, 1983). *Immunoblotting:* see Chapter 26.

References

Achtman, M., Schwuchow, S., Helmuth, R., Morelli, G. and Manning, P. A. (1978). Cell-cell interaction in conjugation *E. coli:* Con-mutants and stabilization of mating aggregates. *Molecular and General Genetics* **164**, 171–183.

Brinton, C. C., Jr. (1959). Non flagellar appendages of bacteria. *Nature (London)* **183**, 782–786.

Brinton, C. C., Jr., Gemski, P., Jr. and Garnahan, I. (1964). A new type of bacterial pilus genetically controlled by the fertility factor of E. coli K-12 and its role in chromosome transfer. *Proceedings of the National Academy of Sciences of The U.S.A.* **52**, 776–783.

Butler, J. E., Feldbush, T. L., McGivern, P. L. and Stewart, N. (1978). The enzyme-linked immunosorbent assay (ELISA): A measure of antibody concentration or affinity? *Immunochemistry* **15**, 131–136.

Duguid, J. P., Smith, I. W., Dempster, G. and Edmunds, P. N. (1955). Non-flagellar filamentous appendages ("fimbriae") and hemagglutinating activity in Bacterium coli. *Journal of Pathology and Bacteriology* **70**, 335–348.

Eisenstein, B. I. and Dodd, D. C. (1982). Pseudocatabolite repression of type 1 fimbriae of Escherichia coli. *Journal of Bacteriology* **151**, 1560–1567.

Eshdat, Y., Speth, V. and Jann, K. (1981). Participation of pili and cell wall adhesin in the yeast agglutinating activity of Escherichia coli. *Infection and Immunity* **34**, 980–986.

Evans, D. G. and Evans, D. J. (1978). New surface-associated heat labile colonization factor antigen (CFA/II) produced by enterotoxigenic E. coli of serogroups 06 and 08. *Infection and Immunity* **21**, 638–647.

Evans, D. G., Evans, D. J., Clegg, S. and Pansley, I. (1979). Purification and characterization of CFA/II antigens of enterotoxigenic E. coli *Infection and Immunity* **25**, 738–748.

Evans, D. J., Evans, D. G., Young, L. S. and Pitt, J. (1980). Hemagglutination typing of Escherichia coli: Definition of seven hemagglutination types. *Journal of Clinical Microbiology* **12**, 235–242.

Gaastra, W. and de Graaf, F. K. (1982). Host-specific fimbrial adhesins of non-invasive enterotoxigenic Escherichia coli strain. *Microbiological Reviews* **46**, 129–161.

Giradeau, J. P., Dubourguier, H. C. and Gouet, P. H. (1982). Inhibition of K99 antigen synthesis by L-alanine enterotoxigenic Escherichia coli. *Journal of General Microbiology* **128**, 463–470.

Isaacson, R. E. (1977). K99 surface antigen of Escherichia coli. Purification and partial characterization. *Infection and Immunity* **15**, 272–279.

Isaacson, R. E. and Richter, P. (1981). Escherichia coli 987P pilus: Purification and partial characterization. *Journal of Bacteriology* **146**, 784–789.

Jann, K., Jann, B. and Schmidt, G. (1981). SDS polyacrylamide gel electrophoresis and seriological analysis of pili from Escherichia coli of different pathogenic origin. *FEMS Microbiology Letters,* **11**, 21–25.

Johnson, R. B., Jr., Libby, R. M. and Nakamura, R. M. (1980). Comparison of glucose oxidase and peroxidase as labels for antibody in enzyme-linked immunosorbent assay. *Journal of Immunoassay* **1**, 27–37.

Kato, K., Hamaguchi, Y., Fukui, H. and Ishikawa, E. (1976). Enzyme-linked immunoassay. Conjugation of rabbit anti-(human immunoglobulin G) antibody with β-galactosidase from *E. coli* and its use for human immunoglobulin G assay. *European Journal of Biochemistry* **62**, 285–292.

Korhonen, T. K., Nurmiaho, E. L., Ranta, H. and Svanborg Edén, C. (1980). New method for isolation of immunologically pure pili from *Escherichia coli. Infection and Immunity* **27**, 569–575.

Larsen, J. C., Ørskov, F., Ørskov, I., Schmidt, M. A., Jann, B. and Jann, K. (1980). Crossed immunoelectrophoresis and chemical structural analysis used for characterization of two varieties of *Escherichia coli* K2 polysaccharide antigen. *Medical Microbiology and Immunology* **168**, 191–200.

Leffler, H. and Svanborg Edén, C. (1981). Glycolipid receptors for uropathogenic *E. coli* binding to human erythrocytes and uroepithelial cells. *Infection and Immunity* **34**, 920–929.

Ofek, I. and Beachey, E. H. (1980). General concepts and principles of bacterial adherence in animals and man. *In* "Bacterial Adherence" (Ed. E. H. Beachey), pp. 3–29. Chapman & Hall, London.

Ørskov, I. and Ørskov, F. (1983). Serology of *Escherichia coli* fimbriae. *Progress in Allergy* **33**, 80–105.

Salit, I. E. and Gotschlich, E. C. (1977). Hemagglutination by purified type 1 *Escherichia coli* pili. *Journal of Experimental Medicine* **146**, 1169–1181.

Stirm, S., Ørskov, I. and Ørskov, F. (1966). K88 an episome-determined protein antigen of *Escherichia coli. Nature (London)* **209**, 507–508.

Stirm, S., Ørskov, F., Ørskov, I. and Mansa, B. (1967). Episome-carried surface antigen K88 of *Escherichia coli*. II. Isolation and chemical analysis. *Journal of Bacteriology* **93**, 731–739.

Weeke, B. (1973a). Immunoelectrophoresis and crossed immunoelectrophoresis I. General remarks on principles, equipment, reagents and procedures. *Scandinavian Journal of Immunology* **2**, Supplement 1, 15–35.

Weeke, B. (1973b). Crossed immunoelectrophoresis. *Scandinavian Journal of Immunology* **2**, Supplement 1, 47–56.

Wevers, P., Picken, R., Schmidt, G., Jann, B., Jann, K., Golecki, J. R. and Kist, M. (1980). Characterization of pili associated with *Escherichia coli* O18ac. *Infection and Immunity* **29**, 685–691.

Yolken, R. H. and Leister, F. J. (1981). Staphylococcal protein A-enzyme immunoglobulin conjugates: Versatile tools for enzyme immunoassays. *Journal of Immunological Methods* **43**, 208–218.

18

Polystyrene Tube Bioluminescence Assay for Bacterial Adherence

M. J. HARBER, RUTH MACKENZIE AND A. W. ASSCHER

Department of Renal Medicine, Welsh National School of Medicine, KRUF Institute, Royal Infirmary, Cardiff, Wales

Introduction

The ability of bacteria to adhere to mucosal surfaces appears to be an important determinant of virulence and yet the possession of type 1 (mannose-sensitive) fimbriae may be disadvantageous for pathogens since these organelles also facilitate attachment to urinary mucus (Ørskov *et al.*, 1980; Chick *et al.*, 1981) and to professional phagocytes (Silverblatt *et al.*, 1979; Mangan and Snyder, 1979). Uropathogenic *Escherichia coli* have been found to be non-fimbriate on initial isolation from urine (Harber *et al.*, 1982), but it is possible that fimbriae are expressed by bacteria in the renal parenchyma or other micro-environments *in vivo*, where they could influence bacteria–host cell interactions. Other factors that confer surface hydrophobicity may also promote bacterial interactions with host cells. Even flagella may play a hitherto unsuspected role in this regard, since there is evidence that these organelles directly promote bacterial attachment to HeLa cells (Jones *et al.*, 1981) and to murine macrophages (Tomita *et al.*, 1981). It is, therefore, of great interest to study and characterize the mechanisms by which bacteria can adhere to different surfaces.

Attempts to investigate bacterial adherence are frequently frustrated by the variable nature of host cell preparations and some workers have preferred to use glass or polystyrene as a medium for attachment (Fletcher, 1976; Rutter and Abbott, 1978) since these materials provide more stable and better defined surfaces. We have found great variability in the capacity of different strains of *E. coli* to adhere to polystyrene, strains possessing both type 1 fimbriae and flagella (but not either alone) adhering strongly in a mannose-sensitive manner (Harber *et al.*, 1983). This model system is useful for assessing the role of different surface properties of bacteria which might influence their interaction with host tissues, and it could possibly be adapted to allow the study of bacterial interaction with

389

physiological surfaces by coating the polystyrene with appropriate 'receptor' substances.

The basic method for measuring bacterial adherence to polystyrene is presented in this chapter. Hydrophobic polystyrene cuvettes are used as the attachment surface, and all operations are performed in a single cuvette. A very short incubation period (10 min) is used, while the number of adherent bacteria is quantified by firefly bioluminescence ATP analysis which provides a rapid, sensitive and accurate detection system (Harber, 1982).

Method

A bacterial culture in nutrient broth is diluted 1/10 in phosphate-buffered saline (PBS), pH 6.8, to give a suspension containing approximately 1×10^8 cfu/ml. Duplicate 300-μl samples from this suspension are placed into polystyrene

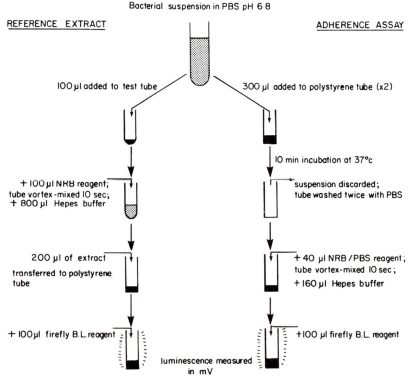

Fig. 1. Flow diagram of the polystyrene tube–bioluminescence adherence assay. (From Harber *et al.*, 1983.)

cuvettes and incubated in a water-bath at 37°C for 10 min. The bacterial suspensions are then aspirated from the cuvettes with a water pump fitted with a trap containing disinfectant, and the cuvettes are washed twice with 1 ml of PBS delivered from a repeating dispenser to remove all non-adherent bacteria. If adherence assays are set up and the polystyrene cuvettes subsequently washed at 1-min intervals it is possible to test 10 bacterial cultures, each in duplicate, in a single experiment.

ATP is extracted from the adherent bacteria by adding 40 µl of NRB–PBS reagent to each cuvette and vortex-mixing for 10 s, followed by the addition of 160 µl of HEPES buffer. ATP is also extracted from a 100-µl sample of the original bacterial suspension in PBS by vortex-mixing with 100 µl of NRB reagent for 10 s, followed by addition of 800 µl of HEPES buffer. A 200-µl aliquot from this reference extract is placed into a cuvette ready for ATP analysis. The ATP in this sample, and in the extracts of the adherent bacteria, is measured by adding 100 µl of firefly luciferin–luciferase reagent to each cuvette and recording the light emission in millivolts with an LKB luminometer 1250. An adherence ratio, which gives an estimate of the number of adherent bacterial cells per thousand of the population in the original PBS suspension, may then be calculated as: mean light emission (mV) from extracts of adherent bacteria × 1000, divided by light emission (mV) from reference extract of PBS suspension × 15 (dilution factor).

A flow diagram of the adherence assay is presented in Fig. 1. Adherence ratios

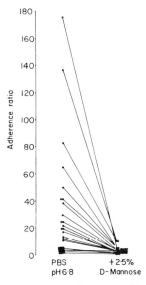

Fig. 2. Inhibitory effect of D-mannose on the adherence of 34 strains of *E. coli* to polystyrene. (Reprinted from Harber, 1982, p. 194, by Courtesy of Marcel Dekker, Inc.)

obtained with 34 urinary isolates of *E. coli* in the presence and absence of 2.5% (w/v) D-mannose are illustrated in Fig. 2. Bacterial strains with an adherence ratio $\geqslant 10$ may be classified as strongly adherent and strains with an adherence ratio <10 as weakly adherent.

Materials

Buffers

The buffer used for the adherence assay is PBS, pH 6.8, which contains, in grams per litre: KH_2PO_4, 3.45; Na_2HPO_4, 4.45; NaCl, 5.0; KCl, 0.2. D-Mannose should be incorporated into this buffer at a concentration of 2.5% (w/v) when required.

The buffer for the bioluminescence assay is 25 mM HEPES, pH 7.75, containing 2 mM EDTA.

ATP Extraction and Assay

ATP is best extracted from bacterial suspensions in PBS with an equal volume of Lumac NRB reagent (nucleotide releasing reagent for bacteria; Sterilin, UK). When ATP is extracted from adherent bacteria in the polystyrene cuvettes a mixture containing equal volumes of NRB and PBS should be used in order to maintain the same ratio of NRB to PBS that is present when extracting ATP from bacterial suspensions in PBS.

Extracted ATP is measured with LKB ATP-monitoring reagent or any equivalent commercial preparation of firefly luciferin–luciferase, which gives an essentially constant light emission in the presence of ATP. The light emission may be quantified with an LKB luminometer 1250 connected to a digital display unit.

Polystyrene Cuvettes

These are the hydrophobic (non-wettable) polystyrene cuvettes designed for use with the LKB luminometers (Clinicon, UK). A cross-hatching in the base of the cuvettes conveniently increases the internal surface area.

Bacteria

The adherence assay was established with strains of *E. coli* possessing both type 1 fimbriae and flagella. Several subcultures may be required to achieve maximum expression of type 1 fimbriation; five consecutive daily subcultures in static Oxoid nutrient broth no. 2 at 37°C is recommended. Silicon-treated glassware should be used for the subcultures to minimize attachment of the organisms to the walls of the containers.

The method is potentially applicable to the study of adhesion with all Gram-negative bacteria which possess type 1 fimbriae and flagella, and it could also be used to investigate the adherence characteristics of other bacteria that have the ability to attach to polystyrene, such as pathogenic oral streptococci (Rutter and Abbott, 1978). The technique has been extended recently in this laboratory to quantify bacterial adhesion to segments of urinary catheters using strains of *Pseudomonas* and *Proteus* isolated from the urine of catheterized patients with spinal injuries.

References

Chick, S., Harber, M. J., Mackenzie, R. and Asscher, A. W. (1981). Modified method for studying bacterial adhesion to isolated uroepithelial cells and uromucoid. *Infection and Immunity* **34**, 256–261.

Fletcher, M. (1976). The effects of proteins on bacterial attachment to polystyrene. *Journal of General Microbiology* **94**, 400–404.

Harber, M. J. (1982). Applications of luminescence in medical microbiology and haematology. *In* "Clinical and Biochemical Luminescence" (Eds. L. J. Kricka and T. J. N. Carter), pp. 189–218. Dekker, New York.

Harber, M. J., Chick, S., Mackenzie, R. and Asscher, A. W. (1982). Lack of adherence to epithelial cells by freshly isolated urinary pathogens. *Lancet* **1**, 586–588.

Harber, M. J., Mackenzie, R. and Asscher, A. W. (1983). A rapid bioluminescence method for quantifying bacterial adhesion to polystyrene. *Journal of General Microbiology* **129**, 621–632.

Jones, G. W., Richardson, L. A. and Uhlman, D. (1981). The invasion of Hela cells by *Salmonella typhimurium:* reversible and irreversible bacterial attachment and the role of bacterial motility. *Journal of General Microbiology* **127**, 351–360.

Mangan, D. F. and Snyder, I. S. (1979). Mannose-sensitive interaction of *Escherichia coli* with human peripheral leukocytes *in vitro*. *Infection and Immunity* **26**, 520–527.

Ørskov, I., Ferencz, A. and Ørskov, F. (1980). Tamm-Horsfall protein or uromucoid is the normal urinary slime that traps type 1 fimbriated *Escherichia coli*. *Lancet* **1**, 887.

Rutter, P. R. and Abbott, A. (1978). A study of the interaction between oral streptococci and hard surfaces. *Journal of General Microbiology* **105**, 219–226.

Silverblatt, F. J., Dreyer, J. S. and Schauer, S. (1979). Effect of pili on susceptability of *Escherichia coli* to phagocytosis. *Infection and Immunity* **24**, 218–223.

Tomita, T., Blumenstock, E. and Kanegasaki, S. (1981). Phagocytic and chemiluminescent responses of mouse peritoneal macrophages to living and killed *Salmonella typhimurium* and other bacteria. *Infection and Immunity* **32**, 1242–1248.

19

Laboratory Tests for Enterotoxin Production, Enteroinvasion and Adhesion in Diarrhoeagenic *Escherichia coli*

SYLVIA M. SCOTLAND, R. J. GROSS AND B. ROWE

Division of Enteric Pathogens, Central Public Health Laboratory, London, UK

Introduction

Enterotoxigenic *Escherichia coli* (ETEC) may produce heat-labile (LT) or heat-stable (ST) enterotoxins. Tissue culture methods are widely used for the detection of LT and that using the Y1 mouse adrenal cell line will be described here. Other cell lines used include Chinese hamster ovary (CHO) and Vero monkey kidney. Antisera to LT can be prepared in animals and a wide range of immunological tests has been described for the detection of LT. These include passive immune haemolysis, enzyme-linked immunosorbent assay (ELISA), solid-phase radio-immunoassay and a precipitin test (the 'Biken' test) performed directly on cultures growing on a special agar medium. We shall describe an ELISA method and the Biken test. So far it has proved impossible to develop tissue culture tests for the two types of ST. ST_B is detected in ligated pig intestinal loops, but as this test is used only in specialized laboratories it will not be described here. The most widely used method to detect ST_A is the infant mouse test and this is described. The non-antigenic nature of ST has delayed the development of immunological methods. Antiserum to ST_A has been prepared by coupling the toxin to a bovine albumin carrier or by polymerization of the toxin with glutaraldehyde. The antiserum has been used to develop a radio-immunoassay and an ELISA method will be reported in the near future. Tests for adhesive factors or colonization factors in ETEC are described elsewhere.

Entcroinvasive *E. coli* (EIEC) share with *Shigella* the ability to penetrate the intestinal epithelial cells and to multiply intracellularly. This enteroinvasive potential can be detected in laboratory tests using the guinea-pig eye (Serény test) or tissue culture. Both methods will be described.

The classical infantile enteropathogenic *E. coli* (EPEC) cause diarrhoea by a

THE VIRULENCE OF ESCHERICHIA COLI

mechanism that remains obscure. Nevertheless, some strains produce a cytotoxin detectable by its action on Vero monkey kidney cells. In addition, many EPEC strains adhere to HEp-2 cells in tissue culture, even in the presence of mannose. The significance of these properties *in vivo* is not yet known but since this is a particularly active area of research both methods will be described.

Methods

Y1 Mouse Adrenal Cell Test for LT

The method is that of Donta *et al.* (1974); the presence of LT is indicated by a morphological change in the cells grown in tissue culture.

Preparation of culture supernatant. Trypticase soy broth, 10 ml, in a 250-ml flask is inoculated with the bacterial strain to be tested. The flask is incubated at 37°C for 18–24 h with shaking (~ 130 oscillations/min). The culture is then centrifuged and the supernatant sterilized by filtration through a Millipore filter (pore size 0.45 μm). This filtrate is used directly for LT tests. A sample of the culture supernatant is also heated at 100°C for 15 min.

Maintenance of Y1 cells. Monolayers are washed twice with Dulbecco's phosphate-buffered saline without calcium and magnesium (DPBS). One millilitre of 0.125% (w/v) trypsin in 0.02% versene (w/v) buffer is added and poured off after 1 min. The monolayers are then incubated at 37°C until the cells begin to detach (~ 5 min). Five millilitres of Ham's F10 growth medium is added and the cells are resuspended. Resuspended cells, 1.5 ml, are then added to 12 ml of growth medium in a tissue culture flask (75 cm²) and incubated at 37°C. This procedure is repeated once weekly.

Test procedure with culture filtrates. A monolayer of Y1 cells, which has been growing for 7 days, is resuspended after trypsin–versene treatment in the usual manner and a portion used for further cultures. For the LT test some of the remaining resuspended cells are diluted in growth medium to obtain a final concentration of ~ 2.5×10^5 cells/ml. Counting in a haemocytometer chamber is recommended but in our laboratory a 1/20 dilution is usually satisfactory. A 0.2-ml sample of this diluted suspension is distributed in each well of a 96-well tissue culture plate (Falcon Micro Test II). The plate is sealed with pressure-sensitive film suitable for tissue culture plates and incubated at 37°C until growth is confluent (usually 3 days). A CO_2 incubator is not essential. Before the test the tissue culture medium is replaced with 0.2 ml growth medium. A 0.025-ml sample of the culture supernatant to be tested is added to each of two wells; 0.025 ml of the heated preparation is tested in a similar way. The plate is sealed and reincubated for 24 h at 37°C. The tissue culture medium is then removed and the cells fixed with methanol for 5 min. After removing the methanol, Giemsa stain (5% v/v) is added. After 45 min the monolayer is washed with distilled water and dried. The cells are examined microscopically for rounding, which indicates the presence of LT in the supernatant. It is usually not feasible to count the cells, but the test is very clear, with over 90% of the cells rounded in a positive test. The effect of the heated preparation should be compared to show that the rounding is due to a heat-labile factor. When LT is not present the Y1 monolayer is unchanged.

Test procedure with bacterial cultures. The method is that of Sack and Sack (1975). A monolayer of Y1 cells growing in a 96-well tissue culture plate is prepared as described above. Before the test the growth medium is removed and replaced with 0.2 ml growth medium. The bacterial strain to be tested has been grown previously at 37°C for 18 h without shaking in 0.5 ml syncase–glucose broth in a tightly closed universal bottle. A 0.05-ml portion of this live bacterial culture is added to a test well. After 5–10 min the medium and bacteria are removed and the cells washed once with phosphate-buffered saline (PBS). A 0.2-ml portion of complete tissue culture medium with penicillin, streptomycin and gentamicin (40 μg/ml) is then added. The plate is resealed and incubated at 37°C for 24 h. The medium is then removed and the monolayer fixed with methanol for 5 min. The methanol is removed and replaced with Giemsa stain (5% v/v). After 45 min the monolayer is washed with distilled water and dried. The cells are examined microscopically for rounding. When the bacterial strain produces LT at least 50% of the cells are rounded (usually 100%); when the strain does not produce LT the monolayer is unchanged.

GM_1–ELISA Test for LT

The method described is an adaptation of the methods of Svennerholm and Holmgren (1978) and Sack *et al.* (1980) and depends on the ability of the GM_1 ganglioside to bind LT.

A 0.1-ml portion of 1 μg/ml solution of ganglioside GM_1 (Supelco) in PBS is added to micro ELISA plates (Dynatech Immulon) and incubated at room temperature overnight. The plates are then washed three times with PBS containing 0.05% (v/v) Tween 20 (PBST). Extra binding sites are blocked by adding 0.2 ml of 1% (w/v) bovine serum albumin (Sigma) dissolved in PBS. After 30 min at 37°C, the plates are again washed three times with PBST. A 0.1-ml portion of the test filtrate, prepared as for tissue culture tests for LT, is added and the plates incubated at room temperature for 18 h. After washing three times with PBST, 0.1 ml of an antiserum prepared against purified cholera toxin or LT is added. The antiserum is diluted in PBST; the appropriate dilution should be determined by titration but may be of the order of 1/200 or 1/400. The plates are incubated at room temperature for 2 h, then washed again three times with PBST. A 0.1-ml portion of anti-rabbit immunoglobulin G (IgG) alkaline phosphate conjugate (Sigma) at a dilution of 1/500 in PBST is added and the plates incubated at room temperature overnight. After washing three times with PBST, 0.2 ml p-nitro-phenyl phosphate (1 mg/ml in carbonate buffer, pH 9.8) is added. After 100 min at room temperature the reaction is stopped with 25 μl NaOH (1 M) and the extinction read at 405 nm. The optical density of the unknown sample (P) is compared with the optical density of a known LT-negative control strain (N). A known LT-positive control strain and medium controls are also included in the test. Samples giving a P/N ratio >2 are considered positive.

Biken (Modified Elek) Test for LT

In this test described by Honda *et al.* (1981), bacteria growing on agar release LT which reacts with a specific antiserum placed in a well to give a line of precipita-tion. The specific antiserum prepared against purified LT is obtained by immu-noaffinity chromatography. The use of antiserum not so treated, even if prepared with purified LT or cholera toxin, may result in the development of non-specific precipitation caused by antibodies against antigens other than LT which may be present in normal rabbit serum.

The strains to be tested are inoculated on Biken agar plates to obtain four large 'colonies' grouped about a central site where a well is to be punched. The inner edge of the final growth should be about 4 mm from the central well position. After 48 h incubation at 37°C the central well is punched and a disc of polymixin B (500 units) is placed on top of each bacterial 'colony'. The plates are reincu-bated for 5–6 h and then 20 μl of antiserum is placed in the central well. After incubation for a further 24 h, the plates are examined against a black background for the development of a precipitation line between the bacterial growth and the central well. They should be re-examined after 24 h.

Infant Mouse Test for ST_A

This test was first described by Dean *et al.* (1972). Four-day-old mice are used and three animals are used for each test. The mice are best kept with their mothers until the test as they are more likely to have milk-filled stomachs and this facilitates the intragastric injections.

The culture filtrate prepared for the Y1 adrenal cell test is used with the addition of 0.04 ml 2% (w/v) Pontamine Blue to 0.5 ml filtrate (this further facilitates accurate intragastric injection). A 0.1-ml portion of filtrate containing Pontamine Blue is injected directly into the stomach, through the abdominal wall. After 4 h at 30°C the mice are killed with chloroform and the intestine examined for distension. The entire intestine distal to the stomach is then removed. The intestines from each experimental group are weighed and the remaining bodies are also weighed. The ratio of gut weight to remaining body weight is then calculated. A ratio greater than 0.1 is considered a positive test for ST_A, ratios less than 0.08 are negative and intermediate values are considered doubtful and the tests repeated. Tests are also repeated if the ratio fails to confirm the visual observations.

Several studies have indicated that ST_A production by *E. coli* strains isolated from animals, other than humans, may not be maximal in trypticase soy broth. An alternative casamino acids–yeast extract medium is described in the Materials section (Evans *et al.*, 1973; Alderete and Robertson, 1977).

Guinea-Pig Eye (Serény) Test for Invasiveness

This is a simple test based on the method of Serény (1957). Cultures are grown on nutrient agar slopes and washed off in physiological saline to give a suspension of approximately 10^{10} organisms per millilitre. This can be obtained by using slopes in $6 \times \frac{5}{8}$ in tubes and suspending the growth in 1 ml saline. A single drop from a '50 dropper' is allowed to fall into the eye while the eyelids are held apart. The eye is examined after 24 h for signs of conjunctivitis. Negative tests are examined daily for 3 days but EIEC and *Shigella* usually give a clear reaction after 24 h.

In the original description the guinea-pigs were observed for a longer period until keratitis developed. We have not found this to be necessary and experiments are terminated as soon as the signs of conjunctivitis appear.

Tissue Culture Test for Invasiveness

The method is based on those of Mehlman *et al.* (1977) and Day *et al.* (1981) and detects penetration of HEp-2 cells by invasive *E. coli*.

Maintenance of HEp-2 cells. Monolayers are washed twice with DPBS. Two millilitres of 0.25% (w/v) trypsin is added and poured off after 1 min. The monolayers are then incubated at 37°C until cells begin to detach ($\sim$ 5 min). Five millilitres of basal medium eagle (BME) growth medium is then added and the cells resuspended. A 1.25-ml portion of suspended cells is then added to 12 ml growth medium in a tissue culture flask (75 cm²) and incubated at 37°C. This procedure is repeated weekly. When the cells are required for invasion or adhesion tests, they are resuspended in growth medium *without added antibiotics* after trypsin treatment.

Test procedure. Two days before the test, a monolayer of HEp-2 cells which has been growing for 7 days is resuspended after trypsin treatment in BME growth medium without antibiotics. A portion is used for further cultures. For the HEp-2 invasiveness test, some of the remaining suspended cells are diluted in the complete tissue culture growth medium without antibiotics to obtain a final concentration of $\sim 10^6$ cells/ml (a dilution of 1/20 is usually satisfactory). Two millilitres of this diluted suspension is distributed into each 4-cm petri dish (tissue culture grade) containing a 22-mm sterile glass coverslip. The dishes are incubated at 37°C for 48 h. When a CO_2 incubator is not available the dishes can be enclosed in a small sealed container.

The bacterial strain to be tested is grown overnight at 37°C in nutrient broth without shaking. Immediately before the test an infection medium is prepared consisting of 70 ml Earle's balanced salts solution (EBSS), 10 ml brain heart infusion broth and 20 ml heat-inactivated foetal calf serum (60°C, 2 h). Each overnight bacterial culture to be tested is diluted 1/20 in this medium, usually 0.25 ml suspension in 5.0 ml of infection medium. The HEp-2 cell monolayer in each dish is washed once with EBSS, 2 ml of the bacteria in the infection medium is added and the dishes are reincubated for 2 h at 37°C. This is the infection period.

After the infection period each cell monolayer is washed thoroughly twice with EBSS and covered with 2 ml of intracellular growth medium. This is prepared with 45 ml complete growth medium without antibiotics, 0.5 ml gentamicin solution (2.0 mg/ml) and 5 ml lysozyme solution (3.0 mg/ml). The cells are reincubated for 3 h at 37°C. This is the intracellular growth period.

After the intracellular growth period, the monolayers are washed thoroughly twice with EBSS, covered with methanol and left for 5–10 min. The methanol is removed and replaced with newly prepared 10% (v/v) Giemsa stain for 30–45 min. The coverslips are then removed, washed twice with Giemsa diluent and mounted on glass slides after passing through acetone, acetone–xylene (50/50 v/v), acetone–xylene (33/66 v/v) and, finally, xylene. When dry and the mounting medium hard, the coverslips can be viewed under oil immersion at ×1000 magnification. At least 300 healthy cells are examined and each cell containing more than one bacterium is counted as positive. In practice, for *E. coli* and *Shigella* the cells contain at least 50 bacteria. The proportion of infected cells may be as low as 1% but is usually 5–80%. If fewer than 60% of the cells appear healthy the test is not recorded and is repeated.

Test for Vero Cytotoxin (VT)

The method is based on those of Konowalchuk *et al.* (1977) and Scotland *et al.* (1980).

Maintenance of vero cells. Monolayers are washed twice with DPBS. Two millilitres of 0.25% (w/v) trypsin in 0.02% (w/v) versene buffer is added and poured off after 1 min. The monolayers are then incubated at 37°C until cells begin to detach (5–15 min). Five millilitres of 199 growth medium is then added and the cells resuspended. Resuspended cells, 1.5 ml, are added to 12 ml of growth medium in a tissue culture flask (75 cm^2) and incubated at 37°C. This procedure is repeated weekly.

Test procedure. A monolayer of Vero cells which has been growing for 7 days is resuspended after trypsin–versene treatment in the usual manner and a portion used for further cultures. For the VT test some of the remaining resuspended cells are diluted in the complete tissue culture medium to obtain a final concentration of $\sim$ 5 $\times$ 10^4 cells/ml. Counting in a haemocytometer is recommended but usually a 1/4 dilution is satisfactory. A 0.2-ml portion of the diluted suspension is distributed in each well of a 96-well tissue culture plate (Falcon Micro Test II). The plate is sealed with pressure-sensitive film suitable for tissue culture plates and incubated until the cell monolayer is confluent (usually 3 days). A CO$_2$ incubator is not essential. The bacterial strain to be tested is grown in trypticase soy broth and a sterile filtrate prepared as for LT testing (see Y1 adrenal cell test). A 0.02-ml portion of the test filtrate is added to a well without changing the medium. Tests are usually performed in duplicate and heated (100°C, 15 min) preparations are compared to ensure that any cytotoxic effect is heat-labile. After the addition of the filtrate the plates are resealed and incubated at 37°C for 4 days. The medium is then removed and the cells fixed with methanol for 5 min. The methanol is removed and Giemsa stain (5% w/v) added. After 45 min the monolayer is washed with distilled water and dried. The cells are examined microscopically. In the presence of the cytotoxin the entire monolayer becomes detached and few cells are visible. In the absence of the cytotoxin the monolayer remains intact. However, Vero cells are also sensitive to LT and if this toxin is present the cells round up and become partially detached from each other.

Tissue Culture Test for Adhesion

The method is based on that of Cravioto *et al.* (1979) and it tests the ability of bacteria to adhere to monolayers of HEp-2 cells. D-Mannose is present during the test to prevent attachment to the tissue culture cells or other surfaces due to type 1 fimbriae, which may be expressed by the bacteria.

Maintenance of HEp-2 cells. Cells are maintained as described in the test for invasiveness.

Test procedure. Monolayers of HEp-2 cells grown for 3 days on glass coverslips in petri dishes are prepared as previously described. The bacterial strains to be tested are grown overnight without shaking at 37°C in peptone water with added D-mannose (1% w/v). Before the test the HEp-2 monolayers are washed twice with EBSS. Each overnight bacterial culture to be tested is diluted 1/50 in an attachment medium of BME medium without antibiotics containing D-mannose (1% w/v) to obtain about 10^7–10^8 bacteria/ml. One millilitre of diluted bacterial culture is added to each washed monolayer and the dishes reincubated at 37°C for 3 h. The monolayers are then washed three times with EBSS and 2 ml of the attachment medium is added. After a further 3-h incubation period the monolayers are washed thoroughly three times with EBSS. The monolayers on glass coverslips are then prepared for viewing under oil immersion at ×1000 magnification by the methods given in the test for invasiveness. *E. coli* strains are considered positive for HEp-2 adhesion when at least 40% of the HEp-2 cells have at least 10 attached bacteria; usually the numbers of bacteria attached are too large to be counted. Some strains give a good positive result after the first 3-h period but for others a 6-h attachment period is necessary for an unequivocal result. Strains that are negative for HEp-2 adhesion show no attached bacteria or 1–5 bacteria attached to less than 5% of the cells.

Materials

1. Trypticase Soy Broth (BBL 11768)

Make up solution of 30 g/litre. Autoclave at not more than 10 lb for 10 min.

	(g/litre):
Trypticase peptone	17.0
Phytone peptone	3.0
Sodium chloride	5.0
K_2HPO_4	2.5
Dextrose	2.5
Final pH 7.3	

Catalogue formula

2. Y1 Mouse Adrenal Cells

Cell line Y1, mouse adrenal cortex tumour, ATCC No. CCL79. *Growth medium:* Ham's F10. To 100 ml add:

12.5 ml horse serum
2.5 ml foetal bovine serum
2.0 ml penicillin/streptomycin (5000 units/ml)
0.5 ml glutamine (200 mM).
Amphotericin B (250 µg/ml) may also be added.

All reagents may be obtained from Flow Laboratories Ltd., Irvine, Scotland.

3. Syncase–Glucose Broth

Na_2HPO_4	5.0 g
K_2HPO_4	5.0 g
Glucose	5.0 g
NH_4Cl	1.18 g
Na_2SO_4	0.089 g
$MgCl_2·6H_2O$	0.042 g
$MnCl_2·4H_2O$	0.004 g
$FeCl_3·6H_2O$	0.005 g
Casamino acids (Difco)	10.0 g
Distilled water	1 litre

Add ingredients to flask in order indicated, followed by water. Shake until clear. Dispense and autoclave at 15 lb for 15 min.

4. Carbonate Buffer

$$Na_2CO_3 \quad 10.6 \text{ g} \quad (0.05 \text{ } M)$$
$$MgCl_2·6H_2O \quad 0.406 \text{ g} \quad (0.001 \text{ } M)$$

Dissolve salts separately. Add distilled water to 2 litres. Adjust pH to 9.8.

5. Biken Agar Plates

Casamino acids (Difco)	2 g
Yeast extract (Difco)	1 g
NaCl	0.25 g
K_2HPO_4	1.5 g
Glucose	0.5 g
Trace salts solution	0.05 ml
Noble agar	1.5 g
Distilled water	100 ml

Adjust pH to 7.5 and autoclave at 15 lb for 15 min. Trace salts solution is 5% (w/v) $MgSO_4$, 2% (w/v) $CoCl_2·6$ H_2O and 0.5% (w/v) $FeCl_3$. A solution of lincomycin (Upjohn or Sigma) of 2.7 mg/ml is also needed; this should not be kept longer than 3 wk. To pour plates, melt agar and cool to 50–60°. Warm lincomycin to 50–60°C and put 0.5 ml in sterile, 85-mm-diameter petri dish. Add 15 ml cooled Biken agar and mix well by rotating plate. Dry plates before use.

6. HEp-2 Cells

Cell line HEp-2, carcinoma of larynx, ATCC No. CCL23.
Growth medium: basal medium eagle with Hanks salts. To 100 ml add:

> 15 ml foetal bovine serum
> 2 ml penicillin–streptomycin (5000 units/ml)
> 0.5 ml glutamine (200 mM)
> Amphotericin B (250 μg/ml) may also be added.

All reagents may be obtained from Flow Laboratories Ltd., Irvine, Scotland.

7. Vero Cells

Cell line Vero, African green monkey kidney, ATCC No. CCL81.
Growth medium: Medium 199 with Earle's salts. To 100 ml add:

> 10 ml foetal bovine serum
> 2 ml penicillin–streptomycin (5000 units/ml)
> 0.5 ml glutamine
> Amphotericin B (250 μg/ml) may also be added.

All reagents may be obtained from Flow Laboratories Ltd., Irvine, Scotland.

8. Casamino Acids–Yeast Extract Medium

Casamino acids (Difco)	20.0 g
Yeast extract (Difco)	6.0 g
NaCl	2.5 g
K_2HPO_4	8.7 g
Trace salts solution	1 ml

Add ingredients to distilled water in order given. Adjust pH to 7.5 with NaOH (5 N). Then adjust to final volume of 1 litre with distilled water. Autoclave at 15 lb for 15 min. Trace salts solution is 5 g $MgSO_4$, 0.5 g $MnCl_2$ and 0.5 g $FeCl_3$ dissolved in 100 ml H_2SO_4 (0.001 N).

References

Alderete, J. F. and Robertson, D. C. (1977). Nutrition and enterotoxin synthesis by enterotoxigenic strains of *Escherichia coli:* Defined medium for production of heat-stable enterotoxin. *Infection and Immunity* **15**, 781–788.
Cravioto, A., Gross, R. J., Scotland, S. M. and Rowe, B. (1979). An adhesive factor found in

strains of *Escherichia coli* belonging to the traditional infantile enteropathogenic serotypes. *Current Microbiology* **3**, 95–99.

Day, N. P., Scotland, S. M. and Rowe, B. (1981). Comparison of an HEp-2 tissue culture test with the Sereny test for detection of enteroinvasiveness in *Shigella* spp. and *Escherichia coli*. *Journal of Clinical Microbiology* **13**, 596–597.

Dean, A. G., Ching, Y. C., Williams, R. G. and Harden, L. B. (1972). Test for *Escherichia coli* enterotoxin using infant mice: Application in a study of diarrhea in children in Honolulu. *Journal of Infectious Diseases* **125**, 407–411.

Donta, S. T., Moon, H. W. and Whipp, S. C. (1974). Detection of heat-labile *Escherichia coli* enterotoxin with the use of adrenal cells in tissue culture. *Science* **183**, 334–336.

Evans, D. G., Evans, D. J. and Gorbach, S. L. (1973). Identification of enterotoxigenic *Escherichia coli* and serum antitoxin activity by the vascular permeability factor assay. *Infection and Immunity* **8**, 731–735.

Honda, T., Taga, S., Takeda, Y. and Miwatani, T. (1981). Modified Elek test for detection of heat-labile enterotoxin of enterotoxigenic *Escherichia coli*. *Journal of Clinical Microbiology* **13**, 1–5.

Konowalchuk, J., Speirs, J. I. and Stavric, S. (1977). Vero response to a cytotoxin of *Escherichia coli*. *Infection and Immunity* **18**, 775–779.

Mehlman, I. J., Eide, E. L., Sanders, A. C., Fishbein, M. and Alusio, C. C. G. (1977). Methodology for recognition of invasive potential of *Escherichia coli*. *Journal of the Association of Official Analytical Chemists* **60**, 546–562.

Sack, D. A. and Sack, R. B. (1975). Test for enterotoxigenic *Escherichia coli* using Y1 adrenal cells in miniculture. *Infection and Immunity* **11**, 334–336.

Sack, D. A., Huda, S., Neogi, P. K. B., Daniel, R. R. and Spira, W. M. (1980). Microtiter ganglioside enzyme-linked immunosorbent assay for *Vibrio* and *Escherichia coli* heat-labile enterotoxins and antitoxin. *Journal of Clinical Microbiology* **11**, 35–40.

Scotland, S. M., Day, N. P. and Rowe, B. (1980). Production of a cytotoxin affecting Vero cells by strains of *Escherichia coli* belonging to traditional enteropathogenic serogroups. *FEMS Microbiology Letters* **7**, 15–17.

Serény, B. (1957). Experimental keratoconjunctivitis Shigellosa. *Acta Microbiologica Academiae Scientiarum Hungaricae* **4**, 367–376.

Svennerholm, A. M. and Holmgren, J. (1978). Identification of *Escherichia coli* heat-labile enterotoxin by means of a ganglioside immunosorbent assay (GM1 ELISA) procedure. *Current Microbiology* **1**, 19–23.

20

Demonstration of Enterotoxigenic *Escherichia coli* in Paraffin-embedded Tissue Sections by an Immunoperoxidase Technique

J. BAILEY, G. A. H. WELLS AND D. J. SHEEHAN

Pathology Department, Central Veterinary Laboratory, Weybridge, Surrey, UK

Introduction

The fluorescent antibody technique (FAT) is widely used for the demonstration of the presence of *Escherichia coli* in unfixed cryostat tissue sections. This technique makes possible visualization of the antigen but little or no histological detail of the host tissues. A technique which determines not only the location but the detailed distribution of the antigen in relation to the histopathological changes in a single section offers obvious advantages over FAT.

The immunoperoxidase technique (IPX) is also an immunohistological method used for the demonstration of a wide range of antigens and can be employed on paraffin-embedded fixed tissue sections in much the same way as tinctorial methods for routine diagnostic histopathology. The method utilizes labelled or unlabelled antibodies and the stable enzyme horseradish peroxidase. The substrate most widely used is diaminobenzidine (DAB), which polymerizes in the presence of peroxidase and hydrogen peroxide to form an insoluble brown polymer that is deposited at the site of the antigen–antibody reaction.

In the direct method the antibody conjugate has specificity directed against the antigen under study. In the indirect method, a primary antiserum having specificity against the antigen is applied initially, followed by a peroxidase–antibody conjugate from a second species, which is directed against the immunoglobulin components of the primary antiserum. The indirect method has advantages over the direct method of greater versatility and increased sensitivity.

This communication describes the use of the indirect method for the histological demonstration of enterotoxigenic *E. coli* in the small intestine. It is further shown that specified antisera can be used to identify virulence factors which have increasing diagnostic importance in the choice of prophylactics.

407

ISBN 0-12-677520-6

Methods

The primary antiserum, to the specific antigen of the *E. coli,* is raised in the rabbit and is unlabelled. The secondary antiserum, goat anti-rabbit IgG, is conjugated to horseradish peroxidase.

Production of antisera. Specific antisera to *E. coli* are raised in rabbits, as described by Sojka (1965).

Titration of antisera. The working dilution of each antiserum is determined by the use of the serum at doubling dilutions from 1/20 to 1/2560 on tissue sections of gut from hysterotomy-derived, colostrum-deprived gnotobiotic piglets, previously inoculated orally with 5×10^8 to 5×10^9 *E. coli.*

Antiserum to rabbit IgG, raised in goat, is conjugated to horseradish peroxidase (Miles Scientific Division, Miles Laboratories Limited, Slough, England) by the method of Avrameas and Ternynck (1971) and is used at a dilution of 1/320. It has been our experience that use of this antiserum at lower dilutions increases the degree of non-specific background reaction.

Preparation of paraffin sections. Small samples of piglet intestine previously infected with enterotoxigenic *E. coli* are collected via laparotomy under halothane general anaesthesia. The samples were fixed in phosphate-buffered 10% neutral formalin, trimmed, processed routinely and embedded in paraffin wax (melting point 56°C). Care is taken to orientate tissue blocks to give longitudinal sections of the villi for microscopy. Sections are cut at 5 μm on a base sledge microtome, floated out on a heated water-bath and collected on grease-free 76 × 26 mm glass slides, without the use of adhesives. The sections are dried overnight in an oven at 37°C and are subsequently transferred to an oven at 60°C to melt the wax.

Immunoperoxidase Technique—Indirect (adapted from Mepham et al., 1979)

1. Dewax sections and take to methanol.
2. Inhibit endogenous peroxidase by treating sections with freshly prepared 0.5% hydrogen peroxide (100 vols) in methanol—10 min.
3. Wash sections in tap water—10 min.
4. Warm sections in distilled water at 37°C to equilibrate the temperature of the slides—10 min.
5. Treat sections with freshly prepared 0.1% trypsin in 0.1% calcium chloride, pH 7.8 at 37°C—10 min.
6. Wash sections in tap water—10 min.

7. Rinse sections with 0.5 *M* Tris-buffered saline pH 7.6 (TBS).
8. Treat sections with rabbit anti-*E. coli* serum—30 min.
9. Wash sections with TBS—3 × 10 min.
10. Treat sections with peroxidase-conjugated goat anti-rabbit IgG anti-serum—30 min.
11. Wash sections with TBS—30 min.
12. Treat sections with DAB—10 min.
13. Wash sections in tap water—10 min.
14. Counterstain with Mayer's haemalum—20 s.
15. 'Blue' sections in running tap water—10 min.
16. Dehydrate, clear and mount sections in DPX.

Procedures 4 and 5 are carried out in a water-bath at 37°C. Procedures 8, 10 and 12 are carried out in a moist chamber at room temperature.

Reagents

1. *Trypsin.* Type II. Crude (Cat. No. T.8128, Sigma Chemical Co. Ltd., Poole, Dorset, England). Freshly prepared 0.1% trypsin in 0.1% $CaCl_2$. Pre-heat 100 ml of deionized water to 37°C, add 0.1 g $CaCl_2$. While stirring constantly take pH readings and adjust to between pH 7.8 and 8.0 by the addition of 0.1 *N* NaOH. Add 0.1 g of trypsin and re-adjust pH to 7.8, again with 0.1 *N* NaOH. Use immediately in a water-bath at 37°C.
2. *DAB* [*3,3'-Diaminobenzidine tetrahydrochloride* (pfs) Grade II (Cat. no. D.5637, Sigma Chemical Co. Ltd., Poole, Dorset, England)]. Dissolve 5 mg of DAB in 10 ml of Tris-HCl buffer, pH 7.6. Immediately before use add 0.1 ml of 1% hydrogen peroxide (100 vols).

Discussion

When examined by routine light microscopy, specific organisms stain brown against a clear background. Organisms colonizing the small intestine can be seen attached to the microvillous border of enterocytes on the intestinal villi. Nuclei stain blue with Mayer's haemalum (Fig. 1).

Sections of uninfected gnotobiotic piglet small intestine show no brown staining. *E. coli* not possessing the specific antigens remain unstained by the specific antisera when used at their optimal or working dilution.

Sections of small intestine from four gnotobiotic piglets experimentally infected with *E. coli* possessing essentially different adhesive antigens were stained by the immunoperoxidase technique, using the specific antiserum. The results are shown in Table 1.

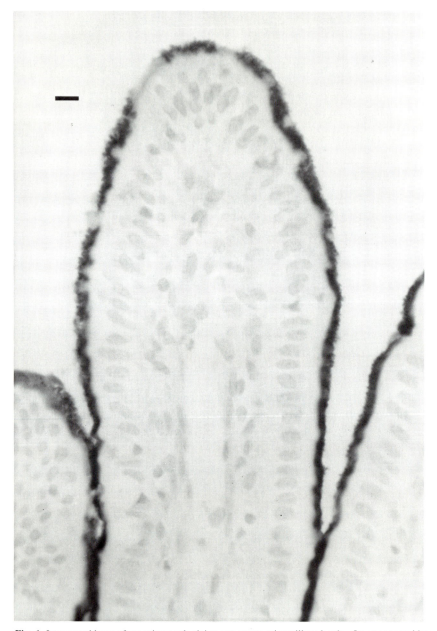

Fig. 1. Interrupted layer of organisms colonizing enterocyte microvillous border. Immunoperoxidase preparation of 5-μm paraffin section of infected piglet intestine. Piglet was infected with *E. coli* strain 1676 (**O**101:K30, F41) and the section was reacted with antiserum to *E. coli* strain 1706 (**O**101:K30, F41). Immunoperoxidase–haematoxylin. Bar marker represents 10 μm.

Table 1. *List of positive results of IPX technique obtained in tissue section of intestine infected with specific E. coli serotypes when reacted with specific antisera*

E. coli	Specific antiserum	Dilution
168-14 (**O8:K87, K88**ab)	OK G7 (**O8:K87, K88**ab)	1/320
B41 (**O101:K99, F41**)	OK B41 and Abs K99	1/320
X177/81 (**O9:K103**, 987P)	OK 987P (**O9:K103**, 987P)	1/640
1706 (**O101:K30, F41**)	OK 1706 (**O101:K30, F41**)	1/1280

Table 2. *Results of IPX technique on tissue section of a field case of piglet diarrhoea from which E. coli Abottstown was isolated*

	Specific antisera	Result[a]
E. coli Abbotstown (**O**149:**K**91, K88ac)	Abs K99 (**K99, F41**)	− ve
	OK 1706 (**O101:K30, F41**)	− ve
	OK 987P (**O9:K103**, 987P)	− ve
	OK G7 (**O8:K87, K88**ab)	+ ve
	Abs K88c	+ ve

[a] +, − Presence or absence, respectively, of antigen-specific staining.

From a field case of piglet diarrhoea from which *E. coli* Abbotstown (**O**149:**K**91, K88ac) was isolated, a range of antisera were reacted in tissue sections of the small intestine. The results are shown in Table 2.

The IPX technique provides the following advantages over the FAT method:

1. It is a permanent record of the location and distribution of specific antigen–antibody reaction.
2. The distribution of the reaction product in relation to both the organism and the host tissues can be visualised by bright-field light microscopy at magnification up to ×1000.
3. Histopathological change in the host tissue attributable to the presence of the antigen can be observed in the same section.
4. When used on field material from cases of enteric disease in domestic food animals the differential diagnostic process can, in the absence of positive results, proceed quickly to alternative histological staining methods.

Acknowledgments

We thank Dr. J. A. Morris and W. J. Sojka for their help and encouragement in the development of the application described here.

References

Avramaes, S. and Ternynck, T. (1971). Peroxidase labelled antibody and Fab conjugates with enhanced cellular penetration. *Immunochemistry* **8**, 1175–1179.

Mepham, B. L., Frater, W. and Mitchell, B. S. (1979). The use of proteolytic enzymes to improve immunoglobulin staining by the PAP technique. *Histochemical Journal* **11**, 345–357.

Sojka, W. J. (1965). *Escherichia coli* in domestic animals and poultry. *Commonwealth Bureau of Animal Health, Review Series* **7**.

21

Growth of *Escherichia coli* under Iron-restricted Conditions

PAULINE STEVENSON AND E. GRIFFITHS

National Institute for Biological Standards and Control, Hampstead, London, UK

Introduction

The amount of iron that is readily available to bacteria in body fluids is now known to be extremely small (Chapter 7). This is because extracellular iron is attached to high-affinity iron-binding glycoproteins, transferrin in serum and lymph and lactoferrin in external secretions and milk. Since this iron-restricted environment induces phenotypic changes in invading bacterial pathogens there is currently much interest in studying the metabolism of such bacteria under iron-restricted conditions.

Several methods have been used to produce iron-restricted bacteria *in vitro*. Some have depended upon the use of media from which iron has been removed by extraction with synthetic co-ordinating agents such as 8-hydroxyquinoline, by co-precipitation with insoluble salts or by use of ion-exchange or chelating resins (Chelex) (Lankford, 1973; Payne and Finkelstein, 1978; O'Brien *et al.*, 1982; Mickelsen and Sparling, 1981). Other methods have depended upon the addition of low-molecular-weight synthetic complexing agents such as α,α'-dipyridyl, ethylenediamine-di-*o*-hydroxyphenylacetic acid or nitrilotriacetate to bind the iron and to limit its availability to bacteria (Braun and Burkhardt, 1982; Archibald and DeVoe, 1979; Williams and Warner, 1980). The natural iron chelators desferrioxime (Desferal, Ciba-Geigy Ltd) and deferriferrichrome A have also been used for this purpose (Norqvist *et al.*, 1978; Mickelsen and Sparling, 1981; Klebba *et al.*, 1982). In each case, however, the iron-chelating agent must be checked empirically to ensure that it does not actually supply iron to the organism used. Media containing high levels of Tris buffer, sometimes in conjunction with succinate, have also been found to induce changes comparable to those observed under conditions of iron restriction (McIntosh and Earhart, 1976; Braun, 1981).

All the methods above suffer from certain disadvantages. In the case of the

413

iron-deficient growth media, there is the danger that the process of removing iron may also remove other essential metals or may introduce undesired contaminants (Lankford, 1973). In addition, the relevance of data obtained with bacteria grown in iron-deficient media, where the quantity of iron is limiting and the organisms starved of iron, to the situation *in vivo* during infection is debatable (Chapter 7). A clear distinction should be made between the quantity of iron present in a medium and its availability to bacteria. There is plenty of iron present in body fluids but it is simply not readily available. The use of synthetic chelating agents to complex iron *in situ* in growth media, although possibly preferable to the use of iron-deficient media, also has its problems (Neilands, 1982). However, where such compounds work, they provide a useful way of producing iron-restricted bacteria.

The methods we describe here for growing *E. coli* under iron-restricted conditions are designed to reflect as closely as possible the situation *in vivo* in host tissues where the iron present is tightly bound to the iron-binding protein transferrin or lactoferrin. Ovotransferrin (conalbumin) can be used instead of lactoferrin or transferrin in synthetic growth media and this is readily obtained commercially and is considerably cheaper than transferrin. Each iron-binding protein has two specific metal-binding sites per molecule and one bicarbonate ion is also required for each Fe^{3+} complexed (Griffiths and Humphreys, 1978). It is worth noting that certain organisms are unable to remove iron equally from transferrin, lactoferrin and ovotransferrin (Mickelsen and Sparling, 1981; Mickelsen *et al.*, 1982). *E. coli*, however, can remove iron from each of these proteins. Body fluids such as human milk, bovine colostrum and horse serum, which contain large quantities of iron-binding proteins, can also be used as growth media. However, these fluids can exert a bacteriostatic effect on certain serotypes of *E. coli* (Griffiths, 1972; Griffiths and Humphreys, 1978). This is thought to be due to the presence of a specific antibody (Chapter 7). It is necessary, therefore, to check whether the particular fluid used as growth medium inhibits the growth of the *E. coli* strain being examined.

Growth of E. coli in Broth Containing Ovotransferrin

Procedure. For a 500-ml culture the following mixture is used; for other volumes the figures are adjusted accordingly. Trypticase soy broth (double strength) 250.0 ml, ovotransferrin solution (4 mg/ml) 62.5 ml, 0.71 M sterile sodium bicarbonate 43.75 ml, sterile water to 500 ml.

The final concentrations of ovotransferrin and $NaHCO_3$ in the medium are 0.5 mg/ml and 0.071 M, respectively. *E. coli* for use as inoculum is grown in brain heart infusion broth (Difco) for 3 h at 37°C. Cultures are harvested by centrifugation and the bacteria resuspended in a 10% (v/v) mixture of brain heart infusion broth in sterile 0.15 M NaCl (10% broth–saline). The total cell count per milli-litre is obtained by use of a colorimeter and a standard graph. Suitable dilutions for inoculations are made in 10% broth–saline and viable counts are made on fresh blood–agar plates. *E. coli* is grown routinely from an inoculum of 10^5–10^6 cells/ml of medium in volumes of 150–500 ml at 37°C in a round-bottom flask fitted with a condenser, through which gas leaves the vessel, and a gas inlet. The gas mixture of 5% CO_2 plus 95% air is sterilized by passage through an in-line gas filter and blown over the surface of the stirred medium at about 200–400 ml/min to maintain the correct pH (pH 7.5–7.8). Bacteria are usually grown for 5–6 h and then collected by centrifugation. Other growth media can be used in place of trypticase soy broth (Griffiths and Humphreys, 1978).

Materials and equipment. OVOTRANSFERRIN. (Conalbumin) (lyophilized, iron-free type I, Sigma Chemical Co. Ltd.). It is important to remove low-molecular-weight chelating agents, presumably used in the preparation of the material, before use. This is done by dialysis. A solution of 4 mg/ml ovotransfer-rin in 0.071 M NaHCO$_3$, 0.15 M NaCl is dialysed for 24 h against about 30 vols of 0.071 M NaHCO$_3$, 0.15 M NaCl. It is sterilized by filtration through a 0.45-μm filter (Millipore Corp.).

TRYPTICASE SOY BROTH. (BBL, Becton, Dickenson & Co.), made up at double strength and sterilized by autoclaving. Quickfit round-bottom flask, condenser, gas inlet and connection to gas cylinder. In-line gas filter (Microflow Pathfinder Ltd.).

Growth of E. coli in Serum, Human Milk and Bovine Colostrum

Procedure. For a 200-ml culture the following mixture is used; for other volumes the figures are adjusted accordingly. Serum, colostrum or human milk 180 ml; 0.71 M sterile sodium bicarbonate 20 ml.

The final concentration of NaHCO$_3$ in the medium is 0.071 M. This is suffi-cient to counteract the iron-mobilizing effect of the citrate normally present in these body fluids (Griffiths and Humphreys, 1977). *E. coli* is routinely grown from an inoculum, prepared as before, of 10^5–10^6 cells per millilitre of growth medium in 150–300 vols at 37°C in the apparatus described above. The pH is maintained at 7.4–7.5 throughout the experiment by passing the sterile gas mixture (5% CO_2 plus 95% air) over the surface of the stirred liquid. Bacteria are grown for 5–6 h and then collected by centrifugation.

Material and equipment. WHOLE NORMAL HORSE SERUM. (Wellcome Diag-nostics, horse serum no. 3 sterile). This is heated at 56°C for 30 min to inactivate complement; other similar sera can be used.

COLOSTRUM. This is obtained from normal cows within 12 h of parturition. Whey is prepared by incubating the colostrum with rennet (0.1% w/v) for 40 min at 37°C and removing the clot by centrifugation at 5°C. Whey is sterilized by filtration first through a 3-μm membrane filter (Millipore Corp.) and then through a 0.45-μm membrane filter and stored at −20°C. The 3-μm filtration step greatly facilitates filtration through the 0.45-μm filter. A protein pre-filter is also used during each filtration.

HUMAN MILK. This is obtained at different stages of lactation, pooled and stored at −20°C. Milk from mothers receiving antibiotics should be excluded. Before use the milk is centrifuged at 56,000 g for 1.5 h and the fluid separated from the fat and sterilized by filtration first through a 3-μm membrane filter and then through an 0.45μm filter as described above.

References

Archibald, F. S. and DeVoe, I. W. (1979). Removal of iron from human transferrin by *Neisseria meningitidis. FEMS Microbiology Letters* **6**, 159–162.

Braun, V. (1981). *Escherichia coli* cells containing the plasmid Col V produce the iron ionophore aerobactin. *FEMS Microbiology Letters* **11**, 225–228.

Braun, V. and Burkhardt, R. (1982). Regulation of the Col V Plasmid-determined Iron (III)-aerobactin transport system in *Escherichia coli. Journal of Bacteriology* **152**, 223–231.

Griffiths, E. (1972). Abnormal phenylalanyl-tRNA found in serum inhibited *Escherichia coli,* strain O111. *FEBS Letters* **25**, 159–164.

Griffiths, E. and Humphreys, J. (1977). Bacteriostatic effect of human milk and bovine colostrum on *Escherichia coli:* Importance of bicarbonate. *Infection and Immunity* **15**, 396–401.

Griffiths, E. and Humphreys, J. (1978). Alterations in tRNAs containing 2-methylthio-N^6-($\triangle^2$-isopentenyl)adenosine during growth of enteropathogenic *Escherichia coli* in the presence of iron-binding proteins. *European Journal of Biochemistry* **82**, 503–513.

Klebba, P. E., McIntosh, M. A. and Neilands, J. B. (1982). Kinetics of biosynthesis of iron-regulated membrane proteins in *Escherichia coli. Journal of Bacteriology* **149**, 880–888.

Lankford, C. E. (1973). Bacterial assimilation of iron. *CRC Critical Reviews in Microbiology* **2**, 273–331.

McIntosh, M. A. and Earhart, C. F. (1976). Effect of iron on the relative abundance of two large polypeptides of the *Escherichia coli* outer-membrane. *Biochimica et Biophysica Acta* **70**, 315–322.

Mickelsen, P. A. and Sparling, P. F. (1981). Ability of *Neisseria gonorrhoeae, Neisseria meningitidis* and commensal Neisseria species to obtain iron from transferrin and iron compounds. *Infection and Immunity* **33**, 555–564.

Mickelsen, P. A., Blackman, E. and Sparling, P. F. (1982). Ability of *Neisseria gonorrhoeae, Neisseria meningitidis* and commensal Neisseria species to obtain iron from lactoferrin. *Infection and Immunity* **35**, 915–920.

Neilands, J. (1982). Microbial envelope proteins related to iron. *Annual Reviews of Microbiology* **36**, 285–309.

Norqvist, A., Davies, J., Norlander, L. and Normark, S. (1978). The effect of iron starvation on the outer membrane protein composition of *Neisseria gonorrhoeae*. *FEMS Microbiology Letters* **4**, 71–75.

O'Brien, A. D., LaVeck, G. D., Thompson, M. R. and Formal, S. B. (1982). Production of *Shigella dysenteriae* Type 1—like cytotoxin by *Escherichia coli*. *Journal of Infectious Diseases* **146**, 763–769.

Payne, S. M. and Finkelstein, R. A. (1978). The critical role of iron in host-bacterial interactions. *Journal of Clinical Investigation* **61**, 1428–1440.

Williams, P. H. and Warner, P. J. (1980). Col V plasmid mediated, Colicin V—independent iron uptake system of invasive strains of *Escherichia coli*. *Infection and Immunity* **29**, 411–416.

22

Detection of Synthesis of the Hydroxamate Siderophore Aerobactin by Pathogenic Isolates of *Escherichia coli*

N. H. CARBONETTI AND P. H. WILLIAMS

Department of Genetics, University of Leicester, Leicester, UK

Introduction

Iron is essential for microbial growth (see Chapter 7). Although it is generally abundant in aerobic conditions, ferric iron forms insoluble aggregates that restrict its availability for microbial assimilation. Thus, aerobic micro-organisms synthesize low-molecular-weight iron-solubilizing compounds (siderophores) to sequester the ferric ions necessary for growth. Enteric bacteria secrete the catechol-type siderophore enterochelin (enterobactin) and the ferric–enterochelin complex formed in the external medium is actively transported back across the bacterial membranes (Rosenberg and Young, 1974). Recently, evidence has accumulated that a significant proportion of *Escherichia coli* isolates from cases of bacteraemia, meningitis and urinary tract infections of man secrete, in addition, the hydroxamate siderophore aerobactin. As far as is known, the enzymes for aerobactin biosynthesis and the specific outer membrane receptor protein for the active transport of ferric aerobactin are products of plasmid genes (Stuart *et al.*, 1980; Braun, 1981; Warner *et al.*, 1981). It has been shown that carriage of an 'aerobactin plasmid' confers upon the host bacterium a significant growth advantage in conditions of iron stress, both *in vitro* and in experimental infections of animals (Williams, 1979). Furthermore, despite the fact that the stability constant of ferric–aerobactin is several orders of magnitude below that of ferric–enterochelin (Harris *et al.*, 1979), the hydroxamate siderophore is apparently better able to sequester iron from its complexed form with transferrin in serum, presumably owing to special, as yet incompletely defined structural features (Konopka *et al.*, 1982). This section describes methods for the detection and quantitation of aerobactin synthesized by clinical isolates of *E. coli*.

THE VIRULENCE OF ESCHERICHIA COLI

Methods

Growth Media

Siderophore synthesis and secretion are induced in conditions of iron starvation. For screening purposes it is usually not appropriate to use media for which attempts have been made either to limit deliberately the addition of, or to remove selectively, iron present as an impurity in constituent chemicals and water. Rather, it is more convenient to use the standard defined or nutrient media on which the isolates to be tested will grow but to reduce the availability of ferric ions present by adding an excess of a specific iron-chelating compound such as transferrin (3 μM) or α,α'-dipyridyl (160 μM). Alternatively, a kind of physiological iron starvation can be induced by growth in medium containing sodium succinate (1% w/v) instead of glucose as the sole carbon source (Warner *et al.*, 1981; Konopka *et al.*, 1982). Several passages of bacterial growth in conditions of iron starvation may be necessary to deplete intracellular iron pools before significant aerobactin induction is observable.

Chemical Detection of Hydroxamate

The presence of aerobactin in culture supernatant fluids can be determined by the method of Csáky (1948). Samples (0.5 ml) are mixed with 0.5 ml 6 M sulphuric acid and autoclaved at 15 p.s.i. (121°C) for 30 min in order to generate free hydroxylamine. After neutralization with 1.5 ml 35% (w/v) sodium acetate, 0.5 ml 1% (w/v) sulphanilic acid and 0.5 ml 1.35% (w/v) iodine, both in 30% (v/v) acetic acid, are added and left to stand for 5 min to allow oxidation of the free hydroxylamine nitrogen to nitrite. Excess iodine is destroyed by adding 1 ml 1% (w/v) sodium arsenate, and nitrite is then estimated colorimetrically (absorbance at 526 nm) after addition of 0.5 ml 0.3% (w/v) 1-naphthylamine in 30% (v/v) acetic acid, and incubation at room temperature for about 30 min to allow development of pink coloration. (*Note:* 1-naphthylamine is thought to be carcinogenic and should be treated with extreme caution.) The concentration of hydroxylamine nitrogen groups present can be roughly quantified from a standard curve constructed with known quantities of hydroxylamine hydrochloride. A 200 μM solution processed as described (omitting the acid hydrolysis step) gives an absorbance of 0.74 at 526 nm. The test is not strictly quantitative for the estimation of aerobactin, possibly owing to the variable release of hydroxylamine by acid hydrolysis and also to the presence of compounds other than aerobactin that will yield hydroxylamine under the same conditions. However, it is certainly adequate for comparison purposes.

Levels of aerobactin secreted into culture media are frequently too low to be measured accurately without prior concentration, for example by passage of the supernatant fraction over an anion exchange resin (Warner *et al.*, 1981). Adsorbed aerobactin elutes from Dowex-1 (Cl⁻) in 0.50–0.55 M ammonium chloride. Thus, siderophore from 250 ml of culture can be quantitatively recovered in as little as 5 ml from a 3.5×1 cm Dowex-1 column either by stepwise (0.1 M intervals) or by gradient (0.4–1.0 M) elution with ammonium chloride. An incidental advantage of this process is that it separates the aerobactin from enterochelin and 2,3-dihydroxybenzoate compounds produced under the same culture conditions.

A quicker, though about 10-fold less sensitive, assay for hydroxamate siderophores involves mixing equal volumes of sample and 5 mM $Fe(ClO_3)_4$–0.1 M $HClO_4$ and measuring the absorbance of the mixture at 455 nm (Atkin and Neilands, 1968). It may be necessary to perform a second centrifugation of the culture supernatant fluid after addition of the ferric perchlorate reagent in order to remove any precipitate formed. This test is specific for aerobactin in mixtures with enterochelin since ferric-catechol complexes do not form in the acidic conditions of the reagent.

Bioassay of Aerobactin

Aerobactin is not utilized by *Arthrobacter flavescens,* the species most commonly used to assay hydroxamate-type siderophores (J. B. Neilands, personal communication), nor, as far as is known, by *E. coli* strains not having the plasmid-specified 74,000-dalton ferric–aerobactin receptor protein in the outer membrane (Grewal *et al.*, 1982). This protein is also the receptor protein for cloacin DF13, a bacteriocin of *Enterobacter cloacae.* Sensitivity to cloacin is a convenient indicator of the presence of the receptor in laboratory strains of *E. coli;* however, its application to screening clinical isolates may be limited by the presence of surface structures which act as barriers to cloacin adsorption. Strains of *E. coli* carrying the mutant plasmid ColV-K30*iuc* (Williams and Warner, 1980) are defective in aerobactin biosynthesis but have normal receptor and are, therefore, able to utilize exogenously supplied aerobactin. A widely used strain for detecting aerobactin production by strains that also secrete enterochelin is the *E. coli* K-12 strain LG1522. This carries ColV-K30*iuc* and so can use ferric–aerobactin but has a chromosomal mutation *fep* (Cox *et al.*, 1970) rendering it defective in ferric–enterochelin uptake, although synthesis of the catechol is normal. Minimal agar, containing 160 μM α,α'-dipyridyl, is seeded with strain LG1522 (approximately 10^6 cells/plate) and inoculated with the bacterial isolates to be tested. Individuals that secrete aerobactin are identified by a halo of growth of the lawn around the inoculum after overnight incubation (Fig. 1a). Because of the rapid rate of diffusion of secreted aerobactin it is normally possible to test only about 10 strains on a single plate. Alternatively, the presence of aerobactin in culture supernatant fluids can be determined by placing aliquots in wells cut in

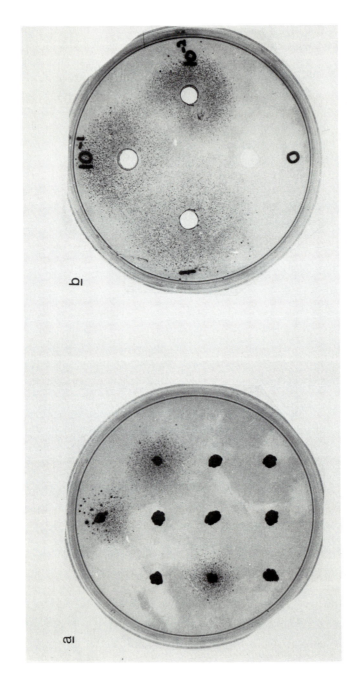

Fig. 1. Bioassay of aerobactin. (a) Production of aerobactin by growing bacteria (in this example, by 3 of 10 clinical isolates tested) or (b) presence of aerobactin in culture medium (here undiluted, 1, or serially diluted, 10^{-1} and 10^{-2}; 0 is a control lacking aerobactin) is indicated by growth of the bacteria in the lawn (strain LG1522) after overnight incubation. In this figure, bacterial growth has been accentuated for photography by addition of 2,3,5-triphenyltetrazolium (50 μg/ml) to the medium.

seeded test plates (Fig. 1b). Indeed, a degree of quantitation is possible if serial dilutions of samples are tested in this way; activity is arbitrarily defined as the reciprocal of the highest dilution of a sample at which growth of the indicator is detectable, and the method is at least 100-fold more sensitive than chemical assays of aerobactin.

Materials

Dowex-1 (chloride form), hydroxylamine hydrochloride (grade 1),α,α'-dipyridyl and transferrin (human) are available from Sigma London Chemical Co. Ltd. Iodine (Analar grade), ferric perchlorate and perchloric acid (70% 'Aristar' ultra-pure grade) are from BDH Chemicals Ltd., and sulphanilic acid (analytical grade) from Fisons Scientific Apparatus, Ltd. We recently used 1-naphthylamine supplied by Fisons for Csáky assays in this laboratory. However, preparations of this reagent have been shown to be carcinogenic (possibly due to contamination with 2-naphthylamine), and its manufacture and large-scale use are now controlled by legislation. Fisons no longer supply 1-naphthylamine, but it is still available from BDH as a 'general purpose reagent' (containing a maximum of 0.5% 2-naphthylamine), and from Sigma either as rather impure 'practical grade' or, more usefully, in sealed bottles to which solvent can be added to yield a 1% solution.

E. coli K-12 strain LG1522 (ara fepA lac leu mtl proC rpsL supE thi tonA trpE xyl ColV-K30iuc) and the Enterobacter cloacae strain which produces cloacin DF13 are available on request from Dr. P. H. Williams.

References

Atkin, C. L. and Neilands, J. B. (1968). Rhodotorulic acid, a diketopiperazine dihydroxamic acid with growth-factor activity. I. Isolation and characterization. *Biochemistry* **7**, 3734–3739.

Braun, V. (1981). *Escherichia coli* cells containing the plasmid ColV produce the iron ionophore aerobactin. *FEMS Microbiology Letters* **11**, 225–228.

Cox, G. B., Gibson, F., Luke, R. J. K., Newton, N. A., O'Brien, I. G. and Rosenberg, H. (1970). Mutations affecting iron transport in *Escherichia coli*. *Journal of Bacteriology* **104**, 219–226.

Csáky, T. Z. (1948). On the estimation of bound hydroxylamine in biological materials. *Acta Chemica Scandinavica* **2**, 450–454.

Grewal, K. K., Warner, P. J. and Williams, P. H. (1982). An inducible outer membrane protein involved in aerobactin mediated iron transport by ColV strains of *Escherichia coli*. *FEBS Letters* **140**, 27–30.

Harris, W. R., Carrano, C. J. and Raymond, K. N. (1979). Co-ordination chemistry of microbial iron transport compounds. Isolation, characterization and formation constants of ferric aerobactin. *Journal of the American Chemical Society* **101**, 2722–2727.

Konopka, K., Bindereif, A. and Neilands, J. B. (1982). Aerobactin-mediated utilization of transferrin iron. *Biochemistry* **21**, 6503–6508.

Rosenberg, H. and Young, I. G. (1974). Iron transport in the enteric bacteria. *In* "Microbial Iron Metabolism" (Ed. J. B. Neilands), pp. 67–82. Academic Press, New York.

Stuart, S. J., Greenwood, K. T. and Luke, R. J. K. (1980). Hydroxamate-mediated transport of iron controlled by ColV plasmids. *Journal of Bacteriology* **143,** 35–42.

Warner, P. J., Williams, P. H., Bindereif, A. and Neilands, J. B. (1981). ColV plasmid-specified aerobactin synthesis by invasive strains of *Escherichia coli*. *Infection and Immunity* **33,** 540–545.

Williams, P. H. (1979). Novel iron uptake system specified by ColV plasmids: An important component in the virulence of invasive strains of *Escherichia coli*. *Infection and Immunity* **26,** 925–932.

Williams, P. H. and Warner, P. J. (1980). ColV plasmid-mediated, colicin V-independent iron uptake system of invasive strains of *Escherichia coli*. *Infection and Immunity* **29,** 411–416.

23

Detection of α-Haemolysin Production by Clinical Isolates of *Escherichia coli*

N. MACKMAN AND P. H. WILLIAMS

Department of Genetics, University of Leicester, Leicester, UK

Introduction

Haemolysins are bacterial toxins that mediate the lysis of erythrocytes. It has been shown that the α-haemolysin of *Escherichia coli*, which is secreted without accompanying lysis of the bacterial cells during the active growth of a haemolytic culture (Springer and Goebel, 1980), plays a significant, though as yet not fully defined, role in the pathogenesis of isolates from extra-intestinal infections of man (Welch *et al.*, 1981). In most of these strains the genes controlling the synthesis and secretion of the active α-haemolysin protein are located on the bacterial chromosome (Hull *et al.*, 1982), while the haemolysins commonly found among animal faecal isolates of *E. coli* are encoded by transmissible plasmids (Goebel *et al.*, 1981).

This section describes the detection and quantitation of *E. coli* α-haemolysin.

Methods

Screening for Haemolysin Production

Strains to be tested are streaked onto nutrient agar plates containing 5% (v/v) defibrinated sheep blood. To avoid haemolysis autoclaved nutrient agar should be allowed to cool to 56°C before addition of an appropriate volume of pre-warmed blood. Haemolytic strains are identified, after overnight incubation, by the presence of a clear zone of erythrocyte lysis where haemolysin secreted during growth has diffused into the agar (Fig. 1).

THE VIRULENCE OF ESCHERICHIA COLI

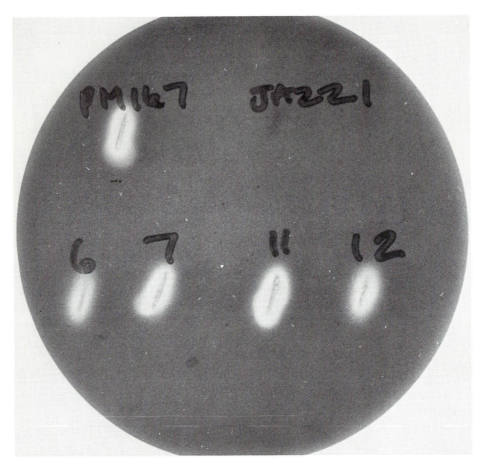

Fig. 1. Detection of *E. coli* α-haemolysin on blood agar. This example shows four haemolytic transconjugants (6, 7, 11, 12) of a mating between a non-haemolytic strain (JA221) and a strain (PM167) carrying a haemolysin plasmid.

Quantitation of Haemolytic Activity

Haemolytic activity in the supernatant fraction of exponentially growing bacterial cultures can be quantified spectrophotometrically from the amount of haemoglobin released from red blood cells under standard incubation conditions. Aliquots (200 μl) of sheep blood are centrifuged for 1 min in an Eppendorf Microfuge and the pelletted erythrocytes resuspended in 700 μl of an isotonic buffer containing 10 mM K$_3$PO$_4$, 155 mM NaCl, 60 mM CaCl$_2$, pH 6.2 (Noegel *et al.*, 1981). Isotonic solutions (200 μl) of haemolysin (e.g., nutrient broth culture supernatants, or dilutions made with the buffer defined above) are added, mixed well and incubated at 37°C for 30 min. Reactions are terminated by rapid removal of unlysed erythrocytes from the mixture by centrifugation for 2 min in a Microfuge. The cell-free fluid (800 μl) is then diluted with 3 ml water and the amount of released haemoglobin determined by measuring absorbance of the solution at 543 nm. A unit of haemolytic activity is arbitrarily defined as the amount of haemolysin protein which causes lysis of all the red blood cells in an assay mixture. Addition of 200 μl distilled water to 700 μl erythrocyte suspension prepared as described above provides a control for complete lysis.

References

Goebel, W., Noegel, A., Rdest, U., Müller, D. and Hughes, C. (1981). Structure and epidemiological spread of the haemolysin determinant of *Escherichia coli*. *In* ''Molecular Biology, Pathogenicity and Ecology of Bacterial Plasmids'' (Eds. S. B. Levy, R. C. Clowes and E. L. Koenig), pp. 43–50. Plenum, New York.

Hull, S. I., Hull, R. A., Minshew, B. H. and Falkow, S. (1982). Genetics of hemolysin of *Escherichia coli*. *Journal of Bacteriology* **151**, 1006–1012.

Noegel, W., Rdest, U. and Goebel, W. (1981). Determination of the functions of hemolytic plasmid pHly152 of *Escherichia coli*. *Journal of Bacteriology* **145**, 233–247.

Springer, W. and Goebel, W. (1980). Synthesis and secretion of hemolysin by *Escherichia coli*. *Journal of Bacteriology* **144**, 53–59.

Welch, R. A., Dellinger, E. P., Minshew, B. and Falkow, S. (1981). Haemolysin contributes to virulence of extra-intestinal *E. coli* infections. *Nature* (*London*) **294**, 665–667.

24

Testing for Carriage of Virulence Factors by Plasmids

C. R. HARWOOD

Department of Microbiology, Medical School, University of Newcastle upon Tyne, Newcastle upon Tyne, UK

I. M. FEAVERS

Department of Microbiology, University of Sheffield, Sheffield, UK

Introduction

The role of plasmids in the carriage and transmission of virulence in *Escherichia coli*, first established by Ørskov and Ørskov (1966), has been reinforced in recent years by experiments that have expanded the range and types of virulence factors that have positively been shown to be plasmid-mediated (Elwell and Shipley, 1980; see also Chapter 8). However, by no means all virulence factors are plasmid-encoded and, as a prerequisite to epidemiological and molecular biological studies, it is often necessary to establish unambiguously the location of virulence genes. The purpose of this chapter, therefore, is to outline a stepwise procedure that may be used to test plasmids for the carriage of virulence factors; the regime is illustrated in Fig. 1. The procedure is designed as a functional approach to the problem of identifying virulence plasmids but it is not a comprehensive catalogue of the methods currently available for the identification, manipulation and analysis of plasmid DNA. For more detailed reviews of these methods the reader is referred to Sherratt (1981) and Timmis (1981). To obtain unambiguous evidence that a plasmid encodes a specific phenotype is not always straightforward. A great deal depends upon the properties of the plasmid, for example its transmissibility, selectability etc., and on the convenience of tests used to detect the phenotype in question. In general, evidence for the involvement of a plasmid in the elaboration of a particular phenotype involves providing affirmative answers to the following questions (Harwood, 1980).

THE VIRULENCE OF ESCHERICHIA COLI

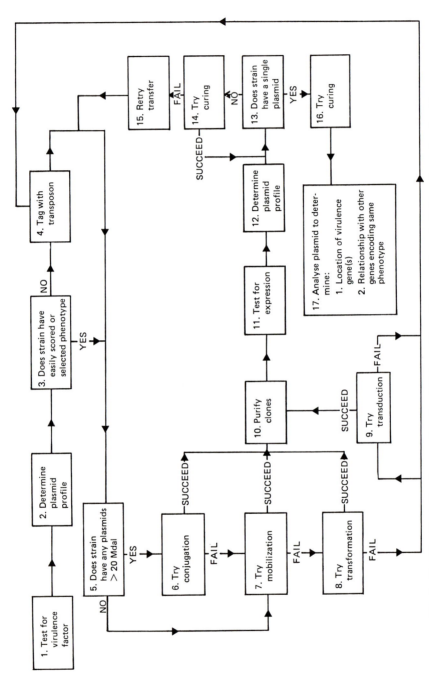

Fig. 1. Algorithm illustrating the procedure for obtaining evidence for the carriage of virulence factors by plasmids.

1. Is the phenotype additional to the characteristics of an otherwise well-defined taxon?
2. Is the phenotype known to be plasmid-mediated in representatives of the same or other bacterial taxa?
3. Is the phenotype lost irreversibly by some of the clones after repeated subculture, or after treatment with agents such as acridine, ultraviolet radiation or elevated temperature, which are known to eliminate plasmids?
4. Is the phenotype transmissible and, if so, is it transmitted in the *absence* of traits known to be chromosomally determined or in the *presence* of traits known to be plasmid-determined?

In the case of virulence factors, questions 1 and 2 are nearly always answered in the affirmative (see Chapter 8). Attempts to eliminate or cure plasmids by subculture, or with the help of curing agents and treatments, will usually answer question 3 with varying degrees of success; some plasmids are cured easily, while others are eliminated at undetectably low frequencies. In some cases it may be difficult to distinguish curing from deletion formation, particularly if the plasmid contains transposable elements. The best evidence for the involvement of a plasmid in determining a specific phenotype is to demonstrate concomitant transmissibility of plasmid and phenotype to a new host (question 4), preferably reinforced by curing data. Even with the evidence of transmissibility two important reservations should be borne in mind. The first is that the plasmid may function to control, rather than mediate, the expression of the phenotype in question, possibly as part of a generalized and not necessarily well-defined pleiotropic effect. The second is that the phenotype may be specified by a gene on a chromosomally located transposon that is only stably acquired by the plasmid prior to transfer.

Methods

The methods involved are outlined in the numbered sections below, which correspond to the stages given in Fig. 1. For the most part, the various methods are adequately described in the literature and only reference is made to these. In the cases of virulence factors with easily detected phenotypes, such as colicin V and haemolysin production, which occur on, or are associated with, conjugative plasmids that mediate efficient transfer, it may be possible to begin at stage 6.

1. Tests for Virulence Factors

Other chapters in this volume describe methods used to detect the various virulence factors associated with *E. coli*. To identify the genetical location of the genes mediating these virulence functions it is important, wherever possible, to select robust tests that are suitable for the screening of large numbers of clones. For example, we have modified the biological assay for aerobactin (Chapter 22) so that culture supernatants can be applied to the surface of the test plates in sterilized fish spines. To prevent growth of test clones on the plate, a nalidixic acid-resistant derivative of the indicator strain LG 1522 is used and nalidixic acid (50 μg ml^{-1}) is incorporated in the test agar. Although there is a slight loss of sensitivity, the modification permits the simultaneous testing of several hundred clones.

In mating experiments, it is important to appreciate that the expression of a virulence phenotype, or indeed of pathogenicity, in an 'unnatural' host may be poor or even undetectable. For example Binns *et al.* (1979) used *E. coli* strain KH933, a strain isolated from a case of septicaemia, rather than strain K-12, for the expression of serum resistance in pathogenicity tests. Similarly, difficulties have been found in obtaining expression of CFA and invasiveness in *E. coli* K-12 (see Chapter 4). Clearly, the advantages of using the genetically well-characterized K-12 strain should be balanced against the possible limitations of expression that may not apply to 'natural' isolates such as KH933.

2. Determination of Plasmid Profiles

Screening for covalently closed circular (ccc) plasmid DNA is, nowadays, straightforward and requires only relatively inexpensive equipment. A wide variety of methods are available; most involve the preparation of lysates from small volumes of starting material ('minilysates'), followed by the electrophoretic separation of the different species of plasmid molecules in agarose. The plasmid molecules separate on the basis of size and molecular configuration, and are visualized by staining with the intercalating dye ethidium bromide (Meyers *et al.*, 1976). A problem with this method is the presence of open circular (oc) forms of the plasmids which migrate independently of the ccc forms to which they are related and may make the interpretation of the plasmid profile difficult. If necessary, oc bands can be identified by electrophoresis of the lysates in two different agarose concentrations or by irradiating the separate bands with short-wave-length (<280 nm) ultraviolet radiation and running in a second direction at right angles to the first (Hintermann *et al.*, 1981). In the latter case, newly formed oc bands of a particular plasmid are detected which, together with the original ccc and oc bands, form the corners of a readily identifiable triangle.

A large number of rapid screening methods is available for determining the plasmid profile of strains of *E. coli*. The most widely used method is that of Birnboim and Doly (1979) in which the majority of proteins and chromosomal DNA are removed by denaturation with alkaline SDS, followed by renaturation in the presence of a high salt concentration. This method is broadly applicable to many Gram-negative and some Gram-positive strains, although in the latter case lysozyme treatment is better carried out at 37°C. Starting material may be 1–2 ml of an overnight broth culture or a colony or two scraped from the surface of an agar plate. In some cases the Birnboim–Doly method may prove unsatisfactory, particularly for high-molecular-mass plasmids, and other methods which have been successfully applied to virulence plasmids may have to be tried (Casse *et al.*, 1979; Eckhardt, 1978; Kado and Liu, 1981). Failure to detect a plasmid by any of the above screening procedures may be due to (1) the physical properties of the plasmid, i.e., binding to the cell envelope or its large size, or (2) its episomal integration into the chromosome of the host.

3. Antibiotic Resistance and Other Easily Detected Phenotypes

The transmissibility of plasmids is extremely variable. This variability is due to a number of factors, one of the most important of which is the mode of transmission. Generally, conjugal transfer is the most efficient and, if a high-frequency transfer (HFT) system can be established, a plasmid may spread epidemically through a population of suitable recipients (Stocker *et al.*, 1963; Falkow, 1975). In many cases, however, the transfer frequency is often as low as 10^{-3} transconjugants per recipient, and may be orders of magnitude lower. With transformation the transfer frequency is likely to be 10^{-4} to 10^{-7} transformants per recipient. Consequently, in order to demonstrate the transfer of a virulence factor whose phenotype does not allow direct selection, it may be necessary to screen several thousand or even tens of thousands of transcipient clones. For virulence factors such as adhesion or iron uptake this can be a mammoth undertaking.

Some virulence factors are often found to be linked with phenotypes that are either selectable or easily detected. For example, the genetical determinants of the K88 antigens are frequently associated with genes determining raffinose utilization (Smith and Parsell, 1975; Schmitt *et al.*, 1979). Raffinose utilization can be selected directly and Raf$^+$ clones tested for K88 production. Similarly, aerobactin-mediated iron uptake is frequently found in association with production of colicin V. In mating experiments, colV-containing strains can be detected with a sandwich-plate method. In other cases the virulence factors may be located on R plasmids; it is, therefore, worthwhile testing the host strains for resistance to antibiotics that are commonly plasmid-mediated, for example ampicillin, chloramphenicol, kanamycin, streptomycin and tetracycline. For methods see Anderson and Threlfall (1974). Resistance to nalidixic acid appears never to be plasmid-mediated and, in combination with a nalidixic acid-resistant recipient, provides a useful agent for counter-selecting donors in mating experiments. Other antibiotics can be used to counter-select donors, as can agents such as colicin E2 (McConnell *et al.*, 1980) and bacteriophages such as T6 (Willetts, 1974).

4. Phenotypic Tagging with Transposons

In strains where plasmid-mediated virulence is suspected but where no antibiotic resistance or other easily detected phenotypes are observed, it is probably worthwhile attempting to tag host cell plasmids phenotypically with a transposon mediating antibiotic resistance. The transposon then provides a selectable phenotype for subsequent transfer experiments. There are a variety of regimes available for transposon tagging and a review of currently available methods is outside the scope of this chapter. However, a few methods that have been successfully applied to virulence factor will be discussed. Most phenotypic tagging methods have been developed for use with *E. coli;* however, some are applicable to other Gram-negative species with very little modification. It should also be possible to develop analogous methods for Gram-positive species.

To tag plasmids phenotypically in their natural host, rather than after transfer to a laboratory strain, two methods are frequently used. One involves the use of transposons located on plasmids that are temperature-sensitive for replication (Robinson *et al.,* 1980); the other involves the use of transposons located on bacteriophage lambda (λ) (Berg, 1977; Berg *et al.,* 1978). We have used a derivative of plasmid RP1, called pMR5 (Robinson *et al.,* 1980), which is temperature-sensitive for replication (Rep$_{ts}$), and developed its potential as a transposon-donating vector. RP1 (Ap, Tc, Km) and its derivatives mediate conjugal transfer between a broad range of Gram-negative species (Jacoby and Shapiro, 1977). Since Tn1 (Ap) of RP1 transposes at a relatively low frequency, we have inactivated the Km resistance gene of pMR5 with Tn7 (Sm, Tp) and introduced Tn5 (Km), which tranposes at a high frequency. The resulting plasmid, pIF21, is a versatile donor of Tn3, Tn5 and Tn7 in a temperature-sensitive (Rep$_{ts}$) background (I. M. Feavers and C. R. Harwood, unpublished data).

The protocol for the use of pIF21 involves its transfer to the virulent strain by patch-mating (see Section 6) at 30°C and selecting transconjugants exhibiting the required transposon phenotype at 42°C. The transconjugants should be sensitive to the other antibiotic resistances determined by pIF21, particularly Tc, which is not transposon-mediated in this plasmid. The newly acquired transposon is not necessarily located on the virulence plasmid and consequently a number of clones should be tested in subsequent transfer experiments.

A number of other plasmids have been used to tag virulence plasmids phenotypically, including F$_{ts}$ 114 *lac*::Tn3/Tn5/Tn10 (Sansonetti *et al.,* 1981), F *proAB lac*::Tn5, F-T::TnA (Smith *et al.,* 1982; McConnell *et al.,* 1981) and R1–19K^{-} (McConnell *et al.,* 1980). In some cases it has been possible to combine phenotypic tagging with mobilization (see Section 7).

Bacteriophage λ vectors have also been used as transposon donors, by means of a regime devised originally by Berg (1977). Phage λ *b*221, *c*I875::Tn5 is used to infect the host, selecting for kanamycin resistance. The kanamycin-resistant clones are tested for lysogeny with λ *b*221,*c*I857. Non-lysogenic, kanamycin-resistant clones are then tested for production of the virulence phenotype and positive clones are used in subsequent transfer experiments (Reis *et al.*, 1980; Stark and Shuster, (1982).

5. *Plasmid Size*

When considering which gene transfer mechanism to try first, it is worth taking into account the range of plasmid sizes found in the virulent strain. The sizes may be estimated from agarose gels if molecular size reference plasmids are run at the same time (Meyers *et al.*, 1976; Macrina *et al.*, 1978). Conjugation between Gram-negative bacteria is mediated by plasmids with molecular sizes in excess of 20 Mdal, consequently strains containing only plasmids smaller than 20 Mdal are unlikely to act as donors in conjugation. Small plasmids are only likely to be transmitted by mobilization (Section 7), transformation (Section 8) or possibly by transduction (Section 8). Many, but by no means all, strains with larger plasmids may act as donors in conjugation, provided the necessary growth conditions are applied (see Section 6). In strains with conjugative plasmids, small coexisting non-conjugative plasmids may be mobilized and transferred by conjugation to a suitable recipient (see Section 7).

6. *Conjugation*

The success of conjugation depends upon a number of factors, including the growth temperature and composition of media, the physiological state of the donor and recipient, the presence of restriction and modification systems and the host range of the mediating plasmid. When designing a conjugal mating experiment particular attention should be paid to the choice of recipient; no universally suitable recipients are available, particularly for the transfer and subsequent expression of virulence factors. In many cases, the recipient of choice is likely to be an F⁻, restriction-deficient derivative of the genetically well-characterized *E. coli* strain K-12. However, as was discussed earlier (Section 1), strain K-12 is not always a suitable host for expressing virulence phenotypes or pathogenicity; therefore, natural isolates like KH933 (Binns *et al.*, 1979) may have to be used. When a non-laboratory strain is used as recipient it is usually necessary first to isolate an antibiotic-resistant derivative so that donor cells can be counter-selected. In the case of nalidixic acid resistance, mutants are usually obtained by spreading 0.2 ml of an overnight culture onto the surface of a nutrient plate containing nalidixic acid (50 μg ml⁻¹) and incubating at 37°C overnight.

With *E. coli* and its relatives, conjugation may occur in liquid cultures or on the surface of an agar plate, depending on the type of conjugative plasmid involved. The protocols for conjugation in liquid cultures tend to vary a little from one group of workers to another. The method we use is to grow both donor and recipient to stationary phase in nutrient broth at 37°C. The donor is then grown to exponential phase by diluting 10-fold into fresh broth and incubating for 90 min at 37°C. Samples (0.5 ml) of donor and recipient are then mixed with 9 ml of pre-warmed broth and incubated at 37°C for 1–16 h. The higher proportion of recipients maximizes contacts between donors and recipients, and reduces unproductive donor-to-donor contacts. When an HFT system is required the method of Meynell and Meynell (1970) may be used. After the mating period the mixture is diluted and inoculated onto selective plates containing an antibiotic to counter-select the donor and, if possible, a second antibiotic to distinguish recipients from transconjugants. In some cases it may be possible to inoculate the mating mixture onto plates that allow this distinction to be made directly, for example blood plates in the case of haemolysin production. Alternative liquid mating procedures are described by Sansonetti et al. (1981) and McConnell et al. (1981).

Many plasmids found in E. coli will mediate conjugation efficiently only on solid surfaces, particularly those mediating rigid sex pili (Bradley et al., 1980). In these cases drops of donor and recipient cultures are mixed together on the surface of a non-selective nutrient plate and incubated overnight. The following day material is scraped from the surface of the mating patch, diluted and then plated onto selective media (Sansonetti et al., 1981).

7. Mobilization

Although not themselves capable of mediating conjugation, many plasmids by virtue of possessing mob (mobilization) genes are capable of conjugal transfer if mobilized by a co-existing conjugative plasmid. Many such non-conjugative virulence plasmids are isolated in strains already possessing conjugative plasmids; however, others can be mobilized with well-characterized conjugative plasmids. This situation lends itself to the establishment of an HFT system (Stocker et al., 1963). A plasmid frequently used for mobilization is R1drd19 (R1–19) (McConnell et al., 1980, 1981; Willshaw et al., 1982). This is a derivative of R1 (Ap, Km, Cm, Su, Sm) derepressed for plasmid transfer (Meynell and Datta, 1967). Since R1–19 is transferred at high frequency (10–100%) most transconjugants will contain this plasmid in addition to any mobilized plasmids. This may not always be desirable, particularly since R1drd19 confers serum resistance (Taylor and Hughes, 1978). Mobilization can also be carried out with pMR5 and pIF21 (see Section 4) and, if necessary, combined with phenotypic tagging. For example, we have introduced pIF21 into a uropathogenic strain and, in a single experiment, achieved mobilization and transposon tagging (I. M. Feavers and C. R. Harwood, unpublished data). Plasmid pIF21 was introduced into the uropathogenic strain by patch-mating at 30°C and purified clones similarly mated with a restriction-deficient K-12 strain. Selection was made at 42°C for Tn1 (Ap) or Tn5 (Km) while counter-selecting the uropathogenic donor. The resulting transconjugants were checked for Tc sensitivity to ensure that the phenotype was not due to loss of Rep_{ts}. All Tc^s clones were found to have acquired plasmids from the uropathogenic strain.

In some instances mobilization results from co-integration of the mobilizing and mobilized plasmid, a situation that was found when the temperature-sensitive conjugative plasmid Rts was used to mobilize a non-conjugative plasmid confering mannose-resistant haemagglutination (MR-HA) (McConnell *et al.*, 1981).

8. Transformation

E. coli and related enterobacteria are not normally transformable; however, Cohen *et al.* (1973) developed a method for inducing *E. coli* to take up molecules of ccc DNA. The method involves heat-shocking cells in the presence of Ca^{2+} to disrupt the cell envelope. This procedure for inducing competence artificially has been studied extensively and variously modified. At best, however, only about 1 in 10^3 recipients is transformed and in most cases the frequency is much lower. Only ccc DNA is effectively transferred and the efficiency of transformation decreases with increasing size, so that molecules greater than 20 Mdal are only rarely transferred. Consequently, transformation is mainly of value for transferring small, easily selected virulence plasmids. However, because competent cells often take up more than a single copy of DNA, co-transformation with a drug resistance plasmid can be used to transfer non-selectable plasmids, as with CFA/II-ST-LT (Penaranda *et al.*, 1980).

Plasmid DNA for transformation can be prepared on a large scale (Maniatis *et al.*, 1982, see pp. 86–96) or can be prepared by a small-scale rapid method like that of Birnboim and Doly (1979). Various modifications to the original method (Cohen *et al.*, 1973) have been devised to improve the transformation frequency (Brown *et al.*, 1979; Kushner, 1978; Maniatis *et al.*, 1982, pp. 249–255). Transformation may also be carried out in conjunction with phenotypic tagging by adapting the methods in Section 7, since transposon-donating plasmids like R1-19, pMR5 and pIF21 are too large to be transformed efficiently.

9. Transduction

Some plasmids may be transferred by bacteriophage-mediated transduction. The efficiency of transduction is low (10^{-5}–10^{-6}) and, since phage coats containing small plasmids may contain other molecules of DNA, some of which may be chromosomal in origin (Ubelaker and Rosenblum, 1978), the results of transduction may be difficult to interpret. Transduction may be of value for the transfer of plasmids between Gram-positive species (Iordanescu, 1977); however, for the transfer of *E. coli* virulence plasmids it is probably not worth considering until at least the second time around the 'circle line' in Fig. 1.

10. Purification of Clones

Transcipients growing on selective plates should be purified before their phenotypes are confirmed. This is done by streaking at least twice for single colonies, under selective conditions where appropriate. However, in at least one purification non-selective medium should be used, to facilitate the detection of contaminants whose growth is prevented under selective conditions (Krieg, 1981).

11. Expression of Virulence Phenotypes in Transcipients

Purified transcipient clones are tested for expression of the virulence phenotype. As discussed in Section 1, the choice of recipient is likely to affect the expression of the virulence phenotypes, and if the results of initial tests prove negative, it may be necessary to use more sensitive tests, when available. Plasmid stability may also be an important factor in the level of expression, and wherever possible the transcipients should be grown under selective conditions. This may not always be possible where the presence of antibiotics may affect the growth of the indicator strains in some of the tests (i.e., aerobactin test; see Chapter 7).

12/13. Plasmid Profiles of Transcipients

Plasmid profiles of purified transcipients are determined by methods outlined in Section 2. Where two or more plasmids are transferred, the profiles of several independent isolates may contain sufficient information to allow preliminary assignment of phenotypes to specific plasmids. Generally, the size of plasmids observed in the transcipients should correlate with those in the donor. Deviations from this correlation can occur in cases of (1) phenotypic tagging, where the size of a tagged plasmid is increased by the size of the transposon, and (2) mobilization, where co-integrates may form between the mobilizing and mobilized plasmids (McConnell *et al.*, 1981). In one case we found a transcipient clone that contained copies of a donor plasmid *and* its transposon-tagged derivative. When selection for the transposon was applied, both plasmids were stably maintained. Clones containing only the transposon-tagged plasmid were obtained by re-transforming, and selecting for the transposon.

14/16. Elimination of Plasmids

Strains that have simultaneously acquired a single plasmid and the virulence phenotype may be analysed directly (Section 17). However, as additional, confirmatory evidence for plasmid carriage it is often worthwhile attempting to cure the plasmid and monitoring concomitant loss of the virulence phenotype. Strains that have acquired more than one plasmid along with the virulence phenotype may require additional treatment to identify the virulence plasmid (see Section 12/13). In this case curing may provide the necessary information by eliminating plasmids not concerned with the phenotype, or by eliminating the virulence plasmid and its phenotype. A wide variety of methods are available for eliminating plasmids; however, in practice results are highly variable from one host–plasmid combination to another. Agents that have been used for curing include ethidium bromide (Bouanchaud et al., 1969), acridine orange (Hohn and Korn, 1969), acriflavine (Mitsuhashi et al., 1961), sodium dodecyl sulphate (SDS) (Inuzuka et al., 1969; Sonstein and Baldwin, 1972) and elevated temperature (Terawaki et al., 1967).

15. Retransfer

In cases where transcipients still have more than one plasmid, it may be necessary to carry out retransfer experiments for subsequent analysis in order to eliminate interfering plasmids. This may be carried out as indicated in Sections 5–8, phenotypically tagging as necessary. Individual plasmids can be physically isolated from mixtures of plasmids by electrophoresis. The plasmid DNA is extracted and is then electrophoresed through agarose to separate the plasmid species. The band containing the required plasmid is cut out and the DNA eluted (Maniatis et al., 1982, see pp. 164–170). The eluted plasmid must be reintroduced by transformation (Section 8).

17. Analysis of Virulence Plasmids

Further confirmation that a plasmid specifies a particular virulence factor can be obtained by detailed analysis of the plasmid.

(i) Location of virulence genes. Transpositional mutagenesis may be used to determine the position of the virulence structural genes and their controlling elements. Methods outlined in Sections 4 and 7 may be used, or the plasmid may be transferred to a strain containing a chromosomally located transposon. In all cases, transposition is detected after retransfer, the transcipients being screened for reduced expression or complete loss of the virulence phenotype. The position of the transposon, and by implication sequences involved in virulence, is deter-

mined by restriction mapping. In the case of large virulence plasmids it may be necessary to clone the virulence genes before transpositional mutagenesis.

(ii) Relationship with other virulence plasmids. Relationships with other plasmids encoding the same virulence phenotype can be determined in a variety of ways. Relationships with similarly sized plasmids may be determined simply by comparing restriction maps. In other cases it may be necessary to carry out hybridization studies. "Dot–blot" hybridization (Kafatos *et al.,* 1979) is a particularly valuable method for examining the relationships of a large number of plasmids. Briefly, the method involves isolating the plasmid DNA from test strains by a rapid method (i.e., Birnboim and Doly, 1979) and immobilizing on a nitrocellulose membrane. The membrane is then treated under hybridizing conditions with labelled probe DNA, and strains containing DNA related to the probe are identified by autoradiography. The specificity of this test depends primarily on the type of probe used. If whole plasmid DNA is used the specificity is rather poor; however, if restriction fragments containing mainly or exclusively DNA from the gene(s) specifying virulence are used, or better still if an oligonucleotide synthesized with DNA sequence data is used, the specificity of hybridization approaches 100%. Consequently, dot–blot hybridization can be used in place of phenotypic tests to screen natural isolates for plasmid-encoded virulence factors.

Discussion

The regime illustrated in Fig. 1 represents a practical, stepwise protocol that can be followed to provide evidence that a particular virulence factor is plasmid-determined. The regime is designed to anticipate many more situations than will apply in individual cases. With many virulence plasmids, successful genetic approaches have apparently been initiated at stages 6 and 7, and it may well be worth trying 'shot-gun' conjugation and/or mobilization as a preliminary.

The regime is designed to provide positive evidence for or against the involvement of plasmids in determining virulence factors. It should be remembered, however, that evidence showing the involvement of chromosomal genes is just as relevant and in some cases may be easier to obtain, either from Hfr studies (Hull *et al.,* 1982) or direct cloning from chromosomal DNA (Hull *et al.,* 1981).

References

Anderson, E. S. and Threlfall, E. J. (1974). The characterisation of plasmids in the enterobacteria. *Journal of Hygiene* **72,** 471–487.
Berg, D. E. (1977). Insertion and excision of the transposable kanamycin resistance determinant

Tn5. *In* "DNA Insertion Elements, Plasmids and Episomes" (Eds. A. I. Bukhari, J. A. Shapiro and S. L. Adhya), pp. 205–212. Cold Spring Harbor Laboratory, Cold Spring Harbor, New York.

Berg, D. E., Jorgensen, R. and Davis, J. (1978). Transposable kanamycin-neomycin resistance determinants. *In* "Microbiology—1978" (Ed. D. Schlessinger), pp. 13–15. American Society for Microbiology, Washington, D.C.

Binns, M. M., Davies, D. L. and Hardy, K. G. (1979). Cloned fragments of the plasmid ColV,I-K94 specifying virulence and serum resistance. *Nature (London)* **279**, 778–781.

Birnboim, H. C. and Doly, J. (1979). A rapid alkaline extraction procedure for screening recombinant plasmid DNA. *Nucleic Acids Research* **1**, 1515–1523.

Bouanchaud, D. H., Scavizzi, M. R. and Chabbert, Y. A. (1969). Elimination by ethidium bromide of antibiotic resistance in enterobacteria and staphylococci. *Journal of General Microbiology* **54**, 417–425.

Bradley, D. E., Taylor, D. E. and Cohen, D. R. (1980). Specification of surface mating systems among conjugative drug resistance plasmids in *Escherichia coli* K-12. *Journal of Bacteriology* **143**, 1466–1470.

Brown, M. G. M., Weston, A., Saunders, J. R. and Humphreys, G. O. (1979). Transformation of *Escherichia coli* C600 by plasmid DNA at different phases of growth. *FEMS Microbiology Letters* **5**, 219–222.

Casse, F., Boucher, C., Julliot, J. S., Michel, M. and Denarie, J. (1979). Identification and characterization of large plasmids in *Rhizobium meliloti* using agarose gel electrophoresis. *Journal of General Microbiology* **113**, 229–242.

Cohen, S. N., Chang, A. C. Y. and Hsu, L. (1973). Non-chromosomal antibiotic resistance in bacteria: Genetic transformation of *Escherichia coli* by R factor DNA. *Proceedings of the National Academy of Sciences of the U.S.A.* **69**, 2110–2114.

Eckhardt, T. (1978). A rapid method for the identification of plasmid deoxyribonucleic acid in bacteria. *Plasmid* **1**, 584–588.

Elwell, L. P. and Shipley, P. L. (1980). Plasmid-mediated factors associated with virulence of bacteria to animals. *Annual Review of Microbiology* **34**, 465–496.

Falkow, S. (1975). "Infectious Multiple Drug Resistance." Pion, London.

Harwood, C. R. (1980). Plasmids. *In* "Microbiological Classification and Identification" (Eds. M. Goodfellow and R. G. Board), pp. 27–53. Academic Press, London.

Hintermann, G., Fischer, H. M., Crameri, R. and Hütter, R. (1981). Simple procedure for distinguishing *ccc, oc* and L forms of plasmid DNA by agarose gel electrophoresis. *Plasmid* **5**, 371–373.

Hohn, B. and Korn, D. (1969). Cosegregation of a sex factor with the *Escherichia coli* chromosome during curing by acridine orange. *Journal of Molecular Biology* **45**, 385–389.

Hull, R. A., Gill, R. E., Hsu, P., Minshew, B. H. and Falkow, S. (1981). Construction and expression of recombinant plasmids encoding type I or D-mannose resistant pili from a urinary tract infection *Escherichia coli* isolate. *Infection and Immunity* **33**, 933–938.

Hull, S. I., Hull, R. A., Minshaw, B. H. and Falkow, S. (1982). Genetics of hemolysin of *Escherichia coli*. *Journal of Bacteriology* **151**, 1006–1012.

Inuzuka, N., Nakamura, S., Inuzuka, M. and Tomoeda, M. (1969). Specific action of sodium dodecyl sulfate on the sex factor of *E. coli* K-12 and Hfr strains. *Journal of Bacteriology* **101**, 827–839.

Iordanescu, S. (1977). Relationship between cotransducible plasmids in *Staphylococcus aureus*. *Journal of Bacteriology* **129**, 71–75.

Jacoby, G. A. and Shapiro, J. A. (1977). Appendix B: Bacterial plasmids *c* Plasmids studied in *Pseudomonas aeruginosa* and other pseudomonads. *In* "DNA Insertion Elements, Plasmids and Episomes" (Eds. A. I. Bukhari, J. A. Shapiro and S. L. Adhya), pp. 539–656. Cold Spring Harbor Laboratory, Cold Spring Harbor, New York.

Kado, C. I. and Liu, S. T. (1981). Rapid procedure for detection and isolation of large and small plasmids. *Journal of Bacteriology* **145**, 1365–1373.

Kafatos, F. C., Jones, C. W. and Efstradiatis, A. (1979). Determination of nucleic acid sequence homologies and relative concentrations by a dot hybridisation procedure. *Nucleic Acids Research* **7**, 1541–1552.

Krieg, N. R. (1981). Enrichment and Isolation. *In* "Manual of Methods for General Bacteriology" (Eds. P. Gerhardt, R. G. E. Murray, R. N. Costilow, E. W. Nester, W. A. Wood, N. R. Krieg and G. B. Phillips), pp. 112–142. American Society for Microbiology, Washington, D.C.

Kushner, S. R. (1978). Improved methods for transformation of *Escherichia coli* with colE1 derived plasmids. *In* "International Symposium on Genetic Engineering" (Eds. H. W. Boyer and S. Micosia), pp. 17–23. North-Holland Publ., Amsterdam.

McConnell, M. M., Smith, H. R., Willshaw, G. A., Scotland, S. M. and Rowe, B. (1980). Plasmids coding for heat-labile enterotoxin production isolated from *Escherichia coli* 078: Comparison of properties. *Journal of Bacteriology* **143**, 158–167.

McConnell, M. M., Smith, W. R., Willshaw, G. A., Field, A. M. and Rowe, B. (1981). Plasmids coding for colonization factor antigen I and heat-stable enterotoxin production isolated from enterotoxigenic *Escherichia coli:* Comparison of their molecular properties. *Infection and Immunity* **32**, 927–936.

Macrina, F. L., Kopecko, D. J., Jones, K. R., Ayers, D. J. and McCowan, S. M. (1978). A multiple plasmid-containing *Escherichia coli* strain: A convenient source of size reference plasmid molecules. *Plasmid* **1**, 417–420.

Maniatis, T., Fritsch, E. F. and Sambrook, J. (1982). "Molecular Cloning: A Laboratory Manual." Cold Spring Harbor Laboratory, Cold Spring Harbor, New York.

Meyers, J. A., Sanchez, D., Elsell, L. P. and Falkow, S. (1976). Simple agarose gel electrophoretic method for the identification and characterisation of plasmid deoxyribonucleic acid. *Journal of Bacteriology* **127**, 1529–1537.

Meynell, E. and Datta, N. (1967). Mutant drug resistance factors of high transmissibility. *Nature (London)* **214**, 885–887.

Meynell, G. G. and Meynell, E. (1970). "Theory and Practice in Experimental Bacteriology," pp. 256–294. Cambridge University Press, London and New York.

Mitsuhashi, S., Harada, K. and Kameda, M. (1961). Elimination of transmissible drug resistance by acriflavine. *Nature (London)* **189**, 947–949.

Ørskov, I. and Ørskov, F. (1966). Episome-carried surface antigen K88 of *Escherichia coli. I.* Transmission of the determinant of the K88 antigen and influence on the transfer of chromosomal markers. *Journal of Bacteriology* **91**, 69–75.

Penaranda, M. E., Mann, M. B., Evans, D. G. and Evans, D. J. (1980). Transfer of an ST:LT:CFA/II plasmid into *Escherichia coli* K-12 strain RR1 by cotransformation with pSC301 plasmid DNA. *FEMS Microbiology Letters* **8**, 251–254.

Reis, M. H. L., Affonso, M. H. T., Trabulsi, L. R., Mazaitis, A. J., Maas, R. and Maas, W. K. (1980). Transfer of a CFA/I-ST plasmid promoted by a conjugative plasmid in a strain of *Escherichia coli* of serotype O128ac:H12. *Infection and Immunity* **29**, 140–143.

Robinson, M. K., Bennett, P. M., Falkow, S. and Dodd, H. M. (1980). Isolation of a temperature-sensitive derivative of RP1. *Plasmid* **3**, 343–347.

Sansonetti, P. J., Kopecko, D. J. and Formal, S. B. (1981). *Shigella sonnei* plasmids: Evidence that a large plasmid is necessary for virulence. *Infection and Immunity* **34**, 75–83.

Schmitt, R., Mattes, R., Schmid, K. and Altenbuchner, J. (1979). Raf plasmids in strains of *Escherichia coli* and their possible role in enteropathogenicity. *Developments in Genetics* **1**, 199–210.

Sherratt, D. (1981). *In vivo* genetic manipulation in bacteria. *Symposium of the Society for General Microbiology* **31**, 35–47.

Smith, H. R., Willshaw, G. A. and Rowe, B. (1982). Mapping of a plasmid coding for colonization factor antigen I and heat-stable enterotoxin production, isolated from an enterotoxigenic strain of *Escherichia coli. Journal of Bacteriology* **149**, 264–275.

Smith, H. W. and Parsell, Z. (1975). Transmissible substrate-utilising ability in enterobacteria. *Journal of General Microbiology* **87**, 129–140.

Sonstein, S. A. and Baldwin, J. N. (1972). Loss of the penicillinase plasmid after treatment of *Staphylococcus aureus* with SDS. *Journal of Bacteriology* **109**, 262–265.

Stark, J. M. and Shuster, C. W. (1982). Analysis of hemolytic determinants of plasmid pHly 185 by Tn5 mutagenesis. *Journal of Bacteriology* **152**, 963–967.

Stocker, B. A. D., Smith, S. M. and Ozeki, H. (1963). High infectivity of *Salmonella typhimurium* newly infected by the ColI factor. *Journal of General Microbiology* **30**, 201–221.

Taylor, P. W. and Hughes, C. (1978). Plasmid carriage and the serum sensitivity of Enterobacteria. *Infection and Immunity* **22**, 10–17.

Terawaki, Y., Takayasy, H. and Akiba, T. (1967). Thermosensitive replication of a kanamycin resistance factor. *Journal of Bacteriology* **94**, 687–690.

Timmis, K. N. (1981). Gene manipulation *in vitro. Symposium of the Society for General Microbiology* **31**, 49–109.

Ubelaker, M. H. and Rosenblum, E. D. (1978). Transduction of plasmid determinants in *Staphylococcus aureus* and *Escherichia coli. Journal of Bacteriology* **133**, 699–707.

Willetts, N. S. (1974). The Kinetics of inhibition of F *lac* transfer by R100 in *E. coli. Molecular and General Genetics* **129**, 123–130.

Willshaw, G. A., Smith, H. R., McConnell, M. M., Barclay, E. A., Krnjulac, J. and Rowe, B. (1982). Genetic and molecular studies of plasmids coding for colonization factor antigen I and heat-stable enterotoxin in several *Escherichia coli* serotypes. *Infection and Immunity* **37**, 858–868.

25

Measurement of the Bactericidal Activity of Serum

P. W. TAYLOR*

*Bayer AG, Pharma-Forschungszentrum, Institut für Chemotherapie, Wuppertal,
Federal Republic of Germany*

Introduction

A variety of specific and non-specific defence mechanisms act together to protect the host against microbial infections. The complement system plays an important role at many stages of the infection process by promoting the inflammatory response through the generation of chemotactic factors and anaphylatoxins, by mediating opsonization and phagocytosis and by direct killing of susceptible Gram-negative bacteria. The importance of the complement system as a component of the host defence is reflected both in its wide distribution within the animal kingdom (Ballow, 1977) and in the frequently observed increased susceptibility to infection of patients congenitally deficient in biosynthesis of individual complement components (Agnello, 1978).

The relative importance of the various complement functions in combating microbial infection is often difficult to assess and, in any case, varies with the type of infection and probably also during the course of any given infection. However, a role for the bactericidal action of complement is suggested by the frequently made observation that Gram-negative bacteria isolated from a variety of human and animal infections are more resistant to serum than comparable isolates from non-infected individuals. Thus, a high incidence of serum resistance has been associated with Gram-negative organisms causing bacteraemia (Fierer *et al.*, 1972; Simberkoff *et al.*, 1976; Vosti and Randall, 1970; Young and Armstrong, 1972), upper urinary tract infection (Gower *et al.*, 1972), disseminated gonococcal infection (Eisenstein *et al.*, 1977) and, in domesticated cattle, bovine mastitis (Carroll and Jasper, 1977).

Killing of susceptible strains is mediated by activated components of either the classical (Inoue *et al.*, 1968) or alternative (Schreiber *et al.*, 1979) complement

*Present address: Molecular Diagnostics, Inc., 400 Morgan Lane, West Haven, Connecticut 06516, U.S.A.

THE VIRULENCE OF ESCHERICHIA COLI

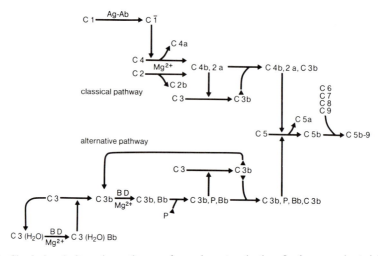

Fig. 1. Classical and alternative pathways of complement activation. In the serum bactericidal reaction, the classical pathway is activated by interaction of antibodies of the IgM or IgG class and antigenic determinants on the surface of the bacterial cell. The Fc portion of the antibody molecule can then combine with the C1q subunit of C1; this complement component is a Ca^{2+}-dependent complex of one C1q, two C1r and two C1s molecules. After binding, critical conformational changes occur in the C1q subunit, leading to conversion of the proenzymes C1r and C1s to active serine proteases C1r̄ and C1s̄. C1s̄ can cleave C4 (M_r 204,000) and C2 (M_r 98,000) to generate the C3-cleaving enzyme C4b,2a. C1s̄ can under certain conditions generate a large number of C4b,2a complexes. C4b, 2a cleaves C3 and the larger cleavage product C3b (M_r 181,000) covalently binds to the activating surface through a short-lived reactive carboxyl group. When this happens in the vicinity of a C4b,2a complex, a new complex, C4b,2a,3b, is formed. The new enzyme function cleaves C5 (M_r 206,000) and this event initiates a self-assembling process, the membrane attack pathway, which results in the formation of a stable supramolecular complex involving C5b, C6 (M_r 128,000), C7 (M_r 121,000), C8 (M_r 153,000) and C9 (M_r 79,000). Activation of the alternative pathway takes place when factor D (M_r 24,000), a serine protease circulating in active form, cleaves factor B (M_r 100,000) molecules in an Mg^{2+}-dependent, transient association with C3 to form small amounts of the C3-cleaving enzyme C3Bb; this complex produces C3b in normal blood or serum at a low rate. This C3b may associate transiently with B, which, when cleared by D, yields C3b,Bb, also a C3b-cleaving enzyme. The two enzyme complexes C3,Bb and C3b,Bb are unstable and the subunits normally dissociate spontaneously. However, C3b,Bb may complex with factor P (properdin M_r 184,000) in the presence of Mg^{2+} to form a stable 'C3 convertase', which may then bind an additional C3b fragment to convert to C3b,P,Bb,C3b, an enzyme with C5-cleaving activity. Under physiological conditions, the alternative route to C5b-9(m) generation is subject to strong negative control by βIH and C3bINA, but on certain bacterial surfaces C3b,Bb,P can readily be formed because some surface configurations protect the enzyme against βIH- and C3bINA-mediated negative control. (From Taylor, 1983.)

pathway (Fig. 1). Activation of the classical pathway is normally effected by an antigen–antibody interaction at the cell surface involving IgM or IgG molecules directed against any of a number of exposed bacterial antigens (Robbins *et al.*, 1965; Schulkind *et al.*, 1972), although certain rough enterobacteria may apparently be killed following antibody-independent classical pathway activation (Betz and Isleker, 1981; Betz *et al.*, 1982). Current evidence suggests that,

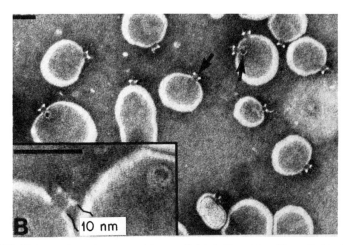

Fig. 2. C5b-9(m) complexes incorporated into artificial lipid vesicles. The cylindrical complex projects 10 nm exterior to the membrane. Scale bar represents 50 nm. Electron micrograph courtesy of Drs. S. Bhakdi, Giessen, German Federal Republic, and J. Tranum-Jensen, Copenhagen, Denmark.

although the alternative pathway may be activated by insoluble immune complexes, in the bactericidal reaction the mechanism of activation is independent of antibody (Lachmann, 1979). Both pathways generate, by the routes indicated in Fig. 1, a supramolecular complex [the C5b-9(m) complex] comprising complement components C5b, C6, C7, C8 and C9 in the ratios 1:1:1:1:12–16 and as a monomeric entity (Bhakdi, 1980; Bhakdi and Tranum-Jensen, 1981). The terminal C5b-9(m) complex (Fig. 2) has been shown to be a hollow cylinder rimmed at one end by an annulus (Bhakdi and Tranum-Jensen, 1978; Tranum-Jensen *et al.*, 1978). The length of the cylinder is approximately 15 nm with a physical internal diameter of 10 nm; a short hydrophobic region of 4 nm in length located at the end of the cylinder distal to the annulus facilitates penetration of the complex into the lipid domain of target membranes (Bhakdi, 1980), resulting in the formation of lesions that either directly or indirectly kill the bacterial cell. Resistant Gram-negative bacteria escape the potentially lethal effects of serum as a result of failure of formed C5b-9(m) complexes to insert into hydrophobic domains of the outer membrane (Joiner *et al.*, 1982a,b).

    This chapter describes a simple and adaptable method for determination of serum bactericidal activity that is suitable for evaluation of the degree of intrinsic resistance of Gram-negative strains to serum and, in addition, provides a methodological basis for studies of the molecular nature of the bactericidal phenomenon.

Methodological Considerations

Measurement of serum bactericidal activity involves exposure of a bacterial population to adequate concentrations of antibody and complement, incubation at the optimum temperature for complement activity and, after suitable periods of time, determination of the absolute concentration of viable cells. The essential components of the system are, therefore, a suspension of viable bacteria, serum and a suitable buffer as diluent. There are frequent reports in the literature of bactericidal assay systems containing a variety of non-essential components that have significant and unpredictable effects on the efficiency of complement killing (DeMatteo et al., 1981; Michael and Braun, 1959) and are therefore best avoided. In order to optimize conditions for complement activation and for deposition of C5b-9(m) complexes onto the Gram-negative cell surface, certain factors must be taken into consideration.

Bacterial growth conditions. The apparent degree of sensitivity of an organism to serum is dependent on energy-yielding functions of the bacterial cell (Griffiths, 1974) and is influenced by the presence of a number of structural components at the cell surface (Taylor and Robinson, 1980). As these parameters may be markedly affected by alterations in the nature of the growth environment, it is not surprising that the serum sensitivity of Gram-negative bacteria has been found to vary during the batch culture cycle (Rowley and Wardlaw, 1958), in relation to the carbon and energy source (Melching and Vas, 1971) and as a result of changes in nutrient limitation (Taylor, 1978). In batch culture, *Escherichia coli* and other enterobacteria are more readily killed by serum when in the early exponential phase than in either the lag or stationary phases (Davis and Wedgwood, 1965; Rowley and Wardlaw, 1958). Similarly, resistance determinants may be completely expressed only when bacteria are grown in nutritionally complex media (Maaloe, 1948a,b). These observations make it essential that particular attention be given to the procedure for inoculum preparation for serum bactericidal assays. Cells in batch culture should be harvested during the early exponential phase of growth from a complex medium that can be formulated with a minimal amount of batch-to-batch variation. We routinely use Müller–Hinton broth (Difco Laboratories, Detroit), which gives excellent and reproducible results.

Harvesting and washing of cells. Complex media based on meat extracts or infusions destroy the biological activity of complement (Muschel and Treffers, 1956) and, therefore, before addition to the bactericidal system, the cells should be washed with buffer. The number of rounds of washing and centrifugation should be kept to a minimum, because bacterial cells that have been extensively washed may show increased susceptibility to serum (Fierer et al., 1974). If cultures are grown to high density in a chemically defined medium that has no anti-complementary activity, it is possible simply to dilute the bacteria to the desired concentration before use (Taylor, 1978). The frequently adopted procedure of centrifugation at 0–4°C during washing is likely to temperature-shock the cells and should be avoided (Wright and Levine, 1981).

Buffer. Unless very high concentrations of serum are used in bactericidal assays, the buffer chosen will contribute significantly to the final pH and ionic environment of the reaction mixture. Serum and plasma have a pH between 7.31 and 7.43 (Documenta Geigy, 1960); complement activity is optimal in the pH range 7.15–7.35 (Mayer, 1960). The pH of stored serum may rise significantly because of the loss of CO_2 and it should be borne in mind that, if relatively high concentrations of serum are used, the strong buffering capacity of serum may limit the ability of a buffer system to maintain an effective pH. The optimum conditions for serum bactericidal activity have not yet been systematically defined but they should not in any case be confused with those for serum bacteriolytic activity that were established by Wardlaw (1962). Gelatin–veronal-buffered saline plus Mg^{2+} and Ca^{2+}, pH 7.35, provides the essential divalent cations for complement activity (Fig. 1) at optimum concentrations, does not affect deleteriously the viability of Gram-negative bacteria and provides an environment in which very high rates of serum killing of susceptible bacteria can be achieved (Schreiber *et al.*, 1979; Wright and Levine, 1981).

Serum. Unless the investigator is specifically interested in the role of the complement system in disease, serum should be obtained from the blood of healthy individuals and used either fresh or after storage of small aliquots at $-20°$ or $-80°C$ for periods up to 1 month. Apparent serum resistance may be due to absence of an activation mechanism for classical pathway functions; this may be particularly important when testing some strains of *E. coli,* because capsular *N*-acetylneuraminic acid (NeuAc)-containing polymers, frequently associated with the surface of this species (Ørskov *et al.*, 1977), prevent effective activation of the alternative pathway (Stevens *et al.*, 1978). Early colonization of the intestinal tract by commensal bacteria ensures that small quantities of antibodies directed against surface antigens of *E. coli* (Cohen and Norins, 1966) and other enterobacteria are present in blood and tissue fluids of man and a variety of animals (Kunin *et al.*, 1962). Antibodies against some overtly pathogenic species are generally absent from non-immune sera and bactericidal systems involving strains of *Vibro cholerae, Salmonella typhi* and others should be supplemented with an antibody source in the form of heated (56°C) immune serum. However, care must be taken to ensure that such systems do not contain concentrations of IgG antibody capable of inhibiting the complement-mediated killing mechanism (Norman *et al.*, 1972). Bactericidal systems containing heterogenous sources of antibody and complement should be avoided wherever possible, as interactions between serum proteins of one animal species and antibodies of another are likely to occur and lead to fixation and deviation of complement from the bactericidal reaction (Rowley, 1973).

The presence of serum antibodies against *E. coli* can be readily estimated by standard procedures (Rose and Friedman, 1980). Complement activation can be monitored by examining the conversion of C3 to C3b (Morrison and Kline,

1977) or formation and deposition of C5b-9(m) complexes on the bacterial surface (Bhakdi *et al.,* 1983).

Sera obtained from different animal species may not be comparable with regard to bactericidal activity. For example, human serum is known to function more efficiently in bactericidal systems than either rabbit or guinea-pig serum (Ogata and Levine, 1980). In a comprehensive comparative study, Schwab and Reeves (1966) found considerable variation in the ability of sera from eight vertebrate species to kill a number of enterobacterial strains. Within a given animal species, however, there is little variation in the bactericidal activity of serum from different healthy individuals; this has been particularly well established for humans (Olling, 1977).

The bactericidal action of serum is demonstrable over a wide range of serum concentrations, although the rate of killing generally increases with increasing serum concentration (Rowley, 1956). For the determination of the serum sensitivity of an *E. coli* isolate, relatively high concentrations of serum should be employed in order to ensure that killing is not limited by availability of essential complement components. In this context, the bactericidal activity of the alternative pathway has been shown to be insignificant at complement concentrations equivalent to a 1:16 dilution of serum (Schreiber *et al.,* 1979). In case of doubt, the bactericidal activity due to alternative pathway activation can be assessed by performing the assay in the presence of 0.01 M magnesium ethylene glycol-bis(β-aminoethyl ether)-N,N'-tetraacetate (Mg-EGTA); this reagent chelates Ca^{2+} and so abrogates classical pathway activity (Fine *et al.,* 1972).

Antibody is needed in such small amounts for complement activation that it is rarely a limiting factor in serum bactericidal reactions involving classical pathway activation, but it has been suggested that it may produce bacterial agglutination during the reaction and thus lead to an underestimate of survival rates (Melching and Vas, 1970). In the author's experience this effect is rarely, if ever, a problem but may be detected by inclusion of a control containing heated (56°C) serum.

Bacterial concentration. Although it has been found the type of response of *E. coli* to human serum does not vary significantly over a bacterial concentration range of 10^4–10^{10} cells/ml (Olling, 1977), it would seem advisable to keep cell concentrations in the range 10^5–10^6 cells/ml. Low cell concentrations ($<10^3$ per millilitre) will favour the occurrence of gross antibody excess and may lead to inhibition of serum bactericidal action.

Incubation and counting. It is essential to incubate assay tubes in a water-bath at the optimum temperature for complement activity (37°C in the case of human serum). Serum complement titres tend to fall after prolonged incubation and, for human serum, the reaction should not be allowed to proceed for more than 3 h; with some animal sera this period should be reduced.

As the killing of Gram-negative bacteria by serum frequently does not proceed at a constant rate, the enumeration of survivors should be made at intervals during the reaction. Viable counting would appear to be the method of choice; the variations inherent in this technique have been extensively discussed by Postgate (1969).

Method

The method presented will enable the bench worker to vary the serum concentration over a very wide range, depending on the type of investigation envisaged. In most cases, however, the aim will be to determine the serum sensitivity of clinical isolates and there is unlikely to be much information available with regard to the presence of cell surface structures that are known to influence the degree of serum resistance. It will, therefore, be of overriding importance to ensure by the use of a high serum concentration that all relevant serum proteins are present in excess.

Inoculate a 50-ml Erlenmeyer flask containing 10 ml Müller–Hinton broth (MHB; Difco) with the strain under investigation and incubate at 37°C overnight with agitation. Transfer 100 μl of the culture to 10 ml of fresh MHB and continue incubating under the same conditions until the $E_{578 \text{ nm}}$ of the culture reaches 0.5. Transfer the culture to a suitable screw-capped tube and sediment the cells at room temperature. Wash the sediment once in 10 ml of GVB^{2+} and resuspend in GVB^{2+} so as to obtain a concentration of 3×10^6 cells/ml; this is done with the aid of a previously prepared calibration curve. Add 250 μl of bacterial suspension to 500 μl of prewarmed (37°C) normal serum and mix well. Immediately remove 10 μl of reaction mixture and pipette into 0.9% NaCl solution for viable counting. Incubate the reaction mixture at 37°C in a water-bath and withdraw 10-μl samples after 1, 2 and 3 h.

With this method, the majority of E. coli strains isolated from the healthy gut or from extra-intestinal infections can be classified into three broad groups (Fig. 3). Completely *resistant* strains survive the 3-h incubation period in serum and generally show a small increase in viable count, but it is not unusual for a slight drop in viability to take place between the second and third hour. Many isolates, and particularly those unable to synthesize or attack the O-specific side-chain moiety of lipopolysaccharide (rough strains), are promptly killed by serum (*promptly sensitive*) and no viable cells are detectable after 1 h of incubation. The majority of smooth, serum-sensitive E. coli strains are killed but only after a lag period of 1 h, and are referred to as *delayed sensitive*. This delay in killing by serum is almost certainly due to the presence at the surface of long and numerous lipopolysaccharide O-side chains (Taylor and Robinson, 1980). Serum responses falling between these categories are encountered; some examples are discussed by Hughes *et al.* (1982).

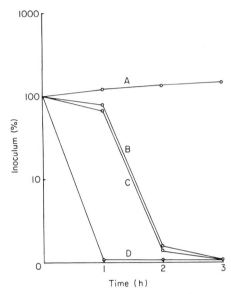

Fig. 3. Response of four *E. coli* urinary tract isolates to normal human serum. Strain A shows complete serum resistance whereas strain D is promptly sensitive. The smooth strains B and C are delayed sensitive.

The method also permits investigation of the mechanism of killing by complement. The amount of serum in the system can be varied by replacement with GVB^{2+} in order to retain the total reaction volume of 750 µl. For example, Fig. 4 shows the killing of urinary isolate LP1092 by 50 µl human serum. Heating the serum at 56°C for 30 min or adding dithiothreitol destroys the bactericidal activity of the serum. Removal of antibodies by adsorption, or addition of Mg-EGTA, considerably reduces the rate of killing, indicating that the major part of the serum bactericidal activity against LP1092 is contributed by the antibody-dependent classical pathway of complement activation.

Materials

Serum. Blood obtained from healthy volunteers should be transferred to a sterile glass or polystyrene container and allowed to clot at room temperature for 1 h. Care should be taken to ensure retraction of the clot from the walls of the container. After centrifugation at 3000 *g* for 20 min, the serum should be removed and divided into a number of small aliquots ready for immediate use or for storage as previously specified. Discard any samples containing amounts of haemoglobin visible to the naked eye.

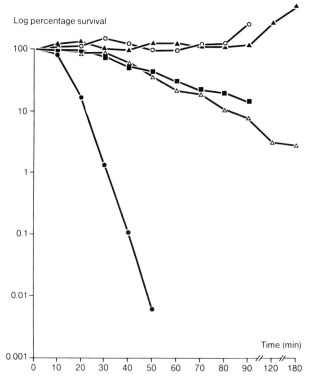

Fig. 4. Killing of *E. coli* LP1092 by 50 μl human serum in a total reaction volume of 750 μl. Symbols: ●, human serum; ○, heated (56°C, 30 min) serum; ▲, serum plus 0.01 *M* dithiothreitol; △, serum plus 0.01 *M* Mg EGTA; ■, serum absorbed twice with heat-killed (100°C, 1 h) LP1092 cells. Each point represents the mean of three independent determinations. (From Taylor and Kroll, 1983.)

Buffer. GVB^{2+} is prepared from the following stock solutions:

Veronal (5×). Dissolve NaCl 41.2 g and sodium 5,5-diethylbarbiturate 5.095 g in 700 ml H$_2$O. Adjust pH to 7.35 with 1 *M* HCl. Bring the volume to 1 litre.

Stock metals. Equal volumes of 2 *M* MgCl$_2$ and 0.3 *M* CaCL$_2$. To prepare the working solution, dissolve 1 g of gelatin in 600 ml of hot H$_2$O. Cool to room temperature, add 200 ml of 5×veronal and 1 ml stock metals. Bring to 1 litre with H$_2$O. Sterilize by filtration (0.45 μm; nitrocellulose membrane).

References

Agnello, V. (1978). Complement deficiency states. *Medicine (Baltimore)* **57**, 1–23.
Ballow, M. (1977). Phylogenetics and ontogenetics of the complement systems. *In* ''Biological

Amplification Systems in Immunology: Comprehensive Immunology," Vol. 2 (Eds. N. K. Day and R. A. Good), pp. 183–204. Plenum, New York.

Betz, S. J. and Isleker, H. (1981). Antibody-independent interactions between *Escherichia coli* J5 and human complement components. *Journal of Immunology* **127**, 1748–1754.

Betz, S. J., Page, N., Estrade, C. and Isleker, H. (1982). The effect of specific antibody on antibody independent interactions between *E. coli* J5 and human complement. *Journal of Immunology* **128**, 707–711.

Bhakdi, S. (1980). On the molecular nature of the complement lesion. *Behring Institut Mitteilungen* **65**, 1–15.

Bhakdi, S. and Tranum-Jensen, J. (1978). Molecular nature of the complement lesion. *Proceedings of the National Academy of Sciences of the U.S.A.* **75**, 5655–5659.

Bhakdi, S. and Tranum-Jensen, J. (1981). Molecular weight of the membrane C5b-9 complex of human complement: Characterization of the terminal complex as a C5b-9 monomer. *Proceedings of the National Academy of Sciences of the U.S.A.* **78**, 1818–1822.

Bhakdi, S., Muhly, M. and Roth, M. (1983). Preparation and isolation of specific antibodies to complement components. *In* "Methods in Enzymology," Vol. 93 (Eds. J. J. Langone and H. Van Vunakis), pp. 409–420. Academic Press, New York.

Carroll, E. J. and Jasper, D. E. (1977). Bactericidal activity of standard bovine serum against coliform bacteria isolated from udders and the environment of dairy cows. *American Journal of Veterinary Research* **38**, 2019–2022.

Cohen, I. R. and Norins, L. C. (1966). Natural human antibodies to Gram-negative bacteria: Immunoglobulins G, A, and M. *Science* **152**, 1257–1259.

Davis, S. D. and Wedgwood, R. J. (1965). Kinetics of the bactericidal action of normal serum on Gram-negative bacteria. *Journal of Immunology* **95**, 75–79.

DeMatteo, C. S., Hammer, M. C., Baltch, A. L., Smith, R. P., Sutphen, N. T. and Michelsen, P. G. (1981). Susceptibility of *Pseudomonas aeruginosa* to serum bactericidal activity. A comparison of three methods with clinical correlations. *Journal of Laboratory and Clinical Medicine* **98**, 511–518.

Documenta Geigy (1960). "Wissenschaftliche Tabellen," 6th ed., p. 521. J. R. Geigy AG, Basel.

Eisenstein, B. I., Lee, T. J. and Sparling, P. F. (1977). Penicillin sensitivity and serum resistance are independent attributes of strains of *Neisseria gonorrhoeae* causing disseminated gonococcal infection. *Infection and Immunity* **15**, 834–841.

Fierer, J., Finley, F. and Braude, A. I. (1972). A plaque assay on agar for detection of Gram-negative bacilli sensitive to complement. *Journal of Immunology* **109**, 1156–1158.

Fierer, J., Finley, F. and Braude, A. I. (1974). Release of ^{51}Cr-endotoxin from bacteria as an assay of serum bactericidal activity. *Journal of Immunology* **112**, 2184–2192.

Fine, D. P., Marney, S. R., Colley, D. G., Sergent, J. S. and Des Prez, R. M. (1972). C3 shunt activation in human serum chelated with EGTA. *Journal of Immunology* **109**, 807–809.

Gower, P. E., Taylor, P. W., Koutsaimanis, K. G. and Roberts, A. P. (1972). Serum bactericidal activity in patients with upper and lower urinary tract infections. *Clinical Science* **43**, 13–22.

Griffiths, E. (1974). Metabolically controlled killing of *Pasteurella septica* by antibody and complement. *Biochimica et Biophysica Acta* **362**, 598–602.

Hughes, C., Phillips, R. and Roberts, A. P. (1982). Serum resistance among *Escherichia coli* strains causing urinary tract infection in relation to O type and the carriage of hemolysin, colicin, and antibiotic resistance determinants. *Infection and Immunity* **35**, 270–275.

Inoue, K., Yonemasu, K., Takamizawa, A. and Amano, T. (1968). Studies on the immune bacteriolysis. XIV. Requirement of all nine components of complement for immune bacteriolysis. *Biken Journal* **11**, 203–206.

Joiner, K. A., Hammer, C. H., Brown, E. J., Cole, R. J. and Frank, M. M. (1982a). Studies on the mechanism of bacterial resistance to complement-mediated killing. I. Terminal complement components are deposited and released from *Salmonella minnesota* S218 without causing bacterial death. *Journal of Experimental Medicine* **155**, 797–808.

Joiner, K. A., Hammer, C. H., Brown, E. J. and Frank, M. M. (1982b). Studies on the mechanism of bacterial resistance to complement-mediated killing. II. C8 and C9 release C5b67 from the surface of *Salmonella minnesota* S218 because the terminal complex does not insert into the bacterial outer membrane. *Journal of Experimental Medicine* **155**, 809–819.

Kunin, C. M., Beard, M. V. and Kalmagyi, N. E. (1962). Evidence for a common hapten– associated with endotoxin fractions of *E. coli* and other Enterobacteriaceae. *Proceedings of the Society for Experimental Biology and Medicine* **111**, 160–166.

Lachmann, P. J. (1979). Complement. *In* "The Antigens," Vol. 5 (Ed. M. Sela), pp. 283–335. Academic Press, New York.

Maaløe, O. (1948a). Pathogenic-apathogenic transformation of *Salmonella typhimurium*. *Acta Pathologica et Microbiologica Scandinavica* **25**, 414–430.

Maaløe, O. (1948b). Pathogenic-apathogenic transformation of *Salmonella typhimurium*. II. Induced change of resistance to complement. *Acta Pathologica et Microbiologica Scandinavica* **25**, 755–766.

Mayer, M. M. (1960). Complement and complement fixation. *In* "Experimental Immunochemistry," 2nd ed. (Eds. E. A. Kabat and M. M. Mayer), pp. 133–240. Thomas, Springfield, Illinois.

Melching, L. and Vas, S. I. (1970). The effects of serum components on the agglutination of Gram-negative bacteria. *Canadian Journal of Microbiology* **16**, 121–124.

Melching, L. and Vas, S. I. (1971). Effects of serum components on Gram-negative bacteria during bactericidal reactions. *Infection and Immunity* **3**, 107–115.

Michael, J. G. and Braun, W. (1959). Modification of bactericidal effects of human sera. *Proceedings of the Society for Experimental Biology and Medicine* **102**, 486–490.

Morrison, D. C. and Kline, L. F. (1977). Activation of the classical and properdin pathways of complement by bacterial lipopolysaccharides (LPS). *Journal of Immunology* **118**, 362–368.

Muschel, L. H. and Treffers, H. P. (1956). Quantitative studies on the bactericidal actions of serum and complement. I. A rapid photometric growth assay for bactericidal activity. *Proceedings of the Society for Experimental Biology and Medicine* **76**, 1–10.

Norman, B., Stendahl, O., Tagesson, C. and Edebo, L. (1972). Characteristics of the inhibition by rabbit immune serum of the bactericidal effect of cattle normal serum on *Salmonella typhimurium* 395 MRO. *Acta Pathologica et Microbiologica Scandinavica, Section B: Microbiology and Immunology* **80B**, 891–899.

Ogata, R. T. and Levine, R. P. (1980). Characterization of complement resistance in *Escherichia coli* conferred by the antibiotic resistance plasmid R100. *Journal of Immunology* **125**, 1494–1498.

Olling, S. (1977). Sensitivity of Gram-negative bacilli to the serum bactericidal activity: A marker of the host-parasite relationship in acute and persisting infections. Scandinavian *Journal of Infectious Diseases, Supplement* **10**, 1–40.

Ørskov, I., Ørskov, F., Jann, B. and Jann, K. (1977). Serology, chemistry, and genetics of O and K antigens of *Escherichia coli*. *Bacteriological Reviews* **41**, 667–710.

Postgate, J. R. (1969). Viable counts and viability. *In* "Methods in Microbiology," Vol. 1 (Ed. J. R. Norris and D. W. Ribbons), pp. 611–628. Academic Press, London.

Robbins, J. B., Kenny, K. and Suter, E. (1965). The isolation and biological activities of rabbit γM ad γG-anti-*Salmonella typhimurium* antibodies. *Journal of Experimental Medicine* **122**, 385–402.

Rose, N. R. and Friedman, H. (1980). "Manual of Clinical Immunology," 2nd ed. American Society for Microbiology, Washington, D.C.

Rowley, D. (1956). Rapidly induced changes in the level of non-specific immunity in laboratory animals. *British Journal of Experimental Pathology* **37**, 223–234.

Rowley, D. (1973). Antibacterial action of antibody and complement. *Journal of Infectious Diseases* **128**, Supplement, 170–175.

Rowley, D. and Wardlaw, A. C. (1958). Lysis of Gram-negative bacteria by serum. *Journal of General Microbiology* **18**, 529–533.

Schreiber, R. D., Morrison, D. C., Podack, E. R. and Müller-Eberhard, H.-J. (1979). Bactericidal activity of the alternative complement pathway generated from eleven isolated plasma proteins. *Journal of Experimental Medicine* **149**, 870–882.

Schulkind, M. L., Kenny, K., Herzberg, M. and Robbins, J. B. (1972). The specific secondary biological activities of rabbit IgM and IgG anti-*Salmonella typhimurium* ''O'' antibodies isolated during the development of the immune response. *Immunology* **23**, 159–170.

Schwab, G. E. and Reeves, P. R. (1966). Comparison of the bactericidal activity of different vertebrate sera. *Journal of Bacteriology* **91**, 106–112.

Simberkoff, M. S., Ricupero, I. and Rahal, J. J. (1976). Host resistance to *Serratia marcescens* infection: Serum bactericidal activity and phagocytosis by normal blood leukocytes. *Journal of Laboratory and Clinical Medicine* **87**, 206–217.

Stevens, P., Huang, S. N.-Y., Welch, W. D. and Young, L. S. (1978). Restricted complement activation by *Escherichia coli* with the K1 capsular serotype: A possible role in pathogenicity. *Journal of Immunology* **121**, 2174–2180.

Taylor, P. W. (1978). The effect of the growth environment on the serum sensitivity of some urinary *Escherichia coli* strains. *FEMS Microbiology Letters* **3**, 119–122.

Taylor, P. W. (1983). Bactericidal and bacteriolytic activity of serum against Gram-negative bacteria. *Microbiological Reviews* **47**, 53.

Taylor, P. W. and Kroll, H. P. (1983). Killing of an encapsulated strain of *Escherichia coli* by human serum. *Infection and Immunity* **39**, 124.

Taylor, P. W. and Robinson, M. K. (1980). Determinants that increase the serum resistance of *Escherchia coli*. *Infection and Immunity* **29**, 278–280.

Tranum-Jensen, J., Bhakdi, S., Bhakdi-Lehnen, B., Bjerrum, O. J. and Speth, V. (1978). Complement lysis: The ultrastructure and orientation of the C5b-9 complex on target sheep erythrocyte membranes. *Scandinavian Journal of Immunology* **7**, 45–56.

Vosti, K. L. and Randall, E. (1970). Sensitivity of serologically classified strains of *Escherichia coli* of human origin to the serum bactericidal system. *American Journal of the Medical Sciences* **259**, 114–119.

Wardlaw, A. C. (1962). The complement-dependent bacteriolytic activity of normal human serum. I. The effect of pH and ionic strength and the role of lysozyme. *Journal of Experimental Medicine* **115**, 1231–1248.

Wright, S. D. and Levine, R. P. (1981). How complement kills *E. coli*. I. Location of the lethal lesion. *Journal of Immunology* **127**, 1146–1151.

Young, L. S. and Armstrong, D. (1972). Human immunity to *Pseudomonas aeruginosa*. I. *In vitro* interaction of bacteria, polymorphonuclear leukocytes, and serum factors. *Journal of Infectious Diseases* **126**, 257–276.

26

Detection of Antibodies and Antigens by Immunoblotting

PAULINE STEVENSON AND E. GRIFFITHS

National Institute for Biological Standards and Control, Hampstead, London, UK

Introduction

The electrophoretic blotting technique devised by Towbin *et al.* (1979) produces replicas on nitrocellulose sheets of proteins separated on polyacrylamide gels. Proteins immobilized on the nitrocellulose can then be used to detect their respective antibodies in sera or other body fluids. Conversely, separated antigens, bound to nitrocellulose, can be detected with specific antibodies. However, it should be mentioned that the nature of the binding of proteins to nitrocellulose sheets is not fully understood and not all proteins bind equally well. The possibility must also be considered that some protein antigens, denatured during sodium dodecyl sulphate (SDS)–polyacrylamide gel electrophoresis, are not renatured on transfer to nitrocellulose and may therefore not be recognized by specific antibodies. The procedures described here are those used to detect serum antibodies directed against outer-membrane proteins of *Escherichia coli*. Essentially, the method involves the electrophoresis of the outer-membrane proteins of *E. coli* in an SDS–polyacrylamide gel, the electrophoretic transfer of the separated proteins to a sheet of nitrocellulose and the use of the electrophoretic 'blot' to detect antibodies. Antibody bound by an antigen immobilized on the nitrocellulose sheet is detected by a second, labelled, antibody directed against the first antibody.

Preparation of Outer-Membrane Proteins

Procedure 1

E. coli (0.2–0.5 g wet weight) are washed in 0.15 *M* NaCl, resuspended in EDTA–Tris buffer (5 ml) and subjected to three freeze–thaw cycles followed by

THE VIRULENCE OF ESCHERICHIA COLI

sonic disruption (2 × 2 min at 0–4°C). Unbroken cells and large fragments are removed by a low-speed centrifugation (2 × 2000 g), the supernatant made up to 1 mM MgCl$_2$ and the membranes pelleted by centrifugation at 196,000 g for 30 min. The membranes are washed in 5 ml 0.1 M Tris–HCl buffer (pH 7.8), 10 mM MgCl$_2$ and the cytoplasmic membrane proteins removed by extraction of the crude envelope preparation with 5 ml Triton X-100, 0.1 M Tris–HCl buffer (pH 7.8), 10 mM MgCl$_2$ at 23°C for 10 min (Schnaitman, 1971). The Triton X-100-extracted membranes, referred to here as 'outer membranes', are collected by centrifugation, washed in 0.1 M Tris–HCl buffer (pH 7.8), 10 mM MgCl$_2$ and resuspended in 0.5–1 ml of 0.06 M Tris–HCl buffer (pH 6.8); protein concentration is estimated by the method of Lowry et al. (1951). The membrane preparation is stored at −20°C. Membrane proteins are solubilized by adding an equal amount of double-strength solubilizing agent and boiling for 5 min. This step is carried out just before applying the protein mixture to the gel for electrophoresis. Solubilized samples can be stored at −20°C and boiled again before application to the gel.

Materials

Ethylenediaminetetraacetic acid (EDTA) (1 mM) in Tris–HCl buffer (0.05 M, pH 7.8).
MgCl$_2$ (10 mM) in Tris–HCl buffer (0.1 M, pH 7.8).
Triton X-100 (2% v/v, Sigma Chemical Co. Ltd.) in Tris–HCl buffer (0.1 M, pH 7.8), MgCl$_2$ (10 mM).
MgCl$_2$ (0.1 M)
NaCl (0.15 M)
Tris–HCl buffer (0.06 M, pH 6.8)

Solubilizing solution (double strength) contains: sodium dodecyl sulphate (4% w/v), Bromphenol Blue (0.002% w/v) in Tris–HCl buffer (0.12 M, pH 6.8), glycerol (20% v/v), 2-mercaptoethanol (10% v/v) (Lugtenberg et al., 1975).

SDS–Polyacrylamide Gel Electrophoresis

Procedure 2

The gel mould is assembled according to the manufacturer's instructions and the running (separating) gel (total acrylamide concentration 11% w/v) prepared by mixing together: stock solution I, 9.38 ml; ammonium persulphate, 0.95 ml; 10% SDS, 0.75 ml; 0.75 M Tris–HCl buffer (pH 8.8), 18.75 ml; H$_2$O, 7.68 ml; $N,N,N,'N'$-tetramethylethylenediamine ($NNN'N'$-TMD), 75 μl.

The running gel is cast first and overlaid with water. When polymerization is complete a sharp interface is seen between the gel and the water layer. After removing the water, the stacking gel (total acrylamide concentration 3% w/v) is cast on top of the running gel. The stacking gel is made by mixing together: stock solution II, 1 ml; ammonium persulphate, 240 μl; 10% SDS, 100 μl; 0.25 M Tris–HCl buffer (pH 6.8), 5.0 ml; H$_2$O, 3.66 ml; NNN'N'-TMD, 20 μl.

Wells are produced in the stacking gel by pushing a plastic well-former into the stacking gel solution before it polymerizes. Care should be taken when removing the well-former. After assembling the electrophoresis apparatus as directed in the manufacturer's instructions, the electrode reservoirs are filled with the electrode buffer. Samples for electrophoresis, prepared as described in procedure 1 and cooled, are then carefully placed into the slots in the stacking gel with an automatic pipette or microsyringe. The quantity to be loaded will depend on the concentration of protein in the sample. The gel is run at a constant current of 50 mA and the progress of electrophoresis monitored by the presence of the bromphenol blue in the sample. The free dye migrates faster than most proteins. Electrophoresis is usually stopped when the dye reaches the bottom of the running gel. However, resolution of proteins of interest can sometimes be increased by running the gel for a longer period. In such cases, electrophoresis is carried out at a constant current of 10 mA overnight and a further 2–3 h at 20 mA once the Bromophenol Blue dye has migrated off the bottom of the gel. At the end of electrophoresis the gel is transferred immediately to the electrophoretic blotting apparatus. Gels which are not being processed in this way can be stained by soaking overnight at room temperature in the staining solution and destained over a 24-h period by washing with the destaining solution. This should be changed frequently and destaining continued until a clear background is obtained. Staining and destaining takes place more rapidly when the gel is shaken gently in the relevant solution on an orbital shaker.

Materials and Equipment

Power pack.
Electrophoresis tank and gel mould (available commercially).
Bulldog clips.
Silicon grease.
Stock solution I: 44% (w/v) acrylamide, 0.8% (w/v) N,N'-bisacrylamide in H$_2$O; this is stable for 6–8 wk when stored in the dark at 4°C.
Stock solution II: 30% (w/v) acrylamide, 0.8% (w/v) N,N'-bisacrylamide in H$_2$O; this is stable for 6–8 wk when stored in the dark at 4°C.
Ammonium persulphate: 1% (w/v) freshly made before use.
Sodium dodecyl sulphate (SDS): 10% (w/v); store at room temperature.

Tris–HCl buffer (0.75 M, pH 8.8).
Tris–HCl buffer (0.25 M, pH 6.8).
$N,N,N,'N'$-tetramethylethylenediamine ($NNN'N'$-TMD).
Electrode buffer: Tris (0.025 M), glycine (0.19 M), SDS (0.1% w/v); pH adjusted to 8.3 with HCl.
Staining solution: Coomassie Blue (0.025% w/v), methanol (50% v/v), acetic acid (5% v/v) in H_2O.
Destaining solution: Methanol (5% v/v), acetic acid (7.5% v/v) in H_2O.

Transfer of Separated Proteins to Nitrocellulose Paper

Procedure 3

A double sheet of 3MM filter paper, folded in half, is soaked in transfer buffer and one half of the folded paper placed on the cathode section of the Electroblot apparatus internal cassette. The polyacrylamide gel is then laid on this paper and carefully overlaid with a sheet of nitrocellulose paper previously soaked in buffer. Care should be taken to expel air bubbles at this stage. The other half of the 3MM filter paper sheet is folded over the nitrocellulose and another piece of 3MM paper, also soaked in buffer, placed on top. Scouring pads are used to fill the remaining space in the cassette and the anode section is then folded over and tightly secured. The cassette is placed in the buffer tank of the Electroblot apparatus, ensuring that this contains sufficient transfer buffer to cover the gel, and the lid fitted. After starting the buffer circulating pump, the power supply is switched on and electrophoresis carried out for $1\frac{1}{2}$ h at a current of 400–500 mA. At the end of this time, the power is switched off and the polyacrylamide gel and nitrocellulose sheet removed. The time required for transfer depends on the acrylamide concentration in the gel and on the molecular weight of the proteins being investigated. Under the conditions used here the polyacrylamide gel contains enough residual protein to produce visible bands on staining with Coomassie Blue (procedure 2). The proteins immobilized on the nitrocellulose sheet are used to detect their respective antibodies in sera (procedure 4).

Materials and Equipment

Electroblot system EC 230: (E–C Apparatus Corporation, St. Petersburg, Florida).
Nitrocellulose paper: cellulose nitrate sheets (0.45 μm); this should be large enough to cover the gel.
Whatman 3MM filter paper.

Transfer buffer: 4.8 litre of Tris (0.025 M), glycine (0.192 M) plus 1.2 litres of methanol.

Scouring pads, nylon.

Proteins electrophoretically separated on an SDS–polyacrylamide gel (procedure 2).

Detection of Serum Antibodies by Electrophoretic Blots

Procedure 4

The nitrocellulose paper, with the separated *E. coli* outer-membrane proteins attached, is placed in a plastic container and covered with haemoglobin–PBS (250 ml) and shaken for 1–2 h at room temperature on an orbital shaker. This process saturates with haemoglobin all protein binding sites on the nitrocellulose sheet not already occupied by the electrophoretically transferred proteins. After the required length of time the haemoglobin–PBS is discarded and replaced by a similar volume of haemoglobin–PBS containing a sample of the serum to be tested and the gentle shaking continued for a further 16 h; 20–200 μl of serum is usually sufficient to allow the detection of antibodies but this volume can be increased if necessary. The haemoglobin–PBS containing the serum under test is then discarded and the nitrocellulose sheet is washed over a period of 30–60 min with 6 × 250 ml of fresh haemoglobin–PBS. Haemoglobin–PBS containing ^{125}I-labelled anti-immunoglobulin (approximately 1×10^6 cpm) is added and the container shaken for 3–4 h. The nitrocellulose paper is finally washed with 6 × 250 ml of PBS–azide over a period of about 30 min and dried in an oven at 45°C for 20–30 min. The dried blot is set up for autoradiography at −70°C, as described by Laskey and Mills (1977), with a cassette fitted with an intensifying screen. The optimum exposure time is found by trial and error. If the film is overexposed, then re-exposure, but for a shorter period, is required; if the image is weak or if no image can be seen, exposure for a longer period is necessary.

Materials and Equipment

Nitrocellulose sheet with attached proteins (procedure 3).

[125]I-labelled anti-immunoglobulin (see procedure 5).

Serum to be tested: this should be centrifuged before use if a clear background is not obtained on autoradiography.

Phosphate-buffered saline containing azide (PBS–azide):
NaCl (0.14 M), Na_2HPO_4 (5.35 mM), KCl (2.7 mM), KH_2PO_4 (1.5 mM), NaN_3 (0.02% w/v).

PBS–azide containing bovine haemoglobin (3% w/v) (haemoglobin–PBS).
Plastic container to hold blot during processing.
Orbital shaker.
Protex X-ray cassettes fitted with fast tungstate screen.
X-o-Mat S film (Kodak).

Iodination of Anti-Immunoglobulin Antibodies for Immunoblotting

Procedure 5

$Na^{125}I$ (100 μCi), immunoglobulin (20 μl) and NaH_2PO_4 buffer (100 μl) are mixed together in a disposable tube. Chloramine-T (10 μl) is added and the mixture incubated at room temperature for 45 s. The reaction is terminated by adding tyrosine in NaH_2PO_4 buffer (50 μl). The mixture is then passed quickly through an equilibrated Dowex column to remove $^{125}I^-$, and the column is washed with bovine serum albumin in NaH_2PO_4 (2 ml). All the column effluent is collected and pooled. This contains the ^{125}I-immunoglobulin; the Dowex resin will become very radio-active and should be discarded. The amount of radio-label covalently bound to immunoglobulin can be measured as follows: 10 μl of ^{125}I-immunoglobulin solution is diluted to 1 ml with NaH_2PO_4 buffer and 20 μl of this solution is added to 200 μl of NaH_2PO_4 buffer containing 20 mg bovine serum albumin. The solutions are mixed and the protein is precipitated by adding 1 ml of trichloroacetic acid (20% w/v). The precipitate is collected by centrifugation at 4000 rpm for 5 min. Both the supernatant and the precipitate are counted in a gamma counter to provide an estimate of the unbound radio-activity and of the bound radio-activity, respectively. The proportion of radio-activity bound to the protein should be at least 90% of the total radio-activity in the sample.

Materials and Equipment

Immunoglobulin, 2–5 mg/ml; species-specific anti-human IgG, anti-mouse IgG or anti-rabbit IgG as required.
NaH_2PO_4 buffer (0.1 M, pH 7.4 adjusted with NaOH).
$Na^{125}I$: 100 μCi (Amersham International).
Chloramine-T, 5 mg/ml in H_2O: freshly made.
Tyrosine (0.4 mg/ml) in NaH_2PO_4 (0.1 M, pH 7.4).
Bovine serum albumin (2 mg/ml) in NaH_2PO_4 (0.1 M, pH 7.4).
Dowex-1–X8(Cl) column: 100–200 mesh standard grade, packed into a Pasteur pipette and equilibrated with 30–40 ml bovine serum albumin–NaH_2PO_4.
Trichloroacetic acid: (20% w/v).

Identification of Antibody–Antigen Pairs

Procedure 6

Although the methods described above allow detection of antibodies directed against antigens bound to the nitrocellulose paper, it is often difficult to align the bands on the autoradiograms of immunoblots with those on the dried Coomassie Blue stained gels. This is due to changes in gel size during processing. The problem can be overcome by radioactively labelling the protein antigens and using the labelled preparations as markers. Here we describe a procedure which is useful when studying outer-membrane proteins that are exposed on the surface of *E. coli*. These proteins are extrinsically labelled with ^{125}I as described in procedure 7 following.

The ^{125}I-labelled proteins are electrophoresed in gels in a lane adjacent to the unlabelled preparation and are transferred to a nitrocellulose sheet along with unlabelled proteins. The sheet is then cut into two, between the two lanes, with pinking shears which produce a serrated edge and the paper carrying the un-labelled proteins processed as described in procedure 4. Both the immunoblot and the nitrocellulose paper carrying the electrophorectically separated ^{125}I-labelled proteins are placed on X-ray film and aligned by means of their serrated edges. In this way a profile of the outer-membrane proteins exposed on the cell surface is produced on the film alongside and aligned with the antibody 'blot' profile.

Iodination of Outer-membrane Proteins of *E. coli*

Procedure 7

Bacterial cells (100–200 mg wet weight) are washed in PBS and resuspended in 0.5 ml of the same solution. The suspension is warmed to 30°C and to it are added 200 μg lactoperoxidase and 1 mCi Na^{125}I. The reaction is started by adding 50 μl H$_2$O$_2$, and further aliquots (50 μl) are added at 3, 6, and 9 min. After a total time of 12 min, the reaction is terminated by adding 20 ml PBI. The suspension is then centrifuged for 10 min at 12,000 g, the supernatant discarded and the cells washed 3 times with PBI (approximately 10 ml each time). The final pellet is stored at -30°C and outer membranes prepared as in procedure 1.

Materials and Equipment

Phosphate buffered saline (PBS): NaCl 0.175 M, KCl 10 mM, NaH$_2$PO$_4$ 10 mM, KH$_2$PO$_4$ 9.2 mM.

Phosphate buffered saline with potassium iodide (PBI): KCl 10 mM, Na$_2$HPO$_4$ 10 mM, KH$_2$PO$_4$ 9.2 mM, KI 50 mM.
Lactoperoxidase (Sigma Chemical Co., London): 200 μg.
Na^{125}I: 1 mCi: (Amersham International).
Hydrogen peroxide (H$_2$O$_2$): 0.1 M.
Water bath at 30°C.

References

Laskey, R. A. and Mills, A. D. (1977). Enhanced autoradiographic detection of ^{32}P and ^{125}I using intensifying screens and hypersensitized film. *FEBS Letters* **82**, 314–316.

Lowry, O. H., Rosebrough, N. J., Farr, L. and Randall, R. J. (1951). Protein measurement with the Folin phenol reagent. *Journal of Biological Chemistry* **193**, 265–275.

Lugtenberg, B., Meijers, J., Peters, R., van der Hoek, P. and van Alphen, L. (1975). Electrophoretic resolution of the major outer membrane protein of *Escherichia coli* K12 into four bands. *FEBS Letters* **58**, 254–258.

Schnaitman, C. A. (1971). Solubilization of the cytoplasmic membrane of *Escherichia coli* by Triton X-100. *Journal of Bacteriology* **108**, 545–552.

Towbin, H., Staehelin, T. and Gordon, J. (1979). Electrophoretic transfer of proteins from polyacrylamide gels to nitrocellulose sheets: Procedure and some applications. *Proceedings of the National Academy of Sciences of the U.S.A.*

Index

The suffix t indicates that the page reference is to a table.

A

N-Acetylneuraminic acid, 167–169
Adenosine monophosphate, cyclic, *see* Cyclic
    adenosine monophosphate
Adenylate cyclase, 13, 181–182
Adherence
    bioluminescence assay for, 389–393
    to epithelial cells, test for, 335
    of extra-intestinal isolates, 126–127
    of intestinal isolates, 125–126
    relationship to haemagglutination, 124–127
Adhesin E8775, 238
Adhesin F41, 50t, 52t, 55, 82t, 87–88
    adhesion properties of, 116
    haemagglutination by, 101–102
    pathogenesis, role in, 138
Adhesin K88, 51–54, 52t–53t, 80, 81t, 82t,
    83–84, 235–236, 273–274
    adhesion properties, 113–116
    haemagglutination by, 100–101
    hydrophobicity, 123
    pathogenicity, role in, 136–137
    receptor, 128–130
Adhesin K99, 50t, 51–55, 80, 81t, 82t, 85–
    87, 236, 274
    adhesion properties, 116
    haemagglutination by, 101
    pathogenesis, role in, 137–138
    receptor, 131
Adhesin 987P, 52t–53t, 54, 81t, 82t, 85, 238
    adhesion properties, 116
    haemagglutination by, 101
    pathogenesis, role in, 137
    receptor, 130
Adhesins, 11, 79–155, 287–288, *see also*
    CFA/I; CFA/II; Colonization factors;
    Mannose-resistant adhesins; Mannose-sen-
    sitive adhesins; specific adhesins
    antigenic properties, 82–96
    characterization, 82–96
    definition, 80

fimbrial, 11, 81t
pathogenesis, role in, 136–143
plasmids for, 235–240
P-specific, 82t, 107–108
purification, 82–96
tissue culture test for, 401–402
X-specific, 107–108
Aerobacter aerogenes, 204
Aerobactin, 204–207
    bioassay, 421–423
    detection of, 419–423
Air-sacculitis, in poultry, 63–65
Antibodies
    anti-α-haemolysin, 209
    anti-K1, 211
    maternal, 272–274
    natural (normal), 30
Antibody synthesis, ontogeny of, 274–275
Antigen, *see* Adhesins; Haemagglutinins; spe-
    cific antigens (e.g. O antigen)
Antigenic pattern and pathogenicity, 160–162
Anti-immunoglobulins, iodination of, 462
Antiserum, anti-O, monospecific, preparation,
    348
Arthritis, calves and lambs, 47
Aspirin, 187
Asymptomatic bacteriuria, 19
ATP-ribosylation, 13

B

Bacteraemia, 50t, 65–66
Bacterium coli commune, 1, 2, 7
Bacterium lactis aerogenes, 2
Bacteriuria
    covert (asymptomatic), 19, 20–21
    haemagglutination by isolates from, 106
    and hypertension, 25
    in pregnancy, 23, 25
Bacteriolysis, by secretory IgA, 273
Bacteroides fragilis, 27
Biken test, for heat-labile enterotoxin, 398

468INDEX

F

FEEC, 81, 90–91, 103–104
  adhesion properties, 120
Ferrichrome, 195, 198–199
Fibrinolysis, activation, 28
Fimbria, 22, 80, *see also* Adhesins; Haemagglutination; Haemagglutinins
  antigenicity, 82
  antiserum preparation, 386
  electron microscopy of, 304–306
  expression in culture, 382–383
  F7 antigen, 92–93
  F8 antigen, 93
  genetics, 252–254
  MR, preparation of, 382–383
  MS, 10, 80, 81t
  P, 10, 23, 81, 81t, 91–93
  P receptor, 131–135
  purification, 91
  receptor specificity, 106–108
  type 1, *see* Fimbria, MS
  vaccines, 32
  X, 10, 95
Fimbrial antigen, 9, 81t, *see also* Adhesins; Haemagglutinins
  F7, 93
  F8, 93–94
  mannose-resistant, 94–95
  P, 93
Flora
  faecal, 10–11
  normal, 10–11

G

Ganglioside GM_1, 13, 182
Gastro-enteritis
  infantile, 15
  Salmonella, 31
β-Glucosaminyl-1,6-glucosamine, 158
Glycoprotein, Tamm-Horsfall, 22
Granulomatous disease of birds, 29
Guanylate cyclase, 183

H

Haemagglutination, 22, 79, 96–111
  adhesion, relationship, 124–127
  extra-intestinal isolates, 126–127
  human isolates, 104t
  indirect, 21

interpretation, 297–301
intestinal isolates, 125–126
meningitis isolates, 105
methodology, 291–296
patterns and fimbrial antigens, 301–304
septicaemia isolates, 105
typing in epidemiology, 306–307
urinary isolates, 105–108
Haemagglutination, mannose-resistant, 22, 98–99t, 100–111, 290–291
  CFA/I, 102–103
  CFA/II, 102–103
  EIEC, 104–105
  EPEC, 103–104
  ETEC
    animal, 100–102
    human, 102–103
  extra-intestinal isolates, 105–108
  F41, 101–102
  FEEC, human, 103–104
  K88, 100–101
  K99, 101
  987P, 101
Haemagglutination, mannose-sensitive, 22, 97–100, 288–290
Haemagglutinins, 287–288, *see also* Haemagglutination
  aeration, effect of, 110
  antibiotics, effect of, 110–111
  electron microscopy of, 304–306
  expression in culture, 105–106
  mannose-resistant, 290–291
    genetics of, 253–254
  mannose-sensitive, 288–290
  medium, effect of, 108–110
  temperature, effect of, 110
Haemolysin, 13, 21, 245–248
  as cytotoxin, 209–210
  genetics, 245–248
  as virulence factor, 208–210
α-Haemolysin, 9, 207–210, 245, 246
  antibodies, 209
  detection, 425–427
β-Haemolysin, 9, 208, 245, 246
Haemolytic uraemic syndrome, 28
Haemorrhagic colitis, 28
Haemorrhagic enteritis, 59–60
H antigen, 9
  identification, 349–352